Carlo Cattani, Hari M. Srivastava, Xiao-Jun Yang (Eds.)
Fractional Dynamics

Carlo Cattani, Hari M. Srivastava, Xiao-Jun Yang
(Eds.)

Fractional Dynamics

Managing Editor: Agnieszka Bednarczyk-Drąg

Language Editor: Nick Rogers

DE GRUYTER
OPEN

Published by De Gruyter Open Ltd, Warsaw/Berlin
Part of Walter de Gruyter GmbH, Berlin/Munich/Boston

ISBN 978-3-11-047208-0
e-ISBN 978-3-11-047209-7

Bibliographic information published by the Deutsche Nationalbibliothek
The Deutsche Nationalbibliothek lists this publication in the Deutsche Nationalbibliografie; detailed bibliographic data are available in the Internet at http://dnb.dnb.de.

Managing Editor: Agnieszka Bednarczyk-Drąg
Language Editor: Nick Rogers

www.degruyteropen.com

Cover illustration: © Thinkstock/3quarks

Contents

Carlo Cattani, H. M. Srivastava, and Xiao-Jun Yang

Fractional Dynamics

This volume is devoted to recent developments in the theory of fractional calculus and its applications. Particular attention is paid to the applicability of this currently popular research field in various branches of pure and applied mathematics. In particular, we have focused on the more recent results in mathematical physics, engineering applications, theoretical and applied physics as quantum mechanics, signal analysis, and in those relevant research fields where nonlinear dynamics occurs and several tools of nonlinear analysis are required. Dynamical processes and dynamical systems of fractional order attract researchers from many areas of sciences and technologies, ranging from mathematics and physics to computer science.

Fractional calculus is an intriguing mathematical theory which dates back to the very foundation of differential calculus when Leibniz (1695) raised some concern about derivatives of fractional order. However, it was Liouville (1832), and nearly contemporarily to him, Riemann (1847) and Fourier (1822), who collected into an independent theory all the basic knowledge about fractional order derivatives and fractional order integrals. Since then and especially in the last century, there has been an increasing interest on fractional calculus and its application in many different fields in Mathematics, Physics, Engineering, Bio-Science, Computer Science, Economics and so on.

In order to explain why fractional calculus is attracting more and more scholars with different scientific and cultural background is due to the following reasons:

1. Although it is almost completely clear what does it mean by fractional derivative and fractional integral and which basic axioms should be taken to define these operators, fractional derivative as operator doesn't fulfill all properties of the ordinary derivative. For this reason, in order to circumvent this problem many additional axioms were assumed, so that, as a consequence, there are different definitions of the same operator, which give rise to different, sometimes competing, fractional derivatives. All definitions are mathematically respectful, but the main problem is that up to now there is not a unique definition of fractional derivative.
2. Since their origins, fractional derivatives were linked to the fractional calculus. There is always a debate on local (for the derivative) and non-local (for derivative) approach to fractional-order operator calculus. For the nonlocal fractional calcu-

Carlo Cattani: School of Engineering (DEIM), Largo dell'Università –Riello, 01100 Viterbo, Italy, E-Mail: cattani@unitus.it
H. M. Srivastava: Department of Mathematics and Statistics, University of Victoria, Victoria, British Columbia V8W 3R4, Canada E-Mail: harimsri@math.uvic.ca
Xiao-Jun Yang: Department of Mathematics and Mechanics, China University of Mining and Technology, Xuzhou, Jiangsu 221008, People's Republic of China, E-Mail: dyangxiaojun@163.com

lus, some boundary conditions are needed and this implies global approach to such problems.

3. The fundamental operators of fractional calculus are based on special functions which, in turn, are strongly depending on efficient computational methods and numerical algorithms. Therefore, the modern advancements on computational and numerical methods have pushed the recent development of fractional calculus and the massive interests of the scientific community.

However, as a consequence of these main concerns, we have reached a situation where there are still some unsolved theoretical problems in fractional calculus whereas there are many different fractional models in applications. In fact, any physical, engineering, and mathematical problems that are being investigated by a fractional calculus might have a variety of fractional models, depending on the local-nonlocal approach and/or on the chosen fractional derivative among the many alternative definitions. Moreover, the many recently developed numerical methods add confusion to the multi-facet fractional models. In fact, different choices of orthogonal functions and bases would lead to different fractional models.

Very recently there have also been several successful attempts to extend local fractional calculus to non differentiable, irregular sets like fractals. The possibility to extend the fractional calculus to self-similar unsmooth objects is opening new frontiers in science, such as, for example, in signal analysis, data sets from complex phenomena, image analysis, nuclear medicine, and so on. Nonlinear analysis of data, collected by modern devices, offer still unsolved analytical problems related not only to complex physics and abstract mathematical theories, but also to nonlinear science. From analytical point of view, these kinds of problems are often leading us to deal with the concepts of scales, fractals, fractional operators, and so on.

In this volume, we have collected some of the main contributions to the fields of local and non-local fractional calculus, aiming to explore the most recent advancement in several branches of science and technology.

This volume consists of a total of 22 chapters which are outlined below.

Chapter 1 (by C. Cattani) deals with the fractional calculus applied to Shannon wavelet theory. This chapter introduces a novel definition of local fractional derivative which makes it possible to compute the derivative of any finite energy function, that is, of a localized function which can be represented by wavelet series.

In Chapter 2, V. E. Tarasov studies three-dimensional discrete dynamical systems with long-range properties and gives some applications for different types of dynamical systems.

In Chapter 3, António M. Lopes and J.A. Tenreiro Machado investigate the statistical distributions of earthquakes in Southern California over the time period from the year 1934 up to the year 2013. The reported results reveal relationships and temporal patterns hidden in the data. Also, the investigated methodology and findings can contribute to a comprehensive explanation of these phenomena and to recognize precursory events for earthquake prediction.

Chapter 4 (by Akira Asada) deals with an integral transform arising from fractional calculus, which is used for investigating the solution of some differential equations.

In Chapter 5, Jordan Hristov fully shows how to obtain approximate solutions to time-fractional models by integral balance approach. This long chapter completely describes the theory of integral balance approach for finding approximate solution of fractional differential equations.

Chapter 6 (by Bashir Ahmad, Ahmed Alsaedi and Hana Al-Hutami) presents a study of sequential fractional q-integro-difference equations with perturbed anti-periodic boundary conditions.

In particular, existence results for the given problem are established by applying Krasnoselskii's fixed point theorem, Leray-Schauder nonlinear alternative for single valued maps and Banach's contraction mapping principle. Moreover, correction terms arising due to the perturbation in the anti-periodic boundary data are highlighted.

Several chapters are devoted to surface processes of sorption and reaction processes in electrolysis: In fact, systems For example, in Chapter 7, by M. K. Lenzi, G. Gonçalves, D. P. Leitoles and E. K. Lenzi, give a new approach to surface processes of sorption and reactions processes. The models address a half space with dynamics of the substances governed by a fractional diffusion equation.

Jocelyn Sabatier in Chapter 8 shows how fractional-order models can be used to capture the dynamical behavior of electrochemical devices such as batteries or super-capacitors, and are also used to build state of charge or state of health estimators. Moreover, in Chapter 9, M. K. Lenzi, G. Gonçalves, F. R. G. B. Silva, R. S. Zola, H. V. Ribeiro, R. Rossato and E. K. Lenzi present a new approach to surface processes of sorption and reactions processes in electrolytic systems. In fact, the electric impedances in electrolytic systems are an old area where fractional derivatives of half-time order are widely applicable. The models address a half-space with dynamics of the substances governed by a fractional diffusion equation to model non-linear phenomena. The processes on the surface are assumed to be of the first order, that is, the kinetic equations are linear and with memory effects which may be connected to an unusual relaxation.

Applications of fractional calculus to epidemiology are given in Chapter 10 by Abdon Atangana by considering the fractional dynamical system which is a generalization of a SIR model in epidemiology. Then, by using some theorems given by the authors, together with the Sumudu transform and the variation iteration model, some significant numerical results for a concrete problem are discussed.

In the category of fractional differential equations there are several chapters. A new numerical method based on Laguerre polynomials and operational matrix for the approximate solution of fractional differential equations on a semi-infinite interval is given in Chapter 11 by A.H. Bhrawy, T.M. Taha, M.A. Abdelkawy and R.M. Hafez.

Chapter 12 (by Hong-Yan Liu and Ji-Huan He) describes some thermal conduction on the wall of a typical Mongolian construction. This problem is summarized by a set of fractional differential equations which are analytically solved and many interesting results are presented. They also find an optimal thickness of the fractal hierarchy of the felt cover.

In Chapter 13, Mohamad Rafi, Segi Rahmat, Dumitru Baleanu and Xiao-Jun Yang present and employ the Cantor-type spherical-coordinate method to solve a problem involving a class of local fractional differential operators on Cantor sets. This method converts differential equations on Cantor sets from Cantorian-coordinate system to Cantor-type spherical-coordinate systems. The capability of the proposed method is confirmed through several examples for some important physical problems namely, the damped wave, Helmholtz and heat conduction equations.

In Chapter 14 on the approximate methods for local fractional differential equations, H. M. Srivastava, J. A. Tenreiro Machado and Xiao-Jun Yang present some analytical techniques for solving a class of local fractional differential equations. They also focus upon the local fractional variational iteration, Adomian decomposition and series expansion methods. In particular, some classical problems such as the Boussinesq and wave equations are considered in the local fractional calculus sense and solved on Cantor sets.

Numerical Solutions for ODEs with local fractional derivative are investigated in Chapter 15 by Xiao-Jun Yang, Dumitru Baleanu and J. A. Tenreiro Machado. In particular, these authors give an efficient numerical algorithm for solving local fractional ODEs by applying the extended differential transform method which employs the generalized local fractional Taylor theorems.

A particular equation arising in fractal heat transfer is studied in Chapter 16 in which Xiao-Jun Yang, Carlo Cattani and Gong-Nan Xie address the homogeneous and non-homogeneous heat conduction, the Poisson and the Laplace equations of fractal heat transfer. The 2D partial differential equations of fractal heat transfer in Cantor-type circle coordinate system are also discussed by these authors.

Fractional partial diferential equations are discussed in Chapter 17 by Hossein Jafari, Hassan Kamil Jassim and Syed Tauseef Mohyud-Din by using the local fractional Laplace decomposition method in order to find the solution of some local fractional differential equations.

Chapter 18 (by Alireza Khalili Golmankhaneh and Dumitru Baleanu) discusses some fundamental principles of fractional calculus and presents a framework for fractional calculus and a calculus on fractals. These authors also present some relations and formulas for calculus on fractal subset of the real-line, F-limit and F- continuity and Taylor series on fractal sets. Some applications on F-calculus in physics and dynamics have also been given.

The solutions of nonlinear fractional differential systems through an implementation of the variational iteration method are given in Chapter 19 by Haci Mehmet Baskonus, Fethi Bin Muhammad Belgacem and Hasan Bulut.

In Chapter 20 (by Tofik Mekkaoui, Zakia Hammouch, Fethi B. M. Belgacem and Ahmed El-Abbassi), the chaotic dynamics of an electrical circuit is analyzed and the corresponding fractional order nonlinear dynamical system is studied.

Chapter 21 (by E. Guariglia) deals with the fractional analysis of the Riemann zeta function by giving some fractional series expansions of the Riemann zeta function and by studying their convergence properties.

A treatment of generalized fractional differential equations: Sumudu transform series expansion solutions and their applications, is presented in Chapter 22 by Fethi Bin Muhammad Belgacem, Vartika Gulati, Pranay Goswami and Abdullah Aljoujiee.

All of the above 22 chapters deal essentially with current problems in the topics within the aim and scope of the present volume. However, they are not an exhaustive representation of the area of fractional calculus and its widespread applications, but they are meant to represent some of the more modern and fascinating fields of research on the subject of this volume. In all of these 22 chapters, the authors have focused on the main aspects of the theory and, although they have proposed some solutions and models, most problems remain still open, thus giving the opportunity to readers for further research and discussions in this field. Thanks to the authors for their excellent contributions addressing all of the key aspects raised in this volume.

Carlo Cattani
H. M. Srivastava
Xiao-Jun Yang

Carlo Cattani

Local Fractional Calculus on Shannon Wavelet Basis

Abstract: In this paper an explicit formula for the local fractional derivative of Shannon wavelets is given and generalized to any $L_2(\mathbb{R})$-functions, as wavelet series based on connection coefficients. Moreover due to the local definition, the fractional derivative can be applied also to nonregular functions, such as localized fractals, of Cantor-type.

Keywords: Shannon wavelets; connection coefficients; local fractional derivative; Cantor set; Cantor function

1 Introduction

Wavelets are special functions, localized in time or frequency, which form a basis for energy bounded functions and in particular for $L_2(\mathbb{R})$-functions. Wavelets allow localized problems to be analyzed easily. Moreover, wavelets allow the multiscale decomposition of problems, thus emphasizing the contribution of each scale. By defining a suitable inner product on the orthogonal family of scaling/wavelet functions, any $L_2(\mathbb{R})$-function can be approximated at a fixed scale, by a truncated series that has the scaling functions and the wavelet functions as a basis. The wavelet coefficients of these series represent the contribution of each scale.

Shannon wavelets are a family of orthogonal wavelets that have many interesting properties (Cattani 2007; Cattani 2008) and which define a multiresolution analysis (Cattani and Rushchitsky 2007; Daubechies 1992). Moreover, it has been shown recently that since they are analytical functions they are infinitely differentiable functions so that their derivatives can be easily computed at any order (Cattani 2003; Cattani 2006; Cattani 2007; Cattani 2008; Cattani 2010; Cattani 2012). It can be also shown that derivatives of any order of the Shannon wavelets are linear combination of themsleves. The coefficients of these linear combination are the so-called connection coefficients. The connection coefficients are the wavelet coefficients of the derivatives of the wavelet basis. These coefficients are a fundamental tool for the approximation of differential operators with respect to the wavelet basis. The theory of connection coefficients was initially given (Latto et Al. 1992; Romine and Peyton 1997) for the compactly supported wavelet families, such as the Daubechies wavelets (Daubechies 1992). The computation of these connection coefficients was based on the recursive equations of wavelet theory and the explicit forms of these coefficients was given only

Carlo Cattani: School of Engineering (DEIM), Largo dell'Università –Riello, 01100 Viterbo, Italy, E-Mail: cattani@unitus.it up to the second order derivatives. In recent papers it was shown that for Shannon wavelets the connection coefficients can be computed for any order derivative and in an explicit analytical form (Cattani 2006; Cattani 2008; Cattani 2010; Cattani 2012). As a consequence, the differential properties of the Shannon wavelets and their corresponding wavelet space are simply defined by these connection coefficients. They are also an expedient tool for the projection of differential operators, and are thus extremely important for computing the wavelet solution of integro-differential equations (Cattani 2005; Deriaz 2007; Latto et Al. 1992; Lin and Zhou 2001; Restrepo and Leaf 1995; Romine and Peyton 1997). In particular, the scaling Shannon function is the *sinc*-function which is an important function in signal analysis (Cattani 2003; Cattani 2005; Cattani and Rushchitsky 2007).

In the following, the local fractional calculus (Yang 2012; Yang 2013; Yang 2014; Yang et Al. 2013) will be defined over a Shannon wavelet basis. The importance of the local fractional calculus is due to the fact that it is possible to extend the local fractional derivative to non-smooth continuous functions, such as fractals or random signals. If these signals are localized then it is possible to represent them by using a wavelet basis. We will define a local fractional derivative on the Shannon wavelet basis using the connection coefficients and this will enable us to compute the local fractional derivative of localized smooth or non-smooth signals.

This chapter is organized as follows: Section 2 gives some preliminary remarks, definitions and properties about Shannon wavelets. Their corresponding connection coefficients are discussed in Section 3. Section 4 deals with some properties of connection coefficients, functional equalities, and errors of approximation. Fractional derivatives of the Shannon scaling function and wavelets are given in Section 5.

2 Preliminary Remarks

In this section, some remarks on Shannon wavelets and connection coefficients are given (see also Cattani 2010).

Shannon wavelet theory (see e.g. Cattani 2006; Cattani 2007; Cattani 2008; Cattani 2010; Deriaz 2007) is based on the scaling function $\varphi(x)$, also known as the sinc function, and the wavelet function $\psi(x)$ respectively defined as

$$\begin{cases} \varphi(x) & = & \operatorname{sinc} x \stackrel{\text{def}}{=} \dfrac{\sin \pi x}{\pi x} = \dfrac{e^{\pi i x} - e^{-\pi i x}}{2\pi i x} \,. \\[2ex] \psi(x) & = & \dfrac{\sin 2\pi(x-\frac{1}{2}) - \sin \pi(x-\frac{1}{2})}{\pi(x-\frac{1}{2})} \\[2ex] & = & \dfrac{e^{-2i\pi x}\left(-i + e^{i\pi x} + e^{3i\pi x} + i\, e^{4i\pi x}\right)}{2\pi\left(x-\frac{1}{2}\right)} \,. \end{cases} \tag{1}$$

The second function can be expressed in terms of the first, as

$$\psi(x) = 2\varphi(2x-1) - \varphi(x-\frac{1}{2}) \tag{2}$$

The corresponding families of translated and dilated instances of the wavelets (Cattani 2006; Cattani 2007; Cattani 2008; Cattani 2010; Deriaz 2007), are

$$\begin{aligned}
\varphi^n_k(x) &= 2^{n/2}\varphi(2^n x - k) = 2^{n/2}\frac{\sin\pi\left(2^n x - k\right)}{\pi\left(2^n x - k\right)} \\
&= 2^{n/2}\frac{e^{\pi i(2^n x - k)} - e^{-\pi i(2^n x - k)}}{2\pi i\left(2^n x - k\right)}\,, \\
&= \frac{2^{n/2}}{2\pi i\left(2^n x - k\right)}\sum_{s=0}^{\infty}\frac{\pi^s i^s}{s!}\left[1-(-1)^s\right]\left(2^n x - k\right)^s \\
&= \frac{2^{n/2}}{2\pi i\left(2^n x - k\right)}\sum_{s=0}^{\infty}\frac{\pi^s i^s}{s!}(1-e^{\pi s})\left(2^n x - k\right)^s \\
&= 2^{n/2-1}\sum_{s=1}^{\infty}\frac{\pi^{s-1} i^{s-1}}{s!}\left[1-(-1)^s\right]\left(2^n x - k\right)^{s-1}\,.
\end{aligned} \tag{3}$$

By using the Mittag-Leffler function,

$$E_{\alpha,\beta}(z) \overset{\text{def}}{=} \sum_{s=0}^{\infty}\frac{z^s}{\Gamma(\alpha s + \beta)} \quad , \quad \alpha, \beta \in \mathbb{C}$$

it is

$$E_{1,1}(z) \overset{\text{def}}{=} \sum_{s=0}^{\infty}\frac{z^s}{s!}$$

so that

$$\varphi^n_k(x) = 2^{n/2-1}\left[E_{1,1}\left(\pi i(2^n x - k)\right) - E_{1,1}\left(-\pi i(2^n x - k)\right)\right]$$

Analogously, it is

$$\begin{aligned}
\psi^n_k(x) &= 2^{n/2}\frac{\sin 2\pi(2^n x - k - \frac{1}{2}) - \sin\pi(2^n x - k - \frac{1}{2})}{\pi(2^n x - k - \frac{1}{2})}\,, \\
&= \frac{2^{n/2}}{2\pi(2^n x - k - \frac{1}{2})}\sum_{r=1}^{2} i^{1+r}e^{r\pi i(2^n x - k)} - i^{1-r}e^{-r\pi i(2^n x - k)}\,.
\end{aligned} \tag{4}$$

or, by taking into account (2)

$$\psi^n_k(x) = 2\varphi^{n+1}_k(x) - \varphi^n_k(x - \frac{1}{2})$$

and eq. (3), it is

$$\begin{aligned}\psi_k^n(x) &= 2^{n/2}\sum_{s=1}^{\infty}\frac{\pi^{s-1}i^{s-1}}{s!}\left[1-(-1)^s\right]\left(2^n x-k\right)^{s-1}\\ &- 2^{n/2-1}\sum_{s=1}^{\infty}\frac{\pi^{s-1}i^{s-1}}{s!}\left[1-(-1)^s\right]\left(2^n(x-\frac{1}{2})-k\right)^{s-1}.\end{aligned}$$

In particular, we denote

$$\varphi_0^0(x)=\varphi(x)\,,\ \psi_0^0(x)=\psi(x)\,,\ \varphi_k^0(x)=\varphi_k(x)=\varphi(x-k)\,,\ \psi_k^0(x)=\psi_k(x)=\psi(x-k)\,.$$

For the basic functions $\varphi(x)$, $\psi(x)$, according to (3),(4) it is

$$\left\{\begin{aligned}\varphi(x) &= \frac{1}{2}\sum_{s=1}^{\infty}\frac{\pi^{s-1}i^{s-1}}{s!}\left[1-(-1)^s\right]x^{s-1}\\ \psi(x) &= \sum_{s=1}^{\infty}\frac{\pi^{s-1}i^{s-1}}{s!}\left[1-(-1)^s\right](2x-1)^{s-1}-\frac{1}{2}\sum_{s=1}^{\infty}\frac{\pi^{s-1}i^{s-1}}{s!}\left[1-(-1)^s\right](x-\frac{1}{2})^{s-1}\end{aligned}\right. \tag{5}$$

and in particular

$$\boxed{\varphi(x)=\frac{1}{2}\sum_{s=1}^{\infty}\left[\left(1-(-1)^s\right)(\pi i)^{s-1}\right]\frac{x^{s-1}}{s!}} \tag{6}$$

2.1 Shannon Wavelets in the Fourier Domain

Let

$$\widehat{f}(\omega)=\widehat{f(x)}\overset{\text{def}}{=}\frac{1}{2\pi}\int_{-\infty}^{\infty}f(x)e^{-i\omega x}dx\quad,\quad f(x)=\int_{-\infty}^{\infty}\widehat{f}(\omega)e^{i\omega x}d\omega$$

be the Fourier transform of the function $f(x)\in L_2(\mathbb{R})$, and its inverse transform respectively. The Fourier transform of (1) give us (Cattani 2008)

$$\left\{\begin{aligned}\widehat{\varphi}(\omega) &= \frac{1}{2\pi}\chi(\omega+3\pi)=\left\{\begin{array}{ll}1/(2\pi) & ,\ -\pi\le\omega<\pi\\ 0 & ,\ \text{elsewhere}\end{array}\right.\\ \widehat{\psi}(\omega) &= \frac{1}{2\pi}e^{i\omega/2}\left[\chi(2\omega)+\chi(-2\omega)\right]\end{aligned}\right.$$

with

$$\chi(\omega)=\left\{\begin{array}{ll}1 & ,\quad 2\pi\le\omega<4\pi\\ 0 & ,\quad \text{elsewhere}\,.\end{array}\right.$$

Analogously for the dilated and translated instances of the scaling/wavelet function in the frequency domain, it is

$$\left\{\begin{array}{lcl}\widehat{\varphi}^n_k(\omega) & = & \dfrac{2^{-n/2}}{2\pi}e^{i\omega k/2^n}\chi(\omega/2^n+3\pi)\\ \widehat{\psi}^n_k(\omega) & = & \dfrac{2^{-n/2}}{2\pi}e^{i\omega(k+1/2)/2^n}\left[\chi(\omega/2^{n-1})+\chi(-\omega/2^{n-1})\right] .\end{array}\right. \tag{7}$$

Both families of Shannon scaling and wavelet are $L_2(\mathbb{R})$-functions, therefore for each $f(x) \in L_2(\mathbb{R})$ and $g(x) \in L_2(\mathbb{R})$, the inner product is defined as

$$\langle f, g\rangle \stackrel{\text{def}}{=} \int_{-\infty}^{\infty} f(x)\overline{g(x)}dx = 2\pi \int_{-\infty}^{\infty} \widehat{f}(\omega)\overline{\widehat{g}(\omega)}d\omega = 2\pi \left\langle \widehat{f}, \widehat{g}\right\rangle , \tag{8}$$

where the bar stands for the complex conjugate.

Shannon wavelets fulfill the following orthogonality properties (for the proof see e.g. Cattani 2008; Cattani 2010)

$$\langle \psi^n_k(x), \psi^m_h(x)\rangle = \delta^{nm}\delta_{hk}, \left\langle \varphi^0_k(x), \varphi^0_h(x)\right\rangle = \delta_{kh}, \left\langle \varphi^0_k(x), \psi^m_h(x)\right\rangle = 0 , \qquad m \geq 0,$$

δ^{nm}, δ_{hk} being the Kroenecker symbols.

Let $\text{Ḅ} \subset L_2(\mathbb{R})$ the set of functions $f(x)$ in $L_2(\mathbb{R})$ such that the integrals

$$\left\{\begin{array}{lcl}\alpha_k & \stackrel{\text{def}}{=} & \langle f(x), \varphi_k(x)\rangle \stackrel{(8)}{=} \displaystyle\int_{-\infty}^{\infty} f(x)\overline{\varphi^0_k(x)}dx\\ \beta^n_k & \stackrel{\text{def}}{=} & \langle f(x), \psi^n_k(x)\rangle \stackrel{(8)}{=} \displaystyle\int_{-\infty}^{\infty} f(x)\overline{\psi^n_k(x)}dx ,\end{array}\right. \tag{9}$$

exist with finite values. It can be shown (Cattani 2008; Cattani 2010; Cattani and Rushchitsky 2007; Daubechies 1992) that the series

$$f(x) = \sum_{h=-\infty}^{\infty} \alpha_h \varphi_h(x) + \sum_{n=0}^{\infty}\sum_{k=-\infty}^{\infty} \beta^n_k \psi^n_k(x) , \tag{10}$$

converges to $f(x)$.

According to (8) the coefficients can be also computed in the Fourier domain (Cattani 2010) so that

$$\left\{\begin{array}{lcl}\alpha_k & = & \displaystyle\int_{-\pi}^{\pi} \widehat{f}(\omega)e^{i\omega k}d\omega ,\\ \beta^n_k & = & 2^{-n/2}\left[\displaystyle\int_{2^n\pi}^{2^{n+1}\pi} \widehat{f}(\omega)e^{i\omega(k+1/2)/2^n}d\omega + \int_{-2^{n+1}\pi}^{-2^n\pi} \widehat{f}(\omega)e^{i\omega(k+1/2)/2^n}d\omega\right] .\end{array}\right. \tag{11}$$

In the frequency domain, Eq. (10) gives (Cattani 2010)

$$\begin{aligned}\widehat{f}(\omega) &= \frac{1}{2\pi}\chi(\omega+3\pi)\sum_{h=-\infty}^{\infty}\alpha_h e^{i\omega h}\\ &+ \frac{1}{2\pi}\chi(\omega/2^{n-1})\sum_{n=0}^{\infty}\sum_{k=-\infty}^{\infty}2^{-n/2}\beta_k^n e^{i\omega(k+1/2)/2^n}\\ &+ \frac{1}{2\pi}\chi(-\omega/2^{n-1})\sum_{n=0}^{\infty}\sum_{k=-\infty}^{\infty}2^{-n/2}\beta_k^n e^{i\omega(k+1/2)/2^n}.\end{aligned}$$

When the upper bound for the series of (10), is finite, then we have the approximation

$$f(x) \cong \sum_{h=-K}^{K}\alpha_h\varphi_h(x) + \sum_{n=0}^{N}\sum_{k=-S}^{S}\beta_k^n\psi_k^n(x). \tag{12}$$

The error of the approximation has been estimated in (Cattani 2010).

2.2 Properties of the Shannon Wavelet

According to (3), Shannon wavelets can be easily computed at some special points, in particular

$$\varphi_k(h) = \varphi_h(k) = \varphi(h-k) = \varphi(k-h) = \delta_{kh} \quad , \quad (h,\ k\in\mathbb{Z}) \quad , \tag{13}$$

so that

$$\varphi_k(h) = \delta_{kh} = \begin{cases} 0 \quad , \quad h\neq k \quad , \quad (h,\ k\in\mathbb{Z})\\ 1 \quad , \quad h=k \quad , \quad (h,\ k\in\mathbb{Z})\end{cases}$$

It is also (Cattani 2010)

$$\begin{aligned}\psi_k^n(h) &= (-1)^{2^n h-k}\frac{2^{1+n/2}}{(2^{n+1}h-2k-1)\pi} \quad , \quad (2^{n+1}h-2k-1\neq 0)\\ \psi_k^n(x) &= 0 \quad , \quad x = 2^{-n}\left(k+\frac{1}{2}\pm\frac{1}{3}\right), \quad (n\in\mathbb{N},\ k\in\mathbb{Z})\\ &\lim_{x\to 2^{-n}(h+\frac{1}{2})}\psi_k^n(x) = -2^{n/2}\delta_{hk}.\end{aligned} \tag{14}$$

In particular, it is

$$\psi_k^0(0) = (-1)^{k+1}\frac{2}{(2k+1)\pi}$$

and since $k\in\mathbb{Z}$, $2k+1\neq 0$.

It can be shown (Cattani 2012) that the max and min values are as follows:

- For the scaling function $\varphi_k(x)$, the max values are located at the integers $x = k$

$$\max[\varphi_k(x_M)] = 1 \quad , \quad x_M = k\,, \tag{15}$$

- The max values of $\psi^n_k(x)$ are

$$\max[\psi^n_k(x_M)] = 2^{n/2}\frac{3\sqrt{3}}{\pi} \quad , \quad x_M = \begin{cases} -2^{-n}\left(k+\dfrac{1}{6}\right) \\ \dfrac{2^{-n-1}}{3}(18k+7)\,. \end{cases} \tag{16}$$

Moreover both families of scaling and wavelet functions belong to $L_2(\mathbb{R})$, and thus have a bounded range and (slow) decay to zero

$$\lim_{x\to\pm\infty} \varphi^n_k(x) = 0 \quad , \quad \lim_{x\to\pm\infty} \psi^n_k(x) = 0$$

3 Connection Coefficients

In order to represent the differential operators in wavelet bases, we have to compute the wavelet decomposition of the derivatives. It can be shown (Cattani 2008; Cattani 2010) that the derivatives of the Shannon wavelets are orthogonal functions:

$$\begin{cases} \dfrac{d^\ell}{dx^\ell}\varphi_h(x) &= \displaystyle\sum_{k=-\infty}^{\infty} \lambda^{(\ell)}_{hk}\,\varphi_k(x)\,, \\ \dfrac{d^\ell}{dx^\ell}\psi^m_h(x) &= \displaystyle\sum_{n=0}^{\infty}\sum_{k=-\infty}^{\infty} \gamma^{(\ell)mn}_{\ \ hk}\,\psi^n_k(x)\,, \end{cases} \tag{17}$$

being

$$\lambda^{(\ell)}_{kh} \stackrel{\text{def}}{=} \left\langle \frac{d^\ell}{dx^\ell}\varphi^0_k(x)\,, \varphi^0_h(x)\right\rangle \quad , \quad \gamma^{(\ell)nm}_{\ \ kh} \stackrel{\text{def}}{=} \left\langle \frac{d^\ell}{dx^\ell}\psi^n_k(x)\,, \psi^m_h(x)\right\rangle\,, \tag{18}$$

the connection coefficients (Cattani 2003; Cattani 2005; Cattani 2006; Cattani 2008; Deriaz 2007; Latto and Al. 1992; Lin and Zhou 2001; Restrepo and Leaf 1995; Romine and Peyton 1997).

The computation of connection coefficients can be easily performed in the Fourier domain, thanks to the equalities (8)

$$\lambda^{(\ell)}_{kh} = 2\pi\left\langle \widehat{\frac{d^\ell}{dx^\ell}\varphi_k(x)}, \widehat{\varphi_h(x)}\right\rangle \quad , \quad \gamma^{(\ell)nm}_{\ \ kh} = 2\pi\left\langle \widehat{\frac{d^\ell}{dx^\ell}\psi^n_k(x)}, \widehat{\psi^m_h(x)}\right\rangle\,, \tag{19}$$

and to the eq. (7).

In fact, in the Fourier domain the ℓth-order derivative of the (scaling) wavelet functions are simply

$$\widehat{\frac{d^\ell}{dx^\ell}\varphi_k^n(x)} = (i\omega)^\ell \widehat{\varphi}_k^n(\omega) \quad , \quad \widehat{\frac{d^\ell}{dx^\ell}\psi_k^n(x)} = (i\omega)^\ell \widehat{\psi}_k^n(\omega) \tag{20}$$

and, according to (7),

$$\begin{cases} \widehat{\frac{d^\ell}{dx^\ell}\varphi_k^n(x)} &= (i\omega)^\ell \frac{2^{-n/2}}{2\pi} e^{i\omega k/2^n}\chi(\omega/2^n + 3\pi)\,, \\ \widehat{\frac{d^\ell}{dx^\ell}\psi_k^n(x)} &= (i\omega)^\ell \frac{2^{-n/2}}{2\pi} e^{i\omega(k+1/2)/2^n}\left[\chi(\omega/2^{n-1}) + \chi(-\omega/2^{n-1})\right]\,, \end{cases}$$

or, in terms of the Mittag-Leffler function

$$\begin{cases} \widehat{\frac{d^\ell}{dx^\ell}\varphi_k^n(x)} &= (i\omega)^\ell \frac{2^{-n/2}}{2\pi} E_{1,1}(i\omega k 2^{-n})\chi(\omega/2^n + 3\pi)\,, \\ \widehat{\frac{d^\ell}{dx^\ell}\psi_k^n(x)} &= (i\omega)^\ell \frac{2^{-n/2}}{2\pi} E_{1,1}(i\omega(k+1/2)2^{-n})\left[\chi(\omega/2^{n-1}) + \chi(-\omega/2^{n-1})\right]\,. \end{cases}$$

It has been shown (Cattani 2006; Cattani 2008; Cattani 2010) that the connection coefficients of any order $(18)_1$ of the Shannon scaling functions $\varphi_k(x)$ are

$$\lambda_{kh}^{(\ell)} = \begin{cases} (-1)^{k-h}\frac{i^\ell}{2\pi}\sum_{s=1}^{\ell}\frac{\ell!\pi^s}{s![i(k-h)]^{\ell-s+1}}\left[(-1)^s - 1\right] & , k \neq h \\ \frac{i^\ell \pi^{\ell+1}}{2\pi(\ell+1)}\left[1 + (-1)^\ell\right] & , k = h\,, \end{cases} \tag{21}$$

or, by defining

$$\mu(m) = \text{sign}(m) = \begin{cases} 1 & , \quad m > 0 \\ -1 & , \quad m < 0 \\ 0 & , \quad m = 0\,, \end{cases}$$

shortly as,

$$\begin{aligned} \lambda_{kh}^{(\ell)} &= \frac{i^\ell \pi^\ell}{2(\ell+1)}\left[1 + (-1)^\ell\right](1 - |\mu(k-h)|) + \\ &+ (-1)^{k-h}|\mu(k-h)|\frac{i^\ell}{2\pi}\sum_{s=1}^{\ell}\frac{\ell!\pi^s}{s![i(k-h)]^{\ell-s+1}}\left[(-1)^s - 1\right]\,, \end{aligned}$$

when $\ell \geq 1$. For $\ell = 0$, it is

$$\lambda_{kh}^{(0)} = \delta_{kh}\,.$$

For the proof see (Cattani 2008).

Analogously for the connection coefficients $(18)_2$ we have that the connection coefficients to any order of the Shannon scaling wavelets $\psi_k^n(x)$ are

$$\begin{cases} \gamma^{(\ell)nm}_{\ \ kh} = \mu(h-k)\delta^{nm}\left\{ \sum_{s=1}^{\ell+1} (-1)^{[1+\mu(h-k)](2\ell-s+1)/2} \dfrac{\ell! i^{\ell-s}\pi^{\ell-s}}{(\ell-s+1)!\,|h-k|^s}(-1)^{-s-2(h+k)}2^{n\ell-s-1} \right. \\ \quad \times \left. \left\{ 2^{\ell+1}\left[(-1)^{4h+s}+(-1)^{4k+\ell}\right] - 2^s\left[(-1)^{3k+h+\ell}+(-1)^{3h+k+s}\right]\right\}\right\} & ,k \neq h \\ \gamma^{(\ell)nm}_{\ \ kh} = \delta^{nm}\left[i^\ell \dfrac{\pi^\ell 2^{n\ell-1}}{\ell+1}(2^{\ell+1}-1)(1+(-1)^\ell)\right] & ,k = h \end{cases} \tag{22}$$

and

$$\gamma^{(\ell)nm}_{\ \ kh} = \delta^{nm}\left\{ i^\ell(1-|\mu(h-k)|)\frac{\pi^\ell 2^{n\ell-1}}{\ell+1}(2^{\ell+1}-1)(1+(-1)^\ell) + \right.$$
$$+\mu(h-k)\sum_{s=1}^{\ell+1}(-1)^{[1+\mu(h-k)](2\ell-s+1)/2}\frac{\ell! i^{\ell-s}\pi^{\ell-s}}{(\ell-s+1)!\,|h-k|^s}(-1)^{-s-2(h+k)}2^{n\ell-s-1}\times$$
$$\left.\times\left\{2^{\ell+1}\left[(-1)^{4h+s}+(-1)^{4k+\ell}\right]-2^s\left[(-1)^{3k+h+\ell}+(-1)^{3h+k+s}\right]\right\}\right\}$$

for $\ell \geq 1$, and

$$\gamma^{(0)nm}_{\ \ kh} = \delta_{kh}\delta^{nm} \tag{23}$$

$\ell = 0$ respectively.

For the proof see (Cattani 2008).

As a consequence of eqs. (21),(22) the ℓ-order derivative of the basic functions (6) are

$$\begin{cases} \dfrac{d^\ell}{dx^\ell}\varphi(x) = \sum_{k=-\infty}^{\infty} \lambda^{(\ell)}_{0k}\,\varphi_k(x)\,, \\ \dfrac{d^\ell}{dx^\ell}\psi(x) = \sum_{n=0}^{\infty}\sum_{k=-\infty}^{\infty} \gamma^{(\ell)0n}_{\ \ 0k}\psi_k^n(x) \\ \qquad\qquad = \sum_{k=-\infty}^{\infty}\gamma^{(\ell)00}_{\ \ 0k}\psi_k^0(x)\,, \end{cases} \tag{24}$$

with

$$\lambda^{(\ell)}_{k0} = \begin{cases} (-1)^k \dfrac{i^\ell}{2\pi}\sum_{s=1}^{\ell}\dfrac{\ell!\pi^s}{s![ik]^{\ell-s+1}}\left[(-1)^s-1\right] & ,k \neq h \\ \dfrac{i^\ell\pi^{\ell+1}}{2\pi(\ell+1)}\left[1+(-1)^\ell\right] & ,k = 0\,, \end{cases} \tag{25}$$

3.1 Properties of Connection Coefficients

3.1.1 Recursiveness

The connection coefficients fulfill some recursive formula as follows:

Theorem 1. *The connection coefficients* (21) *are recursively given by*

$$\lambda_{kh}^{(\ell+1)} = \begin{cases} \dfrac{\ell+1}{k-h}\lambda_{kh}^{(\ell)} - (-1)^{k-h}\dfrac{i^{\ell}\pi^{\ell+1}}{k-h}[(-1)^{\ell}+1] & , k \neq h \\ i\pi\dfrac{\ell+1}{\ell+2}\lambda_{kh}^{(\ell)} + \dfrac{(-i)^{\ell+1}\pi^{\ell+1}}{\ell+2} & , k = h , \end{cases} \tag{26}$$

Proof:
Let us show first when $k = h$. From the definition (21) it is

$$\begin{aligned} \lambda_{kk}^{(\ell+1)} &= \frac{i^{\ell+1}\pi^{\ell+2}}{2\pi(\ell+2)}\left[1+(-1)^{\ell+1}\right] , \\ &= i\pi\frac{(\ell+1)}{(\ell+2)}\frac{i^{\ell}\pi^{\ell+1}}{2\pi(\ell+1)}\left[1+(-1)^{\ell+1}+(-1)^{\ell}-(-1)^{\ell}\right] , \\ &= i\pi\frac{(\ell+1)}{(\ell+2)}\frac{i^{\ell}\pi^{\ell+1}}{2\pi(\ell+1)}\left[1+(-1)^{\ell}+2(-1)^{\ell+1}\right] , \end{aligned}$$

from where $(26)_2$ follows. Analogously with simple computation we obtain $(26)_1$. □

After some calculations, Eq. (26) can be written as

$$\lambda_{kh}^{(\ell+1)} = (1-\delta_{kh})\left[\frac{\ell+1}{k-h}\lambda_{kh}^{(\ell)} - (-1)^{k-h}\frac{i^{\ell}\pi^{\ell+1}}{k-h}[(-1)^{\ell}+1]\right] + \delta_{kh}\left[i\pi\frac{\ell+1}{\ell+2}\lambda_{kh}^{(\ell)} + \frac{(-i)^{\ell+1}\pi^{\ell+1}}{\ell+2}\right]$$

that is

$$\lambda_{kh}^{(\ell+1)} = \left[(1-\delta_{kh})\frac{\ell+1}{k-h} + \delta_{kh}i\pi\frac{\ell+1}{\ell+2}\right]\lambda_{kh}^{(\ell)} - (1-\delta_{kh})(-1)^{k-h}\frac{i^{\ell}\pi^{\ell+1}}{k-h}[(-1)^{\ell}+1] + \delta_{kh}\frac{(-i)^{\ell+1}\pi^{\ell+1}}{\ell+2}$$

It is not so easy to derive a similar property for the γ-coefficients as a function of ℓ, however there is a simple rule for the recursiveness of the scale (upper) indices, as follows:

Theorem 2. *The connection coefficients* (22) *are given recursively by the matrix at the lowest scale level:*

$$\gamma^{(\ell)nn}_{\ \ kh} = 2^{\ell(n-1)}\gamma^{(\ell)11}_{\ \ kh} \, . \tag{27}$$

Proof:
As can be seen from Eqs. (22) parameter n appears only in the factor

$$2^{n\ell-1}$$

so that Eq. (27) follows from the identity

$$2^{n\ell-1}=2^{\ell(n-1)}2^{\ell-1}$$

□

Moreover it can be shown also that

$$\gamma^{(2\ell+1)nn}_{\quad kh}=-\gamma^{(2\ell+1)nn}_{\quad hk}\quad,\quad \gamma^{(2\ell)nn}_{\quad kh}=\gamma^{(2\ell)nn}_{\quad hk}\,.$$

4 Differential Properties of $L_2(\mathbb{R})$-functions in Shannon Wavelet Basis

4.1 Taylor Series

By using the connection coefficients, and taking into account that the basic functions are $\mathcal{C}^{\infty}$-functions, it is easy to establish the following theorem:

Theorem 3. *If $f(x)\in B_{\psi}\subset L_2(\mathbb{R})$ and $f(x)\in {}^{S}$ the Taylor series of $f(x)$ in x_0 is*

$$f(x)=f(x_0)+\sum_{r=1}^{\infty}\left[\sum_{h,k=-\infty}^{\infty}\alpha_h\,\lambda_{hk}^{(r)}\varphi_k(x_0)+\sum_{n=0}^{\infty}\sum_{k,s=-\infty}^{\infty}2^{r(n-1)}\beta_k^n\gamma^{(r)11}_{\quad sk}\psi_s^n(x_0)\right]\frac{(x-x_0)^r}{r!}\tag{28}$$

being α_h and β_k^n given by (9), (11).

Proof: From (10), the ℓth-order derivative ($\ell\le S$) is

$$\begin{aligned}
f^{(\ell)}(x) &= \sum_{h=-\infty}^{\infty}\alpha_h\frac{d^\ell}{dx^\ell}\varphi_h(x)+\sum_{n=0}^{\infty}\sum_{k=-\infty}^{\infty}\beta_k^n\frac{d^\ell}{dx^\ell}\psi_k^n(x)\,,\\
&\overset{(17)}{=} \sum_{h=-\infty}^{\infty}\alpha_h\sum_{k=-\infty}^{\infty}\lambda_{hk}^{(\ell)}\varphi_k(x)+\sum_{n=0}^{\infty}\sum_{k=-\infty}^{\infty}\beta_k^n\sum_{m=-\infty}^{\infty}\sum_{s=-\infty}^{\infty}\gamma^{(\ell)mn}_{\quad sk}\psi_s^m(x)\,,\\
&= \sum_{h,k=-\infty}^{\infty}\alpha_h\,\lambda_{hk}^{(\ell)}\varphi_k(x)+\sum_{n,m=0}^{\infty}\sum_{k,s=-\infty}^{\infty}\beta_k^n\gamma^{(\ell)mn}_{\quad sk}\psi_s^m(x)\,,
\end{aligned}$$

so that by taking into account (27) the proof follows. □

Using a suitable choice of the initial point x_0 Eq. (28) can be simplified. For instance, at the integers, $x_0 = h,\ (h \in \mathbb{Z})$, according to Eqs. (13), (27) it is

$$f(x) \cong f(h) + \sum_{r=1}^{S}\left[\sum_{h=-\infty}^{\infty} \alpha_h \lambda_{hh}^{(r)} + \sum_{n=0}^{\infty}\sum_{k,s=-\infty}^{\infty} \frac{2^{r(n-1)+1+n/2}}{(2^{n+1}h-2s-1)\pi}\beta_k^n \gamma^{(r)11}_{sk}\psi_s^n(h)\right]\frac{(x-h)^r}{r!}$$

In particular, for $x_0 = h = 0$, eq. (28) gives

$$\begin{aligned} f(x) &= f(0) + \sum_{r=1}^{\infty}\left[\sum_{h,k=-\infty}^{\infty} \alpha_h \lambda_{hk}^{(r)}\varphi_k(0) + \sum_{n=0}^{\infty}\sum_{k,s=-\infty}^{\infty} 2^{r(n-1)}\beta_k^n \gamma^{(r)11}_{sk}\psi_s^n(0)\right]\frac{x^r}{r!} \\ &= f(0) + \sum_{r=1}^{\infty}\left[\sum_{h=-\infty}^{\infty} \alpha_h \lambda_{h0}^{(r)} + \sum_{n=0}^{\infty}\sum_{k,s=-\infty}^{\infty} 2^{r(n-1)}\beta_k^n \gamma^{(r)11}_{sk}\psi_s^n(0)\right]\frac{x^r}{r!} \end{aligned} \tag{29}$$

and since

$$\psi_s^n(0) = (-1)^s \frac{2^{1+n/2}}{(-2s-1)\pi} \quad , \quad (-2k-1 \neq 0)$$

we get

$$\boxed{f(x) = f(0) + \sum_{r=1}^{\infty}\left[\sum_{h=-\infty}^{\infty} \alpha_h \lambda_{h0}^{(r)} + \sum_{n=0}^{\infty}\sum_{k,s=-\infty}^{\infty} \frac{(-1)^{s+1}2^{n(r+1/2)+1-r}}{(2s+1)\pi}\beta_k^n \gamma^{(r)11}_{sk}\right]\frac{x^r}{r!}} \tag{30}$$

with $\lambda_{h0}^{(r)}$ given by (25) and $\gamma^{(r)11}_{sk}$ by (22) respectively. So that each function $f(x) \in \mathcal{B} \subset L_2(\mathbb{R})$, can be easily expressed as a power series, when the finite values of the wavelet coefficients α_h , β_k^n are given, according to (9),(11).

There exists a Taylor power series for the basic functions $\varphi(x)$, $\psi(x)$:

$$\left\{\begin{aligned} \varphi(x) &= 1 + \sum_{r=1}^{\infty}\left(\sum_{h=-\infty}^{\infty} \alpha_h \lambda_{h0}^{(r)}\right)\frac{x^r}{r!} \\ \psi(x) &= -\frac{2}{\pi} + \sum_{r=1}^{\infty}\left(\sum_{n=0}^{\infty}\sum_{k,s=-\infty}^{\infty} \frac{(-1)^{s+1}2^{n(r+1/2)+1-r}}{(2s+1)\pi}\delta_k^n \gamma^{(r)11}_{sk}\right)\frac{x^r}{r!} \end{aligned}\right. \tag{31}$$

being $\psi(0) = -\dfrac{2}{\pi}$, according to (1). However the Taylor series for the wavelet function $\psi(x)$ can be also easily computed as follows:

$$\begin{aligned}
\psi(x) &= \psi(0) + \sum_{\ell=1}^{\infty}\left(\frac{d^{\ell}\psi(x)}{dx^{\ell}}\right)_{x=0}\frac{x^{\ell}}{\ell!} \\
&\overset{(17),(24)}{=} -\frac{2}{\pi} + \sum_{\ell=1}^{\infty}\left(\sum_{k=-\infty}^{\infty}\gamma^{(\ell)00}_{0k}\psi^{0}_{k}(0)\right)\frac{x^{\ell}}{\ell!} \\
&\overset{(14)}{=} -\frac{2}{\pi} + \sum_{\ell=1}^{\infty}\left(\sum_{k=-\infty}^{\infty}\gamma^{(\ell)00}_{0k}(-1)^{k+1}\frac{2}{(2k+1)\pi}\right)\frac{x^{\ell}}{\ell!}
\end{aligned}$$

that is

$$\boxed{\psi(x) = -\frac{2}{\pi} + \sum_{\ell=1}^{\infty}\left(\sum_{k=-\infty}^{\infty}(-1)^{k+1}\frac{2}{(2k+1)\pi}\gamma^{(\ell)00}_{0k}\right)\frac{x^{\ell}}{\ell!}} \tag{32}$$

4.2 Functional Equations

The connection coefficients fulfill the following:

Theorem 4. *For any $k \in \mathbb{Z}$ and $\ell \in \mathbb{N}$ it is*

$$(i\omega)^{\ell}e^{-i\omega k} = \sum_{h=-\infty}^{\infty}\lambda^{(\ell)}_{kh}e^{-i\omega h} \quad , \quad -\pi \le \omega \le \pi$$

or

$$(i\omega)^{\ell} = \sum_{h=-\infty}^{\infty}\lambda^{(\ell)}_{kh}e^{-i\omega(h-k)} \quad , \quad -\pi \le \omega \le \pi,\ \forall k \in \mathbb{Z}$$

Proof:
From Eq. (17), taking the Fourier transform of both sides and taking into account (20) we get

$$\begin{aligned}
(i\omega)^{\ell}\widehat{\varphi}_k(\omega) &= \sum_{h=-\infty}^{\infty}\lambda^{(\ell)}_{kh}\widehat{\varphi}_h(\omega) \\
(i\omega)^{\ell}e^{-i\omega k}\chi(\omega+3\pi) &\overset{(7)}{=} \sum_{h=-\infty}^{\infty}\lambda^{(\ell)}_{kh}e^{-i\omega h}\chi(\omega+3\pi)
\end{aligned}$$

from where the identity follows.

□

In particular, by assuming, without restrictions, $k = 0$, we have

Corollary 1. *For any $\ell \in \mathbb{N}$ it is*

$$(i\omega)^{\ell} = \sum_{h=-\infty}^{\infty}\lambda^{(\ell)}_{0h}e^{-i\omega h} \quad , \quad -\pi \le \omega \le \pi \tag{33}$$

so that $\lambda_{0h}^{(\ell)}$ are the Fourier coefficients of $(i\omega)^\ell$.

Analogously, from $(17)_2$, it can be shown that

Theorem 5. *For any $k \in \mathbb{Z}$ and $\ell, n \in \mathbb{N}$ it is*

$$(i\omega)^\ell e^{-i\omega(k+1/2)/2^n} = \sum_{h=-\infty}^{\infty} \gamma_{kh}^{(\ell)nn} e^{-i\omega(h+1/2)/2^n} \quad , \quad \omega \in [-2^{n+1}\pi\, , -2^n\pi] \cup [2^n\pi\, , 2^{n+1}\pi]$$

or

$$(i\omega)^\ell = \sum_{h=-\infty}^{\infty} \gamma_{kh}^{(\ell)nn} e^{-i\omega(h-k)/2^n} \quad , \quad \omega \in [-2^{n+1}\pi\, , -2^n\pi] \cup [2^n\pi\, , 2^{n+1}\pi]$$

In particular, with $k = 0$, and taking into account (27), we have

Corollary 2. *For any $\ell, n \in \mathbb{N}$ it is*

$$(i\omega)^\ell = 2^{\ell(n-1)} \sum_{h=-\infty}^{\infty} \gamma_{0h}^{(\ell)11} e^{-i\omega h/2^n} \quad , \quad \omega \in [-2^{n+1}\pi\, , -2^n\pi] \cup [2^n\pi\, , 2^{n+1}\pi]$$

As a consequence of the previous theorems it follows that

Theorem 6. *For any $\ell, n \in \mathbb{N}$*

$$(i\omega)^\ell = \begin{cases} \displaystyle\sum_{h=-\infty}^{\infty} \lambda_{0h}^{(\ell)} e^{-i\omega h} & , \quad -\pi \le \omega \le \pi \\ \displaystyle 2^{\ell(n-1)} \sum_{h=-\infty}^{\infty} \gamma_{0h}^{(\ell)11} e^{-i\omega h/2^n} & , \quad \omega \in [-2^{n+1}\pi\, , -2^n\pi] \cup [2^n\pi\, , 2^{n+1}\pi] \end{cases}$$

It also follows that:

Corollary 3. *The Fourier transform of the derivatives of a function*

$$\widehat{\frac{d^\ell}{dx^\ell} f(x)} = \widehat{f}(\omega) \times \begin{cases} \displaystyle\sum_{h=-\infty}^{\infty} \lambda_{0h}^{(\ell)} e^{-i\omega h} & , \quad -\pi \le \omega \le \pi \\ \displaystyle 2^{\ell(n-1)} \sum_{h=-\infty}^{\infty} \gamma_{kh}^{(\ell)11} e^{-i\omega h/2^n} & , \quad \omega \in [-2^{n+1}\pi\, , -2^n\pi] \cup [2^n\pi\, , 2^{n+1}\pi] \end{cases}$$

If we express $e^{i\omega}$ as a Taylor series we know that

$$e^{i\omega} = \sum_{\ell=0}^{\infty} \frac{(i\omega)^\ell}{\ell!}$$

so that $e^{i\omega}$ with $-\pi \le \omega \le \pi$ is the solution of the functional equation

$$X = \sum_{\ell=0}^{\infty} \sum_{h=-\infty}^{\infty} \frac{1}{\ell!} \lambda_{0h}^{(\ell)} X^{-h}$$

Moreover, the following theorem of moments

$$\int_{\mathbb{R}} x^{\ell} f(x)dx = i^{\ell}\frac{d\widehat{f}(\omega)}{d\omega^{\ell}}$$

can be written as

$$\int_{\mathbb{R}} x^{\ell} f(x)dx = i^{\ell}\widehat{f}(\omega)\times\begin{cases}\sum\limits_{h=-\infty}^{\infty}\lambda_{kh}^{(\ell)}e^{-i\omega h} & , \quad -\pi \le \omega \le \pi \\ \sum\limits_{h=-\infty}^{\infty}\gamma_{kh}^{(\ell)nn}e^{-i\omega(h-k)/2^n} & , \quad \omega \in [-2^{n+1}\pi\,, -2^n\pi] \cup [2^n\pi\,, 2^{n+1}\pi]\end{cases}$$

4.3 Error of the Approximation by Connection Coefficients

For a fixed scale of approximation in the equations (17), it is possible to estimate the error as follows. It should be noticed that the approximations depend on the upper bound of the limits in the sums. Thus:

Theorem 7 (Error of the approximation of scaling functions derivatives). *The error of the approximation in* $(17)_1$ *is given by*

$$\left|\frac{d^{\ell}}{dx^{\ell}}\varphi_h(x) - \sum_{k=-N}^{N}\lambda_{hk}^{(\ell)}\varphi_k(x)\right| \le \left|\lambda_{h(-N-1)}^{(\ell)} + \lambda_{h(N+1)}^{(\ell)}\right|$$

Proof: The error of the approximation $(17)_1$ is defined as

$$\frac{d^{\ell}}{dx^{\ell}}\varphi_h(x) - \sum_{k=-N}^{N}\lambda_{hk}^{(\ell)}\varphi_k(x) = \sum_{k=-\infty}^{-N-1}\lambda_{hk}^{(\ell)}\varphi_k(x) + \sum_{k=N+1}^{\infty}\lambda_{hk}^{(\ell)}\varphi_k(x)$$

Concerning the r.h.s, and according to (15), the following holds

$$\sum_{k=-\infty}^{-N-1}\lambda_{hk}^{(\ell)}\varphi_k(x) + \sum_{k=N+1}^{\infty}\lambda_{hk}^{(\ell)}\varphi_k(x) \le \max_{x\in\mathbb{R}}\left[\sum_{k=-\infty}^{-N-1}\lambda_{hk}^{(\ell)}\varphi_k(x) + \sum_{k=N+1}^{\infty}\lambda_{hk}^{(\ell)}\varphi_k(x)\right]$$

$$= \lambda_{h(-N-1)}^{(\ell)}\varphi_{-N-1}(x) + \lambda_{h(N+1)}^{(\ell)}\varphi_{N+1}(x) \le \lambda_{h(-N-1)}^{(\ell)} + \lambda_{h(N+1)}^{(\ell)}\,.$$

□

Theorem 8 (Error of the approximation of wavelet functions derivatives). *The error of the approximation in* $(17)_2$ *is given by*

$$\left|\frac{d^{\ell}}{dx^{\ell}}\psi_h^m(x) - \sum_{n=0}^{N}\sum_{k=-S}^{S}\gamma_{hk}^{(\ell)mn}\psi_k^n(x)\right| \le \left|2^{\ell(m-1)+m/2}\frac{3\sqrt{3}}{\pi}\left[\gamma_{h(-S-1)}^{(\ell)11} + \gamma_{h(S+1)}^{(\ell)11}\right]\right|$$

Proof: The error of this approximation is

$$\frac{d^\ell}{dx^\ell}\psi_h^m(x) - \sum_{n=0}^{N}\sum_{k=-S}^{S}\gamma^{(\ell)mn}_{\ \ hk}\,\psi_k^n(x) = \sum_{n=N+1}^{\infty}\left[\sum_{k=-\infty}^{-S-1}\gamma^{(\ell)mn}_{\ \ hk}\,\psi_k^n(x) + \sum_{k=S+1}^{\infty}\gamma^{(\ell)mn}_{\ \ hk}\,\psi_k^n(x)\right]$$

If $m < N$ the r.h.s. according to (22) is zero, therefore we assume that $m > N$ so that the last equation becomes

$$\begin{aligned}
\frac{d^\ell}{dx^\ell}\psi_h^m(x) - \sum_{n=0}^{N}\sum_{k=-S}^{S}\gamma^{(\ell)mn}_{\ \ hk}\,\psi_k^n(x) &= \left[\sum_{k=-\infty}^{-S-1}\gamma^{(\ell)mm}_{\ \ hk}\,\psi_k^n(x) + \sum_{k=S+1}^{\infty}\gamma^{(\ell)mm}_{\ \ hk}\,\psi_k^n(x)\right]\\
&\overset{(27)}{=} 2^{\ell(m-1)}\left[\sum_{k=-\infty}^{-S-1}\gamma^{(\ell)11}_{\ \ hk} + \sum_{k=S+1}^{\infty}\gamma^{(\ell)11}_{\ \ hk}\right]\psi_k^m(x)\\
&\le 2^{\ell(m-1)}\max\left\{\left[\sum_{k=-\infty}^{-S-1}\gamma^{(\ell)11}_{\ \ hk} + \sum_{k=S+1}^{\infty}\gamma^{(\ell)11}_{\ \ hk}\right]\psi_k^m(x)\right\}\\
&= 2^{\ell(m-1)}\left[\gamma^{(\ell)11}_{\ \ h(-S-1)}\psi^m_{(-S-1)}(x) + \gamma^{(\ell)11}_{\ \ h(S+1)}\psi^m_{(S+1)}(x)\right]\\
&\le 2^{\ell(m-1)}\left[\gamma^{(\ell)11}_{\ \ h(-S-1)}\max\psi^m_{(-S-1)}(x) + \gamma^{(\ell)11}_{\ \ h(S+1)}\max\psi^m_{(S+1)}(x)\right]\\
&\overset{(16)}{=} 2^{\ell(m-1)}2^{m/2}\frac{3\sqrt{3}}{\pi}\left[\gamma^{(\ell)11}_{\ \ h(-S-1)} + \gamma^{(\ell)11}_{\ \ h(S+1)}\right].
\end{aligned}$$

□

5 Fractional Derivatives of the Wavelet Basis

The simplest way to define the fractional derivative is based on the assumption that the non integer derivative of the exponential function formally coincides with the derivative of integer order so that

$$\frac{d^\nu}{dx^\nu}e^{ax} = a^\nu e^{ax} \qquad \nu \in \mathbb{Q}\,.$$

For negative values of ν this formula still holds true and represents integration.

It is known that the fractional derivative can't be analytically computed except for some special functions, such as (see e.g. Eldred et Al. 1996; Oldham and Spanier 1970; Ross 1975) the following

$$\frac{d^\nu}{dx^\nu}e^{ax} = a^\nu e^{ax}\,,\quad \frac{d^\nu}{dx^\nu}\cos ax = a^\nu\cos\left(ax + \frac{\pi}{2}\nu\right)\,,\quad \frac{d^\nu}{dx^\nu}\sin ax = a^\nu\sin\left(ax + \frac{\pi}{2}\nu\right)\,.$$

In particular for integer powers of x its frational derivative is:

$$\frac{d^\nu}{dx^\nu}x^n = \frac{\Gamma(n+1)}{\Gamma(n+1-\nu)}x^{n-\nu} \qquad (n \ge 0;\ 0 \le \nu \le 1) \tag{34}$$

where with $n = 1$, we get the fractional differential

$$(dx)^{\nu} \overset{\text{def}}{=} d^{\nu}x = \frac{x^{1-\nu}}{\Gamma(2-\nu)} dx^{\nu} \qquad (0 \leq \nu \leq 1) \tag{35}$$

From these classical examples we can see that the fractional derivative can be also interpreted as an interpolating function between derivatives with integer order, so that

$$\frac{d^{\nu}}{dx^{\nu}} f(x) = (1-\nu) f(x) + \nu \frac{d}{dx} f(x) \quad , \quad 0 \leq \nu \leq 1$$

In general, if $f(x)$ is a single-valued real function, then the Riemann-Liouville fractional order derivative is defined as (Oldham and Spanier 1970)

$$\frac{d^{\nu}}{dx^{\nu}} f(x) \overset{\text{def}}{=} \frac{1}{\Gamma(1-\nu)} \frac{d}{dx} \int_0^x \frac{f(\xi)}{(x-\xi)^{\nu}} d\xi \quad , \quad (0 < \nu < 1 \,, x > 0) \,,$$

$\Gamma(\nu)$ being the Gamma-function.

Other equivalent representations were given by Caputo (for a differentiable function)

$$\frac{d^{\nu}}{dx^{\nu}} f(x) \overset{\text{def}}{=} \frac{1}{\Gamma(1-\nu)} \int_0^x \frac{f'(\xi)}{(x-\xi)^{\nu}} d\xi \quad , \quad 0 < \nu < 1 \,,$$

and by Grünwald (Eldred et Al. 1996; Ross 1975)

$$\frac{d^{\nu}}{dx^{\nu}} f(x) = \lim_{N \to \infty} \frac{1}{\Gamma(-\nu)} \left(\frac{x}{N}\right)^{-\nu} \sum_{k=0}^{N-1} \frac{\Gamma(k-\nu)}{\Gamma(k+1)} f\left[\left(1 - \frac{k}{N}\right) x\right] \quad , \quad (0 < \nu < 1 \,, x > 0). \tag{36}$$

One drawback in the Grünwald definition, as well as in that of Riemann-Liouville, is that it can't be computed for negative values of the variable ($x < 0$).

5.1 Complex Shannon Wavelets on Fractal Sets of Dimension ν

Let us start from eq. (6) and define the basic complex functions of the Shannon wavelet theory on fractal set of fractal dimension ν $(0 < \nu \leq 1)$ as follows:

$$\begin{cases} \varphi_{\nu}(z) \overset{(6)}{=} \dfrac{1}{2} \displaystyle\sum_{s=1}^{\infty} \frac{\pi^{\nu s-1} i^{\nu s-1}}{\Gamma(\nu s+1)} \left[1 - (-1)^{\nu s}\right] z^{\nu(s-1)} \overset{(31)}{=} 1 + \sum_{\ell=1}^{\infty} \lambda_{00}^{(\ell)} \frac{z^{\nu \ell}}{\Gamma(\nu \ell + 1)} \\ \psi_{\nu}(z) \overset{(32)}{=} -\dfrac{2}{\pi} + \displaystyle\sum_{\ell=1}^{\infty} \left(\sum_{k=-\infty}^{\infty} (-1)^{k+1} \frac{2}{(2k+1)\pi} \gamma_{0k}^{(\ell)00} \right) \frac{z^{\nu \ell}}{\Gamma(\nu \ell + 1)} \end{cases} \tag{37}$$

or, if we define:

$$\lambda^{\ell} \stackrel{\text{def}}{=} \begin{cases} 1\,,\ \ell = 0 \\ \lambda_{00}^{(\ell)}\,,\ \ell > 0 \end{cases}$$

$$\gamma^{\ell} \stackrel{\text{def}}{=} \begin{cases} -\dfrac{2}{\pi}\,,\ \ell = 0 \\ \displaystyle\sum_{k=-\infty}^{\infty} (-1)^{k+1} \frac{2}{(2k+1)\pi} \gamma_{0k}^{(\ell)00}\,,\ \ell > 0 \end{cases} \tag{38}$$

we have the two complex power series

$$\varphi_{\nu}(z^{\nu}) = \sum_{\ell=0}^{\infty} \frac{\lambda^{\ell}}{\Gamma(\nu\ell+1)} z^{\nu\ell} \quad , \quad \psi_{\nu}(z^{\nu}) = \sum_{\ell=0}^{\infty} \frac{\gamma^{\ell}}{\Gamma(\nu\ell+1)} z^{\nu\ell}\,. \tag{39}$$

So that, when $\nu = 1$ we have the complex Shannon wavelets on regular domain:

$$\varphi_1(z) = \varphi(z) = \sum_{\ell=0}^{\infty} \frac{\lambda^{\ell}}{\ell!} z^{\ell} \quad , \quad \psi_1(z) = \psi(z) = \sum_{\ell=0}^{\infty} \frac{\gamma^{\ell}}{\ell!} z^{\ell} \tag{40}$$

which reduce to the real Shannon wavelets when $z \to x$. While for $\nu < 1$, we have purely complex Shannon wavelets, derived in a continuous non regular domain, which are functions of the variable z^{ν}. The complex number z^{ν} is defined (Yang 2012) as

$$z^{\nu} \stackrel{\text{def}}{=} x^{\nu} + i^{\nu} y^{\nu} \quad , \qquad (x, y \in \mathbb{R};\ 0 < \nu \le 1)$$

By limiting ourselves to the reals, i.e. $z^{\nu} = x^{\nu} = \Re(z^{\nu})\,, \Im(z^{\nu}) = 0$, the variable x^{ν} might be defined in a non-regular domains, such as the Cantor set. Fig. 1 illustrates the functions: $\varphi(x)$, $\Re(\varphi_0(x^0))$, $\Re(\varphi_{1/2}(x^{1/2}))$, $\Re(\varphi_1(x^1))$, while in Fig. 2 the following $\varphi(x)$, $\Im(\varphi_0(x^0))$, $\Im(\varphi_{1/2}(x^{1/2}))$, $\Im(\varphi_1(x^1))$ are shown.

Starting from the Eq. (30), any function $f(x) \in \mathcal{B}_{\psi} \subset L_2(\mathbb{R})$ can be extended to a complex variable z on fractal set of dimension ν by the following Taylor series

$$f(z) = f(0) + \sum_{\ell=1}^{\infty} \left[\sum_{h=-\infty}^{\infty} \alpha_h \lambda_{h0}^{(r)} + \sum_{n=0}^{\infty} \sum_{k,s=-\infty}^{\infty} \frac{(-1)^{s+1} 2^{n(r+1/2)+1-r}}{(2s+1)\pi} \beta_k^n \gamma_{sk}^{(\ell)11} \right] \frac{z^{\nu\ell}}{\Gamma(\nu\ell+1)} \tag{41}$$

In order to define the power series (39) also for nonregular domains, let us remind ourselves of the simplest well known nonregular domain: the Cantor set.

The Cantor function is defined on the closed interval [0, 1] by the recursive law:

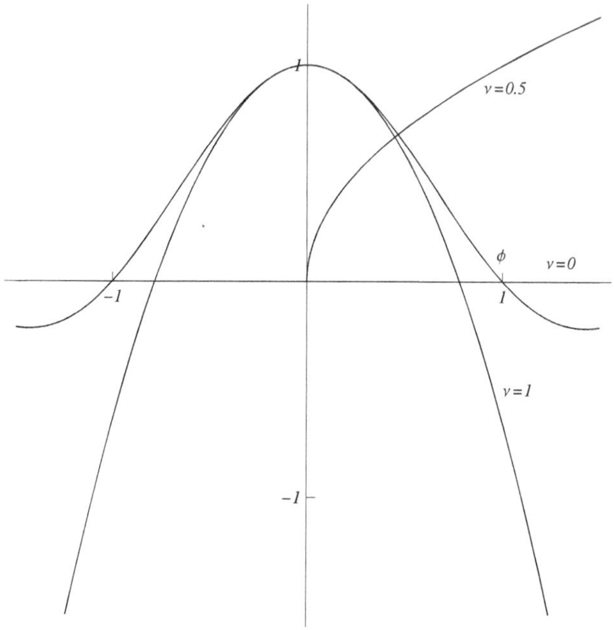

Fig. 1. *The Shannon scaling function $\varphi(x)$ and the real part of the complex function $\varphi_\nu(x)$ with $\nu = 0, 1/2, 1$.*

$$\widetilde{x}^\nu = \begin{cases} f(x,n) = 0 & , \quad x = 0 \wedge n = 0 \\ f(x,n) = 1 & , \quad x = 1 \wedge n = 0 \\ f(x,n) = x & , \quad n = 0 \\ f(x,n) = \frac{1}{2} f(3x, n-1) & , \quad 0 < x \le \frac{1}{3} \\ f(x,n) = \frac{1}{2} & , \quad \frac{1}{3} < x \le \frac{2}{3} \\ f(x,n) = \frac{1}{2} + \frac{1}{2} f(3(x - \frac{2}{3}), n-1) & , \quad \frac{2}{3} < x \le 1 \end{cases} \tag{42}$$

by letting $n \to \infty$.

If we consider $\widetilde{x}^\nu$ as a variable of the Cantor set (42), we can see some pictures of the real part of the complex scaling function $\varphi_\nu(z^\nu)$ on Cantor set. In Fig. 3 the functions: $\varphi(x)$, $\Re(\varphi_{1/2}(x^{1/2}))$, $\Re(\varphi_{1/2}(\widetilde{x}^\nu))$, $\Re(\varphi_1(x^1))$, $\Re(\varphi_1(\widetilde{x}^\nu))$ are shown.

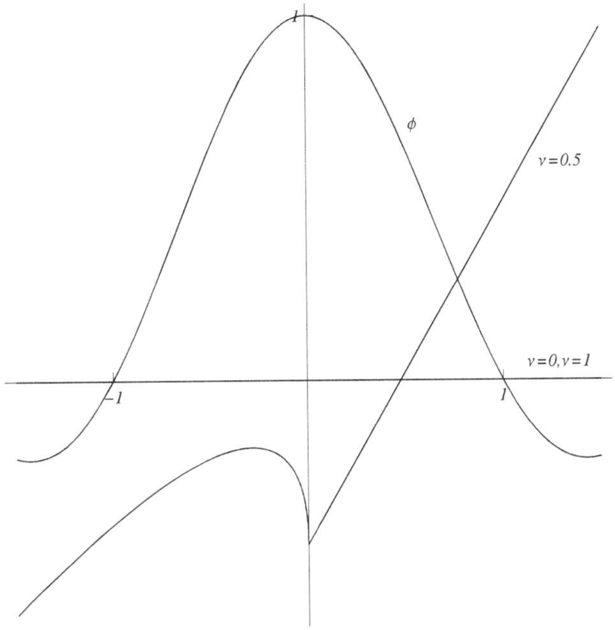

Fig. 2. *The Shannon scaling function $\varphi(x)$ and the imaginary part of the complex function $\varphi_v(x)$ with $v = 0, 1/2, 1$.*

5.2 Local Fractional Derivatives of Complex Functions

In this section we give the local fractional derivative of the complex Shannon wavelets and the local fractional derivative of the functions $f(x) \in \mathcal{B}_\psi \subset L_2(\mathbb{R})$. So that when the wavelet coefficients of $f(x)$ are known we can easily compute their local fractional derivatives. According to the definition given in (Yang 2012), the local fractional derivative of the complex function $f(z)$ in the neighborhood of the point z_0 is defined as:

$$\frac{d^v}{dz^v} f(z)\Big|_{z=z_0} = f^{(v)}(z_0) = \Gamma(1+v) \lim_{z \to z_0} \frac{f(z) - f(z_0)}{(z - z_0)^v} \quad , \qquad 0 < v \le 1$$

According to this definition, it has been shown that almost all of the basic rules of derivatives hold true. Moreover for the elementary functions we have (Yang 2012) that:

$$\frac{d^v}{dz^v} E_{a,b}(z^v) = E_{a,b}(z^v) \quad , \quad \frac{d^v}{dz^v} \sin_v(z^v) = \cos_v(z^v), \quad \frac{d^v}{dz^v} \cos_v(z^v) = -\sin_v(z^v)\,. \tag{43}$$

In particular,

$$\frac{d^v}{dz^v} z^{kv} = \frac{\Gamma(1+kv)}{\Gamma(1+(k-1)v)} z^{(k-1)v} \tag{44}$$

so that for $k = 1$, we have

$$d^v z^v = \Gamma(1+v) dz^v$$

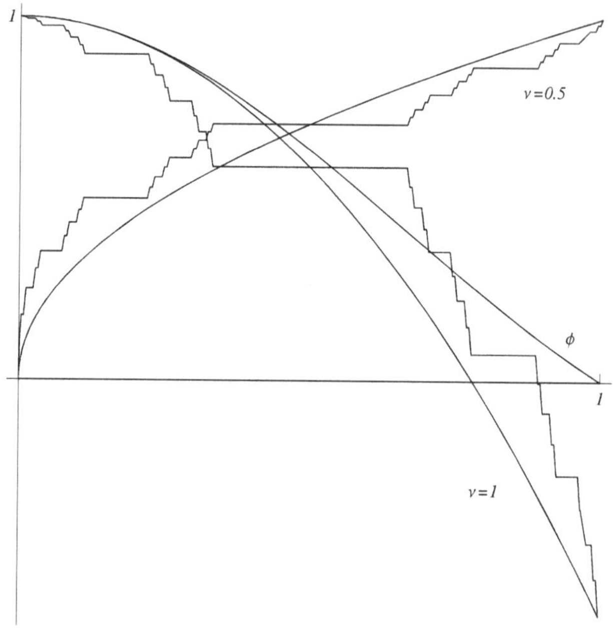

Fig. 3. *The Shannon scaling function $\varphi(x)$ and the real part of the complex function $\varphi_\nu(x)$ with $\nu = 0, 1/2, 1$, on x and on a Cantor set.*

or

$$(dz)^\nu = \Gamma(1+\nu)dz^\nu$$

This value of the differential operator can be used to define several integral transforms on Cantor sets (see e.g. Yang 2012; Yang 2013; Yang 2014; Yang and Srivastava 2013).

We can now define the local fractional derivative of a complex function as follows:

Theorem 9. *The local fractional derivative of a complex function $f(z^\nu) \in \mathcal{B}_{\psi_\nu} \subset L_2(\mathbb{R})$, in the neighborhood of 0 is*

$$\frac{d^\nu}{dz^\nu} f(z) = \sum_{\ell=1}^{\infty} \frac{A_\ell(\alpha,\beta)}{\Gamma(1+(\ell-1)\nu)} z^{(\ell-1)\nu} \tag{45}$$

being

$$A_\ell(\alpha,\beta) \stackrel{\text{def}}{=} \left[\sum_{h=-\infty}^{\infty} \alpha_h \lambda_{h0}^{(\ell)} + \sum_{n=0}^{\infty} \sum_{k,s=-\infty}^{\infty} \frac{(-1)^{s+1} 2^{n(\ell+1/2)+1-\ell}}{(2s+1)\pi} \beta_k^n \gamma^{(\ell)11}_{sk} \right] \tag{46}$$

Proof: It is

$$
\begin{aligned}
\frac{d^{\nu}}{dz^{\nu}}f(z) &= \sum_{\ell=1}^{\infty}\left[\sum_{h=-\infty}^{\infty}\alpha_h\lambda_{h0}^{(\ell)}+\sum_{n=0}^{\infty}\sum_{k,s=-\infty}^{\infty}\frac{(-1)^{s+1}2^{n(\ell+1/2)+1-\ell}}{(2s+1)\pi}\beta_k^n\gamma^{(\ell)11}_{\ \ sk}\right]\\
&\quad\times\frac{1}{\Gamma(\nu\ell+1)}\frac{d^{\nu}}{dz^{\nu}}z^{\nu\ell}\\
&\overset{(44)}{=} \sum_{\ell=1}^{\infty}\left[\sum_{h=-\infty}^{\infty}\alpha_h\lambda_{h0}^{(\ell)}+\sum_{n=0}^{\infty}\sum_{k,s=-\infty}^{\infty}\frac{(-1)^{s+1}2^{n(\ell+1/2)+1-\ell}}{(2s+1)\pi}\beta_k^n\gamma^{(\ell)11}_{\ \ sk}\right]\\
&\quad\times\frac{1}{\Gamma(\nu\ell+1)}\frac{\Gamma(1+\ell\nu)}{\Gamma(1+(\ell-1)\nu)}z^{(\ell-1)\nu}\\
&= \sum_{\ell=1}^{\infty}\left[\sum_{h=-\infty}^{\infty}\alpha_h\lambda_{h0}^{(\ell)}+\sum_{n=0}^{\infty}\sum_{k,s=-\infty}^{\infty}\frac{(-1)^{s+1}2^{n(\ell+1/2)+1-\ell}}{(2s+1)\pi}\beta_k^n\gamma^{(\ell)11}_{\ \ sk}\right]\\
&\quad\times\frac{1}{\Gamma(1+(\ell-1)\nu)}z^{(\ell-1)\nu}
\end{aligned}
\tag{47}
$$

□

5.3 Example: Fractional Derivative of a Gaussian on a Fractal Set

In order to show the efficiency of the proposed method for the computation of a fractional derivative, let us consider the function e^{-x^2}. A good approximation of this function, in terms of Shannon wavelet expansion (10), can be obtained as

$$
\begin{aligned}
e^{-x^2} &\cong \sum_{h=-1}^{1}\alpha_h\varphi_h(x)+\sum_{n=0}^{0}\sum_{h=-1}^{1}\beta_h^n\psi_h^n(x)\\
&\cong \alpha_{-1}\varphi_{-1}(x)+\alpha_0\varphi(x)+\alpha_1\varphi_1(x)+\\
&+ \beta_{-1}^0\psi_{-1}^0(x)+\beta_0^0\psi_0^0(x)+\beta_1^0\psi_1^0(x)
\end{aligned}
$$

where

$$\alpha_{-1}=\alpha_1=0.123,\ \alpha_0=0.30,\ \psi_{-1}^0=\psi_1^0=0.004,\ \psi_0^0=0.001\,.$$

If we neglect the detailed coefficients β_k^n, the approximate Shannon wavelet representation is (see Fig. 4)

$$e^{-x^2}\cong 0.123\ \varphi_{-1}(x)+0.30\ \varphi(x)+0.123\ \varphi_1(x)$$

According to (45),(46) the fractional derivative is

$$\frac{d^{\nu}}{dx^{\nu}}e^{-x^2}\cong\sum_{\ell=1}^{\infty}\frac{A_{\ell}(\alpha,0)}{\Gamma(1+(\ell-1)\nu)}x^{(\ell-1)\nu}$$

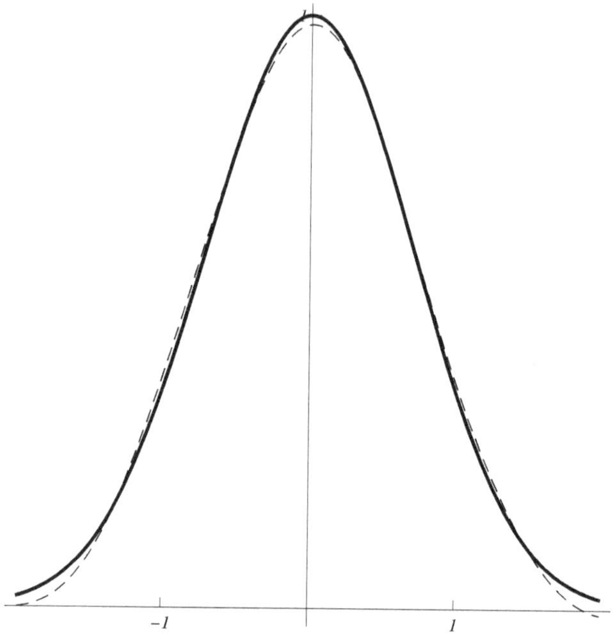

Fig. 4. *The function* e^{-x^2} *and its Shannon wavelet representation (dashed) at the scale* $n = 0$, $k = -1, 0, 1$.

with

$$A_\ell(\alpha, 0) = \sum_{h=-1}^{1} \alpha_h \lambda_{h0}^{(\ell)}$$

From (21) it is

$$\lambda_{-1,0}^{(\ell)} = -\lambda_{1,0}^{(\ell)}, \; \lambda_{0,0}^{(\ell)} = \frac{i^\ell \pi^{\ell+1}}{2\pi(\ell+1)}[1 + (-1)^\ell],$$

so that

$$\frac{d^\nu}{dx^\nu} e^{-x^2} \cong \sum_{\ell=1}^{\infty} 0.30 \frac{i^\ell \pi^{\ell+1}}{2\pi(\ell+1)}[1 + (-1)^\ell] \frac{1}{\Gamma(1 + (\ell - 1)\nu)} x^{(\ell-1)\nu}.$$

On the other hand, by taking into account the properties of local fractional derivatives (Yang 2012) and (43), we have

$$\frac{d^\nu}{dx^\nu} e^{-x^2} = -2x^\nu e^{-\nu x^2}.$$

We can notice, by comparing the wavelet approximation with the analytical results that we have a good approximation even with a very low upper value in the series expansion (Fig. 5).

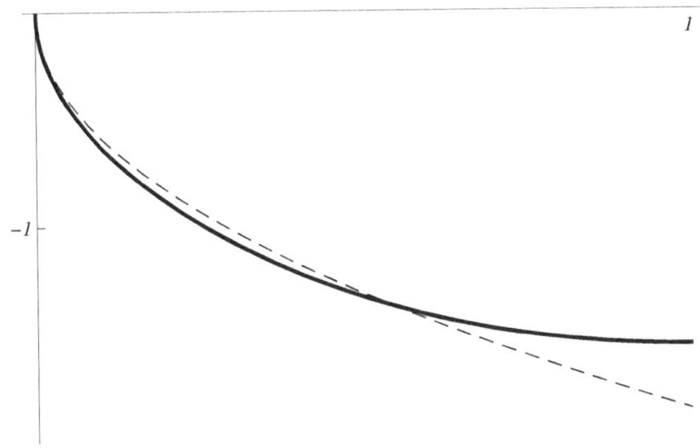

Fig. 5. *The fractional derivative of the function* e^{-x^2} *and its Shannon wavelet series approximation at the scale* $n = 0$, $k = -1, 0, 1$, *as given by (45),(46).*

Conclusion

In this paper the local fractional calculus has be re-derived by using Shannon wavelets. Local fractional derivatives of the Shannon scaling/wavelet functions, based on connection coefficients, are explicitly computed.

Bibliography

Cattani C., Harmonic Wavelet Solutions of the Schrödinger Equation, International Journal of Fluid Mechanics Research, 5, (2003) 1–10.

Cattani C. , Harmonic Wavelets towards Solution of Nonlinear PDE, Computers and Mathematics with Applications, 50, 8-9, (2005) 1191–1210.

Cattani C., Connection Coefficients of Shannon Wavelets, Mathematical Modelling and Analysis, 11 (2), (2006), 1–16.

Cattani C., Shannon Wavelet Analysis, Proceedings of the International Conference on Computational Science, ICCS 2007, May 27-30, 2007, Beijing, China, Y. Shi et Al. (Eds.),Lecture Notes in Computer Science, LNCS 4488, Part II, Springer-Verlag Berlin Heidelberg (2007) 982-989.

Cattani C., Shannon Wavelets Theory, Mathematical Problems in Engineering, vol. 2008, Article ID 164808, 24 pages, (2008), 1–24, doi:10.1155/2008/164808.

Cattani C., Shannon Wavelets for the Solution of Integrodifferential Equations, Mathematical Problems in Engineering, vol. 2010, Article ID 408418, 22 pages, (2010), 1–24, doi:10.1155/2010/408418.

Cattani C., Fractional calculus and Shannon Wavelet, Mathematical Problems in Engineering, vol. 2012, Article ID 502812, 26 pages, (2012).

Cattani C., Rushchitsky J.J., *Wavelet and Wave Analysis as applied to Materials with Micro or Nanostructure*, Series on Advances in Mathematics for Applied Sciences, World Scientific, Singapore, 74 (2007).

Deriaz E., Shannon wavelet approximation of linear differential operators, IM PAN preprints (2007).
Daubechies I., *Ten lectures on wavelets*. Philadelphia: CBMS-NSF Regional Conference Series in Applied Mathematics, SIAM (1992).
Eldred L.B., Parker W.P. and Palazzotto A.N., Numerical application of fractional derivative model constitutive relations for viscoelastic materials, Computers & Structures,vol. 60, No. 6, (1996) 875-882.
Oldham K.B. and Spanier J., *The Fractional Calculus*. Academic Press, London (1970).
Latto A., Resnikoff H.L., and Tenenbaum E., The evaluation of connection coefficients of compactly supported wavelets, Proc. of the French-USA workshop on wavelets and turbulence, Y. Maday ed., Springer-Verlag, (1992) 76-89.
Lin E.B. and Zhou X., Connection coefficients on an interval and wavelet solutions of Burgers equation, Journal of Computational and Applied Mathematics, 135, (2001) 63–78.
Restrepo J. and Leaf G.K., Wavelet-Galerkin discretization of hyperbolic equations, Journal of Computational Physics, 122, (1995) 118–128.
Romine C.H. and Peyton B.W., Computing connection coefficients of compactly supported wavelets on bounded intervals, Computer Science and Mathematics Division, Mathematical Sciences Section, Oak Ridge National Laboratory, Oak Ridge, Technical Report ORNL/TM-13413, (1997) 23 pages. http://www.ornl.gov/ webworks/cpr/rpt/91836.pdf
Ross B., A brief history and exposition of the fundamental theory of fractional calculus, fractional calculus and applications, Lecture Notes in Mathematics, Springer, Berlin, vol 457 (1975), 1–36.
Yang X.J., Advanced local fractional calculus and its applications. World Science, New York, NY, USA, (2012).
Yang X.J., Baleanu D., Tenreiro Machado J.A., Systems of Navier-Stokes equations on Cantor sets, Mathematical Problems in Engineering, (2013), Article ID 769724, 8 pages.
Yang A.M., Cattani C., Zhang J. et al., Local fractional Fourier series solutions for nonhomogeneous heat equations arising in fractal heat flow with local fractional derivative, Mathematical Problems in Engineering, Advances in Mechanical Engineering, (2014), Article ID 514639, 5 pages.
Yang X.J., Srivastava H.M., He J.H., Cantor-type cylindrical-coordinate method for differential equations with local fractional derivatives, Physics Letters A, 377(28), (2013) 1696-1700.

Vasily E. Tarasov

Discretely and Continuously Distributed Dynamical Systems with Fractional Nonlocality

Abstract: In this chapter, we will discuss three-dimensional discrete dynamical systems with long-range properties. The continuum limits for equations of discrete dynamical systems are used to obtain equations for distributed systems that exibit a power-law non-locality. Some applications for different types of dynamical systems are discussed. Fractional generalizations of the nonlinear equations such as sine-Gordon, Burgers, Korteweg-deVries, Kadomtsev-Petviashvili, Boussinesq, and Navier-Stokes equations for 3D lattice with long-range properties and nonlocal continuum are considered.

Keywords: Fractional dynamics, lattice models, fractional derivative, long-range interaction

1 Introduction

Discretely and continuously distributed dynamical systems can be described by lattice and continuum approaches respectively. Continuum models can be considered to be the continuous limits of lattice models, where the length-scales of any continuum element are much larger than the distances between the lattice sites. In modern physics, an important role is played by nonlocal continuum systems and their corresponding discrete (lattice) systems with long-range properties. Applyong generalized infinite series with integer-order derivatives to describe a nonlocality is difficult. This problem can be solved for power-law type of nonlocality by using derivatives of non-integer orders. This procedure is based on the fact that the use of the fractional-order derivatives is equivalent to using an infinite number of integer-order derivatives, which can have arbitrarily large values (for example, see Lemma 15.3 in (Samko 1993)). The theory of derivatives and integrals of non-integer orders (Samko 1993, Kilbas et al. 2006) has a wide application in different areas of physics and mechanics (Carpinteri et al. 1997, Sabatieret al. 2007, Mainardi 2010, Tarasov 2011, 2013a, Atanackovic et al. 2014). The fractional-order derivatives and integrals allow us to describe fractional generalizations of models of continuously distributed systems with nonlocality of power-law type. The first time a connection of fractional nonlocal continuous models and discrete (lattice) models with long-range properties was proved in (Tarasov 2006a,b). Recently, lattice models with long-range properties for continuum with power-law type of nonlocality have been considered in papers (Tarasov 2013b, 2014a,b,c,d, 2015a, 2014f).

Vasily E. Tarasov: Skobeltsyn Institute of Nuclear Physics, Lomonosov Moscow State University, Moscow 119991, Russia, E-mail: tarasov@theory.sinp.msu.ru

A general approach to three-dimensional discrete and continuous nonlocal systems has been suggested in (Tarasov 2014g, 2015b).

Fractional calculus is a powerful tool to describe processes in continuously distributed media with nonlocal properties of power-law type. As was shown in (Tarasov 2006a,b), differential equations with fractional derivatives can be derived from equations for lattice with long-range properties by a special lattice-continuum transform operation. The one-dimensional lattice models for fractional nonlocal continuous system and the corresponding continuum equations have been suggested in (Tarasov 2013b, 2014a,b,c). In this Chapter, we consider three-dimensional models of lattices with long-range properties and their corresponding continuum with power-law nonlocality by using the lattice fractional vector calculus suggested in (Tarasov 2014g). We describe three-dimensional lattice models for continuum with fractional nonlocality of the Riesz type. In this Chapter, we apply this new mathematical tool to describe physical lattices with long-range properties and correspondent fractional-order differential equations for nonlocal continuum with power-law nonlocality. A new relationship between discrete and continuous fractional-order nonlinear equations, such as sine-Gordon, Burgers, Korteweg-deVries, Kadomtsev-Petviashvili, Boussinesq, and Navier-Stokes equations, is suggested.

2 Lattice with Long-range Properties

As a basic model of a discretely distributed dynamical system, we consider a three-dimensional unbounded physical lattice. Physical lattices are characterized by space periodicity. For unbounded lattices we can use three non-coplanar vectors $\mathbf{a}_1, \mathbf{a}_2, \mathbf{a}_3$, that are the shortest vectors by which a lattice can be displaced to combine with itself. Sites of this lattice can be characterized by the number vector $\mathbf{n} = (n_1, n_2, n_3)$, where n_j $(j = 1, 2, 3)$ are integers. For simplification, we consider a lattice with mutually perpendicular primitive lattice vectors $\mathbf{a}_j$, $(j = 1, 2, 3)$, i.e. we use a primitive orthorhombic Bravais lattice. We choose directions of the axes of the Cartesian coordinate system that coincide with the vector $\mathbf{a}_j$. In this case $\mathbf{a}_j = a_j\,\mathbf{e}_j$, where $a_j = |\mathbf{a}_j| > 0$ and $\mathbf{e}_j$ are the basis vectors of the Cartesian coordinate system. Then the vector $\mathbf{n}$ can be represented as $\mathbf{n} = n_1\mathbf{e}_1 + n_2\mathbf{e}_2 + n_3\mathbf{e}_3$.

Let us consider a coordinate origin at one of the lattice sites. Then the positions of all other site with $\mathbf{n} = (n_1, n_2, n_3)$ are described by the vectors $\mathbf{r}(\mathbf{n}) = n_1\mathbf{a}_1 + n_2\mathbf{a}_2 + n_3\mathbf{a}_3$. The lattice sites are numbered by $\mathbf{n}$, so that the vector $\mathbf{n}$ is a number vector of these sites. We assume that all physical values are connected with the lattice sites $\mathbf{r}(\mathbf{n})$. To describe properties of discrete (lattice) system, we consider a scalar field $u(\mathbf{n}, t) = u(n_1, n_2, n_3, t)$ for simplification. Our analysis can be easily generalized to the case of a vector field $\mathbf{u}(\mathbf{n}, t) = \sum_{j=1}^{3} u_j(\mathbf{n}, t)\,\mathbf{e}_j$, where $\mathbf{e}_j$ is the basis of the Cartesian coordinate system, and $u_j(\mathbf{n}, t) = u_j(n_1, n_2, n_3, t)$ are components of the vector for lattice particle that is defined by the vector $\mathbf{n} = (n_1, n_2, n_3)$.

To describe long-range properties of the lattices, we will define fractional-order lattice operators. that can be considered as a lattice partial derivative of arbitrary positive real order α_j in the direction $\mathbf{e}_j = \mathbf{a}_j/|\mathbf{a}_j|$ in the lattice.

Definition. *Lattice fractional partial derivatives are the operator* $\mathbb{D}_L^+\left[\begin{smallmatrix}\alpha_j\\ j\end{smallmatrix}\right]$ *and* $\mathbb{D}_L^-\left[\begin{smallmatrix}\alpha_j\\ j\end{smallmatrix}\right]$ *such that*

$$\mathbb{D}_L^+\begin{bmatrix}\alpha_j\\ j\end{bmatrix} u = \frac{1}{a_j^{\alpha_j}} \sum_{m_j=-\infty}^{+\infty} K_{\alpha_j}^+(n_j - m_j)\, u(\mathbf{m}), \quad (j = 1, 2, 3), \tag{1}$$

$$\mathbb{D}_L^-\begin{bmatrix}\alpha_j\\ j\end{bmatrix} u = \frac{1}{a_j^{\alpha_j}} \sum_{m_j=-\infty}^{+\infty} K_{\alpha_j}^-(n_j - m_j)\, u(\mathbf{m}), \quad (j = 1, 2, 3), \tag{2}$$

where $\alpha_j \in \mathbb{R}$, $\alpha_j > 0$, $m_j \in \mathbb{Z}$, and the kernels $K_{\alpha_j}^{\pm}(n-m)$ are defined by the equations

$$K_{\alpha_j}^+(n_j - m_j) = \frac{\pi^{\alpha_j}}{\alpha_j + 1}\, {}_1F_2\left(\frac{\alpha_j+1}{2}; \frac{1}{2}, \frac{\alpha_j+3}{2}; -\frac{\pi^2\,(n_j-m_j)^2}{4}\right), \quad \alpha_j > 0, \tag{3}$$

$$K_{\alpha_j}^-(n - m) = -\frac{\pi^{\alpha_j+1}\,(n-m)}{\alpha_j + 2}\, {}_1F_2\left(\frac{\alpha_j+2}{2}; \frac{3}{2}, \frac{\alpha_j+4}{2}; -\frac{\pi^2\,(n-m)^2}{4}\right), \quad \alpha_j > 0, \tag{4}$$

where ${}_1F_2$ is the Gauss hypergeometric function (Erdelyi et al. 1981). The parameters $\alpha_j > 0$ will be called the orders of the lattice derivatives.

The kernels $K_{\alpha_j}^{\pm}(n_j)$ are real-valued functions of integer variable $n_j \in \mathbb{Z}$. Note that the kernel $K_{\alpha_j}^+(n_j)$ is an even real-valued functions of the variable $n_j \in \mathbb{Z}$, and $K_{\alpha_j}^-(n_j)$ is odd function, i.e.,

$$K_{\alpha_j}^+(-n_j) = +K_{\alpha_j}^+(n_j), \quad K_{\alpha_j}^-(-n_j) = -K_{\alpha_j}^-(n_j) \tag{5}$$

hold for all $n_j \in \mathbb{Z}$ and $j = 1, 2, 3$.

The kernels $K_\alpha^{\pm}(n-m)$ are defined in the forms (3) and (4) by the following reasons.

The Fourier series transforms $\hat{K}_{\alpha_j}^+(k_j)$ of the kernels $K_{\alpha_j}^+(n_j)$ in the form

$$\hat{K}_{\alpha_j}^+(k_j) = \sum_{n_j=-\infty}^{+\infty} e^{-ik_j n_j} K_{\alpha_j}^+(n_j) = 2\sum_{n_j=1}^{\infty} K_{\alpha_j}^+(n_j)\cos(k_j n_j) + K_{\alpha_j}^+(0) \tag{6}$$

satisfy the condition

$$\hat{K}_{\alpha_j}^+(k_j) = |k_j|^{\alpha_j}, \quad (\alpha_j > 0). \tag{7}$$

The form (3) of the term $K_{\alpha_j}^+(n_j - m_j)$ is completely determined by the requirement (7). If we use an inverse relation to (6) with $\hat{K}_{\alpha_j}^+(k_j) = |k_j|^{\alpha_j}$ that has the form

$$K_{\alpha_j}^+(n_j) = \frac{1}{\pi}\int_0^{\pi} k_j^{\alpha_j} \cos(n_j\, k_j)\, dk, \quad (\alpha_j \in \mathbb{R}, \quad \alpha_j > 0), \tag{8}$$

then we obtain equation (3) for the kernel $K^{+}_{\alpha_j}(n_j - m_j)$.

The Fourier series transforms $\hat{K}^{-}_{\alpha_j}(k_j)$ of the kernels $K^{-}_{\alpha_j}(n_j)$ in the form

$$\hat{K}^{-}_{\alpha_j}(k_j) = \sum_{n_j=-\infty}^{+\infty} e^{-ik_j n_j} K^{-}_{\alpha_j}(n_j) = -2\,i \sum_{n_j=1}^{\infty} K^{-}_{\alpha_j}(n_j) \sin(k_j n_j) \tag{9}$$

satisfy the condition

$$\hat{K}^{-}_{\alpha_j}(k_j) = i \operatorname{sgn}(k_j)\, |k_j|^{\alpha_j}, \quad (\alpha_j > 0). \tag{10}$$

The form (4) of the kernel $K^{-}_{\alpha_j}(n_j - m_j)$ is completely determined by (7). If we use an inverse relation to (9) with $\hat{K}^{-}_{\alpha_j}(k_j) = i \operatorname{sgn}(k_j)\, |k_j|^{\alpha_j}$ that has the form

$$K^{-}_{\alpha_j}(n_j) = -\frac{1}{\pi} \int_0^{\pi} k_j^{\alpha_j} \sin(n_j\, k_j)\, dk_j \quad (\alpha_j \in \mathbb{R}_{+}, \quad \alpha_j > 0), \tag{11}$$

then we get equation (4) for the kernel $K^{-}_{\alpha_j}(n_j - m_j)$. Note that $K^{-}_{\alpha_j}(0) = 0$.

Let us note that integer-order lattice derivatives can also be defined by equations (1) with (3) and (2) with (4) for $\alpha \in \mathbb{N}$. The exact form of the kernels $\hat{K}^{\pm}_{\alpha}(k)$ for integer positive values $\alpha \in \mathbb{N}$ are simpler than (3) and (4). Equations (3) and (4) for the case $\alpha \in \mathbb{N}$ can be simplified by using equation 2.5.3.5 of (Prudnikov et al. 1986). For integer values $\alpha = 1, 2, 3, 4$, we get the kernels $K^{+}_{\alpha}(n)$ with $n \neq 0$ in the form

$$K^{+}_{1}(n) = -\frac{1 - (-1)^n}{\pi\, n^2}, \quad K^{+}_{2}(n) = \frac{2\,(-1)^n}{n^2}, \tag{12}$$

$$K^{+}_{3}(n) = \frac{3\,\pi\,(-1)^n}{n^2} + \frac{6\,(1 - (-1)^n)}{\pi\, n^4}, \quad K^{+}_{4}(n) = \frac{4\,\pi^2\,(-1)^n}{n^2} - \frac{24\,(-1)^n}{n^4}, \tag{13}$$

where $n \neq 0$, $n \in \mathbb{Z}$, and $K^{+}_{m}(0) = \pi^m/(m+1)$ for all $m \in \mathbb{N}$. For $\alpha = 1, 2, 3, 4$, the kernels $K^{-}_{\alpha}(n)$ with $n \neq 0$ are

$$K^{-}_{1}(n) = \frac{(-1)^n}{n}, \quad K^{-}_{2}(n) = \frac{(-1)^n\,\pi}{n} + \frac{2(1 - (-1)^n)}{\pi\, n^3}, \tag{14}$$

$$K^{-}_{3}(n) = \frac{(-1)^n\,\pi^2}{n} - \frac{6\,(-1)^n}{n^3}, \quad K^{-}_{4}(n) = \frac{(-1)^n\,\pi^3}{n} - \frac{12\,(-1)^n \pi}{n^3} - \frac{24\,(1 - (-1)^n)}{\pi\, n^5}, \tag{15}$$

where $n \neq 0$, $n \in \mathbb{Z}$, and $K^{-}_{m}(0) = 0$ for all $m \in \mathbb{N}$. Note that $(1 - (-1)^n) = 2$ for odd n, and $(1 - (-1)^n) = 0$ for even n. These kernels allow us to consider lattice models for continuously distributed systems with local properties that are described by integer-order differential equations (Tarasov 2014h, 2015c).

Note that in equations (1) and (2) the value $j \in \{1, 2, 3\}$ characterizes the component n_j of the lattice vector $\mathbf{n}$ with respect to which this derivative is taken. This is similar to the variable x_j in the usual partial derivatives for the space $\mathbb{R}^3$. The lattice operators $\mathbb{D}^{\pm}_{L}\begin{bmatrix}\alpha_j \\ j\end{bmatrix}$ are analogous to the partial derivatives of order α with respect to coordinates x_j for continuum model.

To describe isotropic discrete physical systems, we should use the lattice operators ${}^R\mathbb{D}^{\pm}_L \begin{bmatrix} \alpha_j \\ j \end{bmatrix}$ with equal orders $\alpha_j = \alpha$ for all $j = 1, 2, 3$.

For repeated fractional lattice derivatives we can use the notation

$$ {}^R\mathbb{D}^{\pm,\pm}_L \begin{bmatrix} \alpha_i \, \beta_j \\ i \, j \end{bmatrix} = {}^R\mathbb{D}^{\pm}_L \begin{bmatrix} \alpha_i \\ i \end{bmatrix} {}^R\mathbb{D}^{\pm}_L \begin{bmatrix} \beta_j \\ j \end{bmatrix}, \tag{16} $$

where $i, j \in \{1, 2, 3\}$. The action of the operator (16) on the lattice fields $u(\mathbf{m}, t)$ is

$$ {}^R\mathbb{D}^{\pm,\pm}_L \begin{bmatrix} \alpha_i \, \beta_j \\ i \, j \end{bmatrix} u(\mathbf{m}, t) = \sum_{m_i=-\infty}^{+\infty} \sum_{m_j=-\infty}^{+\infty} K^{\pm}_{\alpha_i}(n_i - m_i) \, K^{\pm}_{\beta_j}(n_j - m_j) \, u(\mathbf{m}, t). \tag{17} $$

Analogously, we can define other repeated lattice fractional derivatives such as ${}^R\mathbb{D}^{\pm,\pm,\pm}_L \begin{bmatrix} \alpha_i \, \beta_j \, \gamma_l \\ i \, j \, l \end{bmatrix}$, ${}^R\mathbb{D}^{\pm,\pm,\mp}_L \begin{bmatrix} \alpha_i \, \beta_j \, \gamma_l \\ i \, j \, l \end{bmatrix}$, and other.

3 Lattice Fractional Nonlinear Equations

In order to describe a fractional nonlocality of power-law type, we should use the lattice fractional-order derivatives for discrete models. It was proved in (Tarasov 2014g) that the integer-order partial derivatives correspond to the local continuum derivatives by the equations

$$ \frac{\partial^{2m+1}}{\partial x_j^{2m+1}} = (-1)^m \, {}^R\mathbb{D}^{-}_C \begin{bmatrix} 2m+1 \\ j \end{bmatrix} \longrightarrow (-1)^m \, {}^R\mathbb{D}^{-}_L \begin{bmatrix} 2m+1 \\ j \end{bmatrix}, \tag{18} $$

$$ \frac{\partial^{2m}}{\partial x_j^{2m}} = (-1)^m \, {}^R\mathbb{D}^{+}_C \begin{bmatrix} 2m \\ j \end{bmatrix} \longrightarrow (-1)^m \, {}^R\mathbb{D}^{+}_L \begin{bmatrix} 2m \\ j \end{bmatrix} \tag{19} $$

for integer values of $m \geq 0$.

To give lattice analogs of fractional nonlinear equations, we use the correspondence of the form

$$ \frac{\partial}{\partial x_j} \longrightarrow + {}^R\mathbb{D}^{-}_L \begin{bmatrix} 1 \\ j \end{bmatrix}, \quad \frac{\partial^3}{\partial x_j^3} \longrightarrow - {}^R\mathbb{D}^{-}_L \begin{bmatrix} 3 \\ j \end{bmatrix}, \tag{20} $$

$$ \frac{\partial^2}{\partial x_j^2} \longrightarrow - {}^R\mathbb{D}^{+}_L \begin{bmatrix} 2 \\ i \end{bmatrix}, \quad \frac{\partial^4}{\partial x_j^4} \longrightarrow + {}^R\mathbb{D}^{+}_L \begin{bmatrix} 4 \\ j \end{bmatrix}, \tag{21} $$

and then replace integer values by fractional α_j.

As microstructural models of linear waves in the three-dimensional fractional nonlocal anisotropic continua, we can consider the following two types of the linear lattice fractional wave equations

$$ \frac{d^2 u(\mathbf{n}, t)}{dt^2} - \sum_{i,j=1}^{3} A_{ij} \, {}^R\mathbb{D}^{-,-}_L \begin{bmatrix} \alpha_i \, \alpha_j \\ i \, j \end{bmatrix} u(\mathbf{m}, t) = 0, \tag{22} $$

or

$$\frac{d^2u(\mathbf{n},t)}{dt^2} + \sum_{i=1}^{3} A_i \, {}^R\mathbb{D}_L^{\pm} \begin{bmatrix} \alpha_i \\ i \end{bmatrix} u(\mathbf{m},t) = 0. \tag{23}$$

Note that in the general case, we have the following inequalities

$${}^R\mathbb{D}_L^{-,-} \begin{bmatrix} \alpha_i \, \alpha_i \\ i \, i \end{bmatrix} u(\mathbf{m},t) \neq {}^R\mathbb{D}_L^{-} \begin{bmatrix} 2\alpha_i \\ i \end{bmatrix} u(\mathbf{m},t), \tag{24}$$

$${}^R\mathbb{D}_L^{-,-} \begin{bmatrix} \alpha_i \, \alpha_i \\ i \, i \end{bmatrix} u(\mathbf{m},t) \neq \pm {}^R\mathbb{D}_L^{+} \begin{bmatrix} 2\alpha_i \\ i \end{bmatrix} u(\mathbf{m},t), \tag{25}$$

where

$${}^R\mathbb{D}_L^{-,-} \begin{bmatrix} \alpha_i \, \alpha_i \\ i \, i \end{bmatrix} u(\mathbf{m},t) = \left({}^R\mathbb{D}_L^{-} \begin{bmatrix} \alpha_i \\ i \end{bmatrix} \right)^2 u(\mathbf{m},t). \tag{26}$$

For "sufficiently good" lattice functions $u(\mathbf{m},t)$, we can use the relation

$${}^R\mathbb{D}_L^{-,-} \begin{bmatrix} \alpha_i \, \alpha_i \\ i \, i \end{bmatrix} u(\mathbf{m},t) = - {}^R\mathbb{D}_L^{+} \begin{bmatrix} 2\alpha_i \\ i \end{bmatrix} u(\mathbf{m},t). \tag{27}$$

To describe isotropic discrete systems and physical lattices, we should use the lattice operators ${}^R\mathbb{D}_L^{\pm} \begin{bmatrix} \alpha_i \\ i \end{bmatrix}$ with equal orders $\alpha_j = \alpha$ for all $j = 1, 2, 3$.

As microstructural models of the three-dimensional (3D) fractional equations for nonlocal anisotropic continuum, we can use the following lattice fractional equations.

1. The lattice fractional 3D sine-Gordon equation

$$\frac{d^2u(\mathbf{n},t)}{dt^2} + \sum_{i=1}^{3} A_i \, {}^R\mathbb{D}_L^{\pm} \begin{bmatrix} \alpha_i \\ i \end{bmatrix} u(\mathbf{m},t) + \sin(u(\mathbf{n},t)) = 0, \tag{28}$$

where the integer-order lattice equation is defined by $\alpha_i = 2$. Equation (28) is a nonlinear generalization of the lattice wave equation (23).

2. The lattice fractional 3D Burgers equation

$$\frac{du(\mathbf{n},t)}{dt} + \sum_{i=1}^{3} A_i \, {}^R\mathbb{D}_L^{-} \begin{bmatrix} \alpha_i \\ i \end{bmatrix} u^2(\mathbf{m},t) + \sum_{i=1}^{3} B_i \, {}^R\mathbb{D}_L^{+} \begin{bmatrix} \beta_i \\ i \end{bmatrix} u(\mathbf{m},t) = 0, \tag{29}$$

where the integer-order lattice equation is defined by $\alpha_i = 1$ and $\beta_i = 2$.

3. The lattice fractional 3D Korteweg-de Vries equation

$$\frac{du(\mathbf{n},t)}{dt} - \sum_{i=1}^{3} A_i \, {}^R\mathbb{D}_L^{-} \begin{bmatrix} \alpha_i \\ i \end{bmatrix} u^2(\mathbf{m},t) - \sum_{i=1}^{3} B_i \, {}^R\mathbb{D}_L^{-} \begin{bmatrix} \beta_i \\ i \end{bmatrix} u(\mathbf{m},t) = 0, \tag{30}$$

where the integer-order lattice equation is defined by $\alpha_i = 1$ and $\beta_i = 3$.

4. The lattice fractional 3D Kadomtsev-Petviashvili equation

$${}^R\mathbb{D}_L^{-} \begin{bmatrix} \alpha_x \\ x \end{bmatrix} \dot{u}(\mathbf{m},t) - A_x \, {}^R\mathbb{D}_L^{+} \begin{bmatrix} \beta_x \\ x \end{bmatrix} u^2(\mathbf{m},t) -$$

$$- B_x \, {}^R\mathbb{D}_L^{+} \begin{bmatrix} \gamma_x \\ x \end{bmatrix} u(\mathbf{m}, t) - \sum_{i=2}^{3} C_i \, {}^R\mathbb{D}_L^{-} \begin{bmatrix} \delta_i \\ i \end{bmatrix} u(\mathbf{m}, t) = 0, \tag{31}$$

where $\dot{u}(\mathbf{m}, t) = du(\mathbf{m}, t)/dt$ the integer-order lattice equation is defined by $\alpha_x = 1$, $\beta_x = 2$, $\gamma_x = 4$, and $\delta_y = \delta_z = 2$.

5. The lattice fractional 3D Boussinesq equation

$$\frac{d^2u(\mathbf{n}, t)}{dt^2} + \sum_{i=1}^{3} A_i \, {}^R\mathbb{D}_L^{+} \begin{bmatrix} \alpha_i \\ i \end{bmatrix} u(\mathbf{m}, t) - B_x \, {}^R\mathbb{D}_L^{+} \begin{bmatrix} \beta_x \\ x \end{bmatrix} u^2(\mathbf{m}, t) + C_x \, {}^R\mathbb{D}_L^{+} \begin{bmatrix} \gamma_x \\ x \end{bmatrix} u(\mathbf{m}, t) = 0, \tag{32}$$

where the integer-order lattice equation is defined by $\alpha_i = 2$, $\beta_x = 2$, $\gamma_x = 4$.

To consider a lattice fractional analog of the Navier-Stokes equations, we define the lattice vector operators of non-integer orders. The lattice nabla operator of the Riesz type is

$${}^R\nabla_L^{\alpha,\pm} = \sum_{j=1}^{3} \mathbf{e}_j \, {}^R\mathbb{D}_L^{\pm} \begin{bmatrix} \alpha \\ j \end{bmatrix}. \tag{33}$$

where $\mathbf{e}_j = \mathbf{a}_j/|\mathbf{a}_j|$. The Laplacian in three-dimensional space $\mathbb{R}^3$ for the scalar field $u(\mathbf{m}, t) = \sum_{j=1}^{3} \mathbf{e}_j \, u_j(\mathbf{m}, t)$ is can be defined by two different equations with the repeated lattice derivative of orders α,

$${}^R\Delta_L^{\alpha,\alpha,\pm} \, u(\mathbf{m}, t) = \sum_{i=1}^{3} \sum_{j=1}^{3} \mathbf{e}_j \left({}^R\mathbb{D}_L^{\pm} \begin{bmatrix} \alpha \\ i \end{bmatrix} \right)^2 u_j(\mathbf{m}, t), \tag{34}$$

and by the derivative of the doubled order 2α,

$${}^R\Delta_L^{2\alpha,\pm} \, \mathbf{u}(\mathbf{m}, t) = \sum_{i=1}^{3} \sum_{j=1}^{3} \mathbf{e}_j \, {}^R\mathbb{D}_L^{\pm} \begin{bmatrix} 2\alpha \\ i \end{bmatrix} u_j(\mathbf{m}, t). \tag{35}$$

Using the lattice operators (33) and (35), we can write the equations

$$\left(\frac{\partial \mathbf{v}(\mathbf{n}, t)}{\partial t} + \left(\mathbf{v}(\mathbf{n}, t), \, {}^R\nabla_L^{\alpha,-} \right) \mathbf{v}(\mathbf{m}, t) \right) =$$

$$= - {}^R\,\mathrm{Grad}_L^{\alpha,-} \, p(\mathbf{m}, t) + \mu \, {}^R\Delta_L^{2\alpha,+} \, \mathbf{v}(\mathbf{m}, t) + \mathbf{f}(\mathbf{n}, t), \tag{36}$$

where μ is the scalar dynamic viscosity. In component form this equation (36) is the set of the equations

$$\left(\frac{dv_i(\mathbf{n}, t)}{dt} + \sum_{j=1}^{3} v_j(\mathbf{n}, t), \, {}^R\mathbb{D}_L^{-} \begin{bmatrix} \alpha_j \\ x_j \end{bmatrix} v_i(\mathbf{m}, t) \right) =$$

$$= - {}^R\mathbb{D}_L^{-} \begin{bmatrix} \alpha_j \\ x_j \end{bmatrix} p(\mathbf{m}, t) + \mu \sum_{j=1}^{3} {}^R\mathbb{D}_L^{+} \begin{bmatrix} 2\alpha_j \\ x_j \end{bmatrix} v_i(\mathbf{m}, t) + f_i(\mathbf{n}, t). \tag{37}$$

These equations are equivalent to the Navier-Stokes equations for fluids on the physical lattice with long-range properties of the Riesz type.

4 Continuum Fractional Derivatives of the Riesz Type

Let us give definitions of continuum fractional-order derivatives that allows us to describe materials with nonlocality of power-law type.

The continuum derivative of non-integer order α is defined (Samko 1993, Kilbas et al. 2006) by the equation

$$\mathbb{D}_C^+ \begin{bmatrix} \alpha \\ i \end{bmatrix} u(\mathbf{r}) = \frac{1}{d_1(m,\alpha)} \int\limits_{\mathbb{R}^1} \frac{1}{|z_i|^{\alpha+1}} (\Delta_{z_i}^m u)(\mathbf{r})\, dz_i, \quad (0 < \alpha < m), \tag{38}$$

where $(\Delta_{z_i}^m u)(\mathbf{r})$ is a centered finite difference of order m of a function $u(\mathbf{r})$ with the vector step $\mathbf{z}_i = z_i\,\mathbf{e}_i \in \mathbb{R}^3$ for the point $\mathbf{r} \in \mathbb{R}^3$ that is defined by the equation

$$(\Delta_{z_i}^m u)(\mathbf{z}_i) = \sum_{n=0}^{m} (-1)^n \frac{m!}{n!\,(m-n)!}\, u(\mathbf{r} - (m/2 - n)\, z_i\, \mathbf{e}_i); \tag{39}$$

and the constant $d_1(m,\alpha)$ is defined by

$$d_1(m,\alpha) = \frac{\pi^{3/2} A_m(\alpha)}{2^\alpha \Gamma(1+\alpha/2)\Gamma((1+\alpha)/2)\sin(\pi\alpha/2)},$$

with

$$A_m(\alpha) = 2 \sum_{j=0}^{[m/2]} (-1)^{j-1} \frac{m!}{j!(m-j)!} (m/2 - j)^\alpha .$$

The constants $d_1(m,\alpha)$ are different from zero for all $\alpha > 0$ in the case of an even m and centered difference $(\Delta_i^m u)$ (see Theorem 26.1 in (Samko 1993)). Note that the integral (38) does not depend on the choice of $m > \alpha$. Therefore, we can always choose an even number m so that it is greater than parameter α, and we can use the centered difference (39) for all positive real values of α.

The continuum fractional derivatives $\mathbb{D}_C^+ \left[\begin{smallmatrix} \alpha \\ i \end{smallmatrix}\right] u(\mathbf{r})$ can be considered to be the Riesz derivative of the function $u(\mathbf{r})$ with respect to one component $x_i \in \mathbb{R}^1$ of the vector $\mathbf{r} \in \mathbb{R}^3$, i.e. the operator $\mathbb{D}_C^+ \left[\begin{smallmatrix} \alpha \\ i \end{smallmatrix}\right]$ is a partial fractional derivative of Riesz type. An important property of the Riesz fractional derivatives is the Fourier transform $\mathcal{F}$ in the form

$$\mathcal{F}\left(\mathbb{D}_C^+ \begin{bmatrix} \alpha \\ i \end{bmatrix} u(\mathbf{r})\right)(\mathbf{k}) = |k_i|^\alpha (\mathcal{F}u)(\mathbf{k}). \tag{40}$$

The property (40) is valid for functions $u(\mathbf{r})$ from the space $C^\infty(\mathbb{R}^1)$ of infinitely differentiable functions on $\mathbb{R}^1$ with compact support. It is also holds for the Lizorkin space (see Section 8.1 in (Samko 1993)).

Using that fact that $(-i)^{2m} = (-1)^m$, the Riesz fractional derivatives for even $\alpha = 2m$, where $m \in \mathbb{N}$, are connected with the usual partial derivative of integer orders $2m$ by the relation

$$\mathbb{D}_C^+ \begin{bmatrix} 2m \\ i \end{bmatrix} u(\mathbf{r}) = (-1)^m \frac{\partial^{2m} u(\mathbf{r})}{\partial x_i^{2m}} . \tag{41}$$

The fractional derivatives $\mathbb{D}^{+}_{C}\left[\begin{smallmatrix}2m\\ i\end{smallmatrix}\right]$ for even orders α are local operators. Note that the Riesz derivative $\mathbb{D}^{+}_{C}\left[\begin{smallmatrix}1\\ i\end{smallmatrix}\right]$ cannot be considered as a derivative of first order with respect to x_i, i.e., $\mathbb{D}^{+}_{C}\left[\begin{smallmatrix}1\\ i\end{smallmatrix}\right] u(\mathbf{r}) \neq \partial u(\mathbf{r})/\partial x_i$. All Riesz derivatives for odd orders $\alpha = 2m + 1$, where $m \in \mathbb{N}$, are non-local operators that cannot be considered as usual derivatives $\partial^{2m+1}/\partial x^{2m+1}$.

We can define a continuum fractional integral $\mathbb{I}^{+}_{C}\left[\begin{smallmatrix}\alpha\\ i\end{smallmatrix}\right]$ of the Riesz type as the Riesz potential of order α with respect to x_i by the equation

$$\mathbb{I}^{+}_{C}\begin{bmatrix}\alpha\\ i\end{bmatrix} u(\mathbf{r}) = \int\limits_{\mathbb{R}^1} R_\alpha(x_i - z_i)\, u(\mathbf{r} + (z_i - x_i)\,\mathbf{e}_i)\, dz_i, \quad (\alpha > 0), \tag{42}$$

where $\mathbf{e}_i$ is the basis of the Cartesian coordinate system; the function $R_\alpha(\mathbf{r})$ is the Riesz kernel that is defined by

$$R_\alpha(\mathbf{r}) = \begin{cases} \gamma_3^{-1}(\alpha)|\mathbf{r}|^{\alpha-3} & \alpha \neq 3 + 2n, \\ -\gamma_3^{-1}(\alpha)|\mathbf{r}|^{\alpha-3}\ln|\mathbf{r}| & \alpha = 3 + 2n; \end{cases} \tag{43}$$

and the constant $\gamma_3(\alpha)$ has the form

$$\gamma_3(\alpha) = \begin{cases} 2^{\alpha}\pi^{3/2}\Gamma(\alpha/2)/\Gamma((3-\alpha)/2) & \alpha \neq 3 + 2n, \\ (-1)^{(3-\alpha)/2}2^{\alpha-1}\pi^{3/2}\,\Gamma(\alpha/2)\,\Gamma(1 + [\alpha - 3]/2) & \alpha = 3 + 2n, \end{cases} \tag{44}$$

where $n \in \mathbb{N}$ and $\alpha \in \mathbb{R}_{+}$. An important property of this Riesz fractional integration is its Fourier transform $\mathcal{F}$:

$$\mathcal{F}\left(\mathbb{I}^{+}_{C}\begin{bmatrix}\alpha\\ i\end{bmatrix} u(\mathbf{r})\right)(\mathbf{k}) = |k_i|^{-\alpha}(\mathcal{F}u)(\mathbf{k}). \tag{45}$$

Note that the continuum fractional integral $\mathbb{I}^{+}_{C}\left[\begin{smallmatrix}\alpha\\ i\end{smallmatrix}\right]$ uses $|k_i|^{-\alpha}$ instead of $|\mathbf{k}|^{-\alpha}$ that is used in the Riesz potential. The continuum integral $\mathbb{I}^{+}_{C}\left[\begin{smallmatrix}\alpha\\ i\end{smallmatrix}\right]$ is an integration of $u(\mathbf{r})$ with respect to one variable x_i instead of all variables x_1, x_2, x_3 in the Riesz potential.

If $u(\mathbf{r})$ as a function of x_i belongs to the Lizorkin space, then we have (Samko 1993) the semi-group property

$$\mathbb{I}^{+}_{C}\begin{bmatrix}\alpha\\ i\end{bmatrix}\mathbb{I}^{+}_{C}\begin{bmatrix}\beta\\ i\end{bmatrix} u(\mathbf{r}) = \mathbb{I}^{+}_{C}\begin{bmatrix}\alpha+\beta\\ i\end{bmatrix} u(\mathbf{r}), \tag{46}$$

where $\alpha > 0$ and $\beta > 0$, and the continuum fractional derivative $\mathbb{D}^{+}_{C}\left[\begin{smallmatrix}\alpha\\ i\end{smallmatrix}\right]$ yields an operator inverse to the continuum fractional integration $\mathbb{I}^{+}_{C}\left[\begin{smallmatrix}\alpha\\ i\end{smallmatrix}\right]$ as

$$\mathbb{D}^{+}_{C}\begin{bmatrix}\alpha\\ i\end{bmatrix}\mathbb{I}^{+}_{C}\begin{bmatrix}\alpha\\ i\end{bmatrix} u(\mathbf{r}) = u(\mathbf{r}), \quad (\alpha > 0). \tag{47}$$

Note that the property (47) is also valid for the continuum fractional integration in the frame of L_p-spaces $L_p(\mathbb{R}^1)$ for $1 \leqslant p < 1/\alpha$ (see Theorem 26.3 in (Samko 1993)).

The continuum fractional derivative $\mathbb{D}_C^+ \begin{bmatrix} \alpha \\ j \end{bmatrix}$ with $\alpha = 1$ cannot be considered the same as the usual local derivative of first order with respect to x_j. Therefore we will define new continuum fractional derivative $\mathbb{D}_C^- \begin{bmatrix} \alpha \\ j \end{bmatrix}$ of the Riesz type by the equation

$$\mathbb{D}_C^- \begin{bmatrix} \alpha \\ j \end{bmatrix} = \begin{cases} \dfrac{\partial}{\partial x_j} \mathbb{D}_C^+ \begin{bmatrix} \alpha - 1 \\ j \end{bmatrix} & \alpha > 1 \\ \dfrac{\partial}{\partial x_j} & \alpha = 1 \\ \dfrac{\partial}{\partial x_j} \mathbb{I}_C^+ \begin{bmatrix} 1 - \alpha \\ j \end{bmatrix} & 0 < \alpha < 1. \end{cases} \tag{48}$$

The Fourier integral transform $\mathcal{F}$ of the fractional derivative (48) is given by

$$\mathcal{F}\left(\mathbb{D}_C^- \begin{bmatrix} \alpha \\ j \end{bmatrix} u(\mathbf{r}) \right)(\mathbf{k}) = i\, k_j\, |k_j|^{\alpha-1} (\mathcal{F}u)(\mathbf{k}) = i \operatorname{sgn}(k_j)\, |k_j|^{\alpha} (\mathcal{F}u)(\mathbf{k}). \tag{49}$$

For the odd integer values of α, equations (41) and (48) give the relation

$$\mathbb{D}_C^- \begin{bmatrix} 2m+1 \\ i \end{bmatrix} u(\mathbf{r}) = (-1)^m \frac{\partial^{2m+1} u(\mathbf{r})}{\partial x_i^{2m+1}}, \quad (m \in \mathbb{N}). \tag{50}$$

Equation (50) means that the fractional derivatives $\mathbb{D}_C^- \begin{bmatrix} \alpha \\ i \end{bmatrix}$ of the odd orders α are local operators represented by the usual derivatives of integer orders. Note that the continuum derivative $\mathbb{D}_C^- \begin{bmatrix} 2 \\ j \end{bmatrix}$ cannot be considered to be a local derivative of second order with respect to x_j. The derivatives $\mathbb{D}_C^- \begin{bmatrix} \alpha \\ j \end{bmatrix}$ for even orders $\alpha = 2m$, where $m \in \mathbb{N}$, are non-local operators that are not typical derivatives $\partial^{2m}/\partial x_i^{2m}$.

Using equations (41) and (50), we see that the partial derivatives of integer orders are obtained from the fractional derivatives $\mathbb{D}_C^{\pm} \begin{bmatrix} \alpha \\ j \end{bmatrix}$ for odd values $\alpha = 2j + 1 > 0$ by $\mathbb{D}_C^- \begin{bmatrix} \alpha \\ j \end{bmatrix}$, and for even values $\alpha = 2m > 0$, where $m \in \mathbb{N}$, by $\mathbb{D}_C^+ \begin{bmatrix} \alpha \\ j \end{bmatrix}$ only. The continuum derivatives $\mathbb{D}_C^- \begin{bmatrix} 2m \\ j \end{bmatrix}$ and $\mathbb{D}_C^+ \begin{bmatrix} 2m+1 \\ j \end{bmatrix}$ are nonlocal differential operators of integer orders.

5 From Lattice Equations to Continuum Equations

Using the methods suggested in (Tarasov 2006a,b), we can define the operation that transforms a lattice field $u(\mathbf{n})$ into a continuum field $u(\mathbf{r})$, For this transformation, we will consider the lattice scalar field $u(\mathbf{n})$ to be the Fourier series coefficients of some function $\hat{u}(\mathbf{k})$ for $k_j \in [-k_{j0}/2, k_{j0}/2]$, where $j = 1, 2, 3$. We than apply the continuous limit $\mathbf{k}_0 \to \infty$ to obtain $\tilde{u}(\mathbf{k})$. Finally we apply the inverse Fourier integral transformation to obtain the continuum scalar field $u(\mathbf{r})$. Let us now fully describe these three transformations:

The first transformation: The discrete Fourier series transform

$$u(\mathbf{n}) \to \mathcal{F}_{\Delta}\{u(\mathbf{n})\} = \hat{u}(\mathbf{k})$$

of the lattice scalar field $u(\mathbf{n})$ is defined by

$$\hat{u}(\mathbf{k}) = \mathcal{F}_{\Delta}\{u(\mathbf{n})\} = \sum_{n_1,n_2,n_3=-\infty}^{+\infty} u(\mathbf{n})\, e^{-i(\mathbf{k},\mathbf{r}(\mathbf{n}))}, \tag{51}$$

where $\mathbf{r}(\mathbf{n}) = \sum_{j=1}^{3} n_j\, \mathbf{a}_j$, and $a_j = 2\pi/k_{j0}$ is distance between lattice particle in the direction $\mathbf{a}_j$.

The second transformation: The passage to the limit

$$\hat{u}(\mathbf{k}) \to \mathrm{Lim}\{\hat{u}(\mathbf{k})\} = \tilde{u}(\mathbf{k}),$$

where we use $a_j \to 0$ (or $k_{j0} \to \infty$), allows us to derive the function $\tilde{u}(\mathbf{k})$ from $\hat{u}(\mathbf{k})$. Here $\tilde{u}(\mathbf{k})$ is the Fourier integral transform of the continuum field $u(\mathbf{r})$, and the function $\hat{u}(k)$ is the Fourier series transform of the lattice field $u(\mathbf{n})$, where

$$u(\mathbf{n}) = \frac{(2\pi)^3}{k_{10}k_{20}k_{30}} u(\mathbf{r}(\mathbf{n})),$$

and $\mathbf{r}(\mathbf{n}) \to \mathbf{r}$.

The third transformation: The inverse Fourier integral transform

$$\tilde{u}(\mathbf{k}) \to \mathcal{F}^{-1}\{\tilde{u}(\mathbf{k})\} = u(\mathbf{r})$$

is defined by

$$u(\mathbf{r}) = \mathcal{F}^{-1}\{\tilde{u}(\mathbf{k})\} = \frac{1}{(2\pi)^3} \iiint_{-\infty}^{+\infty} dk_1 dk_2 dk_3\, e^{i\sum_{j=1}^{3} k_j x_j}\, \tilde{u}(\mathbf{k}). \tag{52}$$

The combination $\mathcal{F}^{-1} \circ \mathrm{Lim} \circ \mathcal{F}_{\Delta}$ of the operations $\mathcal{F}^{-1}$, Lim, and $\mathcal{F}_{\Delta}$ define the lattice-continuum transform operation

$$\mathcal{T}_{L\to C} = \mathcal{F}^{-1} \circ \mathrm{Lim} \circ \mathcal{F}_{\Delta} \tag{53}$$

that maps lattice models into the continuum models (Tarasov 2006a,b).

It is important to note that the lattice-continuum transform operation $\mathcal{T}_{L\to C}$ can be applied not only the lattice fields but also to lattice operators. The operation $\mathcal{T}_{L\to C}$ allows us to map of lattice derivatives $\mathbb{D}_L^{\pm}\left[\begin{smallmatrix}\alpha\\ i\end{smallmatrix}\right]$ into continuum derivatives $\mathbb{D}_C^{\pm}\left[\begin{smallmatrix}\alpha\\ i\end{smallmatrix}\right]$.

Let us define two functions $\hat{K}_{\alpha}^{\pm}(k_i)$ and $\tilde{K}_{\alpha}^{\pm}(k_i)$ to consider transformations of lattice operators.

The functions $\hat{K}_{\alpha}^{\pm}(k_i)$ are defined by the discrete Fourier series transform $\mathcal{F}_{\Delta}$ of the kernels of lattice operators, The equation that defines $\hat{K}_{\alpha}^{\pm}(k_i)$ has the form

$$\mathcal{F}_{\Delta}\left(\mathbb{D}_L^{\pm}\left[\begin{matrix}\alpha\\ i\end{matrix}\right] u(\mathbf{m})\right) = \frac{1}{a_i^{\alpha}} \hat{K}_{\alpha}^{\pm}(k_i\, a_i)\, \hat{u}(\mathbf{k}), \tag{54}$$

where $\hat{u}(\mathbf{k}) = \mathcal{F}_{\Delta}\{u(\mathbf{m})\}$, and $\mathcal{F}_{\Delta}$ is an operator notation for the discrete Fourier series transform.

The functions $\tilde{K}^{\pm}_{\alpha}(k_i)$ are defined by the Fourier integral transforms $\mathcal{F}$ of the corresponding continuum derivatives. The equation that defines $\tilde{K}^{\pm}_{\alpha}(k_i)$ is

$$\mathcal{F}\left(\mathbb{D}^{\pm}_{C}\begin{bmatrix}\alpha\\ i\end{bmatrix} u(\mathbf{r})\right) = \tilde{K}^{\pm}_{\alpha}(k_i)\, \tilde{u}(\mathbf{k}), \tag{55}$$

where $\tilde{u}(\mathbf{k}) = \mathcal{F}\{u(\mathbf{r})\}$, and $\mathcal{F}_{\Delta}$ is an operator notation for the discrete Fourier transform. In general, the order of the partial derivative $\mathbb{D}^{\pm}_{C}\left[\begin{smallmatrix}\alpha\\ i\end{smallmatrix}\right]$ is defined by the order of lattice operator $\mathbb{D}^{\pm}_{L}\left[\begin{smallmatrix}\alpha\\ i\end{smallmatrix}\right]$. This order can be an integer or non-integer positive real number.

The following theorem illustrates the connection between the lattice and continuum fractional derivatives of non-integer orders.

Theorem *The lattice-continuum transform operation $\mathcal{T}_{L\to C}$ maps the lattice fractional derivatives*

$$\mathbb{D}^{\pm}_{L}\begin{bmatrix}\alpha\\ j\end{bmatrix} u(\mathbf{m}) = \frac{1}{a_i^{\alpha}} \sum_{m_j=-\infty}^{+\infty} K^{\pm}_{\alpha}(n_j - m_j)\, u(\mathbf{m}), \tag{56}$$

where $K^{\pm}_{\alpha}(n_j - m_j)$ are defined by (3),(4), into the continuum fractional derivatives of order α with respect to coordinate x_i by

$$\mathcal{T}_{L\to C}\left(\mathbb{D}^{\pm}_{L}\begin{bmatrix}\alpha\\ j\end{bmatrix} u(\mathbf{m})\right) = \mathbb{D}^{\pm}_{C}\begin{bmatrix}\alpha\\ j\end{bmatrix} u(\mathbf{r}), \tag{57}$$

where $u(\mathbf{r}) = \mathcal{F}^{-1} \circ \mathrm{Lim} \circ \mathcal{F}_{\Delta}\left(u(\mathbf{m})\right)$.

Proof. The multiplication equation (8) by $\exp(-ik_j\, n_j\, a_j)$, and the sum over n_j from $-\infty$ to $+\infty$ give

$$\sum_{n_j=-\infty}^{+\infty} e^{-ik_j\, n_j\, a_j}\, \mathbb{D}^{\pm}_{L}\begin{bmatrix}\alpha\\ j\end{bmatrix} u(\mathbf{m}) = \frac{1}{a_j} \sum_{n_j=-\infty}^{+\infty} \sum_{m_j=-\infty}^{+\infty} e^{-ik_j\, n_j\, a_j}\, K^{\pm}_{\alpha}(n_j - m_j)\, u(\mathbf{m}). \tag{58}$$

Using (51) and $n_j' = n_j - m_j$, the right-hand side of (58) gives

$$\sum_{n_j=-\infty}^{+\infty} \sum_{m_j=-\infty}^{+\infty} e^{-ik_j\, n_j\, a_j}\, K^{\pm}_{\alpha}(n_j - m_j)\, u(\mathbf{m}) =$$

$$= \sum_{n_j=-\infty}^{+\infty} e^{-ik_j\, n_j\, a_j}\, K^{\pm}_{\alpha}(n_j - m_j) \sum_{m_j=-\infty}^{+\infty} u(\mathbf{m}) =$$

$$= \sum_{n_j'=-\infty}^{+\infty} e^{-ik_j\, n_j'\, a_j} K^{\pm}_{\alpha}(n_j') \sum_{m_j=-\infty}^{+\infty} u(\mathbf{m}) e^{-ik_j\, m_j\, a_j} = \hat{K}^{\pm}_{\alpha}(k_j\, a_j)\, \hat{u}(\mathbf{k}). \tag{59}$$

It allows us to represent equation (58) in the form

$$\mathcal{F}_{\Delta}\left(\mathbb{D}_L^{\pm}\begin{bmatrix}\alpha\\ j\end{bmatrix} u(\mathbf{m})\right) = \frac{1}{a_j^{\alpha}}\hat{K}_{\alpha}^{\pm}(k_j\, a_j)\,\hat{u}(\mathbf{k}), \tag{60}$$

where $\mathcal{F}_{\Delta}$ is an operator notation for the Fourier series transform.

Using the expressions

$$\hat{K}_{\alpha}^{+}(a_j\, k_j) = |a_j\, k_j|^{\alpha}, \tag{61}$$

$$\hat{K}_{\alpha}^{-}(a_j\, k_j) = i\ \mathrm{sgn}(k_j)\,|a_j\, k_j|^{\alpha}, \tag{62}$$

the limit $a_j \to 0$ gives

$$\tilde{K}_{\alpha}^{+}(k_j) = \lim_{a_j\to 0}\frac{1}{a_j^{\alpha}}\,\hat{K}_{\alpha}^{+}(k_j\, a_j) = |k_j|^{\alpha}, \tag{63}$$

$$\tilde{K}_{\alpha}^{-}(k_j) = \lim_{a_j\to 0}\frac{1}{a_j^{\alpha}}\,\hat{K}_{\alpha}^{-}(k_j\, a_j) = i\, k_j\, |k_j|^{\alpha-1}. \tag{64}$$

Then the limit $a_j \to 0$ for equation (60) gives

$$\mathrm{Lim}\circ\mathcal{F}_{\Delta}\left(\mathbb{D}_L^{\pm}\begin{bmatrix}\alpha\\ j\end{bmatrix} u(\mathbf{m})\right) = \tilde{K}_{\alpha}^{\pm}(k_j)\,\tilde{u}(\mathbf{k}), \tag{65}$$

where

$$\tilde{K}_{\alpha}^{+}(k_j) = |k_j|^{\alpha}, \quad \tilde{K}_{\alpha}^{-}(k_j) = i\, k_j\, |k_j|^{\alpha-1}, \quad \tilde{u}(\mathbf{k}) = \mathrm{Lim}\,\hat{u}(\mathbf{k}).$$

The inverse Fourier transform of (65) has the form

$$\mathcal{F}^{-1}\circ\mathrm{Lim}\circ\mathcal{F}_{\Delta}\left(\mathbb{D}_L^{+}\begin{bmatrix}\alpha\\ j\end{bmatrix} u(\mathbf{m})\right) = \mathbb{D}_C^{+}\begin{bmatrix}\alpha\\ j\end{bmatrix} u(\mathbf{r}), \quad (\alpha > 0), \tag{66}$$

$$\mathcal{F}^{-1}\circ\mathrm{Lim}\circ\mathcal{F}_{\Delta}\left(\mathbb{D}_L^{-}\begin{bmatrix}\alpha\\ j\end{bmatrix} u(\mathbf{m})\right) = \frac{\partial}{\partial x_j}\left(\mathbb{D}_C^{+}\begin{bmatrix}\alpha-1\\ j\end{bmatrix} u(\mathbf{r})\right), \quad (\alpha > 1), \tag{67}$$

$$\mathcal{F}^{-1}\circ\mathrm{Lim}\circ\mathcal{F}_{\Delta}\left(\mathbb{D}_L^{-}\begin{bmatrix}\alpha\\ j\end{bmatrix} u(\mathbf{m})\right) = \frac{\partial}{\partial x_j}\mathbb{I}_C^{+}\begin{bmatrix}1-\alpha\\ j\end{bmatrix} u(\mathbf{r}), \quad (0 < \alpha < 1). \tag{68}$$

Here we use the connection between the continuum derivative and integral of the Riesz type of the order α and the corresponding Fourier integrals transform

$$\mathbb{D}_C^{+}\begin{bmatrix}\alpha\\ j\end{bmatrix} u(\mathbf{r}) = \mathcal{F}^{-1}\{|k_j|^{\alpha}\tilde{u}(\mathbf{k})\}, \quad \mathbb{I}_C^{+}\begin{bmatrix}\alpha\\ j\end{bmatrix} u(\mathbf{r}) = \mathcal{F}^{-1}\{|k_j|^{-\alpha}\tilde{u}(\mathbf{k})\}. \tag{69}$$

We have proved that lattice fractional-order derivatives are transformed by (57) into continuum fractional-order derivatives of the Riesz type. □

Using (57), and the independence of n_i and n_j for $i \neq j$, it is easy to see that the continuum limits for the lattice mixed partial derivatives (16) and (17) have the form

$$\mathcal{T}_{L\to C}\left(\mathbb{D}_L^{\pm,\pm}\begin{bmatrix}\alpha_1\,\alpha_2\\ i\,j\end{bmatrix} u(\mathbf{m})\right) = \mathbb{D}_C^{\pm}\begin{bmatrix}\alpha_1\\ i\end{bmatrix}\mathbb{D}_C^{\pm}\begin{bmatrix}\alpha_2\\ j\end{bmatrix} u(\mathbf{r}), \quad (i \neq j), \tag{70}$$

$$\mathcal{T}_{L\to C}\left(\mathbb{D}_L^{\pm,\mp}\begin{bmatrix}\alpha_1\,\alpha_2\\ i\,j\end{bmatrix} u(\mathbf{m})\right) = \mathbb{D}_C^{\pm}\begin{bmatrix}\alpha_1\\ i\end{bmatrix}\mathbb{D}_C^{\mp}\begin{bmatrix}\alpha_2\\ j\end{bmatrix} u(\mathbf{r}), \quad (i \neq j). \tag{71}$$

Similarly, we can get relations for other mixed lattice fractional derivatives.

6 Fractional Continuum Nonlinear Equations

The above Theorem allows us to obtain continuum models from lattice models by using the lattice-continuum transform operation (53). Let us consider the continuous limit of the lattice equations (28) - (32). In the continuum limit ($a_j \to 0$), the lattice equations give the continuum equations for the fractional nonlocal continuum.

Lattice fractional equation (22) gives the fractional differential equation for nonlocal continuum

$$\frac{\partial^2 u(\mathbf{r},t)}{\partial t^2} - \sum_{i,j=1}^{3} A_{ij}\, {}^{R}\mathbb{D}_{C}^{-,-} \begin{bmatrix} \alpha_i\, \alpha_j \\ i\, j \end{bmatrix} u(\mathbf{r},t) = 0. \tag{72}$$

The integer-order differential equation is defined by $\alpha_i = \alpha_j = 1$ and we obtain:

$$\frac{\partial^2 u(\mathbf{r},t)}{\partial t^2} - \sum_{i=1}^{3} A_{ij} \frac{\partial^2 u(\mathbf{r},t)}{\partial x_i \partial x_j} = 0. \tag{73}$$

In the continuum limit, lattice equations (23) gives the fractional differential equations for nonlocal continuum in the form

$$\frac{\partial^2 u(\mathbf{r},t)}{\partial t^2} + \sum_{i=1}^{3} A_i\, {}^{R}\mathbb{D}_{C}^{+} \begin{bmatrix} \alpha_i \\ i \end{bmatrix} u(\mathbf{r},t) = 0. \tag{74}$$

The integer-order differential equation is defined by $\alpha_i = 2$ for (74), and we obtain

$$\frac{\partial^2 u(\mathbf{r},t)}{\partial t^2} - \sum_{i=1}^{3} A_i \frac{\partial^2 u(\mathbf{r},t)}{\partial x_i^2} = 0. \tag{75}$$

The three-dimensional fractional-order differential equations for nonlocal anisotropic continuum, which corresponds to the lattice equations (28) - (32), have the following forms.

1. The continuum fractional 3D sine-Gordon equation

$$\frac{\partial^2 u(\mathbf{r},t)}{\partial t^2} + \sum_{i=1}^{3} A_i\, {}^{R}\mathbb{D}_{C}^{+} \begin{bmatrix} \alpha_i \\ i \end{bmatrix} u(\mathbf{r},t) + \sin(u(\mathbf{r},t)) = 0, \tag{76}$$

where the integer-order differential equation is defined by $\alpha_i = 2$ in the form

$$\frac{\partial^2 u(\mathbf{r},t)}{\partial t^2} - \sum_{i=1}^{3} A_i \frac{\partial^2 u(\mathbf{r},t)}{\partial x_i^2} + \sin(u(\mathbf{r},t)) = 0, \tag{77}$$

2. The continuum fractional 3D Burgers equation

$$\frac{\partial u(\mathbf{r},t)}{\partial t} + \sum_{i=1}^{3} A_i\, {}^{R}\mathbb{D}_{C}^{-} \begin{bmatrix} \alpha_i \\ i \end{bmatrix} u^2(\mathbf{r},t) + \sum_{i=1}^{3} B_i\, {}^{R}\mathbb{D}_{C}^{+} \begin{bmatrix} \beta_i \\ i \end{bmatrix} u(\mathbf{r},t) = 0. \tag{78}$$

The integer-order differential equation is defined by $\alpha_i = 1$ and $\beta_i = 2$ in the form

$$\frac{\partial u(\mathbf{r},t)}{\partial t} + \sum_{i=1}^{3} A_i \frac{\partial u^2(\mathbf{r},t)}{\partial x_i} - \sum_{i=1}^{3} B_i \frac{\partial^2 u(\mathbf{r},t)}{\partial x_i^2} = 0. \tag{79}$$

We note that a special case of 1D fractional Burgers equation is suggested in (Biler et al. 1998). The 1D analog of equation (79) gives the Burgers equation (Burgers 2008) that is a nonlinear partial differential equation of second order:

$$\frac{\partial u(x,t)}{\partial t} + A_x \frac{\partial u^2(x,t)}{\partial x} - B_x \frac{\partial^2 u(x,t)}{\partial x^2} = 0. \tag{80}$$

This equation is used in fluid dynamics as a simplified model for turbulence, boundary layer behavior, shock wave formation, and mass transport.

3. The continuum fractional 3D Korteweg-de Vries equation

$$\frac{\partial u(\mathbf{r},t)}{\partial t} - \sum_{i=1}^{3} A_i \, {}^{R}\mathbb{D}_C^{-} \begin{bmatrix} \alpha_i \\ i \end{bmatrix} u^2(\mathbf{r},t) - \sum_{i=1}^{3} B_i \, {}^{R}\mathbb{D}_C^{-} \begin{bmatrix} \beta_i \\ i \end{bmatrix} u(\mathbf{r},t) = 0. \tag{81}$$

The integer-order differential equation is defined by (81) with $\alpha_i = 1$ and $\beta_i = 3$ in the form

$$\frac{\partial u(\mathbf{r},t)}{\partial t} - \sum_{i=1}^{3} A_i \frac{\partial u^2(\mathbf{r},t)}{\partial x_i} + \sum_{i=1}^{3} B_i \frac{\partial^3 u(\mathbf{r},t)}{\partial x_i^3} = 0, \tag{82}$$

Note that 1D fractional KdV equation has been suggested in (Momani 2005, Miskinis 2005). The 1D analog of equation (82) is the 1D Korteweg-de Vries equation

$$\frac{\partial u(x,t)}{\partial t} - A_x \frac{\partial u^2(x,t)}{\partial x} + B_x \frac{\partial^3 u(x,t)}{\partial x^3} = 0, \tag{83}$$

First formulated as a part of an analysis of shallow-water waves in canals, it has subsequently been found to be involved in a wide range of physics phenomena, especially those exhibiting shock waves, travelling waves and solitons. It is used in fluid dynamics, aerodynamics, and continuum mechanics as a model for shock wave formation, solitons, boundary layer behavior, turbulence, and mass transport.

4. The continuum fractional 3D Kadomtsev-Petviashvili equation

$$\begin{aligned} &{}^{R}\mathbb{D}_C^{-} \begin{bmatrix} \alpha_x \\ x \end{bmatrix} \dot{u}(\mathbf{r},t) - A_x \, {}^{R}\mathbb{D}_C^{+} \begin{bmatrix} \beta_x \\ x \end{bmatrix} u^2(\mathbf{r},t) - \\ &- B_x \, {}^{R}\mathbb{D}_C^{+} \begin{bmatrix} \gamma_x \\ x \end{bmatrix} u(\mathbf{r},t) - \sum_{i=2}^{3} C_i \, {}^{R}\mathbb{D}_C^{-} \begin{bmatrix} \delta_i \\ i \end{bmatrix} u(\mathbf{r},t) = 0, \end{aligned} \tag{84}$$

where $\dot{u}(\mathbf{r},t) = \partial u(\mathbf{r},t)/\partial t$, and the integer-order differential equation is defined by $\alpha_x = 1$, $\beta_x = 2$, $\gamma_x = 4$, and $\delta_y = \delta_z = 2$ i the form

$$\frac{\partial \dot{u}(\mathbf{r},t)}{\partial x} + A_x \frac{\partial^2 u^2(\mathbf{r},t)}{\partial x^2} - B_x \frac{\partial^4 u(\mathbf{r},t)}{\partial x_i^4} + C_y \frac{\partial^2 u(\mathbf{r},t)}{\partial y^2} + C_z \frac{\partial^2 u(\mathbf{r},t)}{\partial z^2} = 0, \tag{85}$$

The Kadomtsev-Petviashvili equations has been obtained as a two-dimensional generalization of the KdV equation in the study of plasmas (Kadomtsev et al. 1970). In 1997, Bouard and Saut obtained the three-dimensional generalization of the KdV equation in (Kadomtsev et al. 1970, Bouard et al. 1997a). The three-dimensional KdV equation with power functions is considered in (Jones 2000).

5. The continuum fractional 3D Boussinesq equation

$$\frac{\partial^2 u(\mathbf{r},t)}{\partial t^2} + \sum_{i=1}^{3} A_i \, {}^R\mathbb{D}_L^+ \begin{bmatrix} \alpha_i \\ i \end{bmatrix} u(\mathbf{r},t) - B_x \, {}^R\mathbb{D}_C^+ \begin{bmatrix} \beta_x \\ x \end{bmatrix} u^2(\mathbf{r},t) + C_x \, {}^R\mathbb{D}_C^+ \begin{bmatrix} \gamma_x \\ x \end{bmatrix} u(\mathbf{r},t) = 0. \tag{86}$$

The integer-order differential equation is defined by $\alpha_i = 2$, $\beta_x = 2$, $\gamma_x = 4$ in the form

$$\frac{\partial^2 u(\mathbf{r},t)}{\partial t^2} - \sum_{i=1}^{3} A_i \frac{\partial^2 u(\mathbf{r},t)}{\partial x_i^2} + B_x \frac{\partial^2 u^2(\mathbf{r},t)}{\partial x^2} + C_x \frac{\partial^4 u(\mathbf{r},t)}{\partial x^4} = 0. \tag{87}$$

In the 1D case, the 3DBoussinesq equation (87) s a nonlinear partial differential equation of fourth order of the form

$$\frac{\partial^2 u(x,t)}{\partial t^2} - A_x \frac{\partial^2 u(x,t)}{\partial x^2} + B_x \frac{\partial^2 u^2(x,t)}{\partial x^2} + C_x \frac{\partial^4 u(x,t)}{\partial x^4} = 0. \tag{88}$$

This equation was formulated as a part of an analysis of long waves in shallow water. It was subsequently applied to problems in the percolation of water in porous subsurface strata.

In the continuum limit ($a_j \to 0$), the lattice equations (36) give the Navier-Stokes equations for the fractional nonlocal continuum in the following form

$$\left(\frac{\partial \mathbf{v}(\mathbf{r},t)}{\partial t} + \left(\mathbf{v}(\mathbf{r},t), \, {}^R\nabla_C^{\alpha,-} \right) \mathbf{v}(\mathbf{r},t) \right) =$$

$$= - {}^R\mathrm{Grad}_C^{\alpha,-} \, p(\mathbf{r},t) + \mu \, {}^R\Delta_C^{2\alpha+} \, \mathbf{v}(\mathbf{r},t) + \mathbf{f}(\mathbf{r},t). \tag{89}$$

where ${}^R\nabla_C^{\alpha,\pm}$, ${}^R\mathrm{Grad}_C^{\alpha,\pm}$, ${}^R\Delta_C^{2\alpha,\pm}$ are defined by the fractional derivatives ${}^R\mathbb{D}_C^{\pm}\begin{bmatrix} \alpha \\ i \end{bmatrix}$. Equations (89) are the Navier-Stokes equations for nonlocal media with power-law nonlocality of the Riesz type. It should be noted that other fractional generalizations of the Navier-Stokes equations have been suggested in (Yang et al. 2013b, Tarasov 2005).

7 Conclusion

In this Chapter the three-dimensional lattice models with long-range properties were described for fractional nonlocal continuum. The proposed lattice model can be considered as a new microstructural basis of a unified description of continuum models with power-law nonlocality. The suggested type of discrete dynamical systems can be considered for integer and non-integer (fractional) values of the parameter α. This allows us to obtain lattice models for the local and nonlocal continuum systems. As an example application, we described a relationship of lattice and continuum fractional-order nonlinear equations, that are generalizations of the sine-Gordon, Burgers, Korteweg-deVries, Kadomtsev-Petviashvili, Boussinesq, and Navier-Stokes equations.

We can note some possible extensions of the suggested lattice approach to formulate generalizations of nonlocal fractal systems. We assume that the suggested lattice approach can be generalized for lattices with the fractal spatial dispersion that are suggested in (Tarasov 2008) (see also (Michelitsch et al. 2009, 2011)), and the continuum limit of these fractal lattice models can leads us to different models of fractal material Yang (2012), Tarasov (2011).

Acknowledgment: The author would like to thanks professor Xiao-Jun Yang and professor Carlo Cattani, for their kind invitation to write a chapter for this edited volume.

Bibliography

Samko, S. G., Kilbas, A. A., Marichev, O. I. (1993). Fractional Integrals and Derivatives Theory and Applications. New York: Gordon and Breach.

Kilbas, A. A., Srivastava, H. M., Trujillo, J. J. (2006). Theory and Applications of Fractional Differential Equations. Amsterdam: Elsevier.

Carpinteri, A., Mainardi, F. (1997). Fractals and Fractional Calculus in Continuum Mechanics. New York: Springer.

Sabatier, J., Agrawal, O. P., Tenreiro Machado, J. A. (2007). Advances in Fractional Calculus. Theoretical Developments and Applications in Physics and Engineering. Dordrecht: Springer.

Mainardi, F. (2010). Fractional Calculus and Waves in Linear Viscoelasticity: An Introduction to Mathematical Models. Singapore: World Scientific.

Tarasov, V. E. (2011). Fractional Dynamics: Applications of Fractional Calculus to Dynamics of Particles, Fields and Media. New York: Springer.

Tarasov, V. E. (2013). Review of some promising fractional physical models. International Journal of Modern Physics B. 27 (9), 1330005 (2013), arXiv:1502.07681

Atanackovic, T. M., Pilipovic, S., Stankovic, S., Zorica, D. (2014). Fractional Calculus with Applications in Mechanics: Wave Propagation, Impact and Variational Principles. London, Hoboken: Wiley-ISTE.

Tarasov, V. E. (2013). Lattice model with power-law spatial dispersion for fractional elasticity. Central European Journal of Physics. 11 (11), 1580-1588. (arXiv:1501.0120)

Tarasov, V. E. (2014). Lattice with long-range interaction of power-law type or fractional nonlocal elasticity. International Journal of Solids and Structures. 51 (15-16), 2900-2907. (arXiv:1502.05492)

Tarasov, V. E. (2014). Fractional gradient elasticity from spatial dispersion law. ISRN Condensed Matter Physics. 2014, 794097. (arXiv:1306.257)

Tarasov, V. E. (2014). Lattice model of fractional gradient and integral elasticity: Long-range interaction of Grünwald-Letnikov-Riesz type. Mechanics of Materials. 70, 106-114. (arXiv:1502.06268)

Tarasov, V. E. (2014). Large lattice fractional Fokker-Planck equation. Journal of Statistical Mechanics: Theory and Experiment. 2014 (9), P09036. (arXiv:1503.03636)
Tarasov, V. E. (2015). Non-linear fractional field equations: weak non-linearity at power-law non-locality. Nonlinear Dynamics. 80 (4), 1665-1672.
Tarasov, V. E. (2014). Fractional quantum field theory: From lattice to continuum. Advances in High Energy Physics. 2014, 957863.
Erdelyi, A., Magnus, W., Oberhettinger, F., Tricomi, F. G. (1981), Higher Transcendental Functions. Volume 1. Melbourne, Florida: Krieeger.
Prudnikov, A. P., Brychkov, Yu. A., Marichev, O. I. (1986). Integrals and Series, Vol. 1: Elementary Functions New York: Gordon and Breach.
Tarasov, V. E. (2006). Map of discrete system into continuous. Journal of Mathematical Physics. 47 (9), 092901. (arXiv:0711.2612)
Tarasov, V. E. (2006). Continuous limit of discrete systems with long-range interaction. Journal of Physics A. 39 (48), 14895-14910. (arXiv:0711.0826)
Tarasov, V. E. (2014). Toward lattice fractional vector calculus. Journal of Physics A. 47 (35), 355204.
Tarasov, V. E. (2015). Lattice fractional calculus. Applied Mathematics and Computation. 257, 12-33.
Uchaikin, V. V. (2012). Fractional Derivatives for Physicists and Engineers. Volume I. Background and Theory. New York, Beijing: Springer, Higher Education Press.
Tarasov, V. E. (2014). General lattice model of gradient elasticity. Modern Physics Letters B. 28 (7), 1450054.
Tarasov, V. E. (2015). Lattice Model with nearest-neighbor and next-nearest-neighbor interactions for gradient elasticity. Discontinuity, Nonlinearity, and Complexity. 4 (1), 11-23.
Kadomtsev B. B., Petviashvili, V. I. (1970). On the stability of solitary waves in weakly dispersing media. Soviet Physics Doklady. 15 (6), 539-541.
de Bouard, A., Saut, J. C. (1997). Solitary waves of generalized Kadomtsev-Petviashvili equations. Annales de l'Institut Henri Poincare (C) Analyse Non Lineaire. 14 (2), 211-236.
de Bouard, A., Saut, J. C. (1997). Symmetries and decay of the generalized Kadomtsev-Petviashvili solitary waves. SIAM Journal on Mathematical Analysis. 28 (5), 1064-1085.
Jones, K. L. (2000). Three-dimensional Korteweg-de Vries equation and traveling wave solutions. International Journal of Mathematics and Mathematical Sciences. 24 (6), 379-384.
Webb, G. M., Zank, G. P. (1990). Painleve analysis of the three-dimensional Burgers equation. Physics Letters A. 150 (1), 14-22.
Shandarin, S. F. (1996). Three-dimensional Burgers equation as a model for the large-scale structure formation in the Universe. In S.A. Molchanov, W.A. Woyczynski (Eds.), The IMA Volumes in Mathematics and its Applications. (pp.401-413) New York: Springer. (arXiv:astro-ph/9507082)
Dai, C. Q., Yu, F. B. (2014). Special solitonic localized structures for the (3+1)-dimensional Burgers equation in water waves. Wave Motion. 51 (1), 52-59.
Korpusov, M. O. (2012). Blowup of solutions of the three-dimensional Rosenau-Burgers equation. Theoretical and Mathematical Physics. 170, 280-286.
Wazwaz, A. M. (2015). A variety of (3+1)-dimensional Burgers equations derived by using the Burgers recursion operator. Mathematical Methods in the Applied Sciences. 38 (12), 2642-2649.
Johnson, R. S. (1996). A two-dimensional Boussinesq equation for water waves and some of its solutions. Journal of Fluid Mechanics. 323, 65-78.
El-Sabbagh, M. F., Ali, A. T. (2005). New exact solutions for (3+1)-dimensional Kadomtsev-Petviashvili equation and generalized (2+1)-dimensional Boussinesq equation. International Journal of Nonlinear Sciences and Numerical Simulation. 6 (2), 151-162.
Yong-Qi, W. (2008). Periodic wave solution to the (3+1)-dimensional Boussinesq equation. Chinese Physics Letters. 25 (8), 2739.

Huan, Z., Bo, T., Hai-Qiang, Z., Tao, M., Xiang-Hua, Wen-Jun, L., Ke-Jie, C. (2008). Periodic wave solutions for (2+1)-dimensional Boussinesq equation and (3+1)-dimensional Kadomtsev-Petviashvili equation. Communications in Theoretical Physics. 50 (5), 1169.

Moleleki, L. D., Khalique, C. M. (2014). Symmetries, traveling wave solutions, and conservation laws of a (3+1)-dimensional Boussinesq equation. Advances in Mathematical Physics. 2014, 672679.

Burgers, J. (2008). The Nonlinear Diffusion Equation. Dordrecht, Amsterdam: Reidel.

Biler, P., Funaki, T., Woyczynski, W. A. (1998). Fractal Burger equation. Journal of Differential Equations. 14 (2), 9-46.

Momani, S. (2005). An explicit and numerical solutions of the fractional KdV equation. Mathematics and Computers in Simulation. 70, 110-118.

Miskinis, P. (2005). Weakly nonlocal supersymmetric KdV hierarchy. Nonlinear Analysis: Modelling and Control. 10 (4), 343-348.

Tarasov, V. E. (2008). Chains with fractal dispersion law. Journal of Physics A. 41 (3), 035101. (arXiv:0804.0607)

Michelitsch, T. M., Maugin, G. A., Nicolleau, F. C. G. A., Nowakowski, A. F., Derogar, S. (2009). Dispersion relations and wave operators in self-similar quasicontinuous linear chains. Physical Review E. 80, 011135. (arXiv:0904.0780)

Michelitsch, T. M., Maugin, G. A., Nicolleau, F. C. G. A., Nowakowski, A. F., Derogar, S. (2011). Wave propagation in quasi-continuous linear chains with self-similar harmonic interactions: Towards a fractal mechanics. in H. Altenbach, G.A. Maugin, V. Erofeev, (Eds.) Mechanics of Generalized Continua: Advanced Structured Materials. (pp.231-244) Berlin, Heidelberg: Springer.

Yang, X. J., Baleanu, D., Tenreiro Machado, J. A. (2013). Systems of Navier-Stokes equations on Cantor sets. Mathematical Problems in Engineering. 2013, 769724.

Tarasov, V. E. (2005). Fractional hydrodynamic equations for fractal media. Annals of Physics. 318 (2), 286-307. (arXiv:physics/0602096)

Yang, X. J. (2012). Advanced Local Fractional Calculus and Its Applications. New York: World Science Publisher.

António M. Lopes* and J.A. Tenreiro Machado

Temporal Patterns in Earthquake Data-series

Abstract: This chapter studies the statistical distributions of earthquakes in Southern California over the time period from 1934 to 2013. We adopt a sliding time window in order to divide the original dataset into 16 time intervals and we then use the Jensen-Shannon divergence to compare the empirical data. Second, we use hierarchical clustering algorithms and multidimensional scaling techniques for data analysis and visualization. The results reveal relationships and temporal patterns hidden in the data. The methodology and findings contribute to a comprehensive explanation of these phenomena and to recognize precursory events for earthquake prediction.

Keywords: Complex systems; Dynamical systems; Fourier transform; Visualization

1 Introduction

Earth's tectonic plates move with respect to each other (Stadler, Gurnis, Burstedde, et al. 2010), exhibiting friction and stick-slip behavior along the fault surfaces (Bhattacharya, Chakrabarti & Kamal 2011, Carlson, Langer & Shaw 1994, De Rubeis, Hallgass, Loreto, et al. 1996). The asperities between the plates increase stress, while strain energy accumulates around the fault surfaces. Whenever the stress is sufficiently high as to break through the asperities, the stored energy is released in a sudden motion of the plates, causing an earthquake (Lopes, Machado, Pinto, et al. 2013). Earthquakes are complex phenomena that exhibit characteristics usually found in systems with memory (Lennartz, Livina, Bunde, et al. 2008). Their complex correlations in space, time and magnitude are characterized by self-similarity and the absence of characteristic length-scale, which means that seismic parameters exhibit power-law (PL) behavior (Lopes, Machado, Pinto, et al. 2013). The Gutenberg-Richter (GR) law describes the overall frequency distribution of earthquakes (Gutenberg & Richter 1944):

$$\log N(m) = q - r \cdot m \tag{1}$$

where $N(m)$ is the frequency of earthquakes with magnitude greater than m, r is a parameter that has regional variation, and is in the range $[0.8; 1.06]$ and $[1.23; 1.54]$ for small and big earthquakes, respectively, and q is a measure of the regional level of seismicity (Christensen & Olami 1992).

***Corresponding Author: António M. Lopes:** Institute of Mechanical Engineering, Faculty of Engineering, University of Porto, Rua Dr. Roberto Frias, 4200-465, Porto, Portugal, E-mail: aml@fe.up.pt
J.A. Tenreiro Machado: Institute of Engineering, Polytechnic of Porto, Rua Dr. António Bernardino de Almeida, 431, 4249-015 Porto, Portugal, E-mail: jtm@isep.ipp.pt

The (modified) Omori law describes the rate of decay of aftershock sequences following a main event (Utsu, Ogata & Matsu'ura 1995):

$$N(t) = C_1(t + C_2)^{-\alpha} \tag{2}$$

where t represents the time after the main shock, C_1 and C_2 are constants, and α is the rate of decay. The classical Omori law is obtained for $\alpha = 1$ and $C_2 = 0$ (Ōmori 1894).

Earthquake data have been studied from the perspective of complex systems (CS), statistics and fractals (Lopes, Machado, Pinto, et al. 2013). In certain models (Hallgass, Loreto, Mazzella, et al. 1997) the asperities on the fault surfaces act like fractals sliding over each other, and this explains the fractal scaling behavior observed in earthquake phenomena. Other authors claim that tectonic plates form a CS owing to interactions among faults (Sarlis & Christopoulos 2012, Turcotte & Malamud 2002, Kanamori & Brodsky 2004). The tectonic plates' motion and strain accumulation processes interact on a wide range of scales and loading rates are not uniform in time (Stein, Liu, Calais & Li 2009). Statistical processing of real data and clustering analysis have been used to detect precursory events and for earthquake forecasting (Geller 1997, Jafari 2010, Kagan & Jackson 2000). However, it should be noted that, at present, reliable short-term prediction of earthquakes is not possible (Lopes, Machado, Pinto, et al. 2014).

In this chapter we study the statistical distributions of earthquakes in Southern California (Hutton, Woessner & Hauksson 2010). The data is from the Southern California Earthquake Data Center (SCEDC) and is freely available at http://www.data.scec.org/. We consider the time period from 1934 to 2013 and we adopt a sliding time window in order to divide the original data into 16 time intervals. First, we prepare the seismic catalog for data analysis. Second, we adopt the Jensen-Shannon divergence to compare the empirical data. Third, we use hierarchical clustering algorithms and multidimensional scaling (MDS) techniques for data analysis and visualization. The results reveal relationships and temporal patterns embedded in the data that may explain the phenomena and may contribute to recognize precursory events for earthquake prediction.

This chapter is organized as follows. Section 2 presents the experimental dataset. Section 3 introduces the main mathematical tools used. Section 4 compares the events using different visualization methods and unveils hidden patterns in the data. Finally, Section 5 outlines the main conclusions.

2 Dataset

In this study we use data from the Southern California Earthquake Data Center (SCEDC), which is freely available at http://www.data.scec.org/. The catalog includes over 600,000 seismic events from 1932 to the present day. Each record contains information about the earthquake magnitude, epicenter location (latitude, longitude and depth), and time (with one-second resolution), among other (Hutton, Woessner & Hauksson 2010). The data was retrieved in October, 2014.

The geographic location and the annual evolution of the total number of reported occurrences is depicted in Figures 1 and 2. The magnitude of the events varies between $M = -1$ and $M = 7.5$ (in log scale). It can be noticed that from the early decades of the twentieth century until the seventies, the number of events per year is remarkably smaller than those observed in recent years. However, this observation does not correspond to a real change in seismic behavior but, in fact, is due to a change in the detection capabilities and reporting of the events over time. In other words, the catalog is incomplete in terms of magnitude, particularly for older dates. Before any data processing, the completeness of the catalog must be addressed.

Magnitude completeness is defined as the lowest magnitude at which all earthquakes in a given space-time volume are detected (Stucchi, Albini, Mirto, et al. 2004). We estimate the magnitude completeness by fitting a Gutenberg-Richter (GR) model to the empirical data distributions for each time window, w_i $(i = 1, \ldots, 16)$, and we determine the minimum magnitude at which the lower ends of the frequency-magnitude distributions diverge from the corresponding GR model (Mignan & Woessner 2012). The magnitude value $M_C = 4.0$ was found as the threshold above which magnitude completeness is achieved for all time windows, meaning that events of magnitude below M_C are discarded. Figure 3, for example, depicts the frequency-magnitude distribution corresponding to w_{11}, as well as the GR model that best fits the data, showing that, for this case, $M_C = 3.71$.

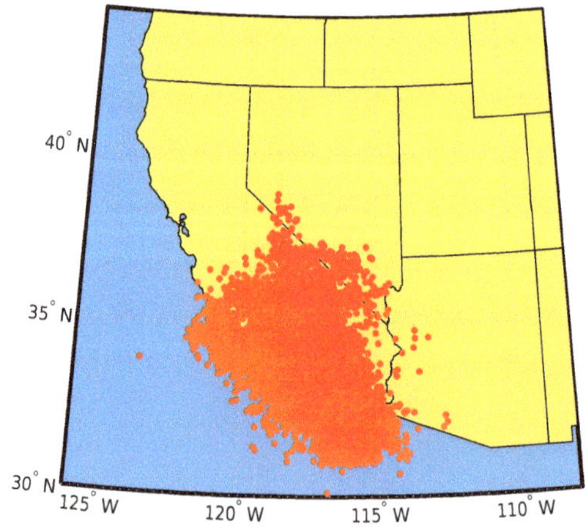

Fig. 1. Geographic location of all reported events occurred in Southern California over the time interval from 1932 up to present.

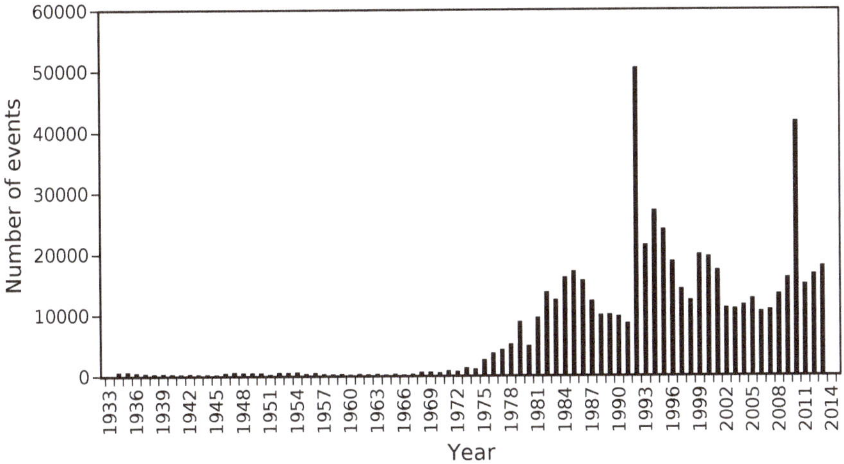

Fig. 2. Annual evolution of the total number of reported seismic events occurred in Southern California over the time interval from 1932 up to present.

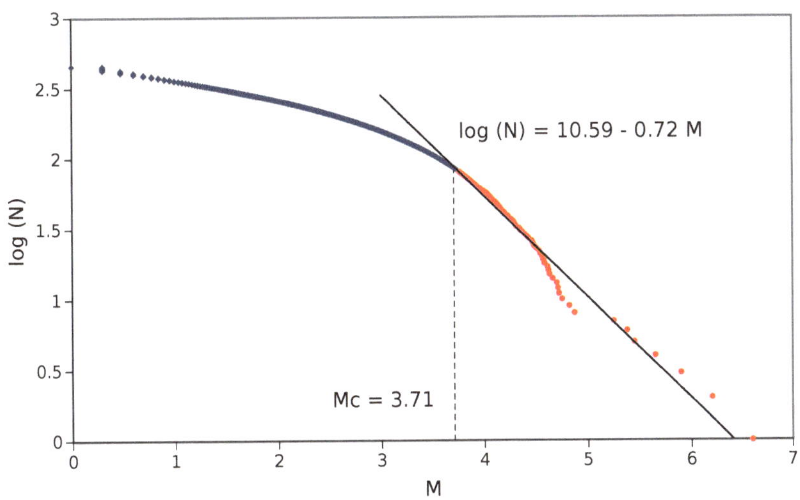

Fig. 3. Frequency-magnitude distribution corresponding to w_{11}, as well as the GR model that best fits the experimental data.

3 Mathematical Tools

This section presents the mathematical tools used to process the data, namely hierarchical clustering and MDS.

3.1 Hierarchichal Clustering

Clustering is a technique for data analysis that has been adopted in many fields including machine learning, pattern recognition, data mining, image analysis, information retrieval and bioinformatics, to name a few (Hartigan 1975). Clustering groups objects that are similar to each other. For hierarchical clustering, two alternative types of algorithms are used to generate a hierarchy of clusters: (i) agglomerative and (i) divisive clustering. In (i) each object starts in its own cluster and, at each step, until there is a single cluster containing all objects, the two most similar clusters are merged. In (ii) all objects start in one cluster and, at each step, until each object is in its own singleton cluster, the algorithm removes the 'outsiders' from the least cohesive cluster. Independent of the choice of algorithm, clustering requires a measure of dissimilarity between clusters, which is achieved by using a metric to quantify the distance between pairs of objects, and a linkage criterion, for quantifying the dissimilarity between clusters as a function of the pairwise distances between objects. By definition, for the clusters R and S, the distance between objects $x_R \in R$ and $x_S \in S$ is expressed as $d(x_R, x_S)$ (Aggarwal, Hinneburg & Keim 2001). Based on that metric, the maximum, minimum and average linkages are commonly used, and are defined, respectively as:

$$d_{max}(R, S) = \max_{x_R \in R, x_S \in S} d(x_R, x_S) \tag{3}$$

$$d_{min}(R, S) = \min_{x_R \in R, x_S \in S} d(x_R, x_S) \tag{4}$$

$$d_{ave}(R, S) = \frac{1}{\| R \| \| S \|} \sum_{x_R \in R, x_S \in S} d(x_R, x_S) \tag{5}$$

The results of hierarchical clustering are presented in a dendrogram or a visualization tree.

3.2 Multidimensional Scaling

MDS is a statistical technique for visualizing data (Cox & Cox 2000). Given s objects and a measure of similarity, an $s \times s$ symmetric matrix, C, of item to item similarities is constructed and passed to the MDS algorithm. MDS assigns points to the items in a multi-dimensional space in order to reproduce the observed similarities. MDS evaluates different configurations for maximizing a goodness-of-fit and arrives at a configuration that best approximates the observed similarities. A common measure that is used to evaluate how a particular configuration reproduces the observed similarities is the raw stress:

$$\mathcal{S} = \left[d_{ij} - f(\delta_{ij})\right]^2 \tag{6}$$

where d_{ij} represents the reproduced similarities, given the respective number of dimensions, δ_{ij} corresponds to the observed similarities and $f(\cdot)$ indicates some type of transformation. The smaller the stress, $\mathcal{S}$, the better is the fit between d_{ij} and δ_{ij}. The final orientation of axes in the space can be chosen in order to produce a clearer visualization.

4 Data Analysys and Pattern Visualization

In this section we adopt the Jensen-Shannon divergence in order to compare the empirical data. In subsection 4.1, we use a hierarchical clustering algorithm for data analysis and visualization, and in subsection 4.2, we apply MDS. In both cases, the results reveal relationships and temporal patterns hidden in the data.

The Jensen-Shannon divergence, D_{JS}, measures the 'distance' between two probability distributions, P and Q (Cover & Thomas 2012), and can be viewed as a symmetrical and smoothed version of the Kullback-Leibler divergence, D_{KL}, which is given by:

$$D_{KL}\left(P \parallel Q\right) = \int_{-\infty}^{+\infty} ln\left[\frac{p(x)}{q(x)}\right] p(x)dx \tag{7}$$

resulting in:

$$D_{JS}\left(P \parallel Q\right) = \frac{1}{2} D_{KL}\left(P \parallel M\right) + \frac{1}{2} D_{KL}\left(Q \parallel M\right) \tag{8}$$

where, $M = (P + Q)/2$ is a mixture distribution.

4.1 Hierarchical Clustering Analysis and Comparison

We calculate a 16×16 matrix $C = [D_{JS_{ij}}]$, $(i, j) = 1, \ldots, 16$, where $D_{JS_{ij}}$ represents the Jensen-Shannon divergence between each time windows' distribution (w_i, w_j). The probabilities are estimated by means of the histograms of relative frequencies, with $N = 350$ bins. We use a hierarchical clustering algorithm based on the successive (agglomerative) clustering and average-linkage method. Figures 4 and 5 depict the dendrogram and visualization tree generated by the software PHYLIP (http://evolution.genetics.washington.edu/phylip.html). We note the emergence of two main clusters: $\mathcal{A} = \{w_2, w_4, w_5, w_{12}, w_{14}, w_{16}\}$ and $\mathcal{B} = \{w_1, w_3, w_6, w_7, w_8, w_9, w_{10}, w_{11}, w_{13}, w_{15}\}$, while smaller groups can be identified in both $\mathcal{A}$ and $\mathcal{B}$.

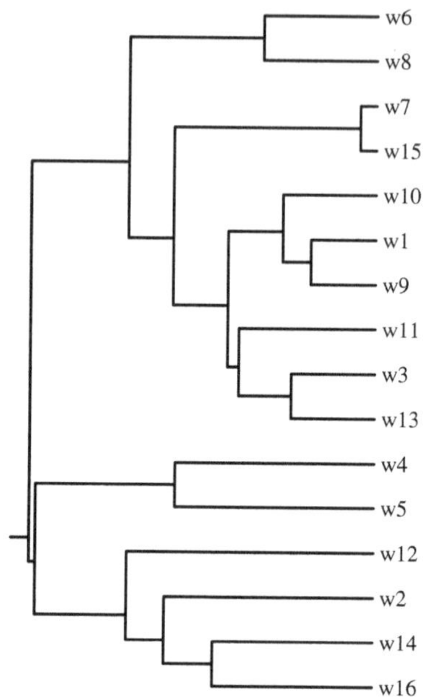

Fig. 4. Dendrogram representing similarities between time windows (w_i, w_j), $(i, j) = 1, \ldots, 16$, distributions, based on Jensen-Shannon divergence.

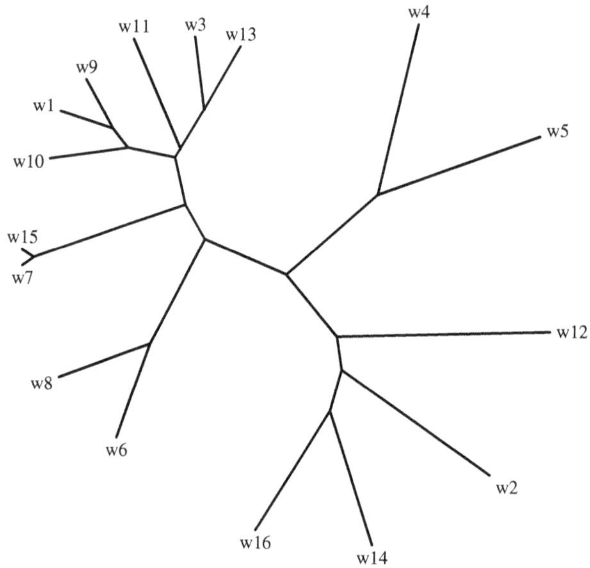

Fig. 5. Visualization tree representing similarities between time windows (w_i, w_j), $(i, j) = 1, \ldots, 16$, distributions, based on Jensen-Shannon divergence.

4.2 MDS Analysis and Visualization

For an alternative visualization, an MDS algorithm is fed with the matrix $C = [D_{JS_{ij}}]$ and the 3-dimensional MDS map is generated (Figure 6). Shorter (larger) distance between two points on the map means that the corresponding objects are more similar (distinct).

Figures 7 and 8 represent the Shepard plot for the MDS 3-dimensional map and the stress graph, which assess the quality of the MDS results. On the former, a scatter of points around the 45 degree line reveals a good correspondence between the original and the reproduced distances. On the latter, the elbow of the curve means that a 3-dimensional map is a good compromise between facility of representation and good visualization.

The results confirm the patterns obtained by the hierarchical clustering algorithm. Both visualization trees and MDS maps allow good visualization of the correlations between the data. Typically, the MDS maps have the advantage of being more intuitive when dealing with large numbers of objects.

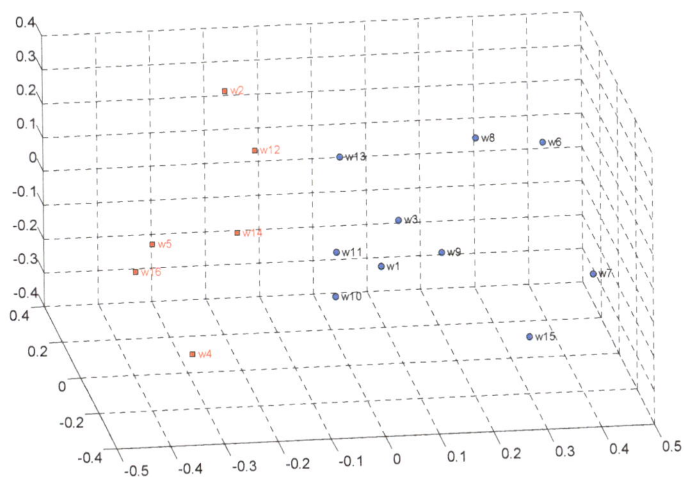

Fig. 6. MDS 3-dimensional map comparing the time windows (w_i, w_j), $(i, j) = 1, \ldots, 16$, distributions, based on Jensen-Shannon divergence.

5 Conclusions

We studied the statistical distributions of earthquakes in Southern California over the time period from 1934 to 2013. First, we adopted the Jensen-Shannon divergence

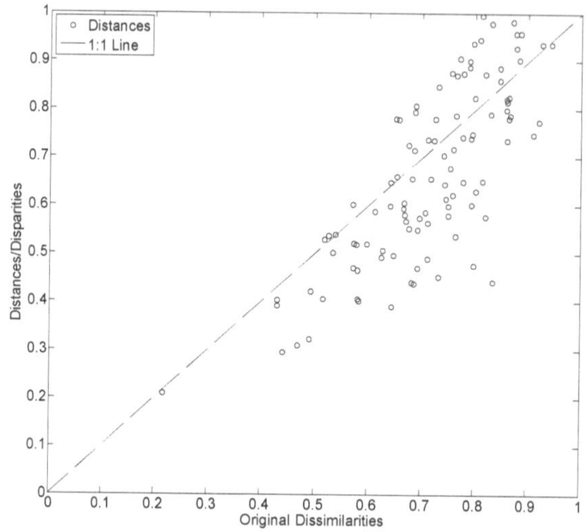

Fig. 7. Shepard plot corresponding to the MDS 3-dimensional maps comparing the time windows (w_i, w_j), $(i, j) = 1, \ldots, 16$, distributions, based on Jensen-Shannon divergence.

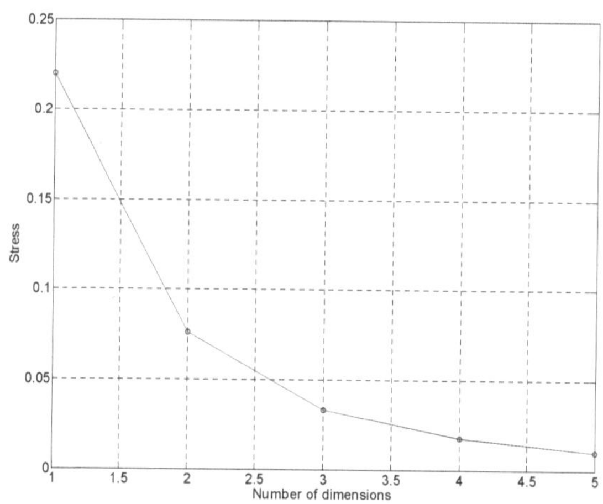

Fig. 8. Stress map for the MDS comparing the time windows (w_i, w_j), $(i, j) = 1, \ldots, 16$, distributions, based on Jensen-Shannon divergence.

to compare the empirical data. Second, we used hierarchical clustering algorithms and multidimensional scaling techniques for data analysis and visualization. Our results reveal relationships and temporal patterns hidden in the data and show that the adopted methodology and findings can contribute to a comprehensive explanation of the phenomena and to recognize precursory events for earthquake prediction.

Acknowledgment: The authors acknowledge the Southern California Earthquake Data Center (SCEDC) for the data (http://www.data.scec.org/).

Bibliography

Aggarwal, C.C., Hinneburg, A., & Keim, D.A. (2001). On the surprising behavior of distance metrics in high dimensional space. Lecture Notes in Computer Science, 1973(1), 420–434.

Jafari, M.A. (2010). Statistical prediction of the next great earthquake around Tehran, Iran. Journal of Geodynamics, 49(1), 14–18.

Bhattacharya, P., Chakrabarti, B.K., & Kamal (2011). A fractal model of earthquake occurrence: Theory, simulations and comparisons with the aftershock data. Journal of Physics: Conference Series, 319, 012004.

Carlson, J.M., Langer, J.S., & Shaw, B.E. (1994). Dynamics of earthquake faults. Reviews of Modern Physics, 66(2), 657–670.

Christensen, K., & Olami, Z. (1992). Variation of the Gutenberg-Richter *b* values and nontrivial temporal correlations in a spring-block model for earthquakes. Journal of Geophysical Research: Solid Earth (1978–2012), 97(B6), 8729–8735.

Cover, T.M., & Thomas, J.A. (2012). Elements of information theory. Hoboken, NJ: John Wiley & Sons.

Cox, T.F., & Cox, M.A. (2000). Multidimensional scaling. Boca Raton, FL: CRC Press.

De Rubeis, V., Hallgass, R., Loreto, V., Paladin, G., Pietronero, L., & Tosi, P. (1996). Self-affine asperity model for earthquakes. Physical review letters, 76(14), 2599–2602.

Geller, R.J. (1997). Earthquake prediction: a critical review. Geophysical Journal International, 131(3), 425–450.

Gutenberg, B., & Richter, C.F. (1944). Frequency of earthquakes in California. Bulletin of the Seismological Society of America, 34(4), 185–188.

Hallgass, R., Loreto, V., Mazzella, O., Paladin, G., & Pietronero, L. (1997). Earthquake statistics and fractal faults. Physical Review E, 56(2), 1346–1356.

Hartigan, J.A. (1975). Clustering algorithms. New York: John Wiley & Sons.

Hutton, K., Woessner, J., & Hauksson, E. (2010). Earthquake monitoring in southern California for seventy-seven years (1932–2008). Bulletin of the Seismological Society of America, 100(2), 423–446.

Kagan, Y.Y., & Jackson, D.D. (2000). Probabilistic forecasting of earthquakes. Geophysical Journal International, 143(2), 438–453.

Kanamori, H., & Brodsky, E.E. (2004). The physics of earthquakes. Reports on Progress in Physics, 67(8), 1429–1496.

Lennartz, S., Livina, V., Bunde, A., & Havlin, S. (2008). Long-term memory in earthquakes and the distribution of interoccurrence times. EPL (Europhysics Letters), 81(6), 69001.

Lopes, A.M., Machado, J.T., Pinto, C., & Galhano, A. (2014). Multidimensional scaling visualization of earthquake phenomena. Journal of seismology, 18(1), 163–179.

Lopes, A.M., Machado, J.T., Pinto, C., & Galhano, A. (2013). Fractional dynamics and MDS visualization of earthquake phenomena. Computers & Mathematics with Applications, 66(5), 647–658.

Mignan, A., & Woessner, J. (2012). Estimating the magnitude of completeness for earthquake catalogs. Community Online Resource for Statistical Seismicity Analysis, doi:10.5078/corssa-00180805.

Ōmori, F. (1894). On the After-shocks of Earthquakes, Journal of the College of Science, Imperial University of Tokyo, 7, 111–200.

Sarlis, N., & Christopoulos, S.R. (2012). Natural time analysis of the centennial earthquake catalog. Chaos: An Interdisciplinary Journal of Nonlinear Science, 22(2), 023123.

Stadler, G., Gurnis, M., Burstedde, C., Wilcox, L.C., Alisic, L., & Ghattas, O. (2010). The dynamics of plate tectonics and mantle flow: From local to global scales. Science, 329(5995), 1033–1038.

Stein, S., Liu, M., Calais, E., & Li, Q. (2009). Mid-continent earthquakes as a complex system. Seismological Research Letters, 80(4), 551–553.

Stucchi, M., Albini, P., Mirto, M., & Rebez, A. (2004). Assessing the completeness of Italian historical earthquake data. Annals of Geophysics, 47(2-3), 659–673.

Turcotte, D.L., & Malamud, B.D. (2002). Earthquakes as a complex system. International Geophysics, 81(Part A), 209–227, IV.

Utsu, T., Ogata, Y., & Matsu'ura, R.S. (1995). The centenary of the Omori formula for a decay law of aftershock activity. Journal of Physics of the Earth, 43(1), 1–33.

Akira Asada*

An Integral Transform arising from Fractional Calculus

Abstract: We introduce an integral transform $\mathcal{R}$;
$\mathcal{R}[f(s)](x) = \int_{-\infty}^{\infty} \frac{x^s}{\Gamma(1+s)} f(s)ds$, such that

$$\frac{d^a}{dx^a}\mathcal{R}[f(s)](x) = \mathcal{R}[\tau_a f(s)](x), \quad \tau_a f(s) = f(s+a).$$

$\mathcal{R}$ changes constant coefficient fractional differential equations to constant coefficients difference equations.
We show that useful domains of $\mathcal{R}$ are spaces of discrete support distributions or generalized functions. We then show $\mathcal{R}$ is a composition of the extended Borel transform $\mathcal{B}$ and a simpler integral transform $\mathcal{N}$; $\mathcal{N}[f(s)](x) = \int_{-\infty}^{\infty} x^s f(s)ds$:

$$\mathcal{R} = \mathcal{B} \circ \mathcal{N}.$$

We describe the usefulness of $\mathcal{R}$ in the study of fractional differential equations.

1 Integral Transform $\mathcal{R}$

We use

$$I^a[f(s)](x) = \frac{1}{\Gamma(a)} \int_0^x (x-s)^{a-1} f(s)ds$$

as the definition of an a-th order indefinite integral. Then we have $I^a[s^c](x) = \frac{\Gamma(1+c)}{\Gamma(1+c+a)} x^{c+a}$, if $\Re c > 0$ and $\Re a > 0$. Since $\frac{\Gamma(1+c)}{\Gamma(1+c-a)}$ is analytic in x, c and a, we use

$$\frac{d^a}{dx^a} x^c = \frac{\Gamma(1+c)}{\Gamma(1+c-a)} x^{c-a},$$

as the definition of fractional order differentiation, where a and c are arbitrary complex numbers.

Let $\mathcal{L}$; $\mathcal{L}[f(s)](y) = \int_{-\infty}^{\infty} e^{ys} f(s)ds$ be the bilateral Laplace transform. If one sets

$$\mathcal{N}[f(s)](x) = \mathcal{L}[f(s)](\log x) = \int_{-\infty}^{\infty} x^s f(s)ds. \tag{1}$$

***Corresponding Author: Akira Asada:** Former; Sinsyu University, E-mail asada-a@poporo.ne.jp then a standard domain of $\mathcal{N}$ is $E_{0+} = \cup_{c>0} E_c$, $E_c = \{f(x) | |f(x)| \le Me^{-c|x|}$, for some $M > 0\}$. To consider $\mathcal{N}[f](x)$ as defined on the positive real axis is natural. But we also consider the cases where $\mathcal{N}[f](x)$ is defined on $\mathbb{C} \setminus \{-\mathbb{R}^+\}$, where $\mathbb{R}^+ = \{x \in \mathbb{R} | x \ge 0\}$, or on the many valued function on $\mathbb{C}^\times = \mathbb{C} \setminus \{0\}$.

If $g(s)$ and $\dfrac{g(s)}{\Gamma(1+s)}$ both belong to E_{0+}, then

$$\frac{d^a}{dx^a}\mathcal{N}[g(s)](x) = \int_{-\infty}^{\infty} \frac{\Gamma(1+s)}{\Gamma(1+s-a)} x^{s-a} g(s)ds.$$

Therefore we have

$$\begin{aligned}\frac{d^a}{dx^a}\mathcal{N}[\frac{g(s)}{\Gamma(1+s)}](x) &= \int_{-\infty}^{\infty} x^{s-a} \frac{g(s)}{\Gamma(1+s-a)} ds \\ &= \int_{-\infty}^{\infty} x^t \frac{g(t+a)}{\Gamma(1+t)} dt, \quad s-a=t. \end{aligned} \tag{2}$$

Definition 1. *We define a particular integral transform $\mathcal{R}$ by*

$$\mathcal{R}[f(s)](x) = \int_{-\infty}^{\infty} \frac{x^s}{\Gamma(1+s)} f(s)ds. \tag{3}$$

By (2), we have (Asada,2011;Asada,2013)

Theorem 1. *If $f(s)$ and $\dfrac{f(s)}{\Gamma(1+s)}$ both belong to E_{0+}, then*

$$\frac{d^a}{dx^a}\mathcal{R}[f(s)](x) = \mathcal{R}[\tau_a f(s)](x), \quad \tau_a f(s) = f(s+a). \tag{4}$$

Note. If a is a complex number $u + iv$, then

$$\frac{d^a}{dx^a}\mathcal{R}[f(s)]ds = \int_{-\infty-iv}^{\infty-iv} \frac{x^s}{\Gamma(1+s)} f(s)ds.$$

But if $|f(s)| \le Me^{-c|s|^2}$ for some $M > 0$ and $c > 0$ if $|\Im s| \le |v|$, then by Cauchy's integral formula, (4) is valid.

We also note that since we have

$$\begin{aligned} x\mathcal{R}[f(s)](x) &= \int_{-\infty}^{\infty} \frac{x^{1+s}}{\Gamma(1+s)} f(s)ds \\ &= \int_{-\infty}^{\infty} \frac{x^t}{\Gamma(1+t)} \frac{\Gamma(1+t)}{\Gamma(t)} f(t-1)dt, \quad t = 1+s, \end{aligned}$$

$$= \int_{-\infty}^{\infty} \frac{x^t}{\Gamma(1+t)} t\tau_{-1} f(t) dt$$

we get

$$x\mathcal{R}[f(s)](x) = \mathcal{R}[s\tau_{-1}f(s)](x), \tag{5}$$

$$x^n\mathcal{R}[f(s)](x) = \mathcal{R}[(s\tau_{-1})^n f(s)](x), \tag{6}$$

where $(s\tau_{-1})^n = s(s-1)\cdots(s-n+1)\tau_{-n}$.

$\mathcal{N}$ and $\mathcal{R}$ are defined for functions of several variables by the same way. One can then say

$$\frac{\partial^{|\mathbf{a}|}}{\partial x_1^{a_1}\cdots\partial x_n^{a_n}}\mathcal{R}[f(\mathbf{s})](\mathbf{x}) = \mathcal{R}[f(\mathbf{s}+\mathbf{a})](\mathbf{x}),$$

where $\mathbf{a} = (a_1, \ldots, a_n)$, $|\mathbf{a}| = a_1 + \cdots + a_n$, and $\mathbf{s} = (s_1, \ldots, s_n)$, $\mathbf{x} = (x_1, \ldots, x_n)$.

Theorem 1 shows that $\{\frac{d^a}{dx^a} | a \in \mathbb{R}\}$ acts as a one-parameter group on F (the space of f such that both f and $\frac{f(x)}{\Gamma(1+x)}$ belong to E_{0+}) and $\mathcal{R}$ maps the dynamical system $\mathbf{d}: \mathbb{R}\times\mathsf{F}; \mathbf{d}(a,f) = \frac{d^a f}{dx^a}$ to the dynamical system τ, $\mathbb{R}\times\mathcal{R}(\mathsf{F})$: $\tau(a,f) = \tau_a f$. Its generating operator is $\log(\frac{d}{dx})$;

$$\log(\frac{d}{dx})f(x) = -\left(\gamma f(x) + \int_0^x \log(x-t)\frac{df_+(t)}{dt}dt\right),$$

where $f_+(x) = \begin{cases} f(x), & x \geq 0, \\ 0, & x < 0, \end{cases}$ and $\frac{df_+(s)}{ds}$ is taken in the sense of distribution. By Theorem 1, we have

$$\log(\frac{d}{dx})\mathcal{R}[f(s)](x) = \mathcal{R}[\frac{df(s)}{ds}](x). \tag{7}$$

Definition 2. *We introduce transforms $\mathcal{N}_+$ and $\mathcal{R}_+$ by*

$$\mathcal{N}_+[f(s)](x) = \int_{-\infty}^{\infty} x_+^s f(s) ds, \tag{8}$$

$$\mathcal{R}_+[f(s)](x) = \int_{-\infty}^{\infty} \frac{x_+^s}{\Gamma(1+s)} f(s) ds. \tag{9}$$

By definition, we have that

$$\mathcal{N}_+[f](x) = \mathcal{N}[f_+](x), \quad \mathcal{R}_+[f](x) = \mathcal{R}[f_+](x),$$

if $x > 0$ (or if $\Re x > 0$, when x is considered a complex number). On the other hand we have

$$\mathcal{N}_+[f](x) = \mathcal{R}_+[f](x) = 0, \ x < 0 \text{ (or } \Re x < 0).$$

As for $\mathcal{R}_+$, we also have

$$\frac{d^a}{dx^a}\mathcal{R}_+[f(s)](x) = \mathcal{R}_+[\tau_a f(s)](x), \tag{10}$$

$$\log(\frac{d}{dx})\mathcal{R}_+[f(s)](x) = \mathcal{R}_+[\frac{df(s)}{ds}](x). \tag{11}$$

$e^{-ax} \notin \mathsf{E}_{0+}$, but $\mathcal{N}_+[e^{-as}](x)$ is defined if $a > 0$, and we have

$$\mathcal{N}_+[e^{-as}](x) = \frac{x^{1/a}}{a}\Gamma(x^{1/a}).$$

We also note if f and g both have compact support and are defined on the positive real axis, then

$$\int_0^\infty \Big(\int_0^\infty x^s f(s)ds\Big) g(x)dx = \int_0^\infty f(s)\Big(\int_0^\infty x^s g(x)dx\Big)ds.$$

Hence $\mathcal{N}_+$ is closely related to the adjoint of the Mellin transform.

2 Dirac's δ-function and $\mathcal{R}$

Let δ be the Dirac function, and let δ_c be $\delta(s-c)$, the Dirac function concentrated at $s = c$.

Lemma 1. *We have*

$$\mathcal{N}[\delta_c] = x^c, \quad \mathcal{R}[\delta_c] = \frac{x^c}{\Gamma(1+c)}. \tag{12}$$

Note. We assume $t > 0$. Then. as distributions, it is known

$$\lim_{t\to\infty} \frac{1}{\sqrt{t\pi}} e^{-t(x-c)^2} = \delta_c.$$

Since we have $x^s e^{-t(s-c)^2} = e^{-t(s-c)^2 + s\log x}$ and

$$-t(s-c)^2 + s\log x = -t(s-c-\frac{\log x}{2t})^2 + c\log x + \frac{(\log x)^2}{4t},$$

we get

$$\frac{1}{\sqrt{t\pi}}\mathcal{N}[e^{-t(s-c)^2}](x) = x^c e^{\frac{(\log x)^2}{4t}}. \tag{13}$$

Hence on $\mathbb{C}^\times = \mathbb{C}\setminus\{0\}$, we have

$$\lim_{t\to\infty} \mathcal{N}[\frac{e^{-t(s-c)^2}}{\sqrt{t\pi}}](x) = x^c(= \mathcal{N}[\delta_c]). \tag{14}$$

Since $\tau_a\delta_b = \delta_{b-a}$ and

$$\frac{d^a}{dx^a}\Big(\frac{x^c}{\Gamma(1+c)}\Big) = \frac{x^{c-a}}{\Gamma(1+c-a)}$$

we have

$$\frac{d^a}{dx^a}\mathcal{R}[\delta_c](x) = \mathcal{R}[\tau_a\delta_c](x). \tag{15}$$

Therefore Theorem 1 is valid when $\mathcal{R}$ acts on the space of discrete delta potential $T = \sum_n c_n\delta_{a_n}$, where $\{a_n\}$ is a discrete subset of $\mathbb{C}$. On the other hand, since $\mathcal{R}[\delta_{-n}] = 0$, we can not establish the formula

$$\frac{d^a}{dx^a}\left(\frac{d^b}{dx^b}\mathcal{R}[\delta_c]\right) = \frac{d^{a+b}}{dx^{a+b}}\mathcal{R}[\delta_c],$$

in general.

We have $\mathcal{N}_+[\delta_c] = x_+^c$ for arbitrary c and $\mathcal{R}_+[\delta_c] = \frac{x_+^c}{\Gamma(1+c)}$ if c is not a negative integer. For example, we have $\mathcal{R}_+[\delta](x) = Y(x)$, the Heaviside function. If $g(x)$ belongs to the Schwartz space $\mathcal{D}_{\mathbb{R}}$, then, since

$$\int_0^\infty \frac{x^{\epsilon-1}}{\Gamma(\epsilon)}g(x)dx = -\frac{1}{\epsilon\Gamma(\epsilon)}\int_0^\infty x^\epsilon g(x)dx.$$

and we have $\lim_{\epsilon\to 0}\frac{x_+^{\epsilon-1}}{\Gamma(\epsilon)} = \delta$, as distributions. Consequently, we have

$$\mathcal{R}_+[\delta_{-n}] = \delta^{(n-1)}\left(= \frac{d^{n-1}}{dx^{n-1}}\delta\right). \tag{16}$$

Therefore we obtain

$$\frac{d^a}{dx^a}\mathcal{R}_+[\delta_c] = \mathcal{R}_+[\tau_a\delta_c], \quad \frac{d^a}{dx^a}\left(\frac{d^b}{dx^b}\mathcal{R}_+[\delta_c]\right) = \frac{d^{a+b}}{dx^{a+b}}\mathcal{R}_+[\delta_c]. \tag{17}$$

Since $\delta_c(x) = \delta(x-c)$, we have

$$\frac{\partial^n}{\partial c^n}\delta_c = (-1)^n\frac{\partial^n}{\partial x^n}\delta_c\left(= -\delta_c^{(n)}\right).$$

Hence by (8), we obtain

Proposition 1. *The following hold*

$$\mathcal{N}[\delta_c^{(n)}](x) = (-1)^n\frac{\partial^n}{\partial c^n}x^c = (-1)^n(\log x)^n x^c. \tag{18}$$

$$\begin{aligned}\mathcal{R}[\delta_c^{(n)}](x) &= (-1)^n\frac{\partial^n}{\partial c^n}\left(\frac{x^c}{\Gamma(1+c)}\right).\\ &= (-1)^n\left(\sum_{k=0}^n\frac{n!}{k!(n-k)!}\frac{\partial^k}{\partial c^k}\left(\frac{1}{\Gamma(1+c)}\right)(\log x)^{n-k}x^c\right).\end{aligned} \tag{19}$$

Since $\tau_a\frac{\partial}{\partial c} = \frac{\partial}{\partial c}\tau_a$, and $\tau_a\delta_c^{(n)} = \delta_{c-a}^{(n)}$, we have

$$\frac{d^a}{dx^a}\mathcal{R}[\delta_c^{(n)}](x) = \mathcal{R}[\tau_a\delta_c^{(n)}](x), \tag{20}$$

$$\log(\frac{d}{dx})\mathcal{R}[\delta_c^{(n)}](x) = \mathcal{R}[\delta_c^{(n+1)}](x). \tag{21}$$

Since $\frac{1}{\Gamma(1+t)} = \sum_{n=1}^{\infty} \alpha_n t^{n-1}$, $\alpha_1 = 1$, $\alpha_2 = \gamma$, and

$$(n-1)\alpha_n = \gamma\alpha_{n-1} - \zeta(2)\alpha_{n-2} + \zeta(3)\alpha_{n-3} + \cdots + (-1)^n\zeta(n-1)\alpha_1,$$

we have

$$\mathcal{R}[\delta^{(n)}](x) = (-1)^n(\sum_{k=0}^{n} \frac{n!}{(n-k)!}\alpha_{k+1}(\log x)^{n-k}). \tag{22}$$

Since $\mathcal{R}[\delta] = 1$ and $\mathcal{R}[\delta'] = -(\log x + \gamma)$, we have $\mathcal{R}[-\delta' - \gamma\delta] = \log x$. Hence we get that

$$\begin{aligned}\frac{d^a}{dx^a}\log x &= -\mathcal{R}[\delta'_{-a} + \gamma\delta_{-a}](x)\\ &= \frac{\Gamma'(1-a) - \gamma\Gamma(1-a)}{(\Gamma(1-a))^2}x^{-a} - \frac{1}{(\Gamma(1-a))}\log x \cdot x^{-a},\end{aligned}$$

if a is not a natural number.

3 Space of Generalized Functions Spanned by $\delta_a^{(n)}$

Let T be a space of generalized functions of the form $T = \sum_n \sum_{a_m} C_{n,m}\delta_{a_m}^{(n)}$, where $n \in \mathbb{N}$ and $\{a_m\}$ is a discrete set of $\mathbb{R}$ (or $\mathbb{C}$), such that τ_a; $a \in \mathbb{R}$ (or $a \in \mathbb{C}$) and $\frac{d^k}{dx^k}$; $k = 0, 1, 2, \ldots$ are defined on T. T may not be a distribution (in the sense of Schwartz). This means that we allow generalized function of the form $T = \sum_{n=0}^{\infty} \delta^{(n)}$, *etc.*

On the other hand, we demand T is realized as a space of generalized function on some function space, especially a space of analytic functions A. Therefore T may be regarded as a kind of analytic functional.

Definition 3. *Let* $f(z) = \sum_{n=0}^{\infty} c_n z^n$, *then we set*

$$f(\delta) = \sum_{n=0}^{\infty} c_n\delta^{(n)}, \quad f(\delta_a) = \sum_{n=0}^{\infty} c_n\delta_a^{(n)}.$$

If $f(z) = e^{\lambda z}$, and $\phi(z)$ is holomorphic on $\{z||z| \le |\lambda|\}$ or on $\{z||z-a| \le |\lambda|\}$, then $f(\delta)[\phi] = \phi(\lambda)$ and $f(\delta_a)[\phi] = \phi(\lambda - a)$. This means the expression of T as a generalized function on A is not unique.

By (16), we have

$$\mathcal{N}[e^{\lambda\delta}](x) = e^{-\lambda\log x} = x^{-\lambda}, \quad \mathcal{N}[e^{\lambda\delta_c}](x) = x^{-\lambda+c}.$$

In general, we have

$$\mathcal{N}[f(\delta)](x) = f(-\log x), \quad \mathcal{N}[f(\delta_c](x) = f(-\log x)x^c. \tag{23}$$

Proposition 2. *Let $f(z)$ be a finite exponential type function. Then $\mathcal{R}[f(\delta)]$ can be defined as a function*

Proof. Let $f(z) = \sum_{n=0}^{\infty} c_n z^n$. Then there are positive numbers M and C such that $|c_n| \le M\frac{C^n}{n!}$. Since

$$\begin{aligned}\sum_{n=k}^{N} |c_n|\frac{n!}{k!}|\alpha_{n-k+1}| &\le \sum_{n=k}^{N} M\frac{C^n}{n!}\frac{n!}{k!}|\alpha_{n-k+1}| \\ &\le \frac{1}{k!}\Big(\sum_{n=k}^{N} MC^n|\alpha_{n-k+1}|\Big) \\ &\le M\frac{C^k}{k!}\Big(\sum_{n=0}^{N}|\alpha_n|C^n\Big),\end{aligned}$$

and since $\frac{1}{\Gamma(1+s)}$ is an entire function, we have

$$\sum_{n=k}^{N} |c_n|\frac{n!}{k!}|\alpha_{n-k+1}| \le M\frac{C^k}{k!}L, \quad L = \sum_{n=0}^{\infty}|\alpha_n|C^n.$$

Hence the following computation is allowed

$$\begin{aligned}\mathcal{R}[f(\delta)] &= \sum_{n=0}^{\infty} c_n\Big((-1)^n\sum_{k=0}^{n}\frac{n!}{k!}\alpha_{n-k+1}(\log x)^k\Big) \\ &= \sum_{k=0}^{\infty}\Big(\sum_{n=k}^{\infty}(-1)^n c_n\frac{n!}{k!}\alpha_{n-k+1}\Big)(\log x)^k.\end{aligned}$$

Here we used (20). Therefore we have established the Proposition, because

$$\sum_{k=0}^{\infty}\Big|\sum_{n=k}^{\infty}(-1)^n c_n\frac{n!}{k!}\alpha_{n-k+1}\Big||\log x|^k \le ML\Big(\sum_{k=0}^{\infty}\frac{C^k}{k!}|\log x|^k\Big).$$

We denote $\mathsf{A}_{\mathrm{ent}}$ and $\mathsf{A}_{\mathrm{exp}}$, the spaces of entire functions and finite exponential type functions. The topology of $\mathsf{A}_{\mathrm{ent}}$ defined via the uniform convergence in the wider sense on $\mathbb{C}$, while a series $\{f_n\}$ of $\mathsf{A}_{\mathrm{exp}}$ converges to f if $|f_n(z)| \le Ae^{B|z|}$ for some positive A, B which are independent from n and converge uniformly on $\mathbb{C}$. We set

$$\mathsf{T}_{\mathrm{ent}} = \{f(\delta)|f \in \mathsf{A}_{\mathrm{ent}}\}, \quad \mathsf{T}_{\mathrm{exp}} = \{f(\delta)|f \in \mathsf{A}_{\mathrm{exp}}\}.$$

Proposition 3. *We have*

$$\mathsf{A}_{\mathrm{ent}}^{\dagger} = \mathsf{T}_{\mathrm{exp}}, \quad \mathsf{A}_{\mathrm{exp}}^{\dagger} = \mathsf{T}_{\mathrm{ent}}. \tag{24}$$

Proof. If A is a space of holomorphic functions containing $\mathbb{C}[z]$ (the algebra of polynomials) as a dense subspace, and $T \in \mathsf{A}^{\dagger}$, then to define T_n by $T_n(z^m) =$

$\begin{cases} T(z^n), n = m, \\ 0, \quad m \neq n, \end{cases}$, we have $T = \sum_{n=0}^{\infty} T_n$ and $T_n(f(z) = (-1)^n \frac{T(z^n)}{n!}\delta^{(n)}$. Hence one can write $T \in \mathsf{A}_{\mathrm{ent}}$ (or $\in \mathsf{A}_{\mathrm{exp}}$) as the form $\sum_{n=0}^{\infty} a_n \delta^{(n)}$. Therefore if $f(z) = \sum_{n=0}^{\infty} b_n z^n$, we have

$$T(f(z)) = \sum_{n=0}^{\infty} (-1)^n n! a_n b_n.$$

This right hand side converges if $|a_n| \leq \frac{MC^n}{n!}$ and $\sum_{n=0}^{\infty} b_n z^n$ converges on $\mathbb{C}$ (or if $\sum_n a_n z^n$ converges on $\mathbb{C}$ and $|b_n| \leq \frac{MC^n}{n!}$). Hence we have proved the Proposition.

Note. δ_a, *etc.*, belongs to $\mathsf{A}_{\mathrm{exp}}^{\dagger}$, *etc.*. But we identify δ_a and $\sum_{n=0}^{\infty} \frac{a^n}{n!}\delta^{(n)}$ as elements of $\mathsf{A}_{\mathrm{exp}}^{\dagger}$, *etc.*.

The correspondence $\mathcal{N}[\delta_n] = x^n \to \mathcal{R}[\delta_n] = \frac{x^n}{n!}$ is known as the Borel transform $\mathcal{B}$;

$$\mathcal{B}[f(\zeta)](z) = \frac{1}{2\pi i} \oint e^{\frac{z}{\zeta}} \frac{f(\zeta)}{\zeta} d\zeta. \tag{25}$$

This suggests $\mathcal{N}$ and $\mathcal{R}$ relate via Borel transform. To justify this observation, we need to extend the Borel transform so that the Borel transform is defined for z^a, where a is not an integer. We discuss this problem in the next Section.

4 Extended Borel Transform

Let $\mathcal{O}$ be the algebra of germs of holomorphic functions at the origin. Here a germ of holomorphic function at the origin means an equivalence class of holomorphic functions at the origin by the equivalence relation $f \sim g$ if and only if $f|U = g|U$ for some neighborhood U of the origin. We can represent a germ by a convergent power series.

Convergence of a series $\{f_n\}$ of elements of $\mathcal{O}$ to f is defined if there is an $\epsilon > 0$ such that each f_n is represented by a holomorphic function on $\{z||z| \leq \epsilon\}$ and $\lim_{n\to\infty} f_n = f$ uniformly on $\{z||z| \leq \epsilon\}$. Then, we have

$$\mathcal{B} : \mathcal{O} \cong \mathsf{A}_{\mathrm{exp}}. \tag{26}$$

As for product structure, we have

$$\mathcal{B}[fg] = \mathcal{B}[f]\sharp\mathcal{B}[g], \quad (u\sharp v)(x) = \frac{d}{dx}\int_0^x u(x-t)v(t)dt \tag{27}$$

We also have

$$z\mathcal{B}[f(\zeta)](z) = \mathcal{B}[\zeta f(\zeta) + \zeta^2 \frac{df(\zeta)}{d\zeta}](z). \tag{28}$$

The right hand side of (23) is meaningful not only for holomorphic function at the origin, but also for analytic functions having isolated singularity at the origin. Let $\tilde{\mathcal{O}}$ be the algebra of germs at the origin of analytic functions having isolated singularities at the origin. Then $\mathcal{B}(\tilde{\mathcal{O}}) = \mathcal{B}(\mathcal{O})$, but z^{-n} acts on $\tilde{\mathcal{O}}$ and we have that

$$\frac{d}{dz}\mathcal{B}[f(\zeta)](z) = \mathcal{B}[\zeta^{-1}f(\zeta)](z). \tag{29}$$

Note. We can define the Borel transform functions of several variables by using same formula as (23). It is an integral on the path $\varsigma = \{|z_1| = \epsilon_1, \dots, |z_n| = \epsilon_n\}$. If $f(z)$ has singularities on $Y; 0 \in Y$, then its Borel transform depends on the local homology class of ς in the local homology group $H_*(\mathbb{C} \setminus Y, \mathbb{C})$ at the origin. For example, since $\frac{x}{x-y} = \begin{cases} 1 + \sum_{n=1}^{\infty} x^{-n}y^n, & |x| > |y|, \\ -\sum_{n=1}^{\infty} x^n y^{-n}, & |x| < |y|, \end{cases}$, its Borel transform is 1, if $\epsilon_1 > \epsilon_2$ and 0, if $\epsilon_1 < \epsilon_2$, in the definition of ς. We denote $\mathcal{B}_\varsigma$, the Borel transform defined by using ς. Then we obtain

$$\frac{\partial}{\partial z_i}\mathcal{B}_\varsigma[f(\zeta)](z) = \mathcal{B}_\varsigma[\zeta_i^{-1}f(\zeta)](z).$$

The inverse of the Borel transform $\mathcal{B}^{-1}$ is given by

$$\mathcal{B}^{-1}[f(t)](x) = \int_0^\infty e^{-t}f(xt)dt. \tag{30}$$

The right hand side of (27) is meaningful some functions that are not of finite exponential type. For example, we have

$$\mathcal{B}^{-1}[t^a](x) = \Gamma(1+a)x^a, \quad \mathcal{B}^{-1}[\log t](x) = \log x - \gamma.$$

Originally, the first equality is valid assuming $\Re a > -1$. But by analytic continuation in a, we consider the first equality to be valid for any $a \in \mathbb{C}$.

Definition 4. *We define*

$$\mathcal{B}[\zeta^a](z) = \frac{z^a}{\Gamma(1+a)}, \quad \mathcal{B}[\log\zeta](z) = \log z + \gamma. \tag{31}$$

By definition, we have $\mathcal{B}[\zeta^a\zeta^b](z) = \mathcal{B}[\zeta^a](z)\sharp\mathcal{B}[\zeta^b](z)$. We also define $\mathcal{B}[(\log\zeta)^n\zeta^a](z)$ by $\frac{\partial^n}{\partial a^n}\mathcal{B}[\zeta^a](z)$. For example, we have

$$\mathcal{B}[(\log\zeta)^n](z) = \sum_{k=0}^{\infty} \frac{n!}{k!(n-k)!}\alpha_{n-k}(\log z)^k (= \mathcal{R}[\delta^{(n)}]).$$

Lemma 2. *We have*

$$\mathcal{B}[(\log\zeta)^n](z) = \overbrace{\mathcal{B}[\log\zeta](z)\sharp\cdots\sharp\mathcal{B}[\log\zeta](z)}^{n}. \tag{32}$$

Proof. Since we have

$$\begin{aligned}\mathcal{B}[(\log\zeta)^n](z) &= \frac{\partial^n}{\partial a^{n-1}\partial b}\mathcal{B}[\zeta^a\zeta^b](z)|_{a=0,b=0}\\ &= \frac{\partial^n}{\partial a^{n-1}\partial b}(\mathcal{B}[\zeta^a](z)\sharp\mathcal{B}[\zeta^b](z))|_{a=0,b=0}\\ &= \mathcal{B}[(\log\zeta)^{n-1}](z)\sharp\mathcal{B}[\log\zeta](z),\end{aligned}$$

we obtain Lemma by induction on n.

Example. We denote $\overbrace{f\sharp\cdots\sharp f}^{n}$ by $f^{\sharp n}$. Then since $\zeta^a = e^{a\log\zeta}$, we have

$$\begin{aligned}\frac{z^a}{\Gamma(1+a)} &= \sum_{n=0}^{\infty}\frac{a^n\mathcal{B}[(\log\zeta)^n](z)}{n!}\\ &= \sum_{n=0}^{\infty}\frac{a^n(\log z+\gamma)^{\sharp n}}{n!} = e^{\gamma a}\Big(\frac{(\log z)^{\sharp n}}{n!}\Big).\end{aligned}$$

Hence we obtain (cf.[1])

$$\sum_{n=0}^{\infty}\frac{a^n(\log z)^{\sharp n}}{n!} = \frac{e^{-\gamma a}}{\Gamma(1+a)}z^a. \tag{33}$$

(30) shows (28) is consistent with the formula $\mathcal{B}[uv] = \mathcal{B}[u]\sharp\mathcal{B}[v]$. We also obtain
Proposition 4. *If $\mathcal{B}[\zeta^{-a}f(\zeta)]$ and $\mathcal{B}[\log\zeta g(\zeta)]$ are defined, then*

$$\frac{d^a}{dz^a}\mathcal{B}[f(\zeta)](z) = \mathcal{B}[\zeta^{-a}f(\zeta)](z), \tag{34}$$

$$\log\left(\frac{d}{dz}\right)\mathcal{B}[g(\zeta)](z) = -\mathcal{B}[\log\zeta g(\zeta)](z). \tag{35}$$

Note. $\mathcal{B}[(\log\zeta)^{-n}](z)$ is computed as follows: Since

$$\int_0^{\infty} x^s\,ds = -\frac{x^a}{\log x},\ |x|<1, \qquad \int_{-\infty}^{a} x^s\,ds = \frac{x^a}{\log x},\ |x|>1,$$

it should be $\mathcal{B}[\frac{\zeta^a}{\log\zeta}](x) = \int_a^{\infty}\frac{x^s}{\Gamma(1+s)}ds$, provided that $\int_a^{\infty}\mathcal{B}[f(s)]ds = \mathcal{B}[\int_a^{\infty}f(s)ds]$. For example, we have

$$\mathcal{B}[(\log\zeta)^{-1}](x) = -\mathcal{R}[Y(s)](x) = -\int_0^{\infty}\frac{x^s}{\Gamma(1+s)}ds. \tag{36}$$

In fact, we have

$$\log\left(\frac{d}{dx}\right)\mathcal{B}\left[\frac{1}{\log\zeta}\right](x) = \mathcal{B}\left[\frac{-\log\zeta}{\log\zeta}\right](x) = -1$$

and

$$-\log\left(\frac{d}{dx}\right)\mathcal{R}[Y(s)](x) = -\mathcal{R}[\delta] = -1$$

by (32).

In general, since

$$\log\left(\frac{d}{dx}\right)\mathcal{B}\left[(\log\zeta)^{-n}\right](x) = -\mathcal{R}[(\log\zeta)^{1-n}](x)$$

and

$$\log\left(\frac{d}{dx}\right)\mathcal{R}[s_+^n](x) = n\mathcal{R}[s_+^{n-1}](x),$$

we know that

$$\begin{aligned}\mathcal{B}[(\log\zeta)^{-n}](x) &= \frac{(-1)^n}{(n-1)!}\mathcal{R}[s_+^{n-1}](x)\\ &= \frac{(-1)^n}{(n-1)!}\int_0^\infty \frac{s^n x^s}{\Gamma(1+s)}ds. \end{aligned} \tag{37}$$

If $f(z) = \sum_{n=0}^\infty c_n z^n$, we can define $\mathcal{B}[f(\log\zeta)]$ by $\sum_{n=0}^\infty c_n\mathcal{B}[(\log\zeta)^n]$. Then since $\mathcal{B}[(\log\zeta)^n] = \mathcal{R}[\delta^{(n)}]$, we can use Proposition 1, and we have

Proposition 5. *If $f(z)$ is a finite exponential type function, then $\mathcal{B}[f(\log\zeta)]$ converges. It is written as $F(\log z)$, where $F(z)$ is a finite exponential type function.*

Similarly, we define $\mathcal{B}[f(z^a)]$ by $\sum_{n=0}^\infty c_n\mathcal{B}[\zeta^a](z)$. It is written as $F(z^a)$, where F is an entire function, if $\Re a > 0$. We say that those transform of functions including z^a or $\log z$ compose *extended Borel transform.*

5 The Transform $\mathcal{R}$ and Extended Borel Transform

By Lemma 1, we have

Theorem 2. *If T is a generalized function of the form $\sum_k\sum_n c_{n,k}\delta_{a_n}^{(k)}$, where $\{a_n\}$ is a discrete set in $\mathbb{C}$, and $\mathcal{R}[T]$ is defined, then*

$$\mathcal{R}[T] = \mathcal{B}(\mathcal{N}[T]), \quad i.e.\ \mathcal{R} = \mathcal{B}\circ\mathcal{N}. \tag{38}$$

For example, if we consider $\mathcal{R}$ to be defined on $\mathsf{T}_{\mathrm{exp}}$, $\mathsf{T}_{\mathrm{ent}}$ or suitable subspace of the space of discrete support distributions, then we can use Theorem 2. Taking these spaces as domains, $\mathcal{N}$ has the inverse $\mathcal{N}^{-1}$ Then we have $\mathcal{B} = \mathcal{R}\circ\mathcal{N}^{-1}$. For example, we have

$$\begin{aligned}\mathcal{B}[\zeta^a] &= \mathcal{R}[\mathcal{N}^{-1}[\zeta^a]] = \mathcal{R}[\delta_a] = \frac{z^a}{\Gamma(1+a)},\\ \mathcal{B}[\log\zeta] &= \mathcal{R}[\mathcal{N}^{-1}[\log\zeta]] = \mathcal{R}[\delta'] = \log z + \gamma.\end{aligned}$$

By (35), we also have

$$\frac{d^a}{dx^a}\mathcal{R}[T] = \frac{d^a}{dx^a}\mathcal{B}[\mathcal{N}[T]] = \mathcal{B}[\zeta^{-a}\mathcal{N}[T]],$$
$$\log(\frac{d}{dx})\mathcal{R}[T] = \log(\frac{d}{dx})\mathcal{B}[\mathcal{N}[T]] = -\mathcal{B}[\log\zeta\mathcal{N}[T]].$$

We also note by (11) and Proposition 5, we have

$$\lim_{t\to\infty}\frac{1}{\sqrt{t\pi}}\mathcal{R}[e^{-t(s-c)^2}](x) = \frac{x^c}{\Gamma(1+c)},$$

on $\mathbb{C}^\times$, where convergence is uniform in wider sense.

Definition 5. *We define a transform $\mathcal{B}_+$ by*

$$\mathcal{B}_+ = \mathcal{R}_+ \circ \mathcal{N}^{-1}. \tag{39}$$

$\mathcal{R}_+[f(s)](x)$ should be considered as a function on $\{x \in \mathbb{R}|x \geq 0\}$, if it is a function, and we have

$$\mathcal{R}_+[f(s)](x) = \mathcal{R}[f(s)]_+(x) = \mathcal{R}[f_+(s)](x),$$

we regard $\mathcal{B}_+[f(\zeta)](z)$ as being defined on positive real axis, and

$$\mathcal{B}_+[f(\zeta)](z) = \mathcal{B}[f(\zeta)]_+(z).$$

Example. Since $\mathcal{N}[\delta_{-n}](x) = x^{-n}$, we have by (14)

$$\mathcal{B}_+[\zeta^{-n}](z) = \mathcal{R}_+[\delta_{-n}] = \delta^{(n-1)}. \tag{40}$$

We set

$$\mathsf{A}_{\mathrm{exp},+} = \{f_+|f \in \mathsf{A}_{\mathrm{exp}}\}.$$

Let $\mathsf{A}_{\mathrm{exp},+}\sharp\mathsf{T}_{\mathrm{ent}}$ be the algebra generated by $\mathsf{A}_{\mathrm{exp}}$ and $\mathsf{T}_{\mathrm{ent}}$ by the $\sharp$-product. Then (37) shows $\mathcal{B}_+$ maps $\hat{\mathcal{O}}$ into $\mathsf{A}_{\mathrm{exp}}\sharp\mathsf{T}_{\mathrm{ent}}$;

$$\mathcal{B}_+ : \hat{\mathcal{O}} \to \mathsf{A}_{\mathrm{exp},+}\sharp\mathsf{T}_{\mathrm{ent}}.$$

The difference of $\mathcal{B}$ (or $\mathcal{R}$) and $\mathcal{B}_+$ (or $\mathcal{R}_+$) can be seen from the following example.

Example. Let $u(z) = z^{a-1}(1-\lambda z^a)^{-1})(= z^{-1}(z^{-a}-\lambda)^{-1})$, then

$$(\frac{d^a}{dz^a} - \lambda)\mathcal{B}[u(\zeta)](z) = \mathcal{B}[\zeta^{-1}](z) = 0, \tag{41}$$
$$(\frac{d^a}{dz^a} - \lambda)\mathcal{B}_+[u(\zeta)](z) = \mathcal{B}_+[\zeta^{-1}](z) = \delta. \tag{42}$$

Hence $\mathcal{B}[u(\zeta)](z)$ is a solution of

$$\frac{d^a U}{dz^a} = \lambda U. \tag{43}$$

While $\mathcal{B}_+[u(\zeta)]$ is a fundamental solution of (40).

By using $\mathcal{R}$ and $\mathcal{R}_+$, this fact can be interpreted as follows: Let $T = \sum_{n=1}^{\infty} \lambda^n \delta_{an-1}$, then since $\tau_a \delta_c = \delta_{c-a}$, we have

$$(\frac{d^a}{dx^a} - \lambda)\mathcal{R}[T](x) = \mathcal{R}[\delta_{-1}](x) = 0, \tag{44}$$

$$(\frac{d^a}{dx^a} - \lambda)\mathcal{R}_+[T](x) = \mathcal{R}_+[\delta_{-1}](x) = \delta. \tag{45}$$

Since $u(z) = z^{-1} + \sum_{n=1}^{\infty} \lambda^n z^{an-1}$, we have $\mathcal{B}[u](x) = \mathcal{R}[T](x)$, which are given by

$$\mathcal{R}[T](x) = \sum_{n=1}^{\infty} \lambda^n \frac{x^{an-1}}{\Gamma(an)} = \lambda x^{a-1} E_{a^{-1}}(\lambda x^a; a), \tag{46}$$

where $E_\rho(x;\mu) = \sum_{k=0}^{\infty} \frac{x^k}{\Gamma(\mu + k/\rho)}$ is the generalized Mittag-Leffler function (cf. Cartwright, 1962).

Note. Let T_m be $\sum_{n=1}^{\infty} \lambda^n \delta_{an-m}$. Then $\mathcal{R}[T_m]$;

$$\mathcal{R}[T_m](x) = \sum_{n=1}^{\infty} \lambda^n \frac{x^{an-m}}{\Gamma(an + 1 - m)} = \lambda x^{a-m} E_{a^{-1}}(\lambda x^a; a + 1 - m)$$

also satisfies (40). If a is a rational number, $\mathcal{R}[T_m]$, $m \in \mathbb{N}$ are not independent, but if a is irrational, then they are independent (cf. Asada, 2014).

Note. The extension of the Borel transform presented in this Section is done in the analytic category. Extension of the Borel transform for non-analytic functions were discussed in Asada, 1977, cf. Asada, 2014.

6 Application of $\mathcal{R}$ to Fractional Differential Equations

In this section, we give some examples of the application of $\mathcal{R}$ and $\mathcal{B}$ to fractional differential equations.

Example 1. *Constant coefficients fractional differential equations of type* $P(\frac{d^a}{dx^a})$.

Let $P(X) = X^m + c_1 X^{m-1} + \cdots + c_m = \prod_i (X - \mu_i)^{k_i}$ be a polynomial. Then $P(\frac{d^a}{dx^a})$ means

$$P(\frac{d^a}{dx^a}) = (\frac{d^a}{dx^a})^m + c_1(\frac{d^a}{dx^a})^{m-1} + \cdots + c_m = \prod_i (\frac{d^a}{dx^a} - \mu_i)^{k_i}. \tag{47}$$

For the simplicity, we assume a is a positive real number. Then a solution of the equation

$$P(\frac{d^a}{dx^a})Y = 0,$$

is given by

$$Y_{i,\ell;\nu}(x) = \mathcal{B}[(\zeta^{-a} - \mu_i)^{-\ell} \zeta^{-\nu}](x), \quad 1 \le \ell \le k_i. \tag{48}$$

ν is an arbitrary integer, if a is an irrational number. If a is a rational number $\frac{q}{p}$, $(p, q) = 1$, then ν is an integer satisfying $1 \le \nu \le q$.

We note that $Y_{i,\ell;\nu}(x)$ may not satisfy the equation

$$(\frac{d^{ma}}{dx^{ma}} + c_1 \frac{d^{(m-1)a}}{dx^{(m-1)a}} + \cdots + c_m)Y = 0.$$

Because we do not have $\frac{d^a}{dx^a}\frac{d^b}{dx^b} = \frac{d^{a+b}}{dx^{a+b}}$ in general.

For example, the equation $(\frac{d^{1/2}}{dx^{1/2}} + 1)(\frac{d^{1/2}}{dx^{1/2}} - 1)Y = 0$ has solutions

$$Y_\pm : (\frac{d^{1/2}}{dx^{1/2}} + 1)Y_+ = 0, \quad (\frac{d^{1/2}}{dx\ 1/2} - 1)Y_- = 0.$$

But since $Y_\pm(x) = \sum_n \frac{(\mp 1)^n x^{n/2}}{\Gamma(n/2)}$, we do not have $(\frac{d}{dx} - 1)Y_\pm = 0$.

Example 2. *Fractional evolution equation*

We consider the fractional evolution equation

$$P(\frac{\partial^a}{\partial t^a}, \frac{\partial}{\partial x}) = \frac{\partial^a U}{\partial t^a} + D_x U = 0, \tag{49}$$

defined on (a subspace of) $\mathbb{R} \times M$, First we assume $M = \mathbb{R}^p$, and $D_x = D(x, \frac{\partial}{\partial x})$ is a partial differential operator with polynomial coefficients on suitable domain in $\mathbb{R}^m$. Then we have

$$((\frac{\partial^a}{\partial t^a} + D_x)\mathcal{B}[v(\tau, \xi)])(t, x) = \mathcal{B}[(\tau^{-a} + D(\vartheta, \xi^{-1})v(\tau, \xi)](t, x),$$

where $\vartheta = (\vartheta_1, \ldots, \vartheta_p)$, $\vartheta_i = \xi_i + \xi_i^2 \frac{\partial}{\partial \xi}$. Hence if

$$(1 + \tau^a D(\vartheta, \xi^{-1})v(\tau, \xi) = \tau^{a-m} g(\xi)$$

where $g(\xi)$ is holomorphic at the origin, we have that

$$(\frac{\partial^a}{\partial t^a} + D_x)\mathcal{B}[v(\tau, \xi)](t, x) = 0.$$

If D_x is a constant coefficient operator, and if $u_{m;g} = \mathcal{B}[(1 + \tau^a D(\xi^{-1}))^{-1}\tau^{a-m} g(\xi)]$ is defined as a function, we have

$$(\frac{\partial^a}{\partial t^a} + D_x)u_{m:g} = 0.$$

To be continuous $u_{m;g}$ at $t = 0$, we need $\Re a - m > -1$, that is;

1. $m = 1$, if $0 < \Re a \le 1$.

2. $m = 1$ and 2, if $1 < \Re a \le 2$.

If $u(x) = \mathcal{B}[g(\xi)](x)$ is a eigenfunction of D_X; $D_X u(x) = \lambda u(x)$, then

$$u_{m;g}(x) = \sum_{n=0}^{\infty}(-1)^n \mathcal{B}[\tau^{(n+1)a-m} D(\xi^{-1})^n g(\xi)](x)$$
$$= \sum_{n=0}^{\infty}(-1)^n I_t^{(n+1)a-m} D_x^n u(x) = \sum_{n=0}^{\infty}(-1)^n I_t^{(n+1)a-m}[\lambda^n u(x)].$$

Therefore we have

$$u_{m;g}(x) = v(t)u(x), \quad \frac{d^a v(t)}{dt^a} + \lambda v(t) = 0. \tag{50}$$

In other words, we can recover the method of Fourier via extended Borel transform.

Example 3. *Relations with elliptic functions and related topics*

We consider the equation

$$\nu\frac{d^\alpha y}{dx^\alpha} - \mu\frac{d^\beta y}{dx^\beta} = 0,$$

where α and β generate a lattice in $\mathbb{C}$. This equation seems to have relations with elliptic function of periods α, β. To see these relations, we consider a discrete support distribution

$$T_{\mu,\nu}^k = \sum_{-\infty<i,j<\infty} \mu^i \nu^j \delta_{i\alpha+j\beta}^{(k)}.$$

If $\mathcal{R}[T_{\mu,\nu}^k]$ is defined, it should be

$$\frac{d^\alpha}{dx^\alpha}\mathcal{R}[T_{\mu,\nu}^k] = \mu\mathcal{R}[T_{\mu,\nu}^k], \quad \frac{d^\beta}{dx^\beta}\mathcal{R}[T_{\mu,\nu}^k] = \nu\mathcal{R}[T_{\mu,\nu}^k].$$

To give meaning to $\mathcal{R}[T_{\mu,\nu}^k]$, the following may be helpful:

If $f(z)$ is holomorphic at $z = c$, we have

$$\delta_c^{(k)}[f] = (-1)^k f^{(k)}(c) = \frac{k!}{2\pi i}\int_{|z-c|=\epsilon} \frac{f(z)}{(z-c)^{k+1}}dz.$$

Hence we may say $\delta_c^{(k)}$ is realized by $\dfrac{k!}{2\pi i}\dfrac{1}{(z-c)^{k+1}}$ as a generalized function (analytic functional) on a suitable space of analytic functions. Especially, $T_{1,1}^2$ may realized as $\wp'(z)$, where $\wp(z)$ is the Weirestrass' $\wp$-function of periods α and β.

To consider an analytic function $\phi(z)$ on $\mathbb{C}$ as a generalized function on a space of analytic functions on $\mathbb{C}$, we set

$$T_{\phi,r,+}[f] = \frac{1}{2\pi i}\int_{|z|=r+0} \phi(z)f(z)dz. \tag{51}$$

Here $\int_{|z|=r+0} g(z)dz$ means $\lim_{\epsilon\downarrow 0}\int_{|z|=r+\epsilon} g(z)dz$, where $g(z)$ is holomorphic on $\{z | r+\epsilon \ge |z| \ge r\}$, for some $\epsilon > 0$. We also set

$$t_{2,\phi}[f] = \lim_{r\to\infty}\frac{1}{\pi r^2} T_{\phi,r,+}[f]. \tag{52}$$

Then if f is an elliptic function of periods α and β (or an additive function of the same periods), then $t_{2,\wp}[f]$ and $t_{2,\wp'}[f]$ are well defined (Note that $t_{2,c}[f] = 0$, c is a constant, if f is an elliptic function). We ask *can we use these generalized functions to treat the equation*

$$\frac{d^\alpha u}{dx^\alpha} - \frac{d^\beta u}{dx^\beta} = 0,$$

and similar type equations?

We also note if $\Re a_n > \Re a_{n-1} \geq \cdots \geq \Re a_0$, the equation

$$\frac{d^{a_n} y}{dx^{a_n}} + f_1(x)\frac{d^{a_{n-1}} y}{dx^{a_{n-1}}} + \cdots + f_n(x)\frac{d^{a_0} y}{dx^{a_0}} = g(x), \tag{53}$$

is changed to a Volterra type equation

$$Y(x) + \int_0^x \Big(\sum_{i=1}^{n-1} f_i(x)\frac{(x-t)^{a_n - a_{n-i}}}{\Gamma(a_n - a_{n-i})}\Big) Y(t)dt$$
$$= g(x), \quad Y = I^{a_n} y, \tag{54}$$

provided $f_1, \ldots, f_n$ are continuous. Since $y(x) = \dfrac{d^{a_n} Y}{dx^{a_n}} + h(x)$, $\dfrac{d^{a_n} h}{dx^{a_n}} = 0$, if a_n is not an integer, we can take $h(x) = x^{a_n - m}$, m to be an arbitrary natural number. So it has infinitely many independent solutions. If we assume $y(x)$ is continuous at $x = 0$, it follows that $\Re a_n \leq m$, as Example 2.

Example 4. *Fractional analogue of regular singularity*

As an example of variable coefficients fractional differential equations, we state on a fractional analogue of regular singularity without proofs. For the details, we refer the reader to Asada,2014.

We say $L = \sum_i \sum_j ci, jx^{a_{ij}}(\frac{d^{a_i}}{dx^{a_i}})^j$ to be a fractional Euler equation. Its characteristic function $\chi_L(q)$ is defined by

$$\chi_L(q) = \sum_i \prod_j c_{i,j}\frac{\Gamma(1 + q - (j-1)a_i)}{\Gamma(1 + q - ja)}.$$

Then $Lx^q = 0$ if and only if $\chi_L(q) = 0$.

Definition 6. *Let* $\Re a > 0$ *and let* K *be*

$$K = g_0(x)(\frac{d^a}{dx^a})^n + g_1(x)(\frac{d^{a-1}}{dx^{a-1}})^{n-1} + \cdots + g_n(x). \tag{55}$$

Here $g_m(x) = x^{a(n-m)}h_m(x^a)$, *where* $h_m(x)$ *is holomorphic at the origin,* $0 \leq m \leq n$ *and* $h_0(0) \neq 0$. *Then we say* L *has regular singularity at the origin*

We set $K_0 = h_0(0)(\frac{d^a}{dx^a})^n + h_1(0)\frac{d^a}{dx^a})^{n-1} + \cdots + h_n(0)$, and define $\chi_K(q)$ by $\chi_K(q) = \chi_{K_0}(q)$. Then we can show

Theorem 3. *If* $\chi_K(q) = 0$ *and* $\chi_K(q + ka) \neq 0$, $k = 1, 2, \ldots$, *then* K *has a solution* y *of the form* $y = x^q f(x^a)$, *where* $f(x)$ *is holomorphic at the origin and* $f(0) \neq 0$.

Bibliography

Asada, A. (1977). Some extension of Borel transformation, J. Fac. Sci. Shinshu Univ.9, 71-89. Borel transformation in non-analytic category, J. Fac.Sci.Shinshu Univ.12, 1-35.

Asada, A. (2011). Lie algebra generated by logarithm of differentiation and logarithm. Balkan J.Geom. and Its Appl.16, 1-11.

Asada, A. (2013). Integral transform arose from fractional calculus and discrete delta potential. In M.R. Adhikari (Ed), Contemporary Topics in Mathematics and Statics with Applications (Chap.3. 29pp), Kolkata,

Asada, A. (2014). Extended Borel transform and fractiona calculus. To appear in R.Daou, M..Xavier (Eds) Fractional Calculus:History, Theory and Applications, Nova Publishers.

Cartwright, M.L. (1962). Integral Functions, Cambridge,

Jordan Hristov*

Approximate Solutions to Time-fractional Models by Integral-balance Approach

Abstract: The chapter is an attempt to collate results on approximate integral-balance solutions of evolution equations involving fractional-time derivatives in the sense of Riemann-Liouville or Caputo. The examples encompass a subdiffusion equation of arbitrary order and transient flows of viscoelastic fluids. The approximate solutions are in closed form and allow clearly see the retarding effect of the fractional order (affecting the power-law memory kernel in the fractional derivative) on the development of the diffusion processes.

Keywords: Integral-balance method, subdiffusion, viscoelastic fluid, memory kernel, approximate solution

1 Introduction

The idea to develop simple, even approximate, solutions to models involving time-fractional derivatives by integral balance method was conceived a couple of years ago (Hristov, 2010a) in a simple exercise on the time-fractional heat equation of order 1/2. The method has also been applied to a number of problems (Hristov, 2010a; 2010b; 2011a; 2011b; 2011c; 2012a;2013) in its simple version known as Heat-balance integral method (HBIM) as well as its close relative termed the Frozen Front Approach (FFA). Details about these solution technologies are presented in this chapter. The integral-balance is a well-known approximate method (Goodman, 1958; Hristov, 2009) widely applied to a number of practically important problems (see Hristov (2009) and the references therein), but never used to solve equations of fractional order before (Hristov, 2010a). Now, we will try to present the main idea of the method, its versions (as integration techniques) and to demonstrate it by solutions of several simple problems. For sake of clarity in the explanations as well as for better performance of the method we focus the efforts on two main problems involving time-fractional derivatives: i) Subdiffusion and ii) Transient flows of complex fluids.

***Corresponding Author: Jordan Hristov:** Department of Chemical Engineering, University of Chemical Technology and Metallurgy, 1756 Sofia, 8 Kl. Ohridsky Blvd., Bulgaria,
E-mail: jordan.hristov@mail.bg

1.1 Subdiffusion

Subdiffusion is associated with evolving Brownian motion through a medium in which the mean square displacement is lower-than-normal time scale, that is $\langle X^2 \rangle \sim t^{\mu}$ (Nigmatullin, 1986). Subdiffusion has been encountered in many real physical systems such as highly ramified media in porous systems (Nigmatullin, 1986; Li et al., 2006; Berkowitz and Sher, 2009), anomalous diffusion in fractals (Korabel et al, 2007), diffusion in thick membranes (Kosztołowicz, 2008), anomalous drug absorption and disposition processes (Dokoumetzidis and Macheras, 2009) and heat transfer close to equilibrium (Li and Wang, 2003). Subdiffusion occurs in a variety of applications as discussed (Metzler and Klafter, 2000; 2004; Ardelean et al., 2007).

Commonly there are two forms of the time-fractional diffusion equation

$$\frac{\partial u(x,t)}{\partial t} = {}_{RL}D_t^{1-\mu}\left[D_\mu \frac{\partial^2 u(x,t)}{\partial x^2}\right], \quad x \in \Re \;\; t \in \Re, \;\; u\left(x, 0^+\right) = u_0(x), \mu \in (0,1) \quad (1)$$

$$ {}_C D_t^{\mu} u(x,t) = D_\mu \frac{\partial^2 u(x,t)}{\partial x^2}. \quad (2)$$

In 1 ${}_tD^{1-\mu} = {}_tD^1 J^{\mu}$ denotes the Riemann-Liouville (RL) fractional derivative of order $1-\mu$ with $m = 1$ (see further in section 2). In eq. 2 ${}_C D_t^{\mu} u(x,t) = J^{\mu} D_t^1$ denotes time-fractional derivative in the Caputo sense (see section 2).

1.2 Time-Fractional Derivatives in Rheology

Start-up flows of viscoelastic fluids are intensively modeled in the literature under different boundary conditions, flow geometries and constitutive equations (Naraim and Joseph,1983; Tan and Xu, 2002) by analytical and numerical methods. There are very few cases in which their exact analytical solutions can be obtained mainly due to the non-linearity imposed by the non-Newtonian constitutive equations of the fluid rheology. The second grade fluid is the common non-Newtonian viscoelastic fluid in industrial fields, such as polymer solution (Momoniat, 2008), emulsions (Derkach, 2009), crude oils (Siginer and Letelier, 2011), blood flow (Schmitt et al., 2011) and magnetohydrodynamic flows with heat and mass transfer (Hsiao, 2009).

Fractional calculus allows, by replacing the time derivative of integer order with the fractional order time-derivative (in the Riemann-Liouville sense) one, to incorporate memory effects in the constitutive equations (Schmitt et al., 2011). The second grade generalized fluids with fractional derivatives under start-up flow conditions have been intensively studied in an attempt to develop exact analytical solutions (Tan and Xu, 2002; Qi and Xu, 2009). In this context, the stress tensor component of the generalized second grade viscoelastic fluid relevant to one-dimensional transient flow (u is the velocity component along the x) is

$$\mathbf{T}_{xy} = \mu_f \frac{\partial u}{\partial y} + \alpha_1 D_t^{\mu} \frac{\partial u}{\partial y} \quad (3)$$

$T_{xx} = T_{yy} = T_{zz} = T_{xz} = T_{yz} = 0$, and $T_{xy} = T_{yx}$ are components of the stress tensor.

Commonly, the constitutive equation for a second grade liquid relating the shear stress $\tau(t)$ and the strain $\varepsilon(t)$ is expressed in two forms (Hayat et al, 2000; Tan et al., 2002)

$$\tau(t) = \mu_f \varepsilon(t) + E\frac{d\varepsilon(t)}{dt} \tag{4a}$$

or

$$\tau(t) = \mu_f \varepsilon(t) + \alpha_1 D_t^{\mu}[\varepsilon(t)]\,. \tag{4b}$$

The coefficient $\alpha_1$in 3 is the first normal stress modulus, while the Riemann-Liouville operator D_t^{μ} defines this as a model of a general second grade fluid. Further, α_1 in Eq. 3 and Eq. 4b equals the viscoelastic coefficient E (see Eq. 4a) with a time divisor c (matching constant), $\alpha_1 = E/c$, where $c = \left[\rho_f y^2/\mu_f\right]^{\mu-1}$. Therefore, for an initially stagnant fluid, the equation of motion due to a sudden change of the boundary condition at $x = 0$ from $u(0,t) = 0$ to $u(0,t) = U_0$ becomes

$$\rho\frac{\partial u}{\partial t} = \mu_f \frac{\partial^2 u}{\partial y^2} + \alpha_1 D_t^{\mu}\frac{\partial^2 u}{\partial y^2} \tag{5}$$

with the following boundary and initial conditions

$$u(y,0) = 0, \quad y > 0; \quad u(0,t) = U_0, \quad t > 0; \quad u \to 0, \quad y \to \infty. \tag{6}$$

1.3 Common Methods of Solutions Involving Time-Fractional Derivatives

The current literature is plenty of articles solving fractional equations relevant to a broad spectrum of problems. We will avoid the approach typical for review articles and will briefly mention some common methods of solutions which allow one to see the existing “mathematical environment” .

The common methods for solving fractional-order equations are purely mathematical (Nakagawa et al., 2010) even tough they are approximate in nature, some examples include: in terms of Mittag–Leffler function (Narahari and Hanneken,2004), similarity solutions (Djordjevic and Atanackovic, 2008), Green’s function (Pskhu, 2003; Langlands, 2006), operational calculus (Luchko and Srivastava,1995), numerical methods (Diethelm et al, 2005), variational iteration method (He, 1998; Ghorbani, 2008), differential transformations (Al-Rabtah et al., 2009), etc.

Most of these methods use mathematical techniques which are correct, of course, but lack physical intuition. This is not a drawback if we look for solutions only, though the physical interpretation of the abovementioned methods fail. When the first attempt to apply the integral-balance solution to a time-fractional model (Hristov, 2010a) was made, the main idea was: *Is it possible some of the classical analytical methods can be applied and without lost of exactness to develop simple, easily understandable, as well as physically motivated solutions*? The integral-balance method is a technique involving old *physically based idea* that allows to develop simple closed-form solutions that are applicable and allow post-solution analysis. After this preamble we will address more details of this method of solution by examples of some common problems widely solved in the literature.

2 Preliminaries Necessary Mathematical Background

We start with the mathematical background necessary to demonstrate the solution method at issue. We will briefly introduce some information about the fractional integral and derivatives of Riemann-Liouville and Caputo, even though this information is available in any book and chapter devoted to fractional calculus. Further, we introduce the basic principle of the integral-balance method.

2.1 Time-Fractional Integral and Derivatives

2.1.1 Fractional Integral

In accordance with the Riemann-Liouville approach the fractional integral of order $\mu > 0$ is a natural result of the Cauchy's formula reducing calculations of the m-fold primitive of a function $f(t)$ and results in a single integral of convolution type (Gorenflo and Mainardi, 1997) for arbitrary positive number $\mu > 0$, namely

$$ {}_0I^{\mu} f(t) := \frac{1}{\Gamma(\mu)} \int_0^t (t-\tau)^{\mu-1} f(\tau)\, d\tau, \quad t > 0, \quad n \in \mathrm{R}^+ \tag{7}$$

For sake of convenience, we will use also the notation ${}_0D^{-\mu} f(t)$ for ${}_0I^{\mu} f(t)$. Further, the law of exponents for fractional integrals means

$$ {}_0D^{-\mu}\, {}_0D^{-\gamma} f(t) = {}_0D^{-\mu-\gamma} f(t) = {}_0D^{-\gamma}\, {}_0D^{-\mu} f(t)\,. $$

The Laplace transform of the fractional integral is defined by the convolution theorem as

$$ L\left[{}_0D_t^{-\mu} f(t)\right] = L\left[\frac{t^{\mu-1}}{\Gamma(\mu)}\right] L\left[f(t); s\right] = s^{-\mu} F(s)\,, \tag{8}$$

where $\Re(s) > 0$, $\Re(\mu) > 0$ and $F(s)$ is the Laplace transform of $f(t)$

2.1.2 Riemann-Liouville fractional derivative

Therefore, we may define the fractional derivative $D^n f(t)$ with $n \in N$ by the relations (Gorenflo and Mainardi, 1997) as: ${}_0D^{\mu}{}_0I^{\mu} = I$ but ${}_0I^{\mu}{}_0D^{\mu} \neq I$. Therefore, D^n is the left-inverse, but not right-inverse, to the integral operator I^{μ}. Hence, introducing a positive integer m such that $m-1 < \mu \leq m$ the natural definition of the Riemann-Liouville (left-sided) fractional derivative of order $\mu > 0$ is

$$ {}_0D^{\mu} f(t) := {}_0D^m {}_0I^{m-\mu} f(t) = \frac{1}{\Gamma(m-\mu)} \frac{d^m}{dt^m} \int_0^t \frac{f(t)}{(t-\tau)^{\mu+1-m}} \quad , \quad m-1 < \mu \leq m, \quad m \in N. \tag{9} $$

Consequently, it follows that $D^0 = I^0 = I$, that is $D^{\mu} I^{\mu} = I$ for $\mu \geq 0$. In addition, the fractional derivative of power-law function and a constant, frequently used in this chapter, are

$$ {}_0D^{\mu} t^{\beta} = \frac{\Gamma(\beta+1)}{\Gamma(\beta+1-\mu)} t^{\beta-\mu} \text{ and } {}_0D^{\mu} C = C \frac{t^{-\mu}}{\Gamma(1-\mu)} \quad \mu > 0, \beta > -1, t > 0. \tag{10} $$

Similarly, the fractional integrals of the power-law function and a constant are

$$ {}_0D^{-\mu} t^{\beta} = \frac{\Gamma(\beta+1)}{\Gamma(\beta+1+\mu)} t^{\mu+\beta} \quad , \tag{11a} $$

$$ {}_0D^{-\mu} C = \frac{C}{\Gamma(1+\mu)} t^{\mu}, \quad \mu \neq 1, 2, \ldots \tag{11b} $$

The Laplace transform of the Riemann-Liouville fractional derivative for $m \in N$ is

$$ L\left[\frac{d^m}{dt^m} f(t); s\right] = s^m F(s) - \sum_{k=0}^{m-1} s^{m-k-1} f^{(m)}(0+) = s^m F(s) - \sum_{k=1}^{m} s^{k-1} D_t^{\mu-k} f(0+), \tag{12} $$

where $\Re(s) > 0$, $\Re(\mu) > 0$ and $m-1 \leq \mu < m$.

2.1.3 Caputo fractional derivative

The Caputo derivative of a casual function $f(t)$, i.e. $f(t) = 0$ for $t < 0$, is defined (Caputo, 1969; Gorenflo and Mainardi, 1997) as

$$ {}_CD_t^{\mu} f(t) = {}_0I^{m-\mu} \frac{d^m}{dt^m} f(t) = {}_0D_t^{-(m-\mu)} f^{(m)}(t) = \frac{1}{\Gamma(m-\mu)} \int_0^t \frac{f^{(m)}(t)}{(\tau-1)^{\mu+1-m}} dt, \tag{13} $$

where $m \in N$ and $m-1 < \mu < m$.

The Laplace transform of Caputo derivative is

$$ L\left[{}_CD_t^{\mu} f(t); s\right] = s^{\mu} F(s) - \sum_{k=0}^{m-1} f^{(m)}(0) s^{\mu-k-1}. \tag{14} $$

The Caputo derivative of a constant is zero, i.e. ${}_C D_t^{\mu} C = 0$ that matches the common knowledge we have from the integer order calculus and because of that the of Riemann–Liouville derivative is the preferred fractional derivative among mathematicians, while Caputo fractional derivative is the preferred one among engineers (Podlubny, 1999).

If $f(0) = f'(0) = f''(0) = \dots = f^{(m-1)}(0) = 0$, then both Riemann-Liouville and Caputo derivatives coincide. In particular for $\mu \in (0.1)$ and $f(0) = 0$ one has ${}_C D_t^{\mu} f(t) = {}_{RL} D_t^{\mu} f(t)$. Further in this chapter will use the notations ${}_{RL} D_t^{\mu} f(t)$ and ${}_C D_t^{\mu} f(t)$ to discriminate the effects on the solutions when both derivatives are used. In addition in some situations we will use also the notation $\partial^{\mu} f(t)/\partial t^{\mu}$. meaning a time-fractional derivative without specification of the type.

2.2 Integral-Balance Method

2.2.1 The main idea

The integral-balance method comes from the field of approximation solutions of integer order diffusion equations (Goodman, 1958; Hristov, 2009).

$$\frac{\partial \theta}{\partial t} = \frac{\partial}{\partial x}\left[D(\theta)\frac{\partial \theta}{\partial x}\right]. \tag{15}$$

The main solution step is the reduction of the model equation by averaging the field function with respect to the α moment by means of the integral transformations 16a

$$\int_0^{\infty} x^{\alpha}\frac{\partial \theta}{\partial t}dx = \int_0^{\infty}\frac{\partial}{\partial x}\left[D(\theta)\frac{\partial \theta}{\partial x}\right]x^{\alpha}dx \tag{16a}$$

or

$$\frac{d}{dt}\int_0^{\infty} x^{\alpha}\theta dx = \int_0^{\infty}\frac{\partial}{\partial x}\left[D(\theta)\frac{\partial \theta}{\partial x}\right]x^{\alpha}dx. \tag{16b}$$

Applying the Leibniz rule to the left-hand side of 16a we get the common expression of the integral-balance relation 16b. For $\alpha = 0$ we have the well-known Heat-balance integral of Goodman (1958) which is precisely a zero moment equation.

The integral-balance method requires the distribution $\theta(x, t)$ to satisfy the soft condition of the moment relationships 16a and 16b but not the original diffusion equation 15. Further, the principle concept of the integral-balance method is the introduction of a finite penetration depth δ of the diffusant (mass, heat or momentum) thus correcting *ad hoc* the unphysical unlimited speed of propagation of the disturbances (caused by the imposed boundary conditions) exhibited by the original parabolic model 15. With the concept of finite penetration depth the boundary conditions at infinity, that are commonly used to solve 15 for semi-infinite media, can be re-formulated as

$$\theta(\delta, t) = 0 \quad , \tag{17a}$$

$$\frac{\partial \theta(\delta, t)}{\partial x} = 0 \tag{17b}$$

Hence, we define a sharp front propagation of the diffusant and the distribution $\theta(x, t)$ within the range $0 \le x \le \delta$ that can be expressed in a profile dependent on the relative coordinate x/δ. The choice of the assumed profile is a principle step of the integral-balance method and, as a rule it is based on some preliminary information of the steady-state solution of 15.

2.2.2 Principle integration techniques: classical approach

After the choice of the assumed profile $\theta_a(x, \delta)$ and application of the boundary condition at $x = 0$ (specific to the problem at issue), and the Goodman's conditions 17a and 17b we get an approximate profile $\theta_a(x/\delta)$. Further, if replace $\theta(x, t)$ in 16b by $\theta_a(x/\delta)$ and set the upper limit of the integral as δ, we get the classical expression of the Heat-balance integral method (HBIM) (for $\alpha = 0$), namely

$$\frac{d}{dt}\int_0^\delta \theta_a(x, \delta) = -D\frac{\partial \theta_a(0, t)}{\partial x}. \tag{18}$$

The expression 18 has been widely used (see Hristov, 2009 and the references therein) but its principle drawback is that the gradient $\partial \theta_a(0, t)/\partial x$ should be expressed through the assumed profile, which affects the accuracy of the approximation. The principle drawback of HBIM can be avoided by the Double-Integration Method (DIM) where the first step is integration from 0 to x. The second step is integration again from 0 to δ. For eq. 15 and the intermediate result 18, the final equation of DIM is (Myers, 2010)

$$\frac{d}{dt}\int_0^\delta \left(\int_0^x x\theta_a dx\right) dx = D\theta_a(0, t). \tag{19}$$

Therefore, the right-hand side of 19 depends on the boundary condition at $x = 0$ and is independent of the choice of the assumed profile.

These two principle integral relationships will be discussed at large though the examples of this chapter in slightly different, but specific forms; precisely, to the cases where the time-derivative of the evolution equation is of either Riemann-Liouville or Caputo sense.

2.2.3 Assumed profile

The examples of approximate solutions presented in this chapter use an assumed distribution in the form of a generalized parabolic profile

$$\theta_a = A\left(1 - \frac{x}{b_3}\right)^n \Rightarrow \theta_a = \theta_0\left(1 - \frac{x}{\delta}\right)^n \tag{20}$$

where $n > 0$ is an undefined exponent (Hristov, 2009). With the Goodman' boundary conditions and Dirichlet problem ($A = \theta(0, t) = \theta_0$), for instance, we get 20. The main idea to use the profile 20 with unspecified exponent comes from the well-known approach of expressing the solution (with $\mu = 1$) by a series $\theta_a = \sum_0^N a_n (1 - z)^n$, $z = x/\delta$, $0 \le z \le 1$. The profile 20 satisfies all the boundary conditions (the Goodman's ones 17a and 17b and those at $x = 0$) for any n (Hristov, 2009). The advantage of 20 is that *there is freedom to optimize the accuracy of approximation* by application of additional constraints missing in the original integral-balance concept (Goodman, 1958). This is discussed further in this chapter.

2.2.4 Integral-balance with time-fractional derivatives

The models to which the method is applied contain time-dependent terms expressed by time-fractional derivatives. Formally, this means that the left-hand sides of HBIM and DIM relationships should be expressed as

$$\int_0^\delta \frac{\partial^\mu}{\partial t^\mu} dx \text{ for HBIM} \tag{21a}$$

and

$$\int_0^\delta \left(\int_0^x \frac{\partial^\mu}{\partial t^\mu} dx\right) dx \text{ for DIM,} \tag{21b}$$

where $\frac{\partial^\mu}{\partial t^\mu}$ could be either Riemann-Liouville or Caputo fractional derivative.

Now the main problem is: How can the integrals 21a and 21b be approximated when the assumed profile is expressed by 20 or any other approximations such as polynomials. In this context, we have to bear in mind that the classical Leibniz rule for differentiation under the integral sign is no longer valid when the order of the derivative is not integer.

3 Introductory Examples

We start with an introductory example involving a convolution integral which explains the main idea to integral-balance solutions of models involving time-fractional derivatives (Hristov, 2015a). This example we stress the principle steps of the method, among them: How an equation involving a convolution integral can be solved approximately by the integral-balance approach? Three principle integration techniques are introduced: *Heat-balance Integral Method* (HBIM), *Double-integration Method* (DIM) and the *Frozen Front Approach* (FFA).

3.1 Fading Memory in the Diffusion Term

Commonly diffusion phenomena, of heat or mass, are described as a rate law $\partial\theta/\partial t = -\partial J/\partial x$. In the diffusion of heat, for instance, the damping function was conceived by Cattaneo (1948) through a Volterra type integral where the flux J and the relaxation (damping) function $R(x,t)$ are assumed to satisfy the constitutive equation (Curtin and Pipkin, 1968) $J(x,t) = -\int_0^\infty R(x,t)\nabla\theta(x,t-\tau)\,d\tau$. In homogeneous media, $R(x,t)$ *should be time-dependent* and *independent of the space co-ordinate*. With a relaxation function expressed as a power-law kernel $R(x,t) = (t-\tau)^{-\mu}$ we obtain eq. 22a (Dehghan, 2006).

$$\frac{\partial\theta}{\partial t} = a_e \int_0^t \frac{1}{(t-\tau)^{\mu}} \frac{\partial^2\theta(x,\tau)}{\partial x^2}\,d\tau \tag{22a}$$

$$\Rightarrow \frac{1}{\Gamma(\mu)}\frac{\partial\theta}{\partial t} = a_e I_t^{\mu}\left[\frac{\partial^2\theta(x,\tau)}{\partial x^2}\right], \quad 0 < \mu < 1. \tag{22b}$$

This is a model appearing in the viscoelasticity literature (Dehghan, 2006) (see further example 4). The fractional order is $0 < \mu \le 1$, while the elastic diffusivity a_e has a dimension $[m^2/s^{2-\mu}]$. The right-hand side of 22a can be presented as a fractional integral of the Riemann-Liouville sense (Oldham and Spanier, 1974; Gorenflo and Mainardi, 1997) which allows one to obtain the form expressed by eq. 22b.

3.2 Example 1: Diffusion of Momentum with Elastic Effects Only

Therefore, from 22a we have a simple parabolic equation that models rigid heat conductors and relaxing viscoelasticity (Dehghan, M., 2006) with fading memory (expressed by the kernel $(t-s)^{-\mu}$. The Dirichlet problem with zero initial condition imposes at the boundary $x = 0$ the conditions

$$\theta(x,0) = 0, 0 \le x \le \infty; \tag{23a}$$

$$\theta(x,t) = 0, x \to \infty \tag{23b}$$

and

$$\theta(0,t) = \theta_0. \tag{23c}$$

Because the problem 22a is parabolic (Dehghan, 2006), the concept of a finite penetration depth $\delta(t)$ is an *ad hoc* correction of the basic inadequacy related to infinite speed inherent of the classical diffusion model 15 as explained above. Therefore, by applying this concept we can replace the conditions 23b and 23c by $\theta(\delta,t) = 0$ and $\theta_x(\delta,t) = 0$.

3.2.1 HBIM solution

Applying the HBIM approach to 22a, with the Leibniz rule to the left-side and replacing the function θ by θ_a we have

$$\frac{d}{dt}\int_0^\delta \theta_a(x,\delta)dx = \int_0^\delta \left[a_e \int_0^t \frac{1}{(t-\tau)^\mu}\frac{\partial^2\theta_a(x,\delta)}{\partial x^2}d\tau\right]dx \tag{24a}$$

$$\Rightarrow \frac{d}{dt}\left(\frac{\delta}{n+1}\right) = a_e \int_0^t \frac{1}{(t-\tau)^\mu}\left(\frac{n}{\delta}\right)d\tau. \tag{24b}$$

The integration of 24b (again from 0 to δ) and the consequent solution of 25a with the initial condition $\delta(t=0) = 0$ result in the expression 25b about the penetration depth δ^e_{HBIM}.

$$\frac{1}{2(n+1)}\frac{d}{dt}\delta^2 = a_e n \int_0^t \frac{1}{(t-\tau)^\mu}d\tau \tag{25a}$$

$$\Rightarrow \delta^e_{HBIM} = \sqrt{a_e}\sqrt{t^{2-\mu}}\sqrt{\frac{2n(n+1)}{(1-\mu)(2-\mu)}}. \tag{25b}$$

The subscript e in δ^e_{HBIM} (and further in this chapter) denotes "elastic".

3.2.2 DIM Solution

The first integration of eq. 22a from 0 to x yields

$$\int_0^x \frac{d\theta}{dt}dx = \int_0^x \left[a_e \int_0^t \frac{1}{(t-\tau)^\mu}\frac{\partial^2\theta(x,\tau)}{\partial x^2}d\tau\right]dx. \tag{26}$$

However, the integral from 0 to δ, can be presented as $\int_0^{\delta} f(\bullet)\,dx = \int_0^{x} f(\bullet)\,dx + \int_x^{\delta} f(\bullet)\,dx$. Applying this approach to 22a, then subtracting 26 from it and consequently integrating the result from 0 to δ we get 27

$$\int_0^{\delta}\left(\int_x^{\delta}\frac{d\theta}{dt}dx\right)dx = \int_0^{\delta}\left\{\int_x^{\delta}\left[a_e\int_0^{t}\frac{1}{(t-\tau)^{\mu}}\frac{\partial^2\theta(x,\tau)}{\partial x^2}d\tau\right]dx\right\}dx. \tag{27}$$

The left side of 27 differs from the classical formulations 19 and 21b. The step to derive it was conceived by Hristov (2015b) for clear presentation the idea of DIM; the final result after the double integration matches that of the classical DIM. Now, replacing in 27 θ by the assumed parabolic profile θ_a 20 and applying the double integration we get the DIM solution about the penetration depth (Hristov, 2015a)

$$\delta^{e}_{DIM} = \sqrt{a_e}\sqrt{t^{2-\mu}}\sqrt{\frac{(n+1)(n+2)}{(1-\mu)(2-\mu)}},\quad 0<\mu<1. \tag{28}$$

3.2.3 Frozen-front approach (FFA)

This approach has been applied to the first (HBIM) solutions of the fractional diffusion equation (Hristov, 2010a; 2010b; 2011a; 2011b; 2011c; 2012a). In these early studies it was more physically motivated rather than mathematically proved but this introductory example will fill the gap. The main physical assumption behind the FFA refers the rate of the penetration front *which is extremely slow with respect to the rate of the relaxation process modelled by the fading memory function*. That is, *for a certain time range the value of δ could be considered as a constant*. However, *this assumption is valid only in the evaluation of the convolution integral*. Analyses of this approach and its outcomes are available elsewhere (Hristov, 2015a).

Therefore, following the FFA concept the memory integral in 22a is *a time-convolution* between the memory function $f_M(t) = t^{-\mu}$ and u_{xx}. With the assumption of "frozen" δ the approximate profile *is independent* of τ and consequently of the general form of the memory kernel in $(t-s)^{-\mu}$. Hence, with the assumed profile $\tilde{\theta}_a = \theta_a/\theta_0 = (1-x/\delta)^n$ the integral-balance approximation of 22a can be presented as

$$\frac{d}{dt}\left(\frac{\delta}{n+1}\right) = \int_0^{\delta}\left[a_e\frac{n(n-1)}{\delta^2}\left(1-\frac{x}{\delta}\right)^{n-2}\int_0^{t}\frac{1}{(t-\tau)^{\mu}}d\tau\right]dx,\quad \delta(t=0)=0. \tag{29}$$

The integration of eq. 29, with the FFA concept, yields (Hristov, 2015a)

$$\frac{d\delta^2}{dt} = 2a_e n(n+1)\frac{1}{1-\mu}t^{1-\mu} \tag{30a}$$

$$\Rightarrow \delta_{FFA}^{e} = \sqrt{a_e t^{2-\mu}} \sqrt{\frac{2n\,(n+1)}{(1-\mu)\,(2-\mu)}} \quad , \quad 0 < \mu < 1. \tag{30b}$$

The result 30b is the same as that obtained by the HBIM solution. In these two examples, we present the penetration depth as a time dependent function proportional to the term $\sqrt{2n\,(n+1)}$. This was especially done, because in the *viscous case* (see Example 4) with $\mu = 1$ the penetration depth reduces to $\delta\,(t)_{\mu=1} = \sqrt{\nu t}\sqrt{2n\,(n+1)}$ (Hristov, 2009), where $\nu\,\left[m^2/s\right]$ is the *viscous diffusivity of momentum.*

The memory term in the parabolic equation 22a, in fact, accounts for a delay in the propagation speed of the disturbances imposed at the boundary $x = 0$, thus compensating the non-physical infinite speed in the classical integer-order diffusion equation 15 with a constant diffusion coefficient. The concept of a finite penetration depth $\delta\,(t)$ does the same, but this is valid for *large times*, while the weakly singular kernels *represent short-time relaxation effects*. Therefore, applying the integral-balance concept to a model containing a memory term expressed as a Volterra integral, *we overlap two techniques relevant to* ***the concept of finite propagation speed.*** The FFA, however, separates the "objects", i.e. the terms of the model to which these two concepts work: The concept of a final depth $\delta\,(t)$ *works on all terms without memories*, while in the convolution integral δ is assumed to be a constant due to the short-time effect of this term.

The above comments ground the physical basis of FFA, but the principle questions are: 1) *What really we did mathematically with the FFA solution*? 2) What is the secret of the FFA solution? The answers are straightforward (Hristov, 2015a)*: nothing incorrect was done and nothing secret was used.* We will explain this answer briefly. In the correct HBIM solution we integrate twice (from 0 to δ) thus making the convolution integral independent of δ (practically, *it does not matter how the double integration with respect to the space coordinate x is performed*, that is when the assumed profile is inside or outside the convolution integral). Finally, the results from HBIM and FFA are identical. With the DIM solution, the effect of the double integration is obvious, i.e. the convolution integral becomes independent of the assumed profile. Hence, looking at the HBIM solution and the FFA integration technology, we may explain the coinciding results in a simple manner, that is: just move the function in the convolution integral outside it (as a pre-factor), replace by the expression derived through the assumed profile and then integrate twice with respect to the space coordinate x from 0 to δ in the manner described by HBIM. The result is the same as that derived when the function was kept inside the convolution integral.

It is important to mention here that *the correct application* of HBIM and DIM generates fractional-time ordinary equations about $\delta\,(t)$ while, in contrast, FFA results in integer-order ODE about $\delta\,(t)$.

3.2.4 The approximate profile (solution)

With the results developed by the three principle integration techniques we have the following approximate profiles (solutions) of the problem 22a-23a, namely

$$\theta_{HBIM} = \theta_{FFA} = \theta_0 \left(1 - \frac{x}{\sqrt{a_e}\sqrt{t^{2-\mu}}\sqrt{\frac{2n(n+1)}{(1-\mu)(2-\mu)}}}\right)^n, \tag{31a}$$

$$\theta_{DIM} = \theta_0 \left(1 - \frac{x}{\sqrt{a_e}\sqrt{t^{2-\mu}}\sqrt{\frac{(n+1)(n+2)}{(1-\mu)(2-\mu)}}}\right)^n. \tag{31b}$$

The approximate profiles 31a and 31b defines a non-Boltzmann similarity variable $x/\sqrt{a_e t^{2-\mu}}$; for $\mu = 1$ we get $x/\sqrt{a_e t}$. The product $a_e t^{2-\mu}$ has a dimension of $\left[m^2\right]$.

The assumed profile 20 satisfies all boundary conditions imposed on the penetration depth and the solutions 31a and 31b are valid for any value of the exponent $n > 0$ (Hristov, 2009). If the exponent n is stipulated, as in the classical HBIM (Goodman, 1958; Hristov, 2009), then we predetermine the accuracy of approximation. However, since the assumed profile satisfies the soft condition imposed by the integral-balance relation 24a (or 26) but not the governing equation (in this case 22a), this addresses problems related to optimization of the exponent as well as the accuracy of approximation discussed for every particular example analyzed in this chapter.

4 Examples Involving Time-fractional Derivatives

4.1 Example 2: Time-Fractional Subdiffusion Equation

4.1.1 The model and problems emerging when the integral-balance method is applied

Let us consider subdiffusion in a semi-infinite subdiffusive material described by the one-dimensional fractional equation with appropriate boundary and initial conditions

$$\frac{\partial^\mu u(x,t)}{\partial t^\mu} = D_\mu \frac{\partial^2 u(x,t)}{\partial x^2}, \quad 0 < \mu < 1, \ u(0,t) = u_s(t), \ t \geq 0; \ u(x,0) = u_\infty = 0, \ x > 0. \tag{32}$$

In the model 32 the time fractional derivative may be presented as a left Riemann-Liouville or as a left Caputo time-fractional derivative of $u(x,t)$, namely

$$\partial^\mu u/\partial t^\mu = {}_{RL}D_t^\mu u(x,t) = \frac{1}{\Gamma(1-\mu)} \frac{d}{dt} \int_0^t \frac{u(x,\tau)}{(t-\tau)^\mu} d\tau \tag{33}$$

$$\partial^{\mu} u/\partial t^{\mu} = {}_{C}D_{t}^{\mu} u(x,t) = \frac{1}{\Gamma(1-\mu)} \int_{0}^{t} \frac{1}{(t-\tau)^{\mu}} \frac{du(x,\tau)}{dt} d\tau. \quad (34)$$

According to the simplest form of the integral-balance concept applied to 32 one obtains

$$\int_{0}^{\delta} \frac{\partial^{\mu} u}{\partial t^{\mu}} dx = -D_{\mu} \left(\frac{\partial u}{\partial x}\right)_{x=0}. \quad (35)$$

The expression 35 presents the classical **HBIM** approach to the fractional-time subdiffusive equation (Hristov, 2010a; 2011b). The left side of 35 is termed ***fractional-time Heat-balance Integral*** **(FT-HBI)** (Hristov, 2010a).

4.1.2 DIM solution

The first step of DIM is the integration of eq. 32 from 0 to x, namely

$$\int_{0}^{x} \frac{\partial^{\mu} u}{\partial t^{\mu}} dx = D_{\mu} \frac{\partial u(x,t)}{\partial x} - D_{\mu} \frac{\partial u(0,t)}{\partial x}. \quad (36)$$

Now, the integration in the left-hand side of 35 can be represented as $\int_{0}^{\delta} \frac{\partial^{\mu} u}{\partial t^{\mu}} dx = \int_{0}^{x} \frac{\partial^{\mu} u}{\partial t^{\mu}} dx + \int_{x}^{\delta} \frac{\partial^{\mu} u}{\partial t^{\mu}} dx$. Then, subtracting 36 from 35 and integrating the result from 0 to δ (the second step of DIM) we obtain (Hristov, 2015b)

$$\int_{0}^{\delta} \left(\int_{x}^{\delta} \frac{\partial^{\mu} u}{\partial t^{\mu}} dx \right) dx = D_{\mu} u(0,t). \quad (37)$$

Equation 37 is the principle equation of the double integration method when the differential equation is of a fractional order (Hristov, 2015b). Following the terminology established (Hristov, 2010a; 2011b) the left side of eq. 37 is termed ***fractional-time Double-balance Integral*** **(FT-DBI).**

4.1.2.1 FT-DBI with Riemann-Liouville time-fractional derivative

Following eq. 37 we have

$$\int_{0}^{\delta} \left[\int_{x}^{\delta} {}_{RL}D_{t}^{\mu} u_a(x,t)\, dx \right] dx = \frac{1}{\Gamma(1-\mu)} \frac{d}{dt} \int_{0}^{t} \frac{1}{(\tau - t)^{\mu}} \frac{\delta^2}{(n+1)(n+2)} d\tau. \quad (38)$$

Therefore, following the definition of the Riemann-Liouville fractional derivative (see 33) we have (Hristov, 2015b)

$$\int_0^{\delta}\left[\int_x^{\delta} {}_{RL}D_t^{\mu} u_a(x,t)\,dx\right]dx = {}_{RL}D_t^{\mu}\left[N\delta^2\right] = N_{RL}\,{}_{RL}D_t^{\mu}\delta^2, \tag{39a}$$

$$N_{RL} = [(n+1)(n+2)]^{-1}. \tag{39b}$$

4.1.2.2 FT-DBI with Caputo time-fractional derivative

With $du_a(x,t)/dt = (x/\delta^2)\,n\,(1-x/\delta)^{n-1}(d\delta/dt)$ the double integration in LHS of 37 with the Caputo derivative results in

$$\int_0^{\delta}\left[\int_x^{\delta} {}_CD_t^{\mu} u_a(x,t)\,dx\right]dx = \frac{1}{\Gamma(1-\mu)}\int_0^t \frac{1}{(t-\tau)^{\mu}}\frac{1}{(n+1)(n+2)}\frac{d\delta^2}{dt}\,d\tau. \tag{40}$$

Then from definition 34 we obtain (Hristov, 2015b)

$$\int_0^{\delta}\left[\int_x^{\delta} {}_CD_t^{\mu} u_a(x,t)\,dx\right]dx = {}_CD_t^{\mu}\left[\frac{N}{2}\delta^2\right] = N_C\,{}_CD_t^{\mu}\delta^2, \tag{41a}$$

$$N_C = N_{RL} = [(n+1)(n+2)]^{-1}. \tag{41b}$$

4.1.2.3 The Penetration depths of the DIM solution

Therefore, from the principle equation 37 with the results 39b and 41a we obtain two fractional equations about the penetration depth (Hristov, 2015b)

$${}_{RL}D_t^{\mu}\delta^2 = D_{\mu}\,[(n+1)(n+2)] \tag{42a}$$

and

$${}_CD_t^{\mu}\delta^2 = D_{\mu}\,[(n+1)(n+2)]. \tag{42b}$$

The right-hand sides of 42a and 42b are constants because we look for a time-independent exponent $n > 0$. In addition, the integral-balance method imposes a physical condition $\delta(t=0) = 0$. Thus, with respect to the Riemann-Liouville derivative we have :$\lim_{t\to 0}\left[{}_{RL}D_t^{\mu-k-1}\delta^2(t)\right] = \delta_{t=0} = 0$ (see Podlubny (1999)-, p. 139). Therefore, the fractional integration with the Riemann-Liouville derivative results in ${}_{RL}\delta^2 = {}_{RL}I^{\mu}\{D_{\mu}[(n+1)(n+2)]\}$ (Hristov, 2015b). Consequently, the result is

$${}_{RL}\delta^2 = D_{\mu}(n+1)(n+2)\frac{t^{\mu}}{\Gamma(1+\mu)} \tag{43a}$$

$$\Rightarrow {}_{RL}\delta^s_{DIM} = \sqrt{D_{\mu}t^{\mu}}\sqrt{\frac{(n+1)(n+2)}{\Gamma(1+\mu)}}. \tag{43b}$$

For $\mu = 1$, for instance, we get the integer-order relationship $\delta = \sqrt{Dt}\sqrt{(n+1)(n+2)}$ which is the well-known result derived by the Refined Integration Method (RIM) (Myers, 2009; 2010).

Looking for the solution of 42b with the Caputo derivative and applying the Laplace transform we have (Podlubny, 1999)

$$L\left[{}_CD_t^{\mu}\delta^2(t), s\right] = s^{\mu}{}_C\delta^2(s) - \sum_{k=0}^{m-1} s^{\mu-k-1}\frac{d^{(k)}}{dt}\left(\delta^2(0)\right) = \frac{D_\mu(n+1)(n+2)}{s}, m-1 < \mu < m. \tag{44}$$

The physical condition $\delta(t = 0) = 0$ applied to 44 and the inverse Laplace transform lead to

$$_C\delta^2 = D_\mu 2(n+1)(n+2)\frac{t^\mu}{\Gamma(1+\mu)} \tag{45a}$$

$$\Rightarrow {}_C\delta^S_{DIM} = \sqrt{D_\mu t^\mu}\sqrt{\frac{(n+1)(n+2)}{\Gamma(1+\mu)}}. \tag{45b}$$

Therefore, the penetration depth equations are equal. This is an expected result because ${}_{RL}D_t^\mu\delta^2 = {}_CD_t^\mu\delta^2$ with zero initial condition (Podlubny, 1999), since $\delta(t = 0) = 0$.

4.1.3 HBIM solution

At this point, for comparison only, we will demonstrate briefly how the simple HBIM works and what are the differences in the evaluation of the penetration depths. Hence, applying HBIM (see 36) to both sides of the model 32 we obtain

– *With Riemann-Liouville derivative*

$$\int_0^\delta {}_{RL}D_t^\mu u_a(x,t)\,dx = \int_0^\delta\left[\frac{1}{\Gamma(1-\mu)}\frac{d}{dt}\int_0^t\frac{1}{(\tau-t)^\mu}\left(\frac{\delta}{n+1}\right)d\tau\right]dx = \frac{n}{\delta}. \tag{46}$$

Then, following the definition 33 and 36 one obtains

$$_{RL}D_t^\mu\frac{\delta}{n+1} = D_\mu\frac{n}{\delta} \tag{47a}$$

$$\Rightarrow \delta_{RL}D_t^\mu\delta = D_\mu n(n+1) \tag{47b}$$

Applying the Laplace transform solution to 47b we get ${}_{RL}\delta^S_{HBIM} = \sqrt{D_\mu t^\mu}\sqrt{\frac{n(n+1)}{\Gamma(1+\mu)}}$.

– *With Caputo derivative*

Similarly to the previous operations we have

$$\int_0^\delta {}_CD_t^\mu u_a(x,t)\,dx = \frac{1}{n+1}D_t^\mu\delta(t) \tag{48a}$$

$$\Rightarrow {}_C D_t^{\mu}\delta = D_{\mu}\frac{n(n+1)}{\delta}. \tag{48b}$$

The Laplace transform solution to eq. 48b is ${}_C\delta^s_{HBIM} = \sqrt{D_\mu n(n+1)}\frac{t^{\mu/2}}{\sqrt{\Gamma(1+\mu)}}$.

4.1.4 FFA solution

We can replace $u(x,t)$ with the assumed distribution $u_a(x,t)$ and turn to the evaluation of the integral

$$I_{fr} = \int_0^{\delta}\left[\frac{\partial^\mu}{\partial t^\mu}u_a(x,t)\right]dx. \tag{49}$$

For the sake of simplicity we consider the case with $\mu = 1/2$ (Hristov, 2010a) where $\partial^\mu/\partial t^\mu$ denotes Riemann-Liouville fractional derivative and $u_a(x,t)$ is presented by the assumed parabolic profile with unspecified exponent 20. With $\mu = 1/2$ the fractional-time heat balance integral 49 becomes

$$I_{fr(1/2)} = \int_0^{\delta}\left[\frac{1}{\sqrt{\pi}}\frac{d}{dt}\int_0^t\frac{u(x,u)}{\sqrt{t-u}}du\right]dx = \frac{1}{\sqrt{\pi}}\int_0^{\delta}\left[\frac{d}{dt}\int_0^t\frac{u(x,u)}{\sqrt{t-u}}du\right]dx. \tag{50}$$

Denoting $\Phi(x,t) = \int_0^t \frac{u(x,u)}{\sqrt{t-u}}du$ and applying the Leibniz rule to the inner integral of 50, taking into account Goodman's conditions at $x = \delta$, we have $I_{fr(\mu=1/2)} = \frac{1}{\sqrt{\pi}}\frac{d}{dt}\int_0^{\delta}\Phi(x,t)\,dx$. For the Dirichlet problem, with $\mu = 1/2$, and replacing $u(x,t)$ by $u_a(x,\delta)$, we may substitute the function $\Phi(x,t)$ with $\Phi_a(x,t) = \int_0^t u_a(x,u)\frac{1}{\sqrt{t-\tau}}d\tau$. Before applying the next step of the integration, recall that for FFA, $\delta(t)$ *is time-independent when it is inside the convolution integral* and we may move the approximate profile in front of the inner integral (see eq. 50 for instance). The consequent integrations yield (the superscript a in I^a means "approximate") $I^a_{fr(\mu/2)} = \frac{2}{\sqrt{\pi}}\frac{1}{n+1}\frac{d}{dt}\left(\delta\sqrt{t}\right)$. In the general case, for $0 < \mu < 1$, and following the FFA scheme, the same integration technique provides an approximation of $\partial^\mu u_a/\partial t^\mu$ (Riemann-Liouville) with u_a defined by 20 as (Hristov, 2010a; 2011b)

$${}_{RL}D_t^{\mu}u_a(x,t) = \frac{1}{\Gamma(1-\mu)}\left[\frac{1}{(1-\mu)}\frac{1}{n+1}\frac{d}{dt}\left(\delta t^{1-\mu}\right)\right]. \tag{51}$$

Next, applying HBIM we obtain an *ordinary (integer-order) equation* about δ (Hristov, 2010a; 2011b)

$$\frac{d}{dt}\left(\delta^s t^{1-\mu}\right) - D_\mu\frac{N_1}{\delta} = 0, \quad N_1 = n(n+1)(1-\mu)\Gamma(1-\mu)\quad, \tag{52a}$$

$$\delta^s_{FFA} = \sqrt{D_\mu t^\mu}\sqrt{2n(n+1)\frac{\Gamma(2-\mu)}{2-\mu}}. \tag{52b}$$

This result differs from the developed HBIM solution (see the result of 47b) but both of them coincide for the limiting case with $\mu = 1$. The non-Boltzmann scaling of the front propagation $\delta \overset{\text{def}}{=} \sqrt{D_\mu t^\mu}$ is preserved in both expressions, but *the only differences exist in the numerical pre-factors depending on n and μ.*

4.2 Approximate Parabolic Profiles

With the developed expression about the penetration depths δ^s (the superscript denotes "subdiffusion") we get the following approximate profiles of the fractional subdiffusion equation (Dirichlet problem)

$$u^s_a(x,t) = \left(1 - \frac{x}{\sqrt{a_\mu t^\mu F^s_n j^s_\mu}}\right)^{n_s}. \tag{53}$$

- For **HBIM** solutions

$$F^s_{n(HBIM)} = \sqrt{n(n+1)}, \tag{54a}$$

$$j^s_{\mu(HBIM)} = \sqrt{\frac{1}{\Gamma(1+\mu)}}. \tag{54b}$$

- For **DIM** solution

$$F^s_{n(DIM)} = \sqrt{(n+1)(n+2)}, \tag{55a}$$

$$j^s_{\mu(HBIM)} = \sqrt{\frac{1}{\Gamma(1+\mu)}}. \tag{55b}$$

- For **FFA** solutions and Riemann-Liouville derivative

$$F^s_{n(FFA)} = \sqrt{2n(n+1)}, \tag{56a}$$

$$j^s_{\mu(FFA)} = \sqrt{\frac{\Gamma(2-\mu)}{2-\mu}}. \tag{56b}$$

The exponent n of the profile is still undefined because the integral-balance method does not have enough boundary conditions to define it which requires additional constraints to be applied (Hristov, 2009).

4.3 Calibration of the Profile Exponent and Results Thereof

4.3.1 Optimal exponents

In the general case of the fractional subdiffusion equation $0 < \mu < 1$ the approximate solution should satisfy the domain equation, so the mean-squared error of approximation should be minimal, namely

$$E_\mu(t) \overset{\text{def}}{=} \int_0^{\delta(t)} \left[\frac{\partial^\mu}{\partial t^\mu} u_a(x,t) - D_\mu \frac{\partial^2 u_a}{\partial^2 x} \right]^2 dx \geq 0, \quad E_\mu(t) \to \min. \tag{57}$$

The squared error of approximation 57 depends on the method of integration employed to develop the penetration depth relationships. Obviously, with different expressions for $\delta(t)$ we have to obtain different values of the optimal exponent n_{opt}. In general, n_{opt} increases as the fractional order is decreased, i.e. $\mu \to 0$ and *vice versa*. As $\mu \to 1$, the value of n_{opt} approaches the optimal exponents established for the integer order integral-balance solutions (Hristov, 2012b; Myers, 2009; 2010) . The optimization procedures with the DIM solution analyzed by Hristov (2015b) as well as the same procedures with the corresponding FFA solutions (Hristov, 2011a, 2011b) provide optimal exponents summarized in Table 1.

Table 1. Optimal exponents of the approximate profile defined by different methods

Fractional order μ	0.1	0.2	0.3	0.4	0.5	0.6	0.7	0.8	0.9
n (*DIM*) *	241.87	59.492	25.712	13.897	8.423	5.447	3.647	2.468	1.690
n_{opt} (*DIM*) * *	214.71	52.841	22.875	12.403	7.579	4.995	3.504	2.772	2.284
n_{opt} (*FFA*)	1.480	1.478	1.477	1.472	1.472	1.469	1.465	1.456	1.434

n_{opt} (*DIM*) from Hristov (2015b) n_{opt} (*FFA*) from (Hristov, 2011a,b); *defined at $x = 0$ setting the residual function equal to zero; ** defined by an approximation of the time-fractional derivative through the assumed profile (Hristov,2015b).

4.3.2 Comparison with exact solutions

Since in the specific case with zero initial condition, both the Riemann-Liouville and the Caputo derivatives are equal, we may compare the integral balance solutions to the exact solution of eq. 32 represented by eq. 58

$$M(\eta_\mu, \mu) = \sum_{k=0}^{\infty} \frac{(-1)^k \eta_\mu^k}{k! \Gamma\left[-\mu k/2 + (1 - \mu/2)\right]}, \quad 0 < \mu/2 < 1 \tag{58}$$

where $M(\eta_\mu, \mu/2)$ is the Mainardi function (Mainardi, 1996) and $\eta_\mu = x/\sqrt{D_\mu t^\mu}$ $(0 < \mu < 1)$ is a similarity variable.

The tests comparing the integral-balance solutions to the exact ones used truncated series approximating the Mainardi function as $M(\eta_\mu, \mu) = \sum_{k=0}^{N} [k, \eta_\mu, \mu]$ for $N = 250$ (calculations performed with Maple 13) following the general rules for calculation of the Mittag-Leffler function (Gorenflo et al., 2005). The comparative plots in Fig. 1a present the solutions by DIM, FFA and the exact solution 58 for $\mu = 0.1$ as an illustrative example. More plots and details are available in Hristov (2015b). The first plot (Fig. 1a) reveals that it is practically impossible to detect differences in the solutions developed by the integration techniques applied. The calculations of the absolute errors (Fig. 1b), show these are acceptable approximations. The FFA solution goes slightly higher than the exact one, but is has practically acceptable error of approximation; for $\eta_\mu > 1.5$ the error of approximation is higher than that of the DIM solution.

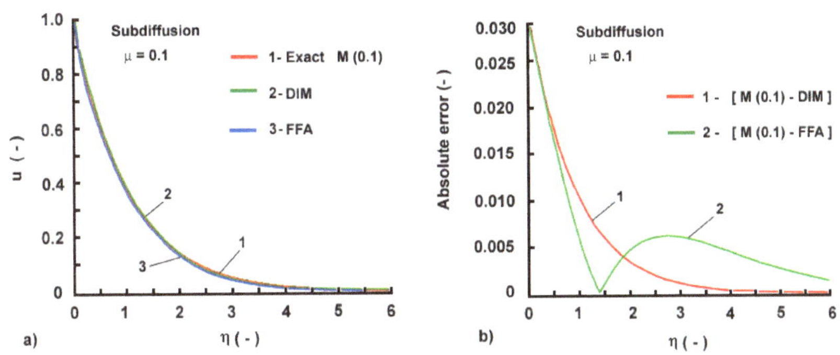

Fig. 1. Comparative plots of the exact solution (Mainardi, 1996) and the approximate solutions developed by DIM and HBIM (i.e. FFA) for $\mu = 0.1$ Distributions against the similarity variable z; b) Absolute errors against the similarity variable η_μ.

Detailed "zoomed" sections of the solutions presented in Fig. 2a, b reveal practically undistinguishable solutions (Fig. 2b, for instance). The best performance of both DIM and FFA solutions are in the range $0 < \eta_\mu < 1.5$. For higher η_μ (see Fig. 2c) the difference between the approximate and the exact solutions increases, but this is an inherent problem of the integral-balance solutions (Hristov, 2010a; 2011a; 2011b; Myers, 2009) because the penetration depth (in fact the point corresponding to $\eta_\mu/\delta(t) = 1$) is shorter than the distance where the exact solution, if it exists, vanishes and approaches zero.

As illustrative examples we demonstrate in Fig. 3 distributions generated by approximate solutions developed by HBIM and FFA (Hristov, 2011b)

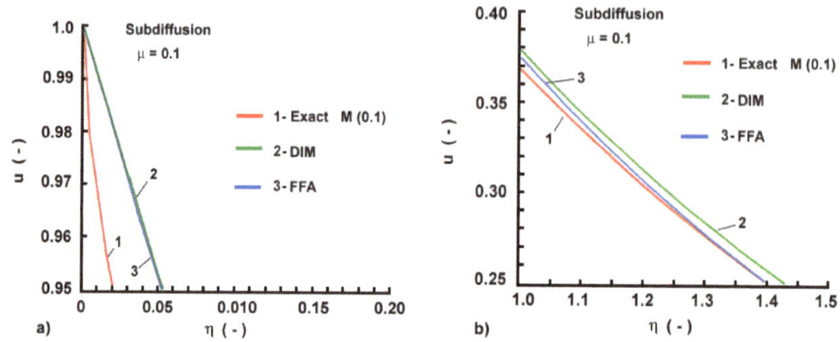

Fig. 2. Detailed zoomed sections of the solutions presented in Fig. 1a.

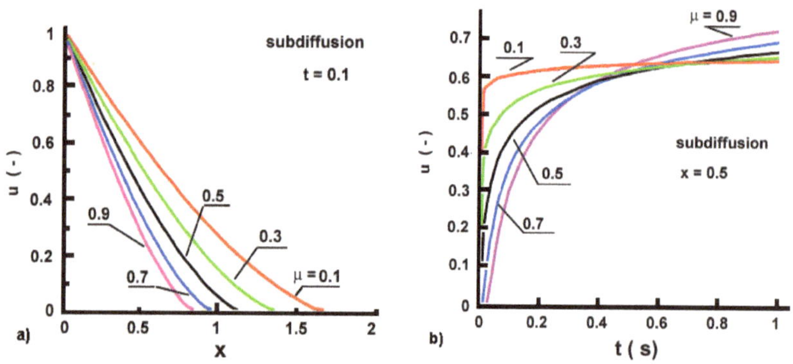

Fig. 3. Short-time distributions of the fractional subdiffusion equation generated by the approximate solutions. For sake of simplicity of calculations it is assumed $a_\mu = 1$. Adapted from Hristov (2011b) by courtesy of *Thermal Science*. a) Approximate solutions as a function of the space coordinate x and a fixed time $t = 0.1$ b) Time-evolution of the distributions at short distances ($x = 0.5$).

4.4 Example 3: Subdiffusion Equation: A Solution by a Weak Approximate Profile

This solution (Hristov, 2010b) used a weak approximate profile (WAP) of a power-law type instead the parabolic one 20, namely

$$\Theta_a(x,t) = 1 - \left(\frac{x}{\delta}\right)^n \tag{59}$$

This profile 59 satisfies the conditions $\Theta_a(0,t) = 1$ and $\Theta_a(\delta,t) = 0$ at any value of the exponent n. Moreover, $\partial\Theta_a(x,t)/\partial x = -n\left(x^{n-1}/\delta^n\right)$ that leads to $\partial\Theta_a(\delta,t)/\partial x = -n/\delta$ and $\partial\Theta_a(0,t)/\partial x = 0$. Hence, this profile does not satisfy the classical boundary conditions $\partial\Theta_a(\delta,t)/\partial x = 0$ and $\partial\Theta_a(0,t)/\partial x \neq 0$ imposed by the integral-balance method. Because of that, is it was termed a *weak approximate profile* (WAP). The condition $\partial\Theta_a(\delta,t)/\partial x = 0$ follows the concept of a sharp front but the experiments on anomalous diffusion Kosztolowicz et al. (2005) do not show a sharp boundary and define a value of the concentration $C(0.05)$ as 5% of C_s $(x = 0)$ as a measurable threshold and corresponding front $\delta_{0.05}(t)$. Hence, the condition $\partial \Theta_a(\delta, t)/\partial x = 0$ can be avoided under the assumption that the other two (i.e. 17a and that at $x = 0$) are exactly obeyed. This approach reduces the number of the conditions imposed on the assumed profile.

The problem with the improper derivatives at the boundaries of the penetration depth can be avoided by a double integration of the subdiffusion equation over the penetration layer, namely

$$\int_0^{\delta}\left\{\int_0^{\delta}\left[\frac{\partial^{\mu}}{\partial t^{\mu}}C(x,t)\right]dx\right\}dx = D_{\mu}C(0,t). \tag{60}$$

Then, with the Riemann-Liouville fractional derivative and applying HBIM and FFA approaches to 60 with WAP one obtains (Hristov, 2010b)

$$\delta_{WAP}^{s} = \sqrt{D_{\mu}t^{\mu}}\sqrt{F_{n(WAP)}^{s}(n)\, j_{\mu(WAP)}^{s}}, \tag{61a}$$

$$F_{n(WAP)}^{s}(n) = 2\frac{(n+1)}{n}, \tag{61b}$$

$$j_{\mu(WAP)}^{s} = \frac{\Gamma(2-\mu)}{(2-\mu)}. \tag{61c}$$

The numerical experiments revealed that the optimal value of n is within the range $1 \le n \le 1.5$ when $0.5 \le \eta \le 1$. The weak profile $\left[1-(x/\delta)^n\right]$ is an adequate approximation for cases when $0.1 < \mu < 0.3$ with $n = 1$. With $n \approx 1.25 \div 1.5$ good approximations for $0.35 < \mu < 0.5$ within the range $0 < \eta < 0.5$ can be obtained. The increase in $\eta \to 1$ and $\mu \to 0.5$ requires the exponent to increase towards $n \approx 2$ and $n \approx 3$. More details about this study are available elsewhere (Hristov, 2010b)

5 Transient Flows of Viscoelastic Fluids

5.1 Example 4: Stokes' First Problem of a Second Grade Fractional (viscoelastic) Fluid

Let us again consider Eq. 5 which can be expressed in a diffusion form as

$$\frac{\partial u}{\partial t} = \nu\frac{\partial^2 u}{\partial y^2} + pD_t^{\mu}\frac{\partial^2 u}{\partial y^2}. \tag{62}$$

The coefficient $p = \alpha_1/\rho_f\ \left[m^2\right]$ can be presented also as $p = \nu\lambda_r$, where λ_r is the relaxation time and $\nu = \rho_f/\mu_f$ is the fluid kinematic viscosity.

The Stokes first problem addresses a transient flow of a fluid, initially at rest, caused by a sudden jump of the surface velocity (at $x = 0$) from 0 to $u(0,t) = U_0$. With the assumed parabolic profile 20 and applying the boundary condition $u(0,t) = U_0$ as well as those defining the sharp penetration front we get: $u_a = U_0 (1 - y/\delta)^n$ or $u_a/U_0 = \overline{u} = (1 - y/\delta)^n$. In this example we stress the attention on the physical adequacy of the integral-balance solutions and the possibility for a *post-solution analysis*. Post-solution analysis refers to the possibility of evaluating the contribution of each term of the closed-form dilution as well as to reduce it to well-known versions when some physical effect would be is neglected.

5.1.1 HBIM solution

Integrating 62 from 0 to δ we get

$$\int_0^\delta \frac{\partial u}{\partial t} dy = \int_0^\delta \nu \frac{\partial^2 u}{\partial y^2} dy + \int_0^\delta p D_t^\mu \frac{\partial^2 u}{\partial y^2} dy. \tag{63}$$

The integration in 63 with application of the Leibnitz rule to the LHS and the FFA concept yields an equation about the time evolution of the penetration depth δ (Hristov, 2012a)

$$\delta_{HBIM}^{\nu+e} = \sqrt{\nu t}\sqrt{2n(n+1)\left[1 + \frac{p}{\nu t^\mu} j_\mu\right]}, \tag{64a}$$

$$\delta_{HBIM}^{\nu+e} = \sqrt{\nu t}\sqrt{2n(n+1)[1 + j_\mu De]}. \tag{64b}$$

Denoting $p/\nu = t_{ev}$ we get the Deborah number $De = t_{ev}/t^\mu = p/\nu t^\mu$ and the penetration depth can be presented as 64b. The approximate velocity profile is a function of two similarity variables: $\xi = y/\sqrt{\nu t}$ and $\chi = \sqrt{p}/\sqrt{\nu t^\mu}$. The former is the Boltzmann similarity variable with $\sqrt{\nu t}$ as a diffusion length scale corresponding to the Newtonian flow while the second one corresponds to the elastic mode of momentum diffusion $(\chi^2 = De)$. From Eq. 64a with $p = 0$ or $p/\nu = \alpha_1/\mu_f \to 0$ we get the Newtonian penetration depth $\delta_N = \sqrt{\nu t}\sqrt{2n(n+1)}$ (Hristov, 2009). Because the integral-balance solution of the Newtonian flow is straightforward, it is possible to extract the term representing only the shear momentum penetration depth δ_N and write the dimensionless ratio $\Delta = \delta_{HBIM}^{\nu+e}/\delta_N = \sqrt{1 + j_\mu De}$. It is evident that we have only viscous (Newtonian) flow when Δ becomes equal to unity, i.e. $De \to 0$ (see Fig. 4a, b). Therefore, the velocity profile is

$$\left.\frac{u_a}{U_0}\right|_{HBIM} = \left(1 - \frac{y}{\sqrt{\nu t}} \frac{1}{\sqrt{2n(n+1)(1 + j_\mu De)}}\right)^n = \tag{65a}$$

$$= \left(1 - \xi \frac{1}{\sqrt{2n(n+1)\left(1 + j_\mu \chi^2\right)}}\right)^n = \tag{65b}$$

$$= \left(1 - \frac{\xi}{F_n R_\mu}\right)^n \tag{65c}$$

$$F_n = \sqrt{2n(n+1)}, \tag{65d}$$

$$R_\mu = 1 + j_\mu De \tag{65e}$$

and

$$R_\mu = 1 + j_\mu \chi^2. \tag{65f}$$

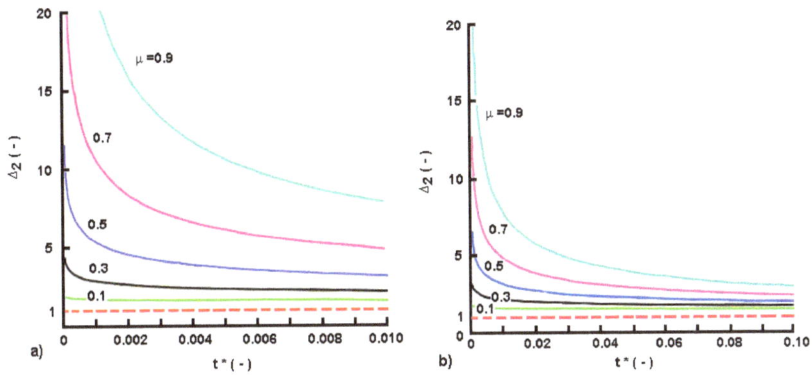

Fig. 4. Relative penetration depth and approximate velocity profiles (Hristov, 2012a) by courtesy of *Thermal Science*; a) Short dimensionless times, b) Medium dimensionless times

5.1.2 The optimal exponent for the HBIM solution

Following the idea used in Example 2, the estimation of the optimal exponent requires minimization of the squared error function in the domain $0 \le y \le \delta$, namely

$$E(y, t, \mu) = \int_0^\delta \left[\frac{\partial \overline{u}}{\partial t} - v\frac{\partial^2 \overline{u}}{\partial y^2} - pD_t^\mu \frac{\partial^2 \overline{u}}{\partial y^2}\right]^2 dy \to \min. \tag{66}$$

The minimization procedure (Hristov, 2012) reveals that $n \approx \frac{1}{\sqrt{2}}\frac{1}{\sqrt{De}}$. Simple tests with $n = 2$and $n = 3$ (Hristov, 2012a) provide $De \approx 0.125$ and $De \approx 0.055$, respectively. Therefore, within the range defined by $O(De) \sim 1$ (that is the viscoelastic flow) (Goodwin and Hughes, 2008), the exponent of the parabolic profile oscillates around $n = 2$, as the value of De varies.

5.1.3 The optimal exponent for the HBIM solution and elastic effects only

The HBIM solutions of Example 1, for instance, lead to following approximate profiles (Hristov, 2015a) following naturally, to some extent, from 65a by setting $\nu = 0$.

$$\left.\frac{u_a}{U_0}\right|_{HBIM}^{(elastic)} = \left(1 - \frac{x}{\sqrt{pt}\sqrt{2n\,(n+1)}\sqrt{\frac{j_\mu}{t^\mu}}}\right)^n. \tag{67}$$

The solution 67 comes from the simplified model 22a where the relaxation is modelled by a fractional integral instead of a fractional derivative. The elastic penetration depth is proportional to $\sqrt{p}$ (modelled by the term with memory) while the viscous one (modelled by terms without memory) is proportional to $\sqrt{\nu}$. The typical relaxation time λ_r for viscoelastic fluids is of order $(1 - 10) \times 10^{-3}s$ for a Newtonian dynamic viscosity (μ_f) range from 2 to 10×10^{-3} $[N/m^2 s]$ (Zilz et al. 2014) (a kinematic viscosity ν of order 10^{-6} $[m^2/s]$). Hence, simply the ratio of the speeds (and the lengths, too) of both penetration depths is proportional to $\sqrt{p/\nu} \sim O\left(10^2\right)$ (Hristov, 2015a).

In the absence of elastic effects, i.e. for $p = 0$ we obtain from 63 the classical diffusion model, then the HBIM solution of the Dirichlet problem has an optimal exponent $n \approx 2.35$ (Myers, 2010). The optimal exponent established for $p = 0$ should be considered to be *the lower limit of exponents modelling the elastic effects only* (Hristov, 2015a). Simple calculations with $n = 2$ and $n = 3$ provide $De \approx 0.125$ and $De \approx 0.055$, respectively. Further, for the optimal Newtonian profile ($p = 0$) with exponent $n \approx 2.35$ we get $De \approx 0.100$. As mentioned above the typical relaxation time λ_r for viscoelastic fluids is in the range $(1 - 10) \times 10^{-3}s$ (Zilz et al. 2014), thus, for observation time of about $1s$, for instance, we get Deborah numbers in the range $(2 - 10) \times 10^{-3}$ and consequently from $n \approx 1/\sqrt{2De}$ the range of optimal exponents from is from $n \approx 15.81$ to $n \approx 7.071$, respectively. Hence, lower Deborah numbers, translate to higher exponents of the parabolic profile for the short-time (with dominating elastic effects) solutions. Examples related to the Stokes first problem with the complete model 63 are available elsewhere (Hristov, 2012a, 2015a).

For $\mu = 1/2$, in the simple case of the model 62 (with $\nu = 0$) applying the Laplace transform in the time domain (Hristov, 2015a) the approximately solution is $u_z(x, t) = u_z(z) \approx 1 - \exp\left(-z/\sqrt{\pi}\right)$; Here $z = x/t^{3/2}$ is the similarity variable $z = x/\sqrt{t^{2-\mu}}$ for $\mu = 1/2$. The plots in Fig. 5a illustrate this almost exact solution compared to approximate one developed by FFA (the same as HBIM) (Hristov, (2015a) where $z = x/t^{3/2}$ is the similarity variable $z = \eta_\mu = x/\sqrt{t^{2-\mu}}$ for $\mu = 1/2$.

5.1.4 HBIM solution with a polynomial profile and elastic effects only

Following the classic HBIM solution (Goodman, 1958) the assumed profile may be a polynomial of 2nd or 3rd order. With the quadratic assumed profile

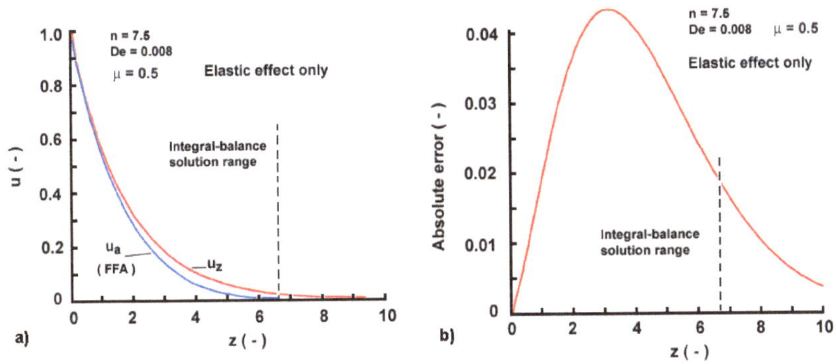

Fig. 5. Solution of the elastic model 62 ($v = 0$) for $n = 7.5$. Left: Comparison of the approximate (FFA=HBIM) integral-balance solution and the almost exact u_z solution; Right: absolute errors. Adapted from Hristov, 2015a) by courtesy of *Thermal Science*

$V_a = \beta_0 + \beta_1 x + \beta_2 x^2$, for instance, and the Goodman's boundary conditions we have $V_a/U_0 = 1 - 2\left(x/\delta\right)^2$. Now, applying the integral-balance method to the model 22a and replacing u by V_a/U_0, we read

$$\frac{d\delta}{dt} = a_e \frac{6}{\delta}\left(\frac{1}{1-\mu}\right) t^{1-\mu} \Rightarrow \delta^e_{poly2} = 2\sqrt{3j^e_\mu}\sqrt{a_e t^{2-\mu}}. \tag{68}$$

Then, the approximate profile of the "elastic" model becomes

$$\left.\frac{V_a}{U_0}\right|^{elastic} = 1-2\left(\frac{x}{2\sqrt{3j^e_\mu}\sqrt{a_e t^{2-\mu}}}\right)+\left(\frac{x}{2\sqrt{3j^e_\mu}\sqrt{a_e t^{2-\mu}}}\right)^2, \quad j^e_\mu = \left[(1-\mu)(2-\mu)\right]^{-1}. \tag{69}$$

The similarity variable $x/\sqrt{a_e t^{2-\mu}}$ is the same as in the solutions using the parabolic profile 20. However, the pre-factor $2\sqrt{3j^e_\mu}$ is predetermined and *there is no freedom to optimize* the approximation with a given fractional order μ. Similarly, when the damping function is presented by a fractional time derivative, i.e., the model 62, the HBIM solution with 2nd order polynomial is

$$\delta^\mu_{ve} = 2\sqrt{vt}\sqrt{1+\frac{1}{4}\frac{1}{(1-\mu)(2-\mu)\Gamma(1-\mu)}\left(\frac{p}{v}\right)t^{-\mu}}. \tag{70}$$

Extracting the *large-time length scale* $\sqrt{vt}$ as a pre-factor we have a better presentation of δ^μ_{ve} in the form 70 with distinguished *large-time* and *short time* (*relaxation*) effects. Precisely, the result is almost the same as that expressed by eq. 64a and 64b with the only differences in the numerical pre-factors strongly dependent on the choice of the assumed profile.

5.1.5 Examples of numerical experiments with the complete approximate profile

The plots in Figure 6, corresponding to the solution 65a, reveal that with increases in $De = \chi^2$ the profile $u/U_0 = f(\xi, \chi = const.)$ becomes almost linear and that with $0 \le \chi \le 0.5$ there is no strong effect of the elasticity on the profile. Hence, as we increase De, the exponent n should decreases from the initial values established for the pure viscous regime as commented in the previous two sections. The short time $(0 \le t \le 0.1)$ profiles at different points from the moving plate in the depth of the fluid presented in Figure 7 are good examples of the effect of magnitude of the ratio p/ν, i.e. the fluid elasticity

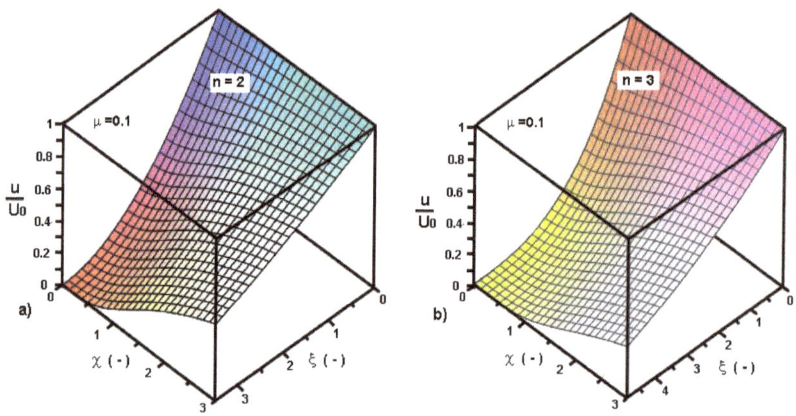

Fig. 6. The velocity profile is expressed by Eq. **??** as a function of two similarity variables: $\xi = y/\sqrt{\nu t}$ and $\chi = \sqrt{p/\nu t^{\mu}}$, a) $n = 2$; b) $n = 3$; Adapted from (Hristov, 2012a) by courtesy of *Thermal Science*

5.1.6 Some comments on the integral-balance solution of the Stokes's first problem

Briefly, we refer to the possibility of developing a post-solution analysis of the approximate closed-form solutions obtained by the integral balance method. The integral-balance method is a simple technique in contrast to the exact methods with cumbersome expressions. Beyond the correctness and the physical adequacy of both the constitutive relationships and the model, the main question arising in modelling after the development of the solution is: *How can the final solutions be used for calculations and analysis of functional relationships between the controlling parameters*? The integral-balance solution explicitly answers this question by clearly defining two dimensionless similarity variables $\xi = y/\sqrt{\nu t}$ and $De = \chi^2 = \sqrt{p/\nu t^{\mu}}$ that are responsible for the viscous and the elastic responses of the fluid to the step jump at the boundary, accordingly. Such findings, for instance are impossible with the exact series solutions (Hayat et al, 2000;Tan an Xu, 2002; Qi and Xu, 2009).

Therefore, the integral-balance solution provides approximate physical information about the interrelations of the effects modelled by different terms of the governing equation. Moreover, the integral balance solution directly shows that:

i) How the term containing De decays in time, thus shifting the solution towards the Newtonian problem controlled by the fluid kinematic viscosity only.

ii) How the weight of the term containing De depends on the fractional order μ?

iii) What is the suitable order of the approximate profile exponent and how does it correspond to the value of De. In this context, the classical $n = 2$ and $n = 3$, used in the simulations (Hristov, 2012a), are adequate because they correspond to two principle conditions: i) to minimize the L_2 of the approximate solution and ii) to match the fluid-like behavior of the viscoelastic medium with low Deborah numbers and elastic solution only (Hristov, 2012a)

5.2 Example 5: Transient Flow of a Generalized Second Grade Fluid Due to a Constant Surface Shear Stress

We will briefly outline the results developed by HBIM (Hristov, 2011c) when the boundary condition at $x = 0$ for the model 62 is defined as

$$\nu \frac{\partial u(0,t)}{\partial y} + \frac{\alpha_1}{\rho} D_t^{\mu} \frac{\partial u(0,t)}{\partial y} = -\frac{\sigma}{\rho} \quad , \quad y = 0 \quad , \quad t > 0 \tag{71}$$

and $u(y, 0) = 0$ for $y > 0$; $t > 0$

The boundary condition 71 refers to a semi-infinite space and a second grade viscoelastic fluid undergoing a transient motion due to tangential shear stress σ exerted at $y = 0$ (parallel to the x axis). That is, the velocity field is $u(y, t)$, where the axis y is normal to plate surface and u is the velocity component along the x axis. The assumed parabolic profile 20 subjected to the boundary condition 71 defines the surface velocity as

$$U_s = \frac{\delta}{n} \frac{\sigma}{\mu_f} \left[1 + \left(\frac{p}{\nu} \right) \frac{1}{(1-\mu)\Gamma(1-\mu)} t^{-\mu} \right]^{-1} . \tag{72}$$

The HBIM (FFA) solution of 62 with the boundary condition 72 defines the penetration depth (see for details in Hristov, 2011c) as $\delta_{HBIM}^{\sigma} = \sqrt{\nu t}\sqrt{n(n+1)}\sqrt{1 + \frac{p}{\nu} j_\mu t^{-\mu}} \overset{\text{def}}{=} \sqrt{1 + j_\mu De}$. The expression for the surface velocity can be presented as $U_s = (\sigma\sqrt{\nu t}/\mu_f)(F_n/nR_\mu)$; the group $\sigma\sqrt{\nu t}/\mu_f$ has a dimension of velocity $[m/s]$. Recall, the problem *has no characteristic time and length scales*. The term $\sqrt{\nu t}$ (the dimension is $[m]$) is like a length scale, as in any other semi-infinite diffusion problems, while σ/μ_f (the dimension is time $[s]$) can be used as a time scale. The minimization of the squared error function $E(y, t, \mu)$ leads to

$n \approx 1/\sqrt{2De}$ (Hristov, 2001c). The exact series solution (Hayat et al., 2000) is quite sensitive with respect to the number of terms in the time-dependent sums. Increase in the number of the terms beyond $m_{max} = 50$ and $k_{max} = 50$ does not affect the exact solution; examples are shown in Figure 7. Further, the exact solution is very sensitive with respect to the value of the fractional order $\mu < 0.5$ and the ratio p/ν (see Hristov, 2011c).

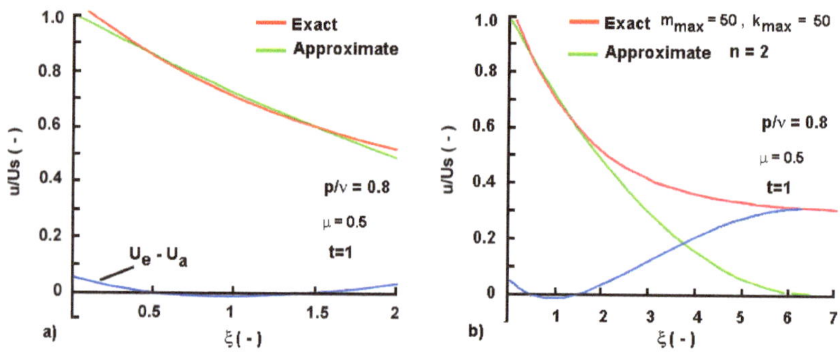

Fig. 7. Numerical tests with the exact solution and the approximate one with a fixed exponent. Adapted from Hristov (2011c). By courtesy of Praise Worthy Prize. U_e-exact solution; U_a-approximate solution; $(U_e - U_a)$-point-wise error

6 Final Comments and Results Outlines

This chapter was written with a single idea of collating results on the integral-balance approach applied to models involving time-fractional derivatives. This analysis could be considered as an attempt to strike the balance between a series of articles utilizing the method, and dissemination of the main achievements, as well as to show emerging problems.

The examples included in the chapter are solved step-by step and demonstrate how this should be done and what the accuracy of approximation is. The main problems are models with Riemann-Liouville derivative, but the case of subdiffusion equation clearly demonstrates that the method works with Caputo derivatives, too.

In addition to the first results developed by the simple Heat-balance integral (HBIM), this chapter collates recently developed results obtained by the Double-Integration method (DIM), working very effective with both Riemann-Liouville and Caputo derivatives.

The chapter evaluates some specific features of the integration technique termed Frozen-Front approach (FFA) used together with HBIM in the first publication on the integral-balance models on solutions of models with time-fractional derivatives. As a general outcome, FFA is practically equivalent to HBIM and in many cases the solutions of these integration techniques coincide.

All the solutions were performed by a general parabolic profile. This profile gives more freedom in the tuning of the approximate solution through search of the optimal exponent in contrast to the case when polynomial profiles (quadratic or cubic) can be used. In the latter cases, the order of the polynomial profile predetermines the accuracy of the approximation.

There are enough comments accompanying the solutions of the examples solved in this chapter, so it would be better to end the discussion and to give a freedom to the readers to estimate what is really the power of the method and what would be further developed on its basis.

I am grateful to all my colleagues reading the manuscript in advance and providing valuable comments.

Bibliography

Al-Rabtah,A., Ertürk,V.S.,& Momani, S. (2009). Solutions of a fractional oscillator by using differential transform method, Computers and Mathematics with Applications, 59, 1356-1362.

Ardelean, I., Farrher, G., & R. Kimmich, R. (2007). Effective diffusion in partially filled nanoscopic and microscopic pores, J. Optoelect. Adv. Mat., 9, 655–660.

Berkowitz, B., Scher, H. (2009). Exploring the nature of non-Fickian transport in laboratory experiments, Advances in Water Resources, 32, 750-755.

Caputo, M. (1969). *Elasticit´a e Dissipazione*, Bologna: Zanichelli.

Cattaneo, C. (1948). On the conduction of heat (in Italian), Atti Sem. Mat. Fis. Universit´a Modena, 3, 83–101.

Curtin, M. E, Pipkin, A.C. (1968) A general theory of heat conduction with finite wave speeds, *Archives of Rational Mathematical Analysis*, 31, 313-332.

Dehghan, M. (2006). Solution of a partial integro-differential equation arising from viscoelasticity, Int. J. Computer Math., 83, 123–129.

Derkach, S.R. (2009). Rheology on the Way from Dilute to Concentrated Emulsions, Int. Rev. Chem. Eng. 2, 465-472.

Diethelm, K., Ford, J.N., Freed, A.D. & Luchko,Y. (2005) Algorithms for the fractional calculus: a selection of numerical methods, Computer Methods in Applied Mechanics and Engineering, 194, 743–773.

Dokoumetzidis, A., Macheras, P. (2009) Fractional kinetics in drug absorption and disposition processes, J Pharmacokinetics and Pharmacodynamics, 36, 165–178.

Djordjevic,V.D., Atanackovic,T.M. (2008). Similarity solutions to nonlinear heat conduction and Burgers/Korteweg–deVries fractional equations, J. Comp. Appl. Math. 222, 701–714.

Ferreira, J.A., de Oliveira, P. (2008). Qualitative analysis of a delayed non-Fickian model, *Applicable* Analysis, 87, 873–886.

Ghorbani, A. (2008). Toward a new analytical method for solving nonlinear fractional differential equations, Computer Methods in Applied Mechanics and Engineering. 197, 4173-4179.

Goodman, T.R. (1958). The heat balance integral and its application to problems involving a change of phase, Transactions of ASME, 80, 335-342.
Goodwin, J.W., Hughes, R.W. (2008). Rheology for chemists: An Introduction, 2nd ed., Cambridge, UK: RSC Publishing.
Gorenflo, R., Mainardi, F. (1997). Fractional Calculus. Integral and differenmtial equations of fractional order. In Carpinteri, A.& Mainardi, F. (Eds.), Fractals and fractional calculus in continuum mechanics, 223-276. Wien: Springer-Verlag Wien.
Gorenflo, R. Loutschko, J., Luchko, Y. (2002). Computation of the Mittag-Leffler function and its derivatives, Fractional Calculus & Applied Analisys, 5, 491–518.
Joseph, D.D., Preciozi, L. (1989). Heat waves, Rev. Mod. Phys. 61, 41-73.
Hayat, T., S. Asghar S., & Siddiqui, A.M. (2000). Some unsteady unidirectional flows of a non-Newtonian fluid, Int. J. Eng. Sci, 38, 337-346.
He, J.H. (1998). Approximate analytical solution for seepage flow with fractional derivatives in porous media, Computer Methods in Applied Mechanics and Engineering, 167, 57-68.
Hristov, J. (2009). The heat-balance integral method by a parabolic profile with unspecified exponent: Analysis and benchmark exercises, Thermal Science, 13, 22-48.
Hristov, J. (2010a). Heat-Balance Integral to Fractional (Half-Time) Heat Diffusion Sub-Model, Thermal Science, 14, 291-316.
Hristov, J. (2010b). A Short-Distance Integral-Balance Solution to a Strong Subdiffusion Equation: Weak Power Law Profile, Int. Rev. Chem. Eng. 2, 555-563.
Hristov J. (2011a). Starting radial subdiffusion from a central point through a diverging medium (a sphere): Heat-balance integral method, Thermal Science, 15, S5-S20.
Hristov J. (2011b). Approximate Solutions to Fractional Subdiffusion Equations: The heat-balance integral method, The Europ. Phys. J.-Special Topics, 193, 229-243.
Hristov, J. (2011c). Transient Flow of a Generalized Second Grade Fluid Due to a Constant Surface Shear Stress: an approximate Integral-Balance Solution, Int. Rev. Chem. Eng., 3, 802-809.
Hristov, J. (2012a). Integral-Balance Solution to the Stokes' First Problem of a Viscoelastic Generalized Second Grade Fluid, *Thermal Science*, 16, 395-410.
Hristov, J. (2012b). The Heat-Balance Integral: 1. How to Calibrate the Parabolic Profile?, CR Mechanique, 340, 485-492.
Hristov, J. (2013). Integral-balance solutions to applied models involving time-fractional derivatives: a scope of the method and some results thereof, Commun. Frac. Calc., 4, 64-104.
Hristov J. (2015a). Diffusion Models with Weakly Singular Kernels in the Fading Memories: How the Integral-Balance Method can be Applied?, *Thermal Science, 19 (in press)* doi: 10.2298/TSCI130803151H
Hristov J. (2015b). Double integral-balance methods to the fractional subdiffusion equation: approximate solutions, optimization problems to be resolved and numerical simulations, J. Vibr. Control, in press
Korabel, N., Klages, R., Chechkin, A.V., Sokolov, I.M. & V. Yu Gonchar, V. Yu (2007). Fractal properties of anomalous diffusion in intermittent maps, Physical Review E, 75, 036213.
Kosztołowicz, T. (2008). Subdiffusion in a system with a thick membrane, J. Memb. Sci., 320,492–499.
Kosztolowicz, T., K. Dworecki, K. & Mrowczinsky,S. (2005). How to measure subdiffusion parameters, Phys. Rev. Lett. 94 (2005) 170602.
Langlands, T. A. M. (2006). Solution of a modified fractional diffusion equation, Physica A 367, 136–144.
Li, Y., Farrher G. R. Kimmich, R. (2006). Sub- and superdiffusion molecular displacement laws in disordered porous media probed by nuclear magnetic resonance, Phys. Rev. E. 74, 066309.

Li, B., Wang, J. (2003). Anomalous Heat Conduction and Anomalous Diffusion in One Dimensional Systems, Physical Review Letters, 91, 044301-1- 4.
Luchko, Y., Srivastava, H.M. (1995). The exact solution of certain differential equations of fractional order by using operational calculus, Computers and Math. Application. 29, 73-85.
Kang, J., Xu, M. (2009). Exact solutions for unsteady unidirectional flows of a generalized second-order fluid through a rectangular conduit, Acta Mech. Sin. 25, 181–186.
Mainardi, F. (1966). The Fundamental Solutions for the Fractional Diffusion-Wave Equation, Appl. Math. Lett., 9, 23-28.
Metzler, R., Klafter, J. (2000). The random walk's guide to anomalous diffusion: A fractional dynamics approach, Physics Report. 39,1–77.
Metzler, R., Klafter, J. (2004). The restaurant at the end of the random walk: Recent developments in the description of anomalous transport by fractional dynamics, Journal of Physics A. 37, 161–208.
Myers, T., (2009). Optimizing the exponent in the hear balance and refined integral methods, Int. Comm. Heat Mass Transfer, 44, 143-147.
Momoniat, E. (2008). A point source solution for unidirectional flow of a viscoelastic fluid, Phys. Lett. A 372, 4041-4044.
Myers, T.G. (2010). Optimal exponent heat balance and refined integral methods applied to Stefan problem, Int. J. Heat Mass Transfer, 53, 1119-1127.
Nakagawa, J., Sakamoto, K., & Yamamoto, M (2010). Overview to mathematical analysis for fractional diffusion equations–new mathematical aspects motivated by industrial collaboration, Journal of Math-for-Industry, 2, 99-108.
Narahari Achar, B.N., Hanneken, J.W. (2004). Fractional radial diffusion in a cylinder, Journal of Molecular Liquids. 114, 147–151.
Naraim, A., D.D. Joseph, D.D. (1983) Remarks about the interpretations of impulsive experiments in shear flows of viscoplastic liquids, Rheologica Acta, 22, 528-538.
Nigmatullin, R.R. (1986). The realization of the generalized transfer equation in a medium with fractal geometry, Physica Status Solidi (B) Basic Research 133, 425–430.
Oldham , K.B. & Spanier, J. (1974). The Fractional calculus, New York: Academic Press.
Podlubny, I. (1999). Fractional Diffrential EquationsSan Diego-Boston-New York –London-Tokyo-Toronto: Academic Press.
Pskhu, A.V. (2003). Solution of Boundary Value Problems for the Fractional Diffusion Equation by the Green Function Method, Differential Equations (Russia), 39, 1509–1513.
Qi, H., Xu, M. (2009). Some unsteady unidirectional flows of a generalized Oldroyd-B fluid with fractional derivative, Appl. Math. Model. 33, 4184–4191.
Schmitt, C., Henni, A.H. & Cloutier, G. (2011). Characterization of blood clot viscoelasticity by dynamic ultrasound elastography and modeling of the rheological behavior, J. Biomech. 44, 622-629.
Siginer, D.A., M.F. Letelier, M.F. (2011). Laminar flow of non-linear viscoelastic fluids in straight tubes of arbitrary contour, Int. J. Heat Mass Transfer. 54, 2188-2202.
Tan, W., Xu, M. (2002). Plane surface suddenly set in motion in a viscoelastic fluid with fractional Maxwell model, Acta Mech. Sin., 18, 342-349.
Zilz, J., Schäfer, C., Wagner, C., Poole, R.J., Alves, M.A. & Lindner, A. (2014). Serpentine channels: micro-rheometers for fluid relaxation times, Lab Chip, 14, 351-358

Bashir Ahmad*, Ahmed Alsaedi*, and Hana Al-Hutami*

A Study of Sequential Fractional q-integro-difference Equations with Perturbed Anti-periodic Boundary Conditions

Abstract: This chapter is devoted to the study of sequential fractional q-integro-difference equations with perturbed fractional q-difference anti-periodic boundary conditions. Existence results for the given problem are established by applying Krasnoselskii's fixed point theorem, Leray-Schauder nonlinear alternative for single valued maps and Banach's contraction mapping principle. Correction terms arising due to the perturbation in the anti-periodic boundary data are highlighted. An illustrative example is also presented.

Keywords: q-difference; fractional integro-differential equations; Sequential; boundary conditions; existence; fixed point

1 Introduction

Nonlinear boundary value problems of fractional differential equations have received considerable attention in the last few decades. One can easily find a variety of results ranging from theoretical analyses to asymptotic behavior and numerical methods for fractional equations in the literature on the topic. The interest in the subject has been mainly due to the extensive applications of fractional calculus mathematical modelling of several real world phenomena occurring in physical and technical sciences, see, for example (Baleanu, 2012; Kilbas 2006; Podlubny, 1999; Sabatier, 2007). An important feature of a fractional order differential operator, distinguishing it from an integer-order differential operator, is that it is nonlocal in nature. This means that the future state of a dynamical system or process based on a fractional operator depends on its current state as well as its past states. Thus, differential equations of arbitrary order are capable of describing memory and hereditary properties of some important and useful materials and processes. This feature has fascinated many researchers and they have shifted their focus to fractional order models from the classical integer-order models. For some recent work on the topic, we refer, for instance, to (Ahmad, 2011; Ahmad, 2014a; Cabada, 2012; Graef, 2014; Henderson, 2013; Kirane, 2014; Liu, 2013; Liu, 2014; O'Regan, 2013; Punzo, 2014; Sudsutad, 2012; Yang, 2014; Zhai, 2014) and the references therein.

***Corresponding Author: Bashir Ahmad, Ahmed Alsaedi, Hana Al-Hutami:** Department of Mathematics, Faculty of Science, King Abdulaziz University, P.O. Box 80203, Jeddah 21589, Saudi Arabia E-mail: bashirahmad-qau@yahoo.com (B. Ahmad), aalsaedi@hotmail.com (A. Alsaedi), hanno.1407@hotmail.com (H. Al-Hutami)

Motivated by the popularity of fractional differential equations, q-difference equations of fractional-order are also attracting a considerable attention. Fractional q-difference equations may be regarded as the fractional analogue of q-difference equations. For earlier work on the topic, we refer to (Agarwal, 1969; Al-Salam, 1966-67), while some recent development of fractional q-difference equations, for instance, can be found in (Ahmad, 2012; Ahmad, 2014b; Ferreira, 2011; Goodrich, 2011; Graef, 2012; Li, 2013). The basic concepts of q-fractional calculus can be found in a recent text (Annaby, 2012).

In this chapter, we consider a boundary value problem of fractional q-integro-difference equations involving two fractional orders with perturbed fractional q-difference anti-periodic conditions given by

$$^cD_q^{\beta}(^cD_q^{\gamma} + \lambda)x(t) = pf(t, x(t)) + kI_q^{\xi}g(t, x(t)), \quad 0 \leq t \leq 1, \; 0 < q < 1, \tag{1}$$

$$x(a) = -x(1), \quad {}^cD_q^{\gamma}x(a) = -{}^cD_q^{\gamma}x(1), \quad 0 < a \ll 1 \tag{2}$$

where $^cD_q^{\beta}$ and $^cD_q^{\gamma}$ denote the fractional q-derivative of the Caputo type, $0 < \beta, \gamma \leq 1$, $I_q^{\xi}(.)$ denotes Riemann-Liouville integral with $0 < \xi < 1$, f, g are given continuous functions, $\lambda \neq 0$ and p, k are real constants.
The chapter is organized as follow. Section 2 deals with some general concepts and results on fractional q-calculus, and an auxiliary lemma for the linear variant of the problem (1)-(2). In Section 3, we show some existence results for the problem (1)-(2) by means of some classical fixed point theorems. The paper concludes with an illustrative example.

2 Preliminaries

First of all, we recall the notations and terminology for q-fractional calculus (Agarwal, 1969; Rajkovic, 2007).

For a real parameter $q \in \mathbb{R}^+ \setminus \{1\}$, a q-real number denoted by $[a]_q$ is defined by

$$[a]_q = \frac{1 - q^a}{1 - q}, \quad a \in \mathbb{R}.$$

The q-analogue of the Pochhammer symbol (q-shifted factorial) is defined as

$$(a; q)_0 = 1, \quad (a; q)_k = \prod_{i=0}^{k-1}(1 - aq^i), \; k \in \mathbb{N} \cup \{\infty\}.$$

The q-analogue of the exponent $(x - y)^k$ is

$$(x - y)^{(0)} = 1, \quad (x - y)^{(k)} = \prod_{j=0}^{k-1}(x - yq^j), \; k \in \mathbb{N}, \; x, y \in \mathbb{R}.$$

The q-gamma function $\Gamma_q(y)$ is defined as

$$\Gamma_q(y) = \frac{(1-q)^{(y-1)}}{(1-q)^{y-1}},$$

where $y \in \mathbb{R} \setminus \{0, -1, -2, \ldots\}$. Observe that $\Gamma_q(y+1) = [y]_q \Gamma_q(y)$.

Definition 10. *(Agarwal, 1969) Let f be a function defined on $[0, 1]$. The fractional q-integral of the Riemann-Liouville type of order $\beta \geq 0$ is $(I_q^0 f)(t) = f(t)$ and*

$$I_q^\beta f(t) := \int_0^t \frac{(t-qs)^{(\beta-1)}}{\Gamma_q(\beta)} f(s) d_q s = t^\beta (1-q)^\beta \sum_{k=0}^{\infty} q^k \frac{(q^\beta; q)_k}{(q; q)_k} f(tq^k), \quad \beta > 0, \quad t \in [0, 1].$$

Observe that $\beta = 1$ in the Definition 10 yields the q-integral

$$I_q f(t) := \int_0^t f(s) d_q s = t(1-q) \sum_{k=0}^{\infty} q^k f(tq^k).$$

For more details on the q and fractional q-integrals, see Section 1.3 and Section 4.2 respectively in (Annaby, 2012).

Remark 11. *The q-fractional integration possesses the semigroup property (Proposition 4.3 (Annaby, 2012):*

$$I_q^\gamma I_q^\beta f(t) = I_q^{\beta+\gamma} f(t); \; \gamma, \beta \in \mathbb{R}^+.$$

Further, it has been shown in Lemma 6 of (Rajkovic, 2007) that

$$I_q^\beta (x)^{(\sigma)} = \frac{\Gamma_q(\sigma+1)}{\Gamma_q(\beta+\sigma+1)} (x)^{(\beta+\sigma)}, \; 0 < x < a, \beta \in \mathbb{R}^+, \sigma \in (-1, \infty).$$

Before giving the definition of fractional q-derivative, we recall the concept of q-derivative.
We know that the q-derivative of a function $f(t)$ is defined as

$$(D_q f)(t) = \frac{f(t) - f(qt)}{t - qt}, \; t \neq 0, \; (D_q f)(0) = \lim_{t \to 0} (D_q f)(t).$$

Furthermore,

$$D_q^0 f = f, \; D_q^n f = D_q(D_q^{n-1} f), \; n = 1, 2, 3, \ldots. \tag{3}$$

Definition 12. *(Annaby, 2012) The Caputo fractional q-derivative of order $\beta > 0$ is defined by*

$${}^c D_q^\beta f(t) = I_q^{\lceil \beta \rceil - \beta} D_q^{\lceil \beta \rceil} f(t),$$

where $\lceil \beta \rceil$ is the smallest integer greater than or equal to β.

Next we recall some properties involving Riemann-Liouville q-fractional integral and Caputo fractional q-derivative (Theorem 5.2 (Annaby, 2012)).

$$I_q^{\beta}\,{}^cD_q^{\beta}f(t) = f(t) - \sum_{k=0}^{\lceil\beta\rceil-1} \frac{t^k}{\Gamma_q(k+1)}(D_q^k f)(0^+),\ \forall\, t \in (0, a],\ \beta > 0; \tag{4}$$

$${}^cD_q^{\beta}\, I_q^{\beta} f(t) = f(t), \quad \forall\, t \in (0, a],\ \beta > 0. \tag{5}$$

In order to define the solution for the problem (1)-(2), we need the following lemma.

Lemma 1. *Let $h \in C([0,1],\mathbb{R})$ be a given function. Then the unique solution of the boundary value problem*

$$\begin{cases} {}^cD_q^{\beta}({}^cD_q^{\gamma} + \lambda)x(t) = h(t), \ \ 0 \le t \le 1,\ 0 < q < 1, \\ x(a) = -x(1),\ \ {}^cD_q^{\gamma}x(a) = -{}^cD_q^{\gamma}x(1). \end{cases} \tag{6}$$

is given by

$$\begin{aligned} x(t) &= \int_0^t \frac{(t-qu)^{(\gamma-1)}}{\Gamma_q(\gamma)}\Big(I_q^{\beta}h(u) - \lambda x(u)\Big)d_q u \\ &+\Big(\frac{(1-2t^{\gamma})}{4\Gamma_q(\gamma+1)} + \frac{a^{\gamma}}{4\Gamma_q(\gamma+1)}\Big)\Big(I_q^{\beta}h(a) + I_q^{\beta}h(1)\Big) \\ &-\frac{1}{2}\Big(\int_0^a \frac{(a-qu)^{(\gamma-1)}}{\Gamma_q(\gamma)}(I_q^{\beta}h(u) - \lambda x(u))\,d_q u + \int_0^1 \frac{(1-qu)^{(\gamma-1)}}{\Gamma_q(\gamma)}(I_q^{\beta}h(u) - \lambda x(u))\,d_q u\Big). \end{aligned} \tag{7}$$

Proof. Applying the operator I_q^{β} on both sides of fractional q-difference equation in (6), we get

$$({}^cD_q^{\gamma} + \lambda)x(t) = I_q^{\beta}h(t) - c_0,$$

which can be written as

$${}^cD_q^{\gamma}x(t) = I_q^{\beta}h(t) - \lambda x(t) - c_0. \tag{8}$$

Now, applying the operator I_q^{γ} on both sides of (8), we get

$$x(t) = \int_0^t \frac{(t-qu)^{(\gamma-1)}}{\Gamma_q(\gamma)}\Big(I_q^{\beta}h(u) - \lambda x(u)\Big)d_q u - \frac{t^{\gamma}}{\Gamma_q(\gamma+1)}c_0 - c_1,\ \ t \in [0,1]. \tag{9}$$

Using the boundary conditions given by (6) in (9), we find that

$$c_0 = \frac{1}{2}\Big(I_q^{\beta}h(a) + I_q^{\beta}h(1)\Big),$$

$$c_1 = \frac{1}{2}\Big(\int_0^a \frac{(a-qu)^{(\gamma-1)}}{\Gamma_q(\gamma)}\big(I_q^\beta h(u) - \lambda x(u)\big)d_qu + \int_0^1 \frac{(1-qu)^{(\gamma-1)}}{\Gamma_q(\gamma)}\big(I_q^\beta h(u) - \lambda x(u)\big)d_qu\Big)$$
$$-\frac{(1+a^\gamma)}{4\Gamma_q(\gamma+1)}\Big(I_q^\beta h(a) + I_q^\beta h(1)\Big).$$

Substituting the values of c_0 and c_1 in (9) yields the solution (7). This completes the proof. □

3 Main Results

Let $\mathcal{C} = C([0,1],\mathbb{R})$ denote the Banach space of all continuous functions from $[0,1]$ into $\mathbb{R}$ endowed with the usual norm defined by $\|x\| = \sup\{|x(t)|, t \in [0,1]\}$.
In view of Lemma 1, we define an operator $\mathcal{G} : \mathcal{C} \to \mathcal{C}$ as

$$\begin{aligned}(\mathcal{G}x)(t) &= \int_0^t \frac{(t-qu)^{(\gamma-1)}}{\Gamma_q(\gamma)}\Big(p\int_0^u \frac{(u-qm)^{(\beta-1)}}{\Gamma_q(\beta)} f(m,x(m))d_qm \\ &+k\int_0^u \frac{(u-qm)^{(\beta+\xi-1)}}{\Gamma_q(\beta+\xi)} g(m,x(m))d_qm - \lambda x(u)\Big)d_qu \\ &+\Big(\frac{(1-2t^\gamma)}{4\Gamma_q(\gamma+1)} + \frac{a^\gamma}{4\Gamma_q(\gamma+1)}\Big)\Big(p\int_0^a \frac{(a-qu)^{(\beta-1)}}{\Gamma_q(\beta)} f(u,x(u))d_qu \\ &+k\int_0^a \frac{(a-qu)^{(\beta+\xi-1)}}{\Gamma_q(\beta+\xi)} g(u,x(u))d_qu\Big)\end{aligned} \tag{10}$$

$$+\Big(\frac{(1-2t^\gamma)}{4\Gamma_q(\gamma+1)} + \frac{a^\gamma}{4\Gamma_q(\gamma+1)}\Big)\Big(p\int_0^1 \frac{(1-qu)^{(\beta-1)}}{\Gamma_q(\beta)} f(u,x(u))d_qu$$
$$+k\int_0^1 \frac{(1-qu)^{(\beta+\xi-1)}}{\Gamma_q(\beta+\xi)} g(u,x(u))d_qu\Big)$$

$$-\frac{1}{2}\int_0^a \frac{(a-qu)^{(\gamma-1)}}{\Gamma_q(\gamma)}\Big(p\int_0^u \frac{(u-qm)^{(\beta-1)}}{\Gamma_q(\beta)} f(m,x(m))d_qm$$
$$+k\int_0^u \frac{(u-qm)^{(\beta+\xi-1)}}{\Gamma_q(\beta+\xi)} g(m,x(m))d_qm - \lambda x(u)\Big)d_qu$$

$$-\frac{1}{2}\int_0^1 \frac{(1-qu)^{(\gamma-1)}}{\Gamma_q(\gamma)}\Big(p\int_0^u \frac{(u-qm)^{(\beta-1)}}{\Gamma_q(\beta)}f(m,x(m))d_qm$$
$$+k\int_0^u \frac{(u-qm)^{(\beta+\xi-1)}}{\Gamma_q(\beta+\xi)}g(m,x(m))d_qm-\lambda x(u)\Big)d_qu.$$

and note that the problem (1)-(2) has solutions only if the operator equation $x = \mathcal{G}x$ has fixed points.

It is worthwhile to note that the terms arising in the operator (10) due to the perturbation in the anti-periodic boundary data are

$$\Big(\frac{(1-2t^{\gamma})}{4\Gamma_q(\gamma+1)}+\frac{a^{\gamma}}{4\Gamma_q(\gamma+1)}\Big)\Big(p\int_0^a \frac{(a-qu)^{(\beta-1)}}{\Gamma_q(\beta)}f(u,x(u))d_qu$$
$$+k\int_0^a \frac{(a-qu)^{(\beta+\xi-1)}}{\Gamma_q(\beta+\xi)}g(u,x(u))d_qu\Big);$$
$$\frac{a^{\gamma}}{4\Gamma_q(\gamma+1)}\Big(p\int_0^1 \frac{(1-qu)^{(\beta-1)}}{\Gamma_q(\beta)}f(u,x(u))d_qu+k\int_0^1 \frac{(1-qu)^{(\beta+\xi-1)}}{\Gamma_q(\beta+\xi)}g(u,x(u))d_qu\Big);$$

and

$$-\frac{1}{2}\int_0^a \frac{(a-qu)^{(\gamma-1)}}{\Gamma_q(\gamma)}\Big(p\int_0^u \frac{(u-qm)^{(\beta-1)}}{\Gamma_q(\beta)}f(m,x(m))d_qm$$
$$+k\int_0^u \frac{(u-qm)^{(\beta+\xi-1)}}{\Gamma_q(\beta+\xi)}g(m,x(m))d_qm-\lambda x(u)\Big)d_qu.$$

In the subsequent sections, we assume that

(A_1) $f, g : [0,1]\times\mathbb{R}\to\mathbb{R}$ are continuous functions such that $|f(t,x)-f(t,y)|\le L_1|x-y|$ and $|g(t,x)-g(t,y)|\le L_2|x-y|$, $\forall t\in[0,1]$, $L_1,\ L_2>0,\ x,y\in\mathbb{R}$.

(A_2) there exist $\vartheta_1,\vartheta_2\in C([0,1],\mathbb{R}^+)$ with $|f(t,x)|\le\vartheta_1(t)$, $|g(t,x)|\le\vartheta_2(t)$, $\forall(t,x)\in[0,1]\times\mathbb{R}$, where $\|\vartheta_i\|=\sup_{t\in[0,1]}|\vartheta_i(t)|$, $i=1,2$.

For the sake of computational convenience, let us set the following notations:

$$\begin{aligned}\varpi_1&=\frac{(3+a^{\beta+\gamma})}{2\Gamma_q(\beta+\gamma+1)}+\frac{M(1+a^{\beta})}{4\Gamma_q(\gamma+1)\Gamma_q(\beta+1)},\\ \varpi_2&=\frac{(3+a^{\beta+\xi+\gamma})}{2\Gamma_q(\beta+\xi+\gamma+1)}+\frac{M(1+a^{\beta+\xi})}{4\Gamma_q(\gamma+1)\Gamma_q(\beta+\xi+1)},\\ \varpi_3&=\frac{(3+a^{\gamma})}{2\Gamma_q(\gamma+1)},\quad M=\max_{t\in[0,1]}\left|1-2t^{\gamma}+a^{\gamma}\right|.\end{aligned}\tag{11}$$

$$\Omega = L\Big[|p|\Big(\frac{M(1+a^{\beta})}{4\Gamma_q(\gamma+1)\Gamma_q(\beta+1)} + \frac{(1+a^{\beta+\gamma})}{2\Gamma_q(\beta+\gamma+1)}\Big) + |k|\Big(\frac{M(1+a^{\beta+\xi})}{4\Gamma_q(\gamma+1)\Gamma_q(\beta+\xi+1)} + \frac{(1+a^{\beta+\xi+\gamma})}{2\Gamma_q(\beta+\xi+\gamma+1)}\Big)\Big] + \frac{|\lambda|(a^{\gamma}+1)}{2\Gamma_q(\gamma+1)}, \tag{12}$$

Our first existence result is based on Krasnoselskii's fixed point theorem.

Lemma 2. *(Krasnoselskii (Smart, 1980)). Let Y be a closed, convex, bounded and nonempty subset of a Banach space X. Let Q_1, Q_2 be the operators such that*
(i) $Q_1x + Q_2y \in Y$ whenever $x, y \in Y$;
(ii) Q_1 is compact and continuous and
(iii) Q_2 is a contraction mapping;
Then there exists $z \in Y$ such that $z = Q_1z + Q_2z$.

Theorem 13. *Let $f, g : [0,1] \times \mathbb{R} \to \mathbb{R}$ be continuous functions satisfying $(A_1) - (A_2)$. Furthermore $\Omega < 1$, where Ω is given by (12)where $L = \max\{L_1, L_2\}$. Then the problem (1)-(2) has at least one solution on $[0,1]$.*

Proof. Consider the set $\widetilde{B}_r = \{x \in \mathcal{C} : \|x\| \le r\}$, where r is given by

$$r \ \ge \ \frac{|p|\|\vartheta_1\|\varpi_1 + |k|\|\vartheta_2\|\varpi_2}{1 - |\lambda|\varpi_3}$$

where $\varpi_1, \varpi_2, \varpi_3$ are given by (11). Define operators $\mathcal{G}_1$ and $\mathcal{G}_2$ on $\widetilde{B}_r$ as

$$(\mathcal{G}_1x)(t) = \int_0^t \frac{(t-qu)^{(\gamma-1)}}{\Gamma_q(\gamma)}\Big(p\int_0^u \frac{(u-qm)^{(\beta-1)}}{\Gamma_q(\beta)} f(m, x(m))d_qm + k\int_0^u \frac{(u-qm)^{(\beta+\xi-1)}}{\Gamma_q(\beta+\xi)} g(m, x(m))d_qm - \lambda x(u)\Big)d_qu, \quad t \in [0,1],$$

$$\begin{aligned}(\mathcal{G}_2x)(t) = {} & \Big(\frac{(1-2t^{\gamma})}{4\Gamma_q(\gamma+1)} + \frac{a^{\gamma}}{4\Gamma_q(\gamma+1)}\Big)\Big(p\int_0^a \frac{(a-qu)^{(\beta-1)}}{\Gamma_q(\beta)} f(u, x(u))d_qu \\ & + k\int_0^a \frac{(a-qu)^{(\beta+\xi-1)}}{\Gamma_q(\beta+\xi)} g(u, x(u))d_qu\Big) \\ & + \Big(\frac{(1-2t^{\gamma})}{4\Gamma_q(\gamma+1)} + \frac{a^{\gamma}}{4\Gamma_q(\gamma+1)}\Big)\Big(p\int_0^1 \frac{(1-qu)^{(\beta-1)}}{\Gamma_q(\beta)} f(u, x(u))d_qu \\ & + k\int_0^1 \frac{(1-qu)^{(\beta+\xi-1)}}{\Gamma_q(\beta+\xi)} g(u, x(u))d_qu\Big)\end{aligned}$$

$$-\frac{1}{2}\int_0^a \frac{(a-qu)^{(\gamma-1)}}{\Gamma_q(\gamma)}\Big(p\int_0^u \frac{(u-qm)^{(\beta-1)}}{\Gamma_q(\beta)} f(m,x(m))d_qm$$
$$+k\int_0^u \frac{(u-qm)^{(\beta+\xi-1)}}{\Gamma_q(\beta+\xi)} g(m,x(m))d_qm-\lambda x(u)\Big)d_qu$$
$$-\frac{1}{2}\int_0^1 \frac{(1-qu)^{(\gamma-1)}}{\Gamma_q(\gamma)}\Big(p\int_0^u \frac{(u-qm)^{(\beta-1)}}{\Gamma_q(\beta)} f(m,x(m))d_qm$$
$$+k\int_0^u \frac{(u-qm)^{(\beta+\xi-1)}}{\Gamma_q(\beta+\xi)} g(m,x(m))d_qm-\lambda x(u)\Big)d_qu, \quad t\in[0,1].$$

For $x, y \in \widetilde{B}_r$, we find that

$$\|\mathcal{G}_1x+\mathcal{G}_2y\| \le |p|\|\vartheta_1\|\varpi_1+|k|\|\vartheta_2\|\varpi_2+|\lambda|r\varpi_3\le r.$$

Thus, $\mathcal{G}_1x+\mathcal{G}_2y\in\widetilde{B}_r$. Continuity of f and g imply that the operator $\mathcal{G}_1$ is continuous. Also, $\mathcal{G}_1$ is uniformly bounded on $\widetilde{B}_r$ as

$$\|\mathcal{G}_1x\|\le \frac{|p|\|\vartheta_1\|}{\Gamma_q(\beta+\gamma+1)}+\frac{|k|\|\vartheta_2\|}{\Gamma_q(\beta+\xi+\gamma+1)}+\frac{|\lambda|r}{\Gamma_q(\gamma+1)}.$$

Now, we prove the compactness of the operator $\mathcal{G}_1$. In view of (A_1), we define

$$\sup_{(t,x)\in[0,1]\times\widetilde{B}_r}|f(t,x)|=f_1, \qquad \sup_{(t,x)\in[0,1]\times\widetilde{B}_r}|g(t,x)|=g_1.$$

Consequently, we have

$$\|(\mathcal{G}_1x)(t_2)-(\mathcal{G}_1x)(t_1)\|$$
$$\le \int_0^{t_1}\frac{(t_2-qu)^{(\gamma-1)}-(t_1-qu)^{(\gamma-1)}}{\Gamma_q(\gamma)}\Big(|p|f_1\int_0^u\frac{(u-qm)^{(\beta-1)}}{\Gamma_q(\beta)}d_qm$$
$$+\ |k|g_1\int_0^u\frac{(u-qm)^{(\beta+\xi-1)}}{\Gamma_q(\beta+\xi)}d_qm+|\lambda|r\Big)d_qu$$
$$+\int_{t_1}^{t_2}\frac{(t_2-qu)^{(\gamma-1)}}{\Gamma_q(\gamma)}\Big(|p|f_1\int_0^u\frac{(u-qm)^{(\beta-1)}}{\Gamma_q(\beta)}d_qm$$
$$+\ |k|g_1\int_0^u\frac{(u-qm)^{(\beta+\xi-1)}}{\Gamma_q(\beta+\xi)}d_qm+|\lambda|r\Big)d_qu$$

which is independent of x and tends to zero as $t_2\to t_1$. Thus, $\mathcal{G}_1$ is relatively compact on $\widetilde{B}_r$. Hence, by the Arzelá-Ascoli Theorem, $\mathcal{G}_1$ is compact on $\widetilde{B}_r$. Now, we shall show

that $\mathcal{G}_2$ is a contraction.
From (A_1) and for $x, y \in \widetilde{B}_r$, we have

$$
\begin{aligned}
&\|\mathcal{G}_2 x - \mathcal{G}_2 y\| \\
&\le \sup_{t\in[0,1]} \Bigg\{ \frac{M}{4\Gamma_q(\gamma+1)} \Bigg(|p| \int_0^a \frac{(a-qu)^{(\beta-1)}}{\Gamma_q(\beta)} \Big| f(u,x(u)) - f(u,y(u)) \Big| d_q u \\
&+ |k| \int_0^a \frac{(a-qu)^{(\beta+\xi-1)}}{\Gamma_q(\beta+\xi)} \Big| g(u,x(u)) - g(u,y(u)) \Big| d_q u \Bigg) \\
&+ \frac{M}{4\Gamma_q(\gamma+1)} \Bigg(|p| \int_0^1 \frac{(1-qu)^{(\beta-1)}}{\Gamma_q(\beta)} \Big| f(u,x(u)) - f(u,y(u)) \Big| d_q u \\
&+ |k| \int_0^1 \frac{(1-qu)^{(\beta+\xi-1)}}{\Gamma_q(\beta+\xi)} \Big| g(u,x(u)) - g(u,y(u)) \Big| d_q u \Bigg) \\
&+ \frac{1}{2} \int_0^a \frac{(a-qu)^{(\gamma-1)}}{\Gamma_q(\gamma)} \Bigg(|p| \int_0^u \frac{(u-qm)^{(\beta-1)}}{\Gamma_q(\beta)} \Big| f(m,x(m)) - f(m,y(m)) \Big| d_q m \\
&+ |k| \int_0^u \frac{(u-qm)^{(\beta+\xi-1)}}{\Gamma_q(\beta+\xi)} \Big| g(m,x(m)) - g(m,y(m)) \Big| d_q m + |\lambda| |x(u) - y(u)| \Bigg) d_q u \\
&+ \frac{1}{2} \int_0^1 \frac{(1-qu)^{(\gamma-1)}}{\Gamma_q(\gamma)} \Bigg(|p| \int_0^u \frac{(u-qm)^{(\beta-1)}}{\Gamma_q(\beta)} \Big| f(m,x(m)) - f(m,y(m)) \Big| d_q m \\
&+ |k| \int_0^u \frac{(u-qm)^{(\beta+\xi-1)}}{\Gamma_q(\beta+\xi)} \Big| g(m,x(m)) - g(m,y(m)) \Big| d_q m + |\lambda| |x(u) - y(u)| \Bigg) d_q u \Bigg\} \\
&\le \sup_{t\in[0,1]} \Bigg\{ \frac{M}{4\Gamma_q(\gamma+1)} \Bigg(|p| \int_0^a \frac{(a-qu)^{(\beta-1)}}{\Gamma_q(\beta)} L_1 |x(u) - y(u)| d_q u \\
&+ |k| \int_0^a \frac{(a-qu)^{(\beta+\xi-1)}}{\Gamma_q(\beta+\xi)} L_2 |x(u) - y(u)| d_q u \Bigg) \\
&+ \frac{M}{4\Gamma_q(\gamma+1)} \Bigg(|p| \int_0^1 \frac{(1-qu)^{(\beta-1)}}{\Gamma_q(\beta)} L_1 |x(u) - y(u)| d_q u \\
&+ |k| \int_0^1 \frac{(1-qu)^{(\beta+\xi-1)}}{\Gamma_q(\beta+\xi)} L_2 |x(u) - y(u)| d_q u \Bigg)
\end{aligned}
$$

$$+\frac{1}{2}\int_0^a \frac{(a-qu)^{(\gamma-1)}}{\Gamma_q(\gamma)}\Big(|p|\int_0^u \frac{(u-qm)^{(\beta-1)}}{\Gamma_q(\beta)}L_1|x(m)-y(m)|d_qm$$

$$+|k|\int_0^u \frac{(u-qm)^{(\beta+\xi-1)}}{\Gamma_q(\beta+\xi)}L_2|x(m)-y(m)|d_qm+|\lambda||x(u)-y(u)|\Big)d_qu$$

$$+\frac{1}{2}\int_0^1 \frac{(1-qu)^{(\gamma-1)}}{\Gamma_q(\gamma)}\Big(|p|\int_0^u \frac{(u-qm)^{(\beta-1)}}{\Gamma_q(\beta)}L_1|x(m)-y(m)|d_qm$$

$$+|k|\int_0^u \frac{(u-qm)^{(\beta+\xi-1)}}{\Gamma_q(\beta+\xi)}L_2|x(m)-y(m)|d_qm+|\lambda||x(u)-y(u)|\Big)d_qu\Big\}$$

$$\leq\Bigg[L\Big[|p|\Big(\frac{M(1+a^{\beta})}{4\Gamma_q(\gamma+1)\Gamma_q(\beta+1)}+\frac{(1+a^{\beta+\gamma})}{2\Gamma_q(\beta+\gamma+1)}\Big)+$$

$$|k|\Big(\frac{M(1+a^{\beta+\xi})}{4\Gamma_q(\gamma+1)\Gamma_q(\beta+\xi+1)}+\frac{(1+a^{\beta+\xi+\gamma})}{2\Gamma_q(\beta+\xi+\gamma+1)}\Big)\Big]+\frac{|\lambda|(a^{\gamma}+1)}{2\Gamma_q(\gamma+1)}\Bigg]\|x-y\|=$$

$$=\Omega\|x-y\|,$$

where we have used (12). Since $\Omega<1$ by our assumption, therefore $\mathcal{G}_2$ is a contraction mapping. Thus all the assumptions of Lemma 3.1 are satisfied. So, by the conclusion of Lemma 3.1, the problem (1) – (2) has at least one solution on $[0, 1]$. □

The second existence result is based on Leray-Schauder Alternative.

Lemma 3. *(Nonlinear alternative for single valued maps (Granas, 2003)). Let E be a Banach space, C a closed, convex subset of E, V an open subset of C and $0\in V$. Suppose that $\mathcal{G}:\overline{V}\to C$ is a continuous, compact (that is, $\mathcal{G}(\overline{V})$ is a relatively compact subset of C) map. Then either*
(i) $\mathcal{G}$ has a fixed point in $\overline{V}$, or
(ii) there is a $x\in\partial V$ (the boundary of V in C) and $\mu\in(0,1)$ with $x=\mu\mathcal{G}(x)$.

Theorem 14. *Let $f, g : [0, 1]\times\mathbb{R}\to\mathbb{R}$ be continuous functions and the following assumptions hold:*
(A_3)there exist functions $v_1, v_2\in C([0,1],\mathbb{R}^+)$, and nondecreasing functions $\psi_1,\psi_2:\mathbb{R}^+\to\mathbb{R}^+$ such that $|f(t,x)|\leq v_1(t)\psi_1(\|x\|)$, $|g(t,x)|\leq v_2(t)\psi_2(\|x\|)$, $\forall(t,x)\in[0,1]\times\mathbb{R}$.
(A_4)There exists a constant $\kappa>0$ such that

$$\kappa>\frac{|p|\|v_1\|\psi_1(\kappa)\varpi_1+|k|\|v_2\|\psi_2(\kappa)\varpi_2}{1-|\lambda|\varpi_3}.$$

where $|\lambda|\neq\frac{1}{\varpi_3}$.

Then the boundary value problem (1) – (2) *has at least one solution on* $[0, 1]$.

Proof. Consider the operator $\mathcal{G} : \mathcal{C} \to \mathcal{C}$ defined by (10). It is easy to show that $\mathcal{G}$ is continuous. We complete the proof in three steps.

(*i*) $\mathcal{G}$ maps bounded sets into bounded sets in $C([0, 1], \mathbb{R})$.
For a positive number ε, let $\widetilde{B}_\varepsilon = \{x \in \mathcal{C} : \|x\| \le \varepsilon\}$ be a bounded set in $C([0, 1], \mathbb{R})$.
Then, we have

$$
\begin{aligned}
&\|(\mathcal{G}x)\| \\
&\le \sup_{t\in[0,1]} \Bigg\{ \int_0^t \frac{(t-qu)^{(\gamma-1)}}{\Gamma_q(\gamma)} \Bigg(|p| \int_0^u \frac{(u-qm)^{(\beta-1)}}{\Gamma_q(\beta)} |f(m, x(m))| d_q m \\
&+ |k| \int_0^u \frac{(u-qm)^{(\beta+\xi-1)}}{\Gamma_q(\beta+\xi)} |g(m, x(m))| d_q m + |\lambda||x(u)| \Bigg) d_q u \\
&+ \frac{M}{4\Gamma_q(\gamma+1)} \Bigg(|p| \int_0^a \frac{(a-qu)^{(\beta-1)}}{\Gamma_q(\beta)} |f(u, x(u))| d_q u \\
&+ |k| \int_0^a \frac{(a-qu)^{(\beta+\xi-1)}}{\Gamma_q(\beta+\xi)} |g(u, x(u))| d_q u \Bigg) \\
&+ \frac{M}{4\Gamma_q(\gamma+1)} \Bigg(|p| \int_0^1 \frac{(1-qu)^{(\beta-1)}}{\Gamma_q(\beta)} |f(u, x(u))| d_q u \\
&+ |k| \int_0^1 \frac{(1-qu)^{(\beta+\xi-1)}}{\Gamma_q(\beta+\xi)} |g(u, x(u))| d_q u \Bigg) \\
&+ \frac{1}{2} \int_0^a \frac{(a-qu)^{(\gamma-1)}}{\Gamma_q(\gamma)} \Bigg(|p| \int_0^u \frac{(u-qm)^{(\beta-1)}}{\Gamma_q(\beta)} |f(m, x(m))| d_q m \\
&+ |k| \int_0^u \frac{(u-qm)^{(\beta+\xi-1)}}{\Gamma_q(\beta+\xi)} |g(m, x(m))| d_q m + |\lambda||x(u)| \Bigg) d_q u \\
&+ \frac{1}{2} \int_0^1 \frac{(1-qu)^{(\gamma-1)}}{\Gamma_q(\gamma)} \Bigg[|p| \int_0^u \frac{(u-qm)^{(\beta-1)}}{\Gamma_q(\beta)} |f(m, x(m))| d_q m \\
&+ |k| \int_0^u \frac{(u-qm)^{(\beta+\xi-1)}}{\Gamma_q(\beta+\xi)} |g(m, x(m))| d_q m + |\lambda||x(u)| \Bigg] d_q u \Bigg\}
\end{aligned}
$$

$$\leq \sup_{t\in[0,1]} \Bigg\{ \int_0^t \frac{(t-qu)^{(\gamma-1)}}{\Gamma_q(\gamma)} \Big(|p| \int_0^u \frac{(u-qm)^{(\beta-1)}}{\Gamma_q(\beta)} v_1(m)\psi_1(\|x\|)d_qm$$

$$+|k| \int_0^u \frac{(u-qm)^{(\beta+\xi-1)}}{\Gamma_q(\beta+\xi)} v_2(m)\psi_2(\|x\|)d_qm + |\lambda||x(u)| \Big) d_qu$$

$$+\frac{M}{4\Gamma(\gamma+1)_q} \Big(|p| \int_0^a \frac{(a-qu)^{(\beta-1)}}{\Gamma_q(\beta)} v_1(u)\psi_1(\|x\|)d_qu$$

$$+|k| \int_0^a \frac{(a-qu)^{(\beta+\xi-1)}}{\Gamma_q(\beta+\xi)} v_2(u)\psi_2(\|x\|)d_qu \Big)$$

$$+\frac{M}{4\Gamma(\gamma+1)_q} \Big(|p| \int_0^1 \frac{(1-qu)^{(\beta-1)}}{\Gamma_q(\beta)} v_1(u)\psi_1(\|x\|)d_qu$$

$$+|k| \int_0^1 \frac{(1-qu)^{(\beta+\xi-1)}}{\Gamma_q(\beta+\xi)} v_2(u)\psi_2(\|x\|)d_qu \Big)$$

$$+\frac{1}{2} \int_0^a \frac{(a-qu)^{(\gamma-1)}}{\Gamma_q(\gamma)} \Big(|p| \int_0^u \frac{(u-qm)^{(\beta-1)}}{\Gamma_q(\beta)} v_1(m)\psi_1(\|x\|)d_qm$$

$$+|k| \int_0^u \frac{(u-qm)^{(\beta+\xi-1)}}{\Gamma_q(\beta+\xi)} v_2(m)\psi_2(\|x\|)d_qm + |\lambda||x(u)| \Big) d_qu$$

$$+\frac{1}{2} \int_0^1 \frac{(1-qu)^{(\gamma-1)}}{\Gamma_q(\gamma)} \Big(|p| \int_0^u \frac{(u-qm)^{(\beta-1)}}{\Gamma_q(\beta)} v_1(m)\psi_1(\|x\|)d_qm$$

$$+|k| \int_0^u \frac{(u-qm)^{(\beta+\xi-1)}}{\Gamma_q(\beta+\xi)} v_2(m)\psi_2(\|x\|)d_qm + |\lambda||x(u)| \Big) d_qu \Bigg\}$$

$$\leq |p|\|v_1\|\psi_1(\|x\|)\varpi_1 + |k|\|v_2\|\psi_2(\|x\|)\varpi_2 + |\lambda|\|x\|\varpi_3,$$

(*ii*) $\mathcal{G}$ maps bounded sets into equicontinuous sets of $C([0,1],\mathbb{R})$.
Let $t_1, t_2 \in [0,1]$ with $t_1 < t_2$ and $x \in \widetilde{B}_\varepsilon$, where $\widetilde{B}_\varepsilon$ is a bounded set of $C([0,1],\mathbb{R})$. Then, we obtain

$$\|(\mathcal{G}x)(t_2) - (\mathcal{G}x)(t_1)\|$$

$$\leq \Bigg| \int_0^{t_1} \frac{(t_2-qu)^{(\gamma-1)} - (t_1-qu)^{(\gamma-1)}}{\Gamma_q(\gamma)} \Big(|p| \int_0^u \frac{(u-qm)^{(\beta-1)}}{\Gamma_q(\beta)} v_1(m)\psi_1(\varepsilon)d_qm$$

$$
\begin{aligned}
&+ \quad |k| \int_0^u \frac{(u-qm)^{(\beta+\xi-1)}}{\Gamma_q(\beta+\xi)} \nu_2(m)\psi_2(\varepsilon) d_q m + |\lambda|\varepsilon \Big) d_q u \\
&+ \quad \int_{t_1}^{t_2} \frac{(t_2-qu)^{(\gamma-1)}}{\Gamma_q(\gamma)} \Big(|p| \int_0^u \frac{(u-qm)^{(\beta-1)}}{\Gamma_q(\beta)} \nu_1(m)\psi_1(\varepsilon) d_q m \\
&+ \quad |k| \int_0^u \frac{(u-qm)^{(\beta+\xi-1)}}{\Gamma_q(\beta+\xi)} \nu_2(m)\psi_2(\varepsilon) d_q m + |\lambda|\varepsilon \Big) d_q u \Big| \\
&+ \quad \frac{(t_2^{\gamma}-t_1^{\gamma})}{2\Gamma_q(\gamma+1)} \Big(|p| \int_0^a \frac{(a-qu)^{(\beta-1)}}{\Gamma_q(\beta)} \nu_1(u)\psi_1(\varepsilon) d_q u \\
&+ \quad |k| \int_0^a \frac{(a-qu)^{(\beta+\xi-1)}}{\Gamma_q(\beta+\xi)} \nu_2(u)\psi_2(\varepsilon) d_q u \Big) \\
&+ \quad \frac{(t_2^{\gamma}-t_1^{\gamma})}{2\Gamma_q(\gamma+1)} \Big(|p| \int_0^1 \frac{(1-qu)^{(\beta-1)}}{\Gamma_q(\beta)} \nu_1(u)\psi_1(\varepsilon) d_q u \\
&+ \quad |k| \int_0^1 \frac{(1-qu)^{(\beta+\xi-1)}}{\Gamma_q(\beta+\xi)} \nu_2(u)\psi_2(\varepsilon) d_q u \Big).
\end{aligned}
$$

Obviously the right hand side of the above inequality tends to zero independently of $x \in \widetilde{B}_\varepsilon$ as $t_2 - t_1 \to 0$. Since $\mathcal{G}$ satisfies the above assumptions, it follows by the Arzelá-Ascoli theorem that $\mathcal{G} : \mathcal{C} \to \mathcal{C}$ is completely continuous.

(*iii*) Let x be a solution and $x = \mu\mathcal{G}x$ for $\mu \in (0, 1)$. Then, for $t \in [0, 1]$, and using the computations in proving that $\mathcal{G}$ is bounded, we have

$$|x(t)| = |\mu(\mathcal{G}x)(t)| \le |p|\|\nu_1\|\psi_1(\|x\|)\varpi_1 + |k|\|\nu_2\|\psi_2(\|x\|)\varpi_2 + |\lambda|\|x\|\varpi_3.$$

Consequently, we have

$$\|x\| \le \frac{|p|\|\nu_1\|\psi_1(\|x\|)\varpi_1 + |k|\|\nu_2\|\psi_2(\|x\|)\varpi_2}{1 - |\lambda|\varpi_3}.$$

In view of (A_4), there exists κ such that $\|x\| \neq \kappa$. Let us set

$$V = \{x \in \mathcal{C} : \|x\| < \kappa\}.$$

Note that the operator $\mathcal{G} : \overline{V} \to C([0, 1], \mathbb{R})$ is completely continuous. From the choice of V, there is no $x \in \partial V$ such that $x = \mu\mathcal{G}(x)$ for some $\mu \in (0, 1)$. Consequently, by the nonlinear alternative of Leray-Schauder type (Lemma 3), we deduce that $\mathcal{G}$ has a fixed point $x \in \overline{V}$ which is a solution of the problem (1) – (2). This completes the proof. □

In the next result, we discuss the uniqueness of solutions for the given problem by means of Banach's contraction principle (Banach fixed point theorem).

Theorem 15. *Suppose that the assumption* (A_1) *holds and that*

$$\overline{\Omega} = (L\Lambda + |\lambda|\varpi_3) < 1, \quad \Lambda = |p|\varpi_1 + |k|\varpi_2 \tag{13}$$

where $\varpi_1, \varpi_2, \varpi_3$ *are given by (11) and* $L = \max\{L_1, L_2\}$. *Then the boundary value problem (1)-(2) has a unique solution on* $[0, 1]$.

Proof. Let us define $N = \max\{N_1, N_2\}$, where N_1, N_2 are finite numbers given by $N_1 = \sup_{t\in[0,1]} |f(t, 0)|$, $N_2 = \sup_{t\in[0,1]} |g(t, 0)|$. Selecting $\overline{\sigma} \geq \dfrac{N\Lambda}{1-\overline{\Omega}}$, we show that $\mathcal{G}\widetilde{B}_{\overline{\sigma}} \subset \widetilde{B}_{\overline{\sigma}}$, where $\widetilde{B}_{\overline{\sigma}} = \{x \in \mathcal{C} : \|x\| \leq \overline{\sigma}\}$. For $x \in \widetilde{B}_{\overline{\sigma}}$, we have

$$\begin{aligned}
&\|(\mathcal{G}x)\| \\
&\leq \sup_{t\in[0,1]} \Bigg\{ \int_0^t \frac{(t-qu)^{(\gamma-1)}}{\Gamma_q(\gamma)} \Bigg(|p| \int_0^u \frac{(u-qm)^{(\beta-1)}}{\Gamma_q(\beta)} |f(m, x(m))| d_qm \\
&+ |k| \int_0^u \frac{(u-qm)^{(\beta+\xi-1)}}{\Gamma_q(\beta+\xi)} |g(m, x(m))| d_qm + |\lambda||x(u)| \Bigg) d_qu \\
&+ \frac{M}{4\Gamma_q(\gamma+1)} \Bigg(|p| \int_0^a \frac{(a-qu)^{(\beta-1)}}{\Gamma_q(\beta)} |f(u, x(u))| d_qu \\
&+ |k| \int_0^a \frac{(a-qu)^{(\beta+\xi-1)}}{\Gamma_q(\beta+\xi)} |g(u, x(u))| d_qu \Bigg) \\
&+ \frac{M}{4\Gamma_q(\gamma+1)} \Bigg(|p| \int_0^1 \frac{(1-qu)^{(\beta-1)}}{\Gamma_q(\beta)} |f(u, x(u))| d_qu \\
&+ |k| \int_0^1 \frac{(1-qu)^{(\beta+\xi-1)}}{\Gamma_q(\beta+\xi)} |g(u, x(u))| d_qu \Bigg) \\
&+ \frac{1}{2} \int_0^a \frac{(a-qu)^{(\gamma-1)}}{\Gamma_q(\gamma)} \Bigg(|p| \int_0^u \frac{(u-qm)^{(\beta-1)}}{\Gamma_q(\beta)} |f(m, x(m))| d_qm \\
&+ |k| \int_0^u \frac{(u-qm)^{(\beta+\xi-1)}}{\Gamma_q(\beta+\xi)} |g(m, x(m))| d_qm + |\lambda||x(u)| \Bigg) d_qu
\end{aligned}$$

$$+\frac{1}{2}\int_0^1 \frac{(1-qu)^{(\gamma-1)}}{\Gamma_q(\gamma)}\bigg(|p|\int_0^u \frac{(u-qm)^{(\beta-1)}}{\Gamma_q(\beta)}|f(m,x(m))|d_qm$$

$$+|k|\int_0^u \frac{(u-qm)^{(\beta+\xi-1)}}{\Gamma_q(\beta+\xi)}|g(m,x(m))|d_qm+|\lambda||x(u)|\bigg)d_qu\bigg\}$$

$$\leq \sup_{t\in[0,1]}\bigg\{\int_0^t \frac{(t-qu)^{(\gamma-1)}}{\Gamma_q(\gamma)}\bigg(|p|\int_0^u \frac{(u-qm)^{(\beta-1)}}{\Gamma_q(\beta)}\Big(|f(m,x(m))-f(m,0)|+|f(m,0)|\Big)d_qm$$

$$+|k|\int_0^u \frac{(u-qm)^{(\beta+\xi-1)}}{\Gamma_q(\beta+\xi)}\Big(|g(m,x(m))-g(m,0)|+|g(m,0)|\Big)d_qm+|\lambda||x(u)|\bigg)d_qu$$

$$+\frac{M}{4\Gamma_q(\gamma+1)}\bigg(|p|\int_0^a \frac{(a-qu)^{(\beta-1)}}{\Gamma_q(\beta)}\Big(|f(u,x(u))-f(u,0)|+|f(u,0)|\Big)d_qu$$

$$+|k|\int_0^a \frac{(a-qu)^{(\beta+\xi-1)}}{\Gamma_q(\beta+\xi)}\Big(|g(u,x(u))-g(u,0)|+|g(u,0)|\Big)d_qu\bigg)$$

$$+\frac{M}{4\Gamma_q(\gamma+1)}\bigg(|p|\int_0^1 \frac{(1-qu)^{(\beta-1)}}{\Gamma_q(\beta)}\Big(|f(u,x(u))-f(u,0)|+|f(u,0)|\Big)d_qu$$

$$+|k|\int_0^1 \frac{(1-qu)^{(\beta+\xi-1)}}{\Gamma_q(\beta+\xi)}\Big(|g(u,x(u))-g(u,0)|+|g(u,0)|\Big)d_qu\bigg)$$

$$+\frac{1}{2}\int_0^a \frac{(a-qu)^{(\gamma-1)}}{\Gamma_q(\gamma)}\bigg(|p|\int_0^u \frac{(u-qm)^{(\beta-1)}}{\Gamma_q(\beta)}\Big(|f(m,x(m))-f(m,0)|+|f(m,0)|\Big)d_qm$$

$$+|k|\int_0^u \frac{(u-qm)^{(\beta+\xi-1)}}{\Gamma_q(\beta+\xi)}\Big(|g(m,x(m))-g(m,0)|+|g(m,0)|\Big)d_qm+|\lambda||x(u)|\bigg)d_qu$$

$$+\frac{1}{2}\int_0^1 \frac{(1-qu)^{(\gamma-1)}}{\Gamma_q(\gamma)}\bigg(|p|\int_0^u \frac{(u-qm)^{(\beta-1)}}{\Gamma_q(\beta)}\Big(|f(m,x(m))-f(m,0)|+|f(m,0)|\Big)d_qm$$

$$+|k|\int_0^u \frac{(u-qm)^{(\beta+\xi-1)}}{\Gamma_q(\beta+\xi)}\Big(|g(m,x(m))-g(m,0)|+|g(m,0)|\Big)d_qm+|\lambda||x(u)|\bigg)d_qu\bigg\}$$

$$\leq |p|(L_1\overline{\sigma}+N_1)\sup_{t\in[0,1]}\bigg\{\int_0^t \frac{(t-qu)^{(\gamma-1)}}{\Gamma_q(\gamma)}\bigg(\int_0^u \frac{(u-qm)^{(\beta-1)}}{\Gamma_q(\beta)}d_qm\bigg)d_qu$$

$$+\frac{M}{4\Gamma_q(\gamma+1)}\int_0^a \frac{(a-qu)^{(\beta-1)}}{\Gamma_q(\beta)}d_qu+\frac{M}{4\Gamma_q(\gamma+1)}\int_0^1 \frac{(1-qu)^{(\beta-1)}}{\Gamma_q(\beta)}d_qu$$
$$+\frac{1}{2}\int_0^a \frac{(a-qu)^{(\gamma-1)}}{\Gamma_q(\gamma)}\Big(\int_0^u \frac{(u-qm)^{(\beta-1)}}{\Gamma_q(\beta)}d_qm\Big)d_qu$$
$$+\frac{1}{2}\int_0^1 \frac{(1-qu)^{(\gamma-1)}}{\Gamma_q(\gamma)}\Big(\int_0^u \frac{(u-qm)^{(\beta-1)}}{\Gamma_q(\beta)}d_qm\Big)d_qu\Bigg\}$$
$$+|k|(L_2\overline{\sigma}+N_2)\sup_{t\in[0,1]}\Bigg\{\int_0^t \frac{(t-qu)^{(\gamma-1)}}{\Gamma_q(\gamma)}\Big(\int_0^u \frac{(u-qm)^{(\beta+\xi-1)}}{\Gamma_q(\beta+\xi)}d_qm\Big)d_qu$$
$$+\frac{M}{4\Gamma_q(\gamma+1)}\int_0^a \frac{(a-qu)^{(\beta+\xi-1)}}{\Gamma_q(\beta+\xi)}d_qu+\frac{M}{4\Gamma_q(\gamma+1)}\int_0^1 \frac{(1-qu)^{(\beta+\xi-1)}}{\Gamma_q(\beta+\xi)}d_qu$$
$$+\frac{1}{2}\int_0^a \frac{(a-qu)^{(\gamma-1)}}{\Gamma_q(\gamma)}\Big(\int_0^u \frac{(u-qm)^{(\beta+\xi-1)}}{\Gamma_q(\beta+\xi)}d_qm\Big)d_qu$$
$$+\frac{1}{2}\int_0^1 \frac{(1-qu)^{(\gamma-1)}}{\Gamma_q(\gamma)}\Big(\int_0^u \frac{(u-qm)^{(\beta+\xi-1)}}{\Gamma_q(\beta+\xi)}d_qm\Big)d_qu\Bigg\}$$
$$+|\lambda|\overline{\sigma}\sup_{t\in[0,1]}\Bigg\{\int_0^t \frac{(t-qu)^{(\gamma-1)}}{\Gamma_q(\gamma)}d_qu+\frac{1}{2}\int_0^a \frac{(a-qu)^{(\gamma-1)}}{\Gamma_q(\gamma)}d_qu+\frac{1}{2}\int_0^1 \frac{(1-qu)^{(\gamma-1)}}{\Gamma_q(\gamma)}d_qu\Bigg\}$$
$$\le (L\overline{\sigma}+N)\Lambda+|\lambda|\overline{\sigma}\varpi_3\le\overline{\sigma},$$

which means that $\mathcal{G}\widetilde{B}_{\overline{\sigma}}\subset\widetilde{B}_{\overline{\sigma}}$.
Now, for $x,y\in\mathcal{C}$, we obtain

$$\|\mathcal{G}x-\mathcal{G}y\|$$
$$\le\sup_{t\in[0,1]}\Bigg\{\int_0^t \frac{(t-qu)^{(\gamma-1)}}{\Gamma_q(\gamma)}\Big(|p|\int_0^u \frac{(u-qm)^{(\beta-1)}}{\Gamma_q(\beta)}\Big|f(m,x(m))-f(m,y(m))\Big|d_qm$$
$$+|k|\int_0^u \frac{(u-qm)^{(\beta+\xi-1)}}{\Gamma_q(\beta+\xi)}\Big|g(m,x(m))-g(m,y(m))\Big|d_qm+|\lambda||x(u)-y(u)|\Big)d_qu$$
$$+\frac{M}{4\Gamma_q(\gamma+1)}\Big(|p|\int_0^a \frac{(a-qu)^{(\beta-1)}}{\Gamma_q(\beta)}\Big|f(u,x(u))-f(u,y(u))\Big|d_qu$$
$$+|k|\int_0^a \frac{(a-qu)^{(\beta+\xi-1)}}{\Gamma_q(\beta+\xi)}\Big|g(u,x(u))-g(u,y(u))\Big|d_qu\Big)$$

$$+\frac{M}{4\Gamma_q(\gamma+1)}\Big(|p|\int_0^1 \frac{(1-qu)^{(\beta-1)}}{\Gamma_q(\beta)}\Big|f(u,x(u))-f(u,y(u))\Big|d_qu$$

$$+|k|\int_0^1 \frac{(1-qu)^{(\beta+\xi-1)}}{\Gamma_q(\beta+\xi)}\Big|g(u,x(u))-g(u,y(u))\Big|d_qu\Big)$$

$$+\frac{1}{2}\int_0^a \frac{(a-qu)^{(\gamma-1)}}{\Gamma_q(\gamma)}\Big(|p|\int_0^u \frac{(u-qm)^{(\beta-1)}}{\Gamma_q(\beta)}\Big|f(m,x(m))-f(m,y(m))\Big|d_qm$$

$$+|k|\int_0^u \frac{(u-qm)^{(\beta+\xi-1)}}{\Gamma_q(\beta+\xi)}\Big|g(m,x(m))-g(m,y(m))\Big|d_qm+|\lambda||x(u)-y(u)|\Big)d_qu$$

$$+\frac{1}{2}\int_0^1 \frac{(1-qu)^{(\gamma-1)}}{\Gamma_q(\gamma)}\Big(|p|\int_0^u \frac{(u-qm)^{(\beta-1)}}{\Gamma_q(\beta)}\Big|f(m,x(m))-f(m,y(m))\Big|d_qm$$

$$+|k|\int_0^u \frac{(u-qm)^{(\beta+\xi-1)}}{\Gamma_q(\beta+\xi)}\Big|g(m,x(m))-g(m,y(m))\Big|d_qm+|\lambda||x(u)-y(u)|\Big)d_qu\Bigg\}$$

$$\leq \overline{\Omega}\|x-y\|$$

which shows that $\mathcal{G}$ is a contraction as $\overline{\Omega} < 1$ by the given assumption. Therefore, it follows by Banach's contraction principle that the problem (1)-(2) has a unique solution. □

4 Example

Consider a boundary value problem of integro-differential equations of fractional order given by

$$\begin{cases} {}^cD_q^{1/2}({}^cD_q^{1/2}+\frac{1}{6})x(t)=\frac{1}{3}f(t,x(t))+\frac{1}{5}I_q^{1/2}g(t,x(t)), \quad 0\leq t\leq 1, \\ x(a)=-x(1), \quad {}^cD_q^{\gamma}x(a)=-{}^cD_q^{\gamma}x(1), \end{cases} \tag{14}$$

where $f(t,x)=\frac{1}{(3+t^2)^2}\Big(\cos t+\frac{|x|}{1+|x|}+|x|\Big)$, $g(t,x)=\frac{1}{3}\tan^{-1}x+t^3$, $q=1/2$ and $a=1/3$. Clearly it follows from the inequalities:

$$|f(t,x)-f(t,y)|\leq \frac{2}{9}|x-y|, \quad |g(t,x)-g(t,y)|\leq \frac{1}{3}|x-y|$$

that $L_1=2/9$ and $L_2=1/3$. So $L=\max\{L_1,L_2\}=\frac{1}{3}$. With the given data, it is found that

$$\overline{\Omega}\simeq 0.717855<1.$$

Thus all the assumptions of Theorem 15 are satisfied. Hence, by the conclusion of Theorem 15, the problem (14) has a unique solution.

Bibliography

Agarwal, R. P. (1969). Certain fractional q-integrals and q-derivatives. Mathematical Proceedings of the Cambridge Philosophical Society, 66, 365-370.

Ahmad, B., & Nieto, J. J. (2011). Riemann-Liouville fractional integro-differential equations with fractional nonlocal integral boundary conditions. Boundary Value Problems, 2011:36, 9 pp.

Ahmad, B., Ntouyas, S. K., & Purnaras, I. K. (2012). Existence results for nonlocal boundary value problems of nonlinear fractional q-difference equations. Advances in Difference Equations, 2012:140, 15 pp.

Ahmad, B., & Nieto, J. J. (2014). Sequential differential equations of fractional order with multi-point boundary conditions. Georgian Mathematics Journal, 21, 243-248.

Ahmad, B., Nieto, J. J., Alsaedi, A., et al. (2014). Existence of solutions for nonlinear fractional q-difference integral equations with two fractional orders and nonlocal four-point boundary conditions. Journal of the Franklin Institute, 351, 2890-2909.

Al-Salam, W. A. (1966-1967). Some fractional q-integrals and q-derivatives. Proceedings of the Edinburgh Mathematical Society, 15, 135-140.

Annaby, M. H., & Mansour, Z. S. (2012). q-Fractional Calculus and Equations: Lecture Notes in Mathematics 2056. Berlin: Springer.

Baleanu, D., Diethelm, K., Scalas, E., et al. (2012). Fractional Calculus Models and Numerical Methods: Series on Complexity, Nonlinearity and Chaos. Boston: World Scientific.

Cabada, A., & Wang, G. (2012). Positive solutions of nonlinear fractional differential equations with integral boundary value conditions. Journal of Mathematical Analysis and Applications, 389, 403-411.

Ferreira, R. A. C. (2011). Positive solutions for a class of boundary value problems with fractional q-differences. Computers & Mathematics with Applications, 61, 367-373.

Goodrich, C. S. (2011). Existence and uniqueness of solutions to a fractional difference equation with nonlocal conditions. Computers & Mathematics with Applications, 61, 191-202.

Graef, J. R., & Kong, L. (2012). Positive solutions for a class of higher order boundary value problems with fractional q-derivatives. Applied Mathematics and Computation, 218, 9682-9689.

Graef, J. R., Kong, L., & Wang, M. (2014). Existence and uniqueness of solutions for a fractional boundary value problem on a graph. Fractional Calculus and Applied Analysis, 17, 499-510.

Granas, A., & Dugundji, J. (2003). Fixed Point Theory. New York: Springer.

Henderson J., & Luca, R. (2013). Positive solutions for a system of nonlocal fractional boundary value problems. Fractional Calculus and Applied Analysis, 16, 985-1008.

Kilbas, A. A., Srivastava, H. M., & Trujillo, J. J. (2006). Theory and Applications of Fractional Differential Equations: North-Holland Mathematics Studies, 204. Amsterdam: Elsevier.

Kirane, M., Ahmad, B., Alsaedi, A., et al. (2014). Non-existence of global solutions to a system of fractional diffusion equations. Acta Applicandae Mathematicae, 133, 235-248.

Li, X., Han, Z., & Sun, S. (2013). Existence of positive solutions of nonlinear fractional q-difference equation with parameter. Advances in Difference Equations, 2013:260, 13 pp.

Liu, X., Jia, M., & Ge, W. (2013). Multiple solutions of a p-Laplacian model involving a fractional derivative. Advances in Difference Equations, 2013:126, 12 pp.

Liu, X., Liu, Z., & Fu, X. (2014). Relaxation in nonconvex optimal control problems described by fractional differential equations. Journal of Mathematical Analysis and Applications, 409, 446-458.

O'Regan, D., & Stanek, S. (2013). Fractional boundary value problems with singularities in space variables. Nonlinear Dynamics, 71, 641-652.

Podlubny, I. (1999). Fractional Differential Equations. San Diego: Academic Press.

Punzo, F., & Terrone, G. (2014). On the Cauchy problem for a general fractional porous medium equation with variable density. Nonlinear Analysis, 98, 27-47.

Rajkovic, P. M., Marinkovic, S. D., & Stankovic, M. S. (2007). On q-analogues of Caputo derivative and Mittag-Leffler function. Fractional Calculus and Applied Analysis, 10, 359-373.

Sabatier, J., Agrawal, O.P., & Machado, J.A.T. (2007). Advances in Fractional Calculus: Theoretical Developments and Applications in Physics and Engineering. Dordrecht: Springer.

Smart, D. R. (1980). Fixed Point Theorems. Cambridge University Press.

Sudsutad, W., & Tariboon, J. (2012). Existence results of fractional integro-differential equations with m-point multi-term fractional order integral boundary conditions. Boundary Value Problems, 2012:94, 11 pp.

Yang, X.-J., Hristov, J., Srivastava, H. M., et al. (2014). Modelling fractal waves on shallow water surfaces via local fractional Korteweg-de Vries equation. Abstract Applied Analysis, Article ID 278672, 10 pages.

Zhai, C., & Xu, L. (2014). Properties of positive solutions to a class of four-point boundary value problem of Caputo fractional differential equations with a parameter. Communications in Nonlinear Science and Numerical Simulation, 19, 2820-2827.

M. K. Lenzi*, G. Gonçalves, D. P. Leitoles, and E. K. Lenzi

Fractional Diffusion Equation, Sorption and Reaction Processes on a Surface

Abstract: We investigate the sorption and reaction processes which may occur on a surface with active sites in contact with a system. We consider the system defined in a half space with dynamics of the substances governed by a fractional diffusion equation. The processes on the surface are assumed to be of first order, i.e., the kinetic equations are linear, and with memory effects which may be connected to a nonusual relaxation. In this scenario, we obtain the behavior of the quantities on the surface where the processes are present and in the bulk.

Keywords: anomalous diffusion; sorption; reaction on surfaces

1 Introduction

Diffusion and reaction play an important role in many situations of interest, e.g., in biology (Benjamin, et al. 2013, Soula, et al. 2013, Köpf, et al. 1996), in engineering (Strizhak 2004, Avnir 1990, Masel 1996), and physics (Lenzi, et al. 2014, Mendez, Campos & Bartumeus 2014, ben-Avraham & Havlin 2005), and, in general, are described in terms of the stochastic differential equations which are connected to the Markovian processes (Crank 1975). However, the influence of the morphology of the surfaces, i.e., fractal characteristics, and the non-Fickian diffusion (Pekalski & Weron 1998) present in these systems have produced evidence of the limitations of the usual approach in describing the large variety of the situations and motivated the researchers to investigate other formalisms in order provide a suitable analysis of these phenomena. For instance, continuous time random walk (Klafter & Sokolov 2011), generalized Langevin equations and fractional diffusion equations (see Refs. Leijnse, et al. (2012), Hilfer, et al. (2004), Metzler & Klafter (2000, 2004), Eliazar & Shlesinger (213), Bressloff & Newby (2013), Condamin, et al. (2008)) have been analyzed and applied in several situations such as molecular diffusion *in vivo* (Leijnse, et al. 2012, Robson, Burrage & Leake 2013), in membrane cells (Caputo, Cametti & Ruggero 2008, Caputo & Cametti 2009), subdiffusion in thin membranes (Kosztolowicz, Dworecki

*Corresponding Author: M. K. Lenzi: Departamento de Engenharia Química, Universidade Federal do Paraná - Curitiba, PR 81531 - 990, Brazil
G. Gonçalves: Departamento de Engenharia Química, Universidade Tecnológica Federal do Paraná - Ponta Grossa, 84016-210 PR, Brazil
D. P. Leitoles: Departamento de Engenharia Química, Universidade Federal do Paraná - Curitiba, PR 81531 - 990, Brazil
E. K. Lenzi: Departamento de Física, Universidade Estadual de Ponta Grossa - Ponta Grossa, PR 87030-900, Brazil

& Lewandowska 2012) and chemotaxis diffusion (Langlands & Henry 2010). In these situations one of the main issues is the nonlinear time dependence exhibited by the mean-square displacement which, in general, is characterized by $(\Delta x)^2(t) \sim t^\alpha$, being the superdiffusion ($\alpha > 1$) commonly related, for example, to active transport (Caspi, Granek & Elbaum 2000, Bruno, et al. 2009, Brangwynne, et al. 2009, Weber, Spakowitz & Theriot 2012), while the subdiffusive behavior ($\alpha < 1$) may be related to the molecular crowding (Sokolov 2012) and fractal structure (Weigel, et al. 2011). In Ref. Jaing, et al. (2013), fractional diffusion equations are used to analyze the desorption process of methane in coal. Thus, understanding the challenges is important in order to establish connections between the models and the experimental results.

Here, we investigate a sorption followed by a reaction process which may occur on a surface with active sites in contact with a system. The system considers a half space and the dynamics are assumed to be governed by fractional diffusion equations. The processes on the surface are also assumed to be of first order, i.e., the kinetic equations are linear, and with memory effects which may be connected to a nonusual relaxation. In this scenario, we obtain the behavior of the quantities on the surface where the processes are present and in the bulk. This development is performed in Sec. II and in Sec. III a summary of the results and our conclusions are presented.

2 Diffusion and Reaction

Let us start our analysis by considering that the particles in the bulk are governed by the fractional diffusion equations

$$\frac{\partial}{\partial t}\rho_a(x,t) = \mathcal{K}_a\, {}_0\mathcal{D}_t^{1-\alpha}\left(\frac{\partial^2}{\partial x^2}\rho_a(x,t)\right) \tag{1}$$

and

$$\frac{\partial}{\partial t}\rho_b(x,t) = \mathcal{K}_b\, {}_0\mathcal{D}_t^{1-\alpha}\left(\frac{\partial^2}{\partial x^2}\rho_b(x,t)\right), \tag{2}$$

where $0 < \alpha \leq 1$, ρ_a and ρ_b are the density of particles (the subindexes a and b represent the two types of particles which may be present in the bulk), and the fractional time derivative present in Eq.(1) is the Riemann-Liouville,

$${}_0\mathcal{D}_t^{\overline{\alpha}}\left(\rho_{a(b)}(x,t)\right) = \frac{1}{\Gamma(n-\overline{\alpha})}\frac{d^n}{dt^n}\int_{t_0}^{t} dt'\,\frac{\rho_{a(b)}(x,t')}{(t-t')^{1+\overline{\alpha}-n}}, \tag{3}$$

as defined in Refs. Podlubny (1999) and Mathai, Saxena & Haubold (2009) with $n-1 < \overline{\alpha} < n$. For the processes which may occur on the surface such as adsorption, desorption, and chemical reactions, we assume that they may satisfy the following

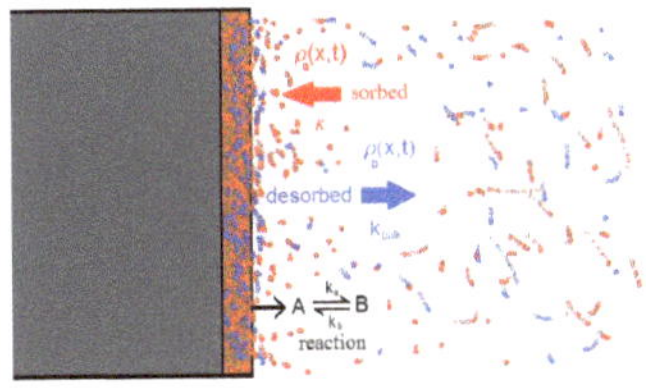

Fig. 1. This figure illustrates the system under consideration where the particles a are sorbed by the surface and by a reaction process the substance b is produced. The substance b may also leave the surface to the bulk with the rate k_{bulk}.

kinetic equations

$$\frac{d}{dt}\Gamma_a(t) = \kappa\rho_a(0,t) - \int_0^t k_a(t-t')\Gamma_a(t')dt' + \int_0^t k_b(t-t')\Gamma_b(t')dt' , \tag{4}$$

$$\begin{aligned}\frac{d}{dt}\Gamma_b(t) &= -\int_0^t k_{bulk}(t-t')\Gamma_b(t')dt' - \int_0^t k_b(t-t')\Gamma_b(t')dt' \\ &+ \int_0^t k_a(t-t')\Gamma_a(t')dt' ,\end{aligned} \tag{5}$$

and the substance, obtained by the reaction process, $\Gamma_b(t)$ may be released to the bulk (see Fig. 1). In this case, we have that

$$\frac{d}{dt}\Gamma_{bulk}(t) = \int_0^t k_{bulk}(t-t')\Gamma_b(t')dt' \tag{6}$$

and, consequently,

$$\mathcal{K}_b\, {}_0\mathcal{D}_t^{1-\alpha}\left(\frac{\partial}{\partial x}\rho_b(x,t)\right)\Bigg|_{x=0} = -\frac{d}{dt}\Gamma_{bulk}(t) \tag{7}$$

where k_{bulk} is connected to the rate at which the substance leaves the surface to join the bulk. In Eqs.(4) and (5), $\Gamma_{a(b)}(t)$ and $\Gamma_b(t)$ are the surface density and κ, k_a, and k_b are the rates connected to the reaction process which may be manifested by the surface. Note that these kinetic equations extend the usual ones to situations characterized by nonusual relaxations, i.e., non-Debye relaxations, where a non-exponential behavior can be obtained depending on the choice of the kernels. From a phenomenological point of view, the choice of the kernels present Eqs. (4), (5), and (6) could be related, for example, to surface irregularities Kopelman (1988) which is an important factor in adsorption–desorption, diffusion, and catalysis processes. In addition, they recover the classical ones for $k_a(t) = k_a\delta(t)$, $k_b(t) = k_b\delta(t)$, and $k_{bulk}(t) = k_{bulk}\delta(t)$.

In order to solve these equations, we initially consider them subject to $\rho_a(x,0) = \varphi(x)$ (with $\int_0^\infty dx\rho_a(x) = 1$), $\rho_b(x,0) = 0$, and $\Gamma_a(0) = \Gamma_b(0) = 0$. We also consider as boundary conditions $\rho_a(\infty,t) = \rho_b(\infty,t) = 0$ and

$$\Gamma_a(t) + \Gamma_b(t) + \Gamma_{bulk}(t) + \int_0^\infty \rho_a(x,t)dx = constant. \tag{8}$$

that imply the conservation of the number of particles present in the system. It is also possible to consider situations with source (or sink) by performing suitable changes in Eq. (8). Equation (8) that may be connected to the following boundary condition

$$\mathcal{K}_a\, {}_0\mathcal{D}_t^{1-\alpha}\left(\frac{\partial}{\partial x}\rho_a(x,t)\right)\Bigg|_{x=0} = \frac{d}{dt}\Gamma_t(t) \tag{9}$$

where $\Gamma_t(t) = \Gamma_a(t) + \Gamma_b(t) + \Gamma_{bulk}(t)$. By using the Laplace transform and the Green function approach, the solution of Eq.(1) subject to the previous boundary conditions can be written as

$$\rho_a(x,s) = -\int_0^\infty \mathcal{G}_a(x,x';s)\varphi(x')dx' \tag{10}$$

with the Green function given by

$$\begin{aligned}\mathcal{G}_a(x,x';t) &= -\frac{1}{2s\sqrt{\mathcal{K}_a/s^\alpha}}\left(e^{-\sqrt{s^\alpha/\mathcal{K}_a}|x-x'|} + e^{-\sqrt{s^\alpha/\mathcal{K}_a}|x+x'|}\right)\\ &+ \frac{2\kappa}{s\sqrt{\mathcal{K}_a/s^\alpha}+\kappa}\frac{1}{2s\sqrt{\mathcal{K}_a/s^\alpha}}e^{-\sqrt{s^\alpha/\mathcal{K}_a}(x+x')}\,.\end{aligned} \tag{11}$$

The last term of the previous equation represents the effect of the surface on the bulk, i.e., how the processes which occur on the surface can modify the evolution of the initial condition of the substance a. An interesting point about this case, by analyzing Eq. (11), is that the reaction processes present on the surface have no influence on the spreading of the substance a into the bulk. The influence of the surface on the substance a is manifested by κ, the sorption rate. For Eq. (2), we have that

$$\rho_b(x,s) = \frac{k_{bulk}(s)}{s\sqrt{\mathcal{K}_b/s^\alpha}}\Gamma_b(s)e^{-\sqrt{s^\alpha/\mathcal{K}_a}x}\,, \tag{12}$$

which in contrast to the solution obtained for $\rho_a(x,t)$, depends on the reaction process present on the surface.

By performing the inverse Laplace transform, we obtain that

$$\rho_a(x,t) = -\int_0^\infty \mathcal{G}_a(x,x';t)\varphi(x')dx'\,, \tag{13}$$

$$\rho_b(x',t) = \int_0^t \frac{dt'}{\sqrt{\mathcal{K}_b t'^\alpha}}\Phi(t-t')\mathrm{H}_{1,1}^{1,0}\left[\frac{x}{\sqrt{\mathcal{K}_a t'^\alpha}}\middle|\begin{matrix}(1-\frac{\alpha}{2},\frac{\alpha}{2})\\(0,\,1)\end{matrix}\right] \tag{14}$$

with $\Phi(t) = \int_0^t k_{bulk}(t-t')\Gamma_b(t')dt'$, and

$$\mathcal{G}_a(x,x';t) = -\frac{1}{\sqrt{4\mathcal{K}_a t^\alpha}}\left(\mathrm{H}_{1,1}^{1,0}\left[\frac{|x-x'|}{\sqrt{\mathcal{K}_a t^\alpha}}\middle|\begin{matrix}(1-\frac{\alpha}{2},\frac{\alpha}{2})\\(0,\,1)\end{matrix}\right]+\mathrm{H}_{1,1}^{1,0}\left[\frac{|x+x'|}{\sqrt{\mathcal{K}_a t^\alpha}}\middle|\begin{matrix}(1-\frac{\alpha}{2},\frac{\alpha}{2})\\(0,\,1)\end{matrix}\right]\right)$$
$$+\int_0^t \frac{\kappa dt'}{\mathcal{K}_a\sqrt{(t-t')^\alpha t'^\alpha}}\mathrm{E}_{\overline{\alpha},\overline{\alpha}}\left(-\frac{\kappa t^{\overline{\alpha}}}{\sqrt{\mathcal{K}_a}}\right)\mathrm{H}_{1,1}^{1,0}\left[\frac{x+x'}{\sqrt{\mathcal{K}_a(t-t')^\alpha}}\middle|\begin{matrix}(1-\frac{\alpha}{2},\frac{\alpha}{2})\\(0,\,1)\end{matrix}\right] \tag{15}$$

where $\overline{\alpha} = 1-\alpha/2$, $\mathrm{E}_{\beta,\alpha}(x)$ is the generalized Mittag-Leffler function Podlubny (1999) and $\mathrm{H}_{p,q}^{m,n}\left[x\middle|\begin{matrix}(a_p,A_p)\\(b_q\ B_q)\end{matrix}\right]$ is the Fox H function Mathai, Saxena & Haubold (2009). The presence of these previous functions is connected to the anomalous relaxation of the system due to the sorption process present on the surface and the fractional time derivative present in Eqs. (1) and (2).

The concentrations, in the Laplace space, of the species a and b on the surface are given by

$$\Gamma_a(s) = \frac{\kappa\left(s+k_b(s)+k_{bulk}(s)\right)\Theta(s)}{(s+k_a(s))(s+k_b(s)+k_{bulk}(s))-k_a(s)k_b(s)}, \tag{16}$$
$$\Gamma_b(s) = \frac{\kappa k_a(s)\Theta(s)}{(s+k_a(s))(s+k_b(s)+k_{bulk}(s))-k_a(s)k_b(s)}, \tag{17}$$
$$\Gamma_{bulk}(s) = \frac{\kappa k_a(s)k_{bulk}(s)\Theta(s)}{s\left[(s+k_a(s))(s+k_b(s)+k_{bulk}(s))-k_a(s)k_b(s)\right]}, \tag{18}$$

with

$$\Theta(s) = \frac{1}{\kappa+s\sqrt{\mathcal{K}_a/s^\alpha}}\int_0^\infty \varphi(x)e^{-\sqrt{s^\alpha/\mathcal{K}_a}x}dx\,. \tag{19}$$

These equations were obtained from Eqs. (4), (5) and (6) by using Eqs.(8) and (1). The boundary and initial conditions for $\rho(x,t)$, $\Gamma_a(t)$, $\Gamma_b(t)$, and $\Gamma_{bulk}(t)$ were also used. Notice the presence of an stationary state when $t \to \infty$ ($s \to 0$) if $k_a(s) \to k_a = constant$, $k_b(s) \to k_b = constant$, and $k_{bulk}(s) = 0$, which results in

$$\Gamma_a(t) \sim \frac{k_b}{k_a+k_b} \quad \text{and} \tag{20}$$
$$\Gamma_b(t) \sim \frac{k_a}{k_a+k_b}\,. \tag{21}$$

Equations (8), (20), and (21) imply that, after some time, the substance, initially present in the bulk, is totally sorbed by the surface where the reversible reaction process is occurring and is not desorbed to the bulk due to the type of the kinetic process

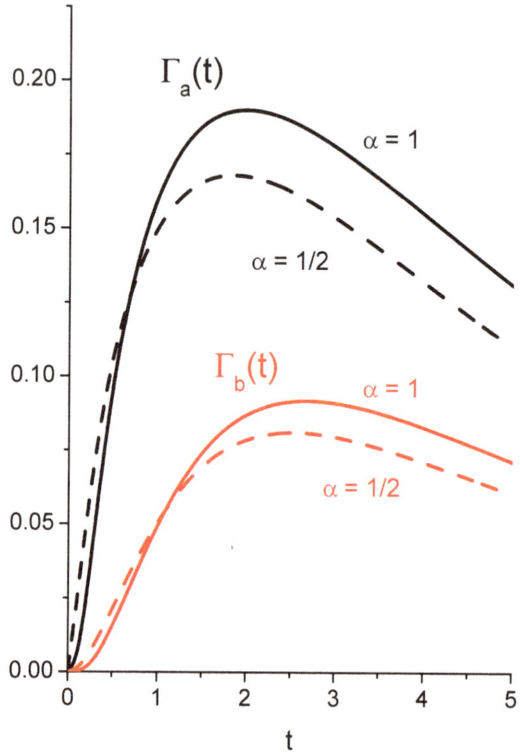

Fig. 2. This figure illustrates the behavior of $\Gamma_a(t)$ and $\Gamma_b(t)$ versus t for two different values of α. We consider, for simplicity, $\kappa = 1$, $k_a(t) = k_b(t) = k_{bulk}(t) = \delta(t)$, and $\mathcal{K}_a = \mathcal{K}_b = 1$.

present on the surface. For the case $t \to \infty$ ($s \to 0$) with $k_a(s) \to k_a = constant$, $k_b(s) \to k_b = constant$, and $k_{bulk}(s) = constant$, we have that $\Gamma_a(t) \to 0$, $\Gamma_b(t) \to 0$, and $\Gamma_{bulk}(t) \to constant$. In addition, by applying the inverse Laplace transform in Eqs. (16), (17) and (18), for $k_a(s) = k_a = constant$, $k_b(s) = k_b = constant$, and $k_{bulk}(s) = k_{bulk} = constant$, we obtain

$$\begin{aligned}\Gamma_a(t) &= \frac{k_b}{\kappa} e^{-k_a t} \int_0^t \Gamma_b(t') e^{k_a t'} dt' \\ &+ e^{-k_a t} \frac{k_b}{\kappa} \int_0^t dt' e^{k_a t'} \int_0^{t'} d\tilde{t}\, \frac{1}{\tilde{t}} \int_0^\infty dx \frac{\varphi(x)}{\sqrt{\mathcal{K}_a (t' - \tilde{t}\,)^\alpha}}\end{aligned}$$

$$\times \quad E_{\bar{\alpha},\bar{\alpha}}\left(-\frac{\kappa}{\sqrt{\mathcal{K}_a}}(t'-\tilde{t})^{-\bar{\alpha}}\right) H^{1,0}_{1,1}\left[\frac{x}{\sqrt{\mathcal{K}_a \tilde{t}^{\alpha}}}\left|\begin{matrix}(0,\frac{\alpha}{2})\\(0,\ 1)\end{matrix}\right.\right],$$

$$\Gamma_b(t) = \int_0^t dt' \Delta(t-t')\frac{1}{t'}\int_0^\infty dx\varphi(x) H^{1,0}_{1,1}\left[\frac{x}{\sqrt{\mathcal{K}_a t'^{\alpha}}}\left|\begin{matrix}(0,\frac{\alpha}{2})\\(0,\ 1)\end{matrix}\right.\right], \tag{22}$$

$$\Gamma_{bulk}(t) = k_{bulk}\int_0^t dt' \Delta(t-t')\int_0^\infty dx\varphi(x) H^{1,0}_{1,1}\left[\frac{x}{\sqrt{\mathcal{K}_a t'^{\alpha}}}\left|\begin{matrix}(1,\frac{\alpha}{2})\\(0,\ 1)\end{matrix}\right.\right], \tag{23}$$

with

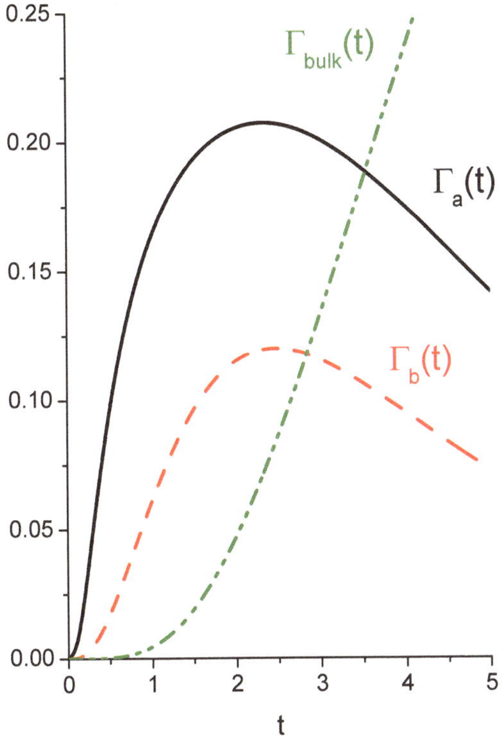

Fig. 3. This figure illustrates the behavior of $\Gamma_a(t)$, $\Gamma_b(t)$, and $\Gamma_{bulk}(t)$ versus t for $\alpha = 1$ and $k_{bulk}(t) = (k'_{bulk}/\tau)\, e^{-t/\tau}\, (k'_{bulk} = 1)$. We consider, for simplicity, $\kappa = 1$, $k_a(t) = k_b(t) = \delta(t)$, $\tau = 1$, and $\mathcal{K}_a = \mathcal{K}_b = 1$.

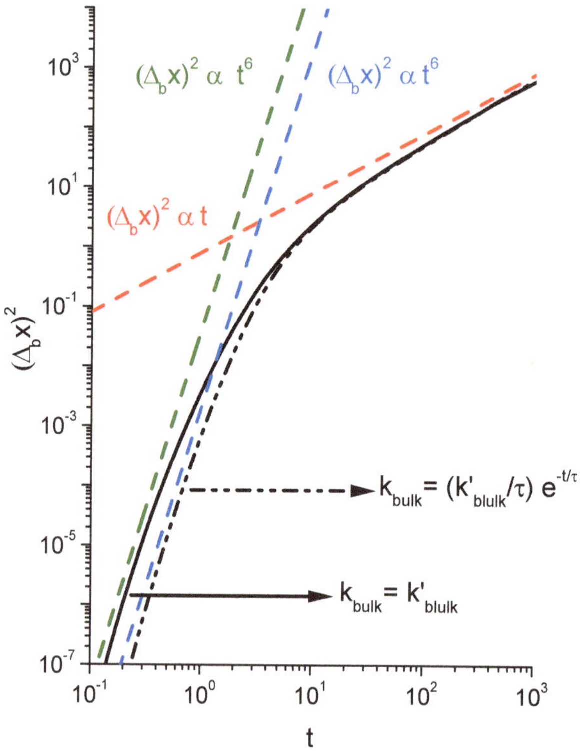

Fig. 4. This figure illustrates the behavior of $(\Delta_b x)^2$ versus t for $\alpha = 1$ and two different $k_{bulk}(t)$. We consider, for simplicity, $\kappa = 1$, $k_a = k_b = k'_{bulk} = 1$, $\tau = 1$. $\mathcal{K}_a = 10$, and $\mathcal{K}_b = 1$. The presence of the straight lines are used to evidence that the asymptotic behavior manifested by the system b for $t \to \infty$ and $t \to 0$ (red and blue lines) is anomalous. For these cases, before reaching the asymptotic limit, the initial spreading of the substance b is very faster.

$$\Delta(t) = 2k_a e^{-\bar{k}t} \int_0^t \frac{\kappa}{\sqrt{\mathcal{K}_a t'^{\alpha}}} \mathrm{E}_{\bar{\alpha},\bar{\alpha}} \left(\frac{\kappa}{\sqrt{\mathcal{K}_a}} t^{\bar{\alpha}} \right) e^{\bar{k}t'} \sinh\left(\delta\left(t - t'\right)\right) dt' , \tag{24}$$

where $\bar{k} = (k_a + k_b + k_{bulk})/2$ and $\delta = \sqrt{\bar{k}^2 - k_a k_{bulk}}$. Figures 1 and 2 show the behavior of $\Gamma_a(t)$, $\Gamma_b(t)$, and $\Gamma_{bulk}(t)$ for $\alpha = 1$ and $\alpha \neq 1$ in order to illustrate the influence of the fractional time derivative and, consequently, the behavior manifested by the solutions of the kinetic equations when $\alpha \neq 1$. Figure 3 illustrates the case characterized by $k_{bulk}(t) = \left(k'_{bulk}/\tau\right) e^{-t/\tau}$ which introduces a memory effect on the desorption process of the substance b. By using Eq.(12) the mean square displacement for the substance

b can be obtained. In particular, after some calculations, one can show that

$$\begin{aligned}(\Delta_b x)^2 &= \langle (x - \langle x \rangle)^2 \rangle_b \\ &= \langle x^2 \rangle_b - (2 - \Gamma_{bulk}(t)) \langle x \rangle_b^2\end{aligned} \tag{25}$$

with

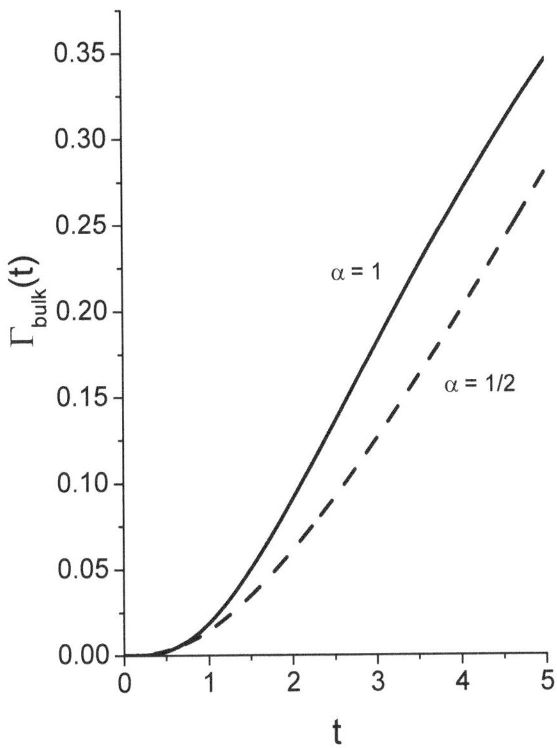

Fig. 5. This figure illustrates the behavior of $\Gamma_{bulk}(t)$ versus t for two different values of α. We consider, for simplicity, $\kappa = 1$, $k_a(t) = k_b(t) = k_{bulk}(t) = \delta(t)$, and $\mathcal{K}_a = \mathcal{K}_b = 1$.

$$\langle x^2 \rangle_b = 2 \frac{\mathcal{K}_b}{\Gamma(1+\alpha)} \int_0^t dt' (t - t')^{\alpha} \Phi(t') \quad \text{and} \tag{26}$$

$$\langle x \rangle_b = \frac{\sqrt{\mathcal{K}_b}}{\Gamma(1+\alpha/2)} \int_0^t dt'(t-t')^{\alpha/2} \Phi(t') \,. \tag{27}$$

These equations in the asymptotic limit ($t \to \infty$) lead us to $(\Delta_b x)^2 \sim t^\alpha$ which corresponds to a subdiffusive behavior for $0 < \alpha < 1$ and for $\alpha = 1$ the usual behavior is manifested. Similar results are found if $k_a(s) \to contant$, $k_b(s) \to contant$, and $k_{bulk}(s) \to contant$ for $t \to \infty$ (see Fig. 3). This result shows that for usual reaction processes on the surface the spreading of the substance released to the bulk is essentially governed by the diffusion equation.

3 Discussion and Conclusions

We have investigated sorption and reaction processes on a surface in contact with a system. We have obtained, in this scenario, solutions for the quantities formed on the surface by a reaction process that is governed by Eqs. (4), (5) and (6) and for the quantities present in the bulk. The results show that a reaction process present on the surface may present an stationary state for $k_{bulk} = 0$ and, consequently, $\rho_b(x,t) = 0$ for $t > 0$ and $\int_0^\infty dx \rho_a(x,t) \to 0$ for $t \to \infty$. This feature implies that the substance is sorbed by the surface where the reaction processes is occurring and is not desorbed. For $k_{bulk} \neq 0$, we have $\int_0^\infty dx \rho_b(x,t) \to constant$ and $\int_0^\infty dx \rho_a(x,t) \to 0$ for $t \to \infty$, implying that the substance a is converted into b by the reaction process present on the surface and released to the bulk, i.e., $\Gamma_a \rightleftharpoons \Gamma_b \to \Gamma_{bulk}$. We have also shown that the asymptotic behavior of the substance b in the bulk obtained by the reaction process may exhibit an anomalous spreading depending on the α value.

Acknowledgment: The authors are thankful to CNPq and CAPES (Brazilian agencies) for financial support.

Bibliography

Benjamin M. R., Vucini D., Domnisoru C., Bartol T. M., Hetzer M. W., Tartakovsky, D. M Sejnowski., Terrence J. (2013). Anomalous Diffusion of Single Particles in Cytoplasm. Biophysical, 104, 1652-1660.

Soula H., Care B., Beslon G., Berry H. (2013). Anomalous versus Slowed-Down Brownian Diffusion in the Ligand-Binding Equilibrium. Biophysical Journal, 105, 2064-2073.

Köpf M., Corinth C., Haferkamp O., Nonnenmachert T. F. (1996). Anomalous diffusion of water in biological tissues. Biophysical Journal, 70, 2950-2958.

Strizhak P. E. (2004). Macrokinetics of Chemical Processes on Porous Catalysts Having Regard to Anomalous Diffusion. Theoretical and Experimental Chemistry, 40, 203-208.

Avnir D. (1990) The Fractal Approach to Heterogeneous Chemistry. New York:Wiley-Interscience.

Masel R. I. (1996). Principles of Adsorption and Reaction on Solid Surfaces. New York: Wiley-Interscience.
Lenzi E. K., Lenzi M. K., Zola R. S., Ribeiro H. V., Zola F. C., Evangelista L. R., Gonçalves G. (2014). Reaction on a solid surface supplied by an anomalous mass transfer source, Physica A, 410, 399-406.
Mendez V., Campos D.l & Bartumeus F. (2014). Stochastic Foundations in Movement Ecology. Heidelberg: Springer.
ben-Avraham D. & Havlin S. (2005). Diffusion and Reactions in Fractals and Disordered Systems. Cambridge: Cambridge University Press.
Crank J. (1975). The Mathematics of Diffusion. England: Oxford Science Publications.
Pekalski A. & Weron K. S. (1998). Anomalous Diffusion: From Basics to Applications, Lecture Notes in Physics (vol. 519). Berlin: Springer.
Klafter J. & Sokolov I. M. (2011). First Steps in Random Walks: From Tools to Applications, Oxford: Oxford University Press.
Leijnse N., Jeon J.-H., Loft S., Metzler R. & Oddershede L. B. (2012). Diffusion inside living human cells. European Physical Journal – Special Topics, 204, 75-84.
Hilfer R., Metzler R., Blumen A. & Klafter J. (2004). Strange Kinetics, Chemical Physics (vol. 284). Amsterdam: Pergamon-Elsevier, .
Metzler R. & Klafter J. (2000). The random walk's guide to anomalous diffusion: a fractional dynamics approach. Physics Report, 339, 1 –77.
Metzler R. & Klafter J. (2004)., The restaurant at the end of the random walk: recent developments in the description of anomalous transport by fractional dynamics. Journal Physics A: Mathematical and General, 37, R361.
Eliazar I. I. & Shlesinger M. F., Fractional Motions, Physics Report, 527, 101 – 129.
Bressloff P. C. & Newby J. M. (2013). Stochastic models of intracellular transport. Reviews of Modern Physics, 85, 135 – 196.
Condamin S., Tejedor V., Voituriez R., Bénichou O. & Klafter J. (2008). Probing microscopic origins of confined subdiffusion by first-passage observables. Proceedings of the National. Academy of Science of the United States of America, 105, 5675.
Robson A., Burrage K. & Leake M. C. (2013). Research article: Inferring diffusion in single live cells at the single-molecule level, Phil. Trans. R. Soc. B, 368, 1471 – 2970.
Caputo M., Cametti C., & Ruggero V. (2008). Time and spatial concentration profile inside a membrane by means of a memory formalism, Physica A, 387, 2010 – 2018.
Caputo M. & Cametti C. (2009). The memory formalism in the diffusion of drugs through skin membrane, Journal of Physics D: Applied Physics, 42, 12505
Kosztolowicz T., Dworecki K., Lewandowska K. D. (2012). Subdiffusion in a system with thin membranes, Physical Review E, 86, 021123.
Langlands T. A. M. & Henry B. I.,(2010). Fractional chemotaxis diffusion equations, Physical Review E, 81, 051102.
Caspi A., Granek R. & Elbaum M. (2000). Enhanced Diffusion in Active Intracellular Transport, Physical Review Letters, 85, 5655.
Bruno L., Levi V., Brunstein M. &Desposito M. A. (2009). Transition to superdiffusive behavior in intracellular actin-based transport mediated by molecular motors, Physical Review E, 80, 011912.
Brangwynne C. P., Koenderink G. H., MacKintosh F. C. & Weitz D. A. (2009). Intracellular transport by active diffusion, Trends in Cell Biology, 19, 423 – 427.
Weber S. C., Spakowitz A. J. & Theriot J. A. (2012). Nonthermal ATP-dependent fluctuations contribute to the in vivo motion of chromosomal loci, Proceedings of the National. Academy of Science of the United States of America, 109, 7338 – 7343 .

Sokolov I. M. (2012). Models of anomalous diffusion in crowded environments. Soft Matter, 8, 9043 – 9052 .

Weigel A. V., Simon B., Tamkun M. M. & Krapf D. (2011). Ergodic and nonergodic processes coexist in the plasma membrane as observed by single-molecule tracking, Proceedings of the National. Academy of Science of the United States of America, 108, 6439 – 6443.

Jaing H., Cheng Y., Tuan L., An F. &Jin. K., (2013). A fractal theory based fractional diffusion model used for the fast desorption process of methane in coal Chaos, 23, 033111.

Podlubny I. (1999). Fractional Differential Equations. San Diego, CA: Academic Press.

Mathai A. M., Saxena R. K. & Haubold H. J. (2009). The H-Function: Theory and Applications, New York: Springer.

Kopelman R.(1988). Fractal Reaction Kinetics. Science, 241, 1620 – 1626.

Jocelyn Sabatier*

Fractional Order Models for Electrochemical Devices

1 Introduction

High automobile pollution standards and rising oil prices have led car manufacturers to design new vehicles (stop and start, hybrid, or electric). The underlying idea of these devices is to reduce exhaust emissions in built-up areas, either by stopping the internal combustion engine when the vehicle is stationary, or by substituting electric fuel for fossil fuel. The latter solution may involve an electric motor and one or more energy storage systems.

These new engines have increased the use of devices such as supercapacitors or batteries on board the vehicle. To keep these new vehicles in good working order, car manufacturers must integrate reliable electrical energy storage management. To achieve this, state of charge (SOC) and state of health (SOH) estimators must be designed. To design these estimators, dynamical models of the supercapacitor and battery pack can be valuable tools.

Electrochemical devices such as ultracapacitors or batteries exhibit long memory behaviors. While the physical origin of these behaviors is still not fully understood, they can be described well by fractional order models. This chapter reviews several applications of fractional order models to the dynamical modeling of electrochemical devices. It also shows that these models can be used to evaluate the SOC or the SOH of the considered devices.

2 Fractional Modeling of Supercapacitors

In several studies available in the literature, long-range phenomena observed during supercapacitor relaxation are taken into account through the introduction of fractional order models (Riu et al., 2004) (Quintana et al., 2006). The voltage dependency of their dynamical behavior however is not addressed in these studies. Dynamic behavior voltage dependency is taken into account in (Behachemi et al., 2000) through the introduction of nonlinear components in the equivalent electrical model. Voltage dependency in component values illustrates bias voltages. But as demonstrated in

***Corresponding Author: Jocelyn Sabatier:** IMS Laboratory, CRONE group, Bordeaux University, 351, Cours de la Libération, 33400 Talence, France, E-mail: Jocelyn.sabatier@ims-bordeaux.fr

(Bertrand et al., 2008), this modeling method can produce a local linear model that does not fit the supercapacitor behavior around an operating voltage.

A large number of the above-mentioned modeling methods are based on impedance spectroscopies or specific time signals that require complex specialized instrumentation and that cannot be done on-line.

The modeling approach proposed in this chapter attempts to solve these problems. It leads to a model that:

- has parameters connected to the physical characteristics of the supercapacitor,
- takes into account dynamic behavior voltage dependency,
- has a small number of parameters due to the introduction of fractional differentiation,
- is simple enough to be embedded,
- requires a simple test for parameter determination, that can be done on-line.

In order to develop an analytical supercapacitor model, it is important to know how the supercapacitor is constituted and the phenomenon which governs its behavior.

A description of a supercapacitor design is given in (Maxwell, 2004) and an explanation of the electrochemical phenomena that take place inside a supercapacitor (charge balances) is available in (Zubieta, 2000). As defined in (Maxwell, 2004) and illustrated by Fig. 1, the supercapacitors considered here are composed of three layers: a carbon powder layer (conducting carbon layer) compressed on the current collector layer and an activated carbon layer added to the conducting carbon.

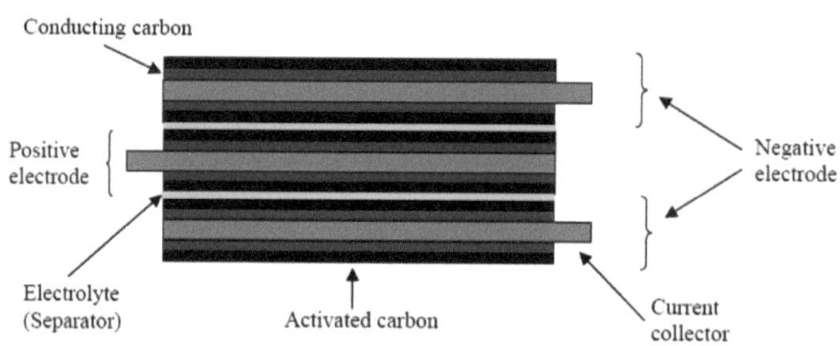

Fig. 1. Supercapacitor structure

In our modeling approach, it was assumed that

- no Faradic process takes place inside the capacitor (no oxydo-reduction reaction at the electrode/electrolyte interface);
- the electrode/electrolyte interface has a rough aspect;
- there is no voltage or temperature dependency in the parameters;
- the ohmic losses generated by the electrode resistance are null;

– the influence of wires and separator is taken into account by R_S (resistance in series with the interface).

Based on the above hypotheses, the equivalent electrical model of the supercapacitor can be represented as in Fig. 2. The resistive components $r_{el}(k)$ are deduced from the electrolyte conductivity. The double layer capacitor $C_{DL}(k)$ is induced by the electrical potential at the electrode/electrolyte interface. The rough aspect of the electrode/electrolyte interface leads to additional electrolyte resistive components $r_I(k)$ in series with the double layer capacitor.

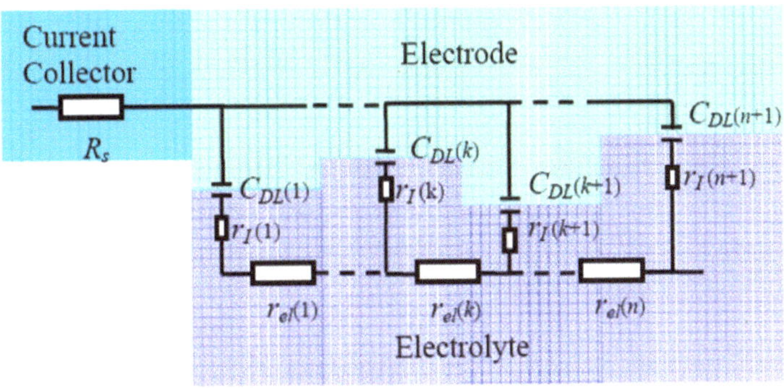

Fig. 2. Supercapacitor equivalent electrical model

Using Görh's method (Görh, 1997), it has been shown (Bertrand et al., 2010a), that the supercapacitor equivalent electrical circuit of Fig. 2 admits the following transfer function:

$$Z_C(s) = R_S + \left(\frac{R_{el}}{Q(s)^{1-\gamma}}\right)^{1/2} \cdot \coth\left[\left(R_{el}Q(s)^{1-\gamma}\right)^{1/2}\right]. \tag{1}$$

In order to validate the supercapacitor model defined by 1, frequency parameter identification was performed. The supercapacitor frequency response was obtained by impedance spectroscopy on a Maxwell 2000F/2.5 V supercapacitor. Note that all the impedance spectroscopies presented here were carried out in a climatic chamber at 25°C, in potentiostatic mode, on the frequency band [1 mHz-100 Hz] with an AC voltage of 20 mV amplitude around a 1 V bias voltage.

The frequency behavior of the identified model is similar to the supercapacitor frequency behavior as presented in Fig. 3.

Other tests were conducted for different operating voltages and at different temperatures, and similar results were obtained. Thus, it is assumed that the linear model 1 represents the supercapacitor dynamic behavior for a given operating voltage and a given temperature.

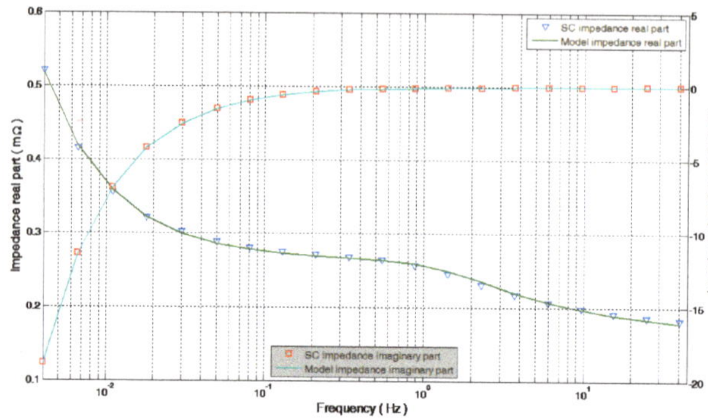

Fig. 3. Comparison between the supercapacitor and the non-integer model frequency responses

This model is accurate but is very complex for a real time implementation due to the *coth* function. To obtain a rational model, the *coth* function was approximated using the Padé approximation method. This method has the advantage of approximating functions in a large frequency band, depending on the degree of Padé approximation chosen.

Here, we obtain an accurate model in the frequency range [1m Hz-10 Hz], and the following model was obtained:

$$\tilde{Z}_C(s) = R + R_1 + \frac{1}{Q_1 s^{1-\gamma}} + \frac{R_2}{R_2 \cdot Q_2 s^{1-\gamma} + 1}. \tag{2}$$

The (R_1, R_2) and (Q_1, Q_2) parameters are respectively proportional to R_{el} and Q as shown by the following equations :

$$R_1 = \frac{R_{el}}{10}, \quad R_2 = 0.233R_{el}, \quad Q_1 = Q, \quad Q_2 = 1,2Q. \tag{3}$$

This implies that only two of the four parameters need to be identified. The remaining two parameters can be deduced from the relations 3.

To validate the model 2, its frequency behavior was compared to that of the supercapacitor in Fig. 4 (Remark - impedance spectroscopy is used here only to validate the model 2. The parameter values of the nonlinear model presented in the next section will be determined through time identification). This figure shows a model frequency behavior similar to that of the supercapacitor within the frequency range [1 mHz-10 Hz].

The model 2 is only valid around an operating voltage. The goal now is therefore to obtain a non-linear supercapacitor model with a voltage dependency of the double layer capacitor.

To reach this goal, it is important to note that the first order Taylor approximation of a nonlinear system can be used to obtain local models, models whose behavior is

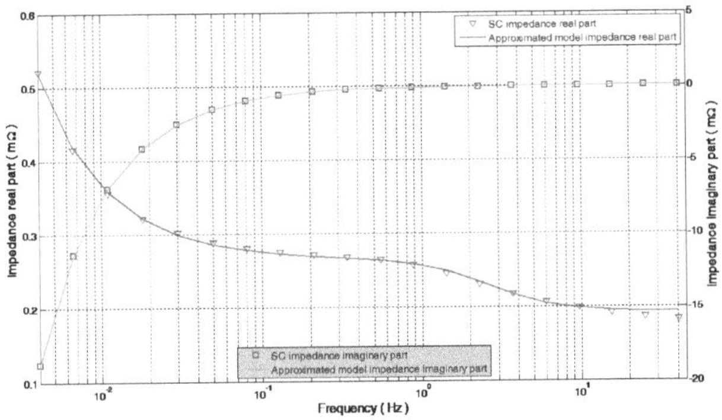

Fig. 4. Comparison between supercapacitor and the approximated model frequency responses

equivalent to their nonlinear counterparts around the operating point selected to compute the first order approximation. A method (denoted pseudo-integration) developed by (Mouyon, 1986) has shown how to obtain a nonlinear system using a set of local models.

This method was applied to the supercapacitor nonlinear model. Only the voltage dependency of the double layer capacitor was taken into account in this nonlinear model. The local model 2 can be written as the following pseudo state space model:

$$\begin{cases} \dfrac{d^{1-\gamma}\delta z_1}{dt^{\gamma}} = \delta z_2 \\ \dfrac{d^{1-\gamma}\delta z_2}{dt^{\gamma}} = -\dfrac{1}{R_2 Q_2 (V_0)}\delta z_2 + \dfrac{1}{R_2 Q_2 (V_0) Q_1 (V_0)}\delta I \end{cases} \tag{4}$$

$$\delta V = \delta z_1 + (Q_2 (V_0) R_2 + Q_1 (V_0) R_2)\,\delta z_2 + (R_S + R_1)\,\delta I$$

Pseudo-integration of relation 4 gives the following nonlinear fractional model:

$$\begin{cases} \dfrac{d^{1-\gamma} z_1}{dt^{\gamma}} = z_2 \\ \dfrac{d^{1-\gamma} z_2}{dt^{\gamma}} = -\dfrac{1}{R_2 Q_2 (V)} z_2 + \dfrac{1}{R_2 Q_2 (V) Q_1 (V)} I \end{cases} \tag{5}$$

$$V = z_1 + (Q_2 (V) R_2 + Q_1 (V) R_2)\, z_2 + (R_S + R_1)\, I$$

$Q_1(V)$ and $Q_2(V)$ are polynomials that fit the variations of Q_1 and Q_2 as the operating voltage V varies. It is easy to check that the Taylor approximation of 5 around the operating voltage V_0 produces the model 4. The non-linear model 5 is of low complexity and can be embedded.

The nonlinear model 5 can be simulated using the Matlab/Simulink diagram of Fig. 5. To embed the model 5, the same diagram can be used, with the $1/s^{1-\gamma}$ fractional integrator approximated by an integer transmittance. The method based on the recursive distribution of poles and zeros presented in (Oustaloup et al., 2000) can be used to compute this approximation.

The most usual method to accurately characterize supercapacitor dynamic behavior is based on impedance spectroscopy. However, this requires specialized instrumentation and cannot be easily implemented onboard for model identification in a large frequency band, in embedded applications. For hybrid electric vehicle applications, the current profile for the identification test must be of low complexity. That is why a specific time identification method was developed (Bertrand et al., 2010b) in order to obtain the supercapacitor model parameters.

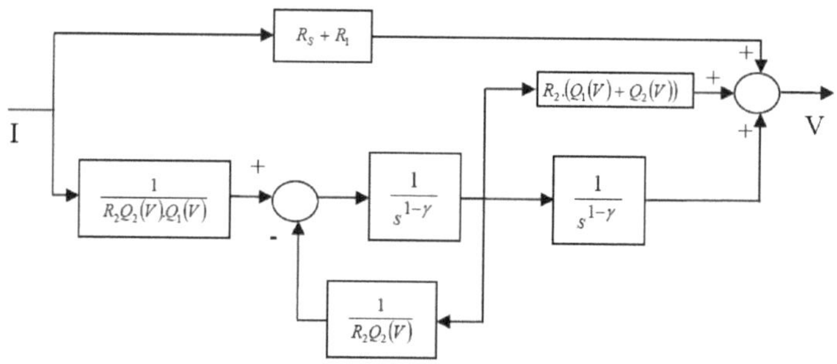

Fig. 5. Matlab/Simulink supercapacitor model (voltage dependency appears in parameters $Q_1(V)$ and $Q_2(V)$)

A first validation test was done on the supercapacitor at ambient temperature with a 10 amp charge/discharge current in order to be consistent with the measures used for the model identification.

The supercapacitor and the model voltage responses are presented in Fig. 6. In order to evaluate the model performance, the relative error between supercapacitor and model voltages is also shown in Fig. 6. The maximum value of the error is about 1%. For this test profile the model gives very accurate results.

The same profile was applied to a supercapacitor with a 100 amp current in order to observe the influence of current magnitude on the cell behavior. The voltage responses of the model and of the supercapacitor are presented in Fig. 7. The durations of rest times are the same as for the previous test.

The maximum relative error between the supercapacitor and the non-linear model voltage responses reached 2% for about one hour of test. The relative error during the test is also presented in Fig. 7.

The result shows that the model gives an accurate voltage response for a 100 amp current level. Given that parameter identification was conducted for a 10 amp current, it can be assumed that there is no current level dependency of the supercapacitor. As for the previous test, the capacitance Q(V') was equal in charge and in discharge.

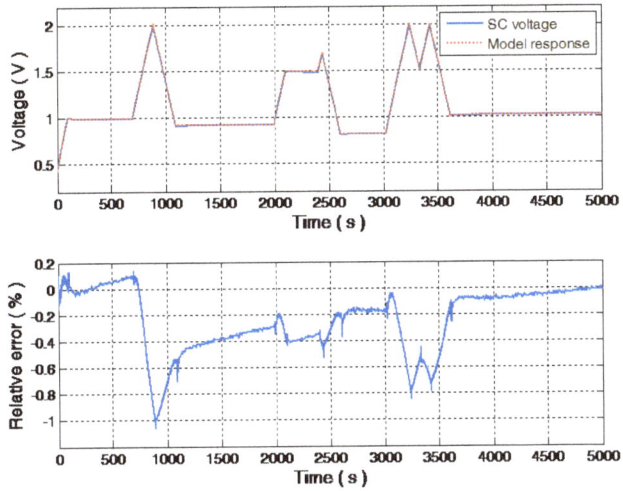

Fig. 6. Validation test with 10 amp current profile

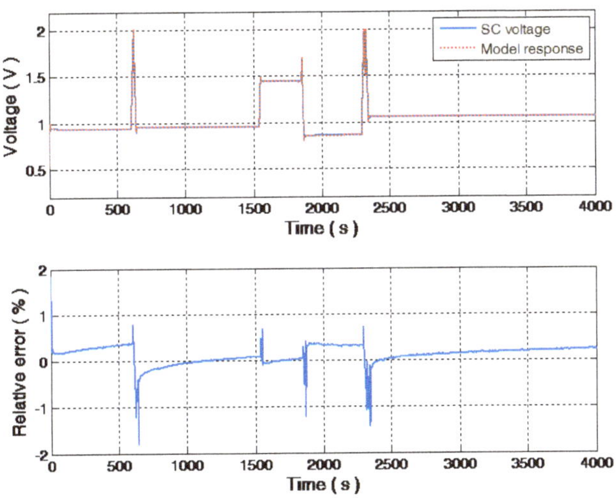

Fig. 7. Validation test with a 100 amp current profile

3 Fractional Modeling of Lead Acid Batteries with Application to State of Charge and State of Health Estimation

The Randles model is frequently used in the literature to model lead acid batteries. This model arises from a simplified resolution of the electrochemical diffusion equation in batteries (Fick's equations) (Sathyanarayana et al., 1979). If $u(t)$ denotes voltage variations in relation to the open circuit voltage and if $i(t)$ denotes the battery current, the Randles model is defined by the electric circuit of Fig. 8.

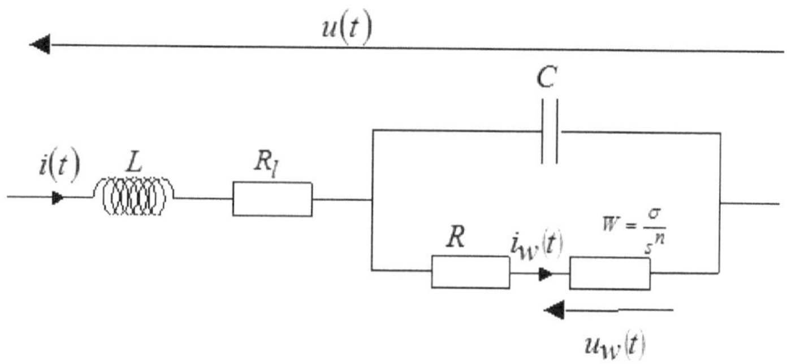

Fig. 8. Randles model of a lead acid battery

The fractional behavior of the Randles model is due to fractional impedance $W(s)$. This impedance is known as the Warburg cell and is a fractional order integrator of order $n = 0.5$.

It was demonstrated in Sabatier et al. (2006) that the low frequency asymptotic behavior of the Randles model (a fractional integrator of order 0.5) is in contradiction with the time behavior of the battery in discharge. This is why another model was proposed. It is defined by the fractional transfer function:

$$H_1(s) = K\left(\frac{1+\frac{s}{\omega_b}}{1+\frac{s}{\omega_r}}\right)^{-n_1}\left(\frac{1+\frac{s}{\omega_r}}{1+\frac{s}{\omega_h}}\right)^{-n_2}. \tag{6}$$

The associated parametric estimation method was defined in Sabatier et al. (2006). Fig. 9 is a comparison of the battery voltage variations with the model response. The input current applied to the battery is a pseudo random binary sequence. It reveals that the model 6 can satisfactorily capture the dynamic behavior of the battery.

Fig. 10 shows the parameter variations of the model in relation 6 as the battery state of charge (SOC) varies. This figure highlights the monotonic variation of some

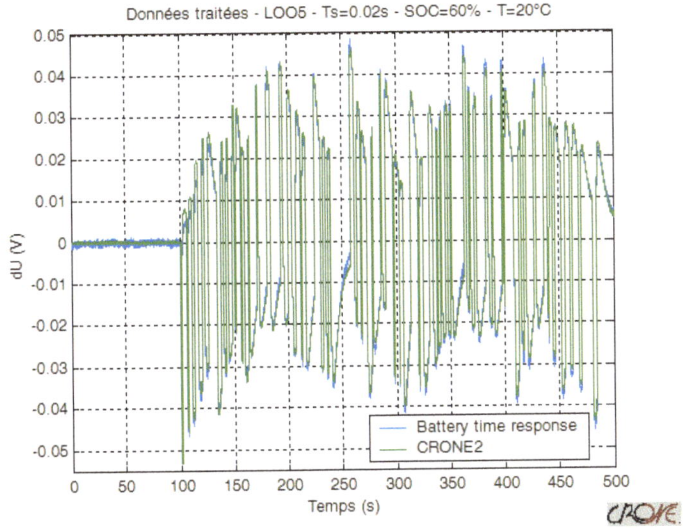

Fig. 9. Comparison of the battery voltage variations with the model response.

parameters such as ω_b, ω_r and n_1 as SOC varies. These variations were used to determine the battery state of charge through battery behavior identification.

The SOC estimator that was designed is described by Fig. 11. As shown in this figure, SOC estimation is based on an estimation function involving variations curves of Fig. 10, which uses parametric variations of $H_1(s)$ models as the battery SOC varies (Sabatier et al., 2006; Ragot et al., 2006a; 2006b).

To validate the proposed state of charge estimation method, two batteries were submitted to a charge and discharge current of a pseudo random binary sequence type, for two temperatures (20°C and 40°C). The battery state of charge was unknown by the estimator users (Sabatier et al., 2006). The SOC estimations had an error close to 2%, thus validating the estimation method but also the battery model proposed (Sabatier et al., 2006; Gregoire et al., 2006; Ragot et al., 2007).

Aging is also an increasing concern in the design of battery state estimators and was taken into account in Cugnet et al. (2009a). For lead-acid batteries, aging mainly involves an increase in resistance (corrosion, sulfation, loss of conductivity at the collector / active material interface). Therefore, the battery resistance measurement is an indicator that is frequently used to estimate age. Also, as demonstrated by Cugnet et al. (2009b), battery available power is correlated to battery resistance (Cugnet et al., 2009d). Given that car cranking mainly requires power (and not energy), battery resistance can be used to evaluate battery cranking capability.

If the lead-acid battery study is limited to the frequency range [8 Hz; 1 kHz] corresponding to the vehicle cranking signal spectrum, a simpler model than the one given by relation 6 can be used to describe its dynamic behavior. It is defined by (Sabatier et

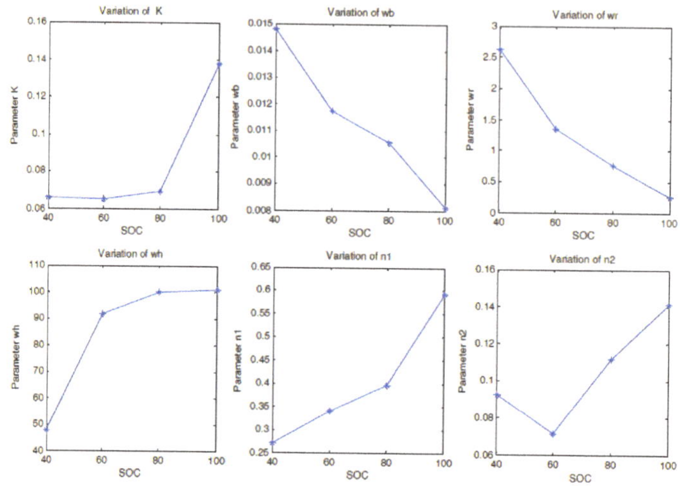

Fig. 10. $H_1(s)$ model parameter variations as state of charge variations

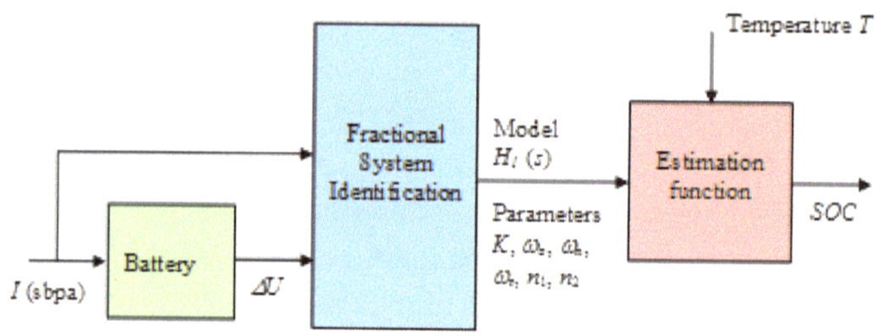

Fig. 11. Functional diagram of SOC estimator based on battery fractional model identification

al., 2010a):

$$H_2(s) = K \left(\frac{1 + \frac{s}{\omega_h}}{1 + \frac{s}{\omega_b}} \right)^{\gamma} . \tag{7}$$

As shown in Fig. 12, which presents a comparison of the battery voltage variation during cranking and of the model response, the model in relation 7 captures the dynamic behavior the battery well during cranking (better than an equivalent complexity integer model).

The frequency responses of several models, obtained from data measured with a battery used at different SOF_C (SOF_C denotes the ratio of the available capacity to the rated capacity), are given in Fig. 13. The battery resistance corresponds to the $H_2(s)$ fractional model high-frequency gain (null imaginary part), and it is thus possible to estimate the battery resistance from a $H_2(s)$ model identification during cranking.

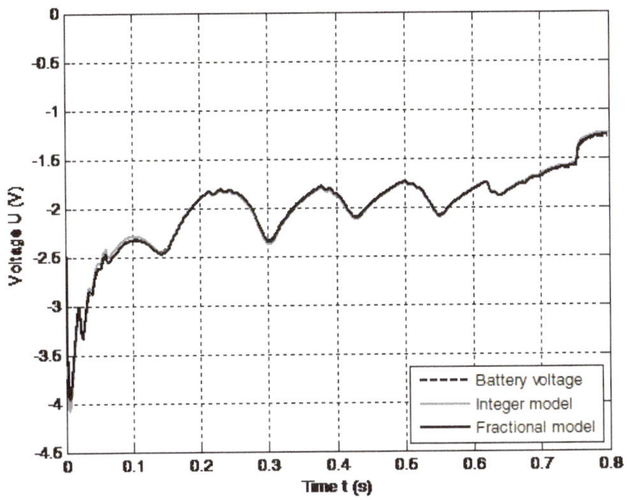

Fig. 12. Comparison of both integer and fractional models simulation performances according to the data measured during vehicle cranking

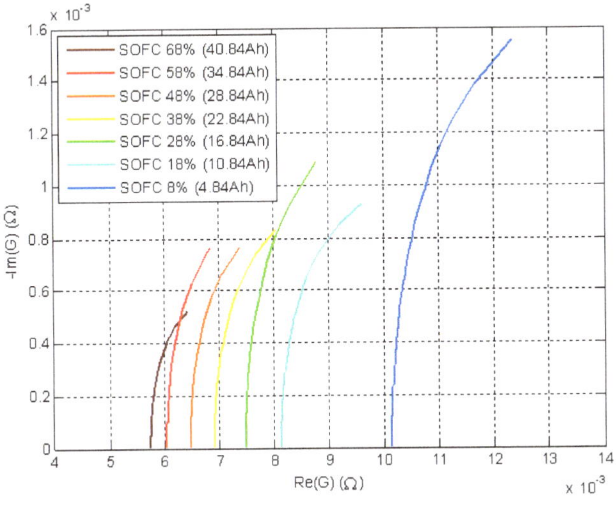

Fig. 13. Nyquist Diagrams of several battery identified models for various SOF_C

In Fig. 13, it can also be seen that the high-frequency gain, and thus the resistance, is directly linked to the SOF_C. The higher the resistance, the lower the SOF_C. It can be concluded from the above that the battery resistance can be determined from an $H_2(s)$ model identification during cranking associated to a SOF_C estimator and a resistance variation law with temperature.

Another studies presented in Cugnet et al. (2010) and Sabatier et al. (2011) showed that a vehicle wil not start when the power is less than 4 kW or when resistance is higher than 8.5 mΩ. The reason for the non-cranking is not the lack of energy but a lack of power induced by the increase in battery resistance. Estimating the resistance thus enables the available power and cranking capability to be estimated. Such a estimation was done in Cugnet et al. (2009c; 2010) using a model invalidation method.

4 Fractional Modeling of Lithium-ion Batteries with Application to State of Charge

As lithium ion-batteries have an excellent stored energy / weight ratio, many manufacturers have become interested in them. Among all the models available in the literature, a well-known electrochemical model for a lithium-ion cell is the pseudo 2D model in Fig. 14.

The representation on Fig. 14 emphasizes the two time scales of Li^+ ions diffusion in the cell. This idea is roughly materialized on the scheme by the following:

- electrodes are seen as an aggregation of spherical particles in which the Li+ ions are inserted;
- the first spatial dimension of this model, represented by variable x, is the horizontal axis;
- the second spatial dimension is the particle radius.

A more exhaustive presentation of this model is available in Newman and Thomas Aleys (2004) and Botte et al. (2000).

This model has a physical meaning, unlike purely electric models proposed in the literature but it is impossible to use it with a car microcontroller. To overcome this problem and thus to reduce the electrochemical model to a simpler one, several simplifying assumptions and some approximations were proposed in (Sabatier et al. (2014a; 2014b). An initial fractional model was obtained in (Sabatier et al., 2014a) and a simpler one in (Sabatier et al., 2014b). The simplest model obtained is in fact a single electrode model. A diagram describing this fractional model is given in Fig. 15.

The main assumptions used to derive this model are the following.

- On a wide SOC range, the most significant contribution to the cell voltage comes from the positive electrode;

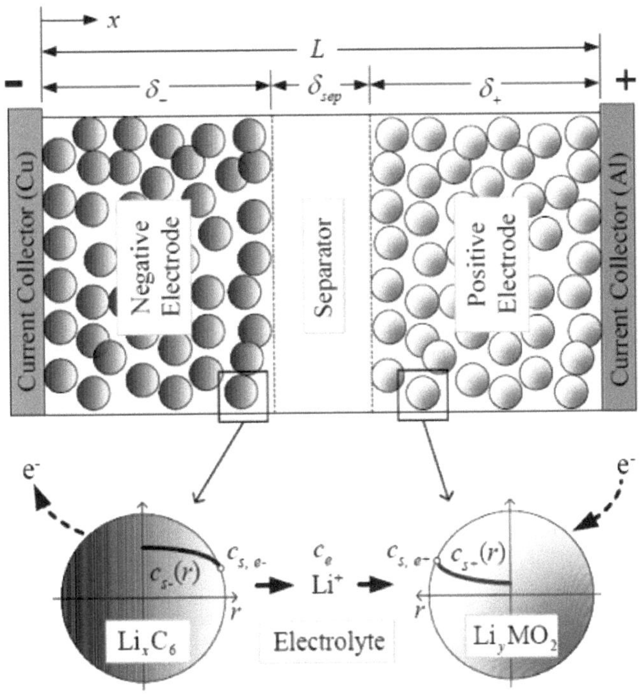

Fig. 14. Pseudo 2D model of a Lithium-ion cell (Newman & Thomas-Alyea, 2004)

- From the former it follows that the Open-Circuit Voltage (OCV) is reduced to the positive electrode equilibrium potential U_p. The latter was modeled using the Redlich-Kister representation of the Gibbs cell free excess energy (Karthikeyan et al., 2008; Hashim Ali, 2005);
- The influence of the electrolyte is neglected.

Thanks to the above assumptions, the model parameters keep a physical meaning:
- K_1 is the inverse of the maximum theoretical capacity of the cell [C^{-1}];
- K_2 is the sum of the internal resistance and contact resistance of the cell [Ω];
- ω_c is proportional to the average solid-phase diffusion coefficient of the positive electrode [s^{-1}];
- OCV accounts for the open-circuit voltage according to the Redlich-Kister expansion (Karthikeyan et al., 2008):

$$OCV \approx U_p = U_p^0 + \frac{RT}{F} \ln\left(\frac{1 - x_p}{x_p}\right) + \sum_{k=0}^{N_p} A_k \left[(2x_p - 1)^{k+1} - \frac{2x_p k(1 - x_p)}{(2x_p - 1)^{1-k}}\right]. \quad (8)$$

The user-defined SOC is linked to the positive electrode insertion rate x_p by:

$$x_p = SOC\left(x_{p,100} - x_{p,0}\right) + x_{p,0} \tag{9}$$

where $x_{p,100}$ and $x_{p,0}$ are respectively the insertion rates at the states SOC=100% and SOC=0%. In Fig. 15, the initial value of the insertion is noted x_p^{init}.

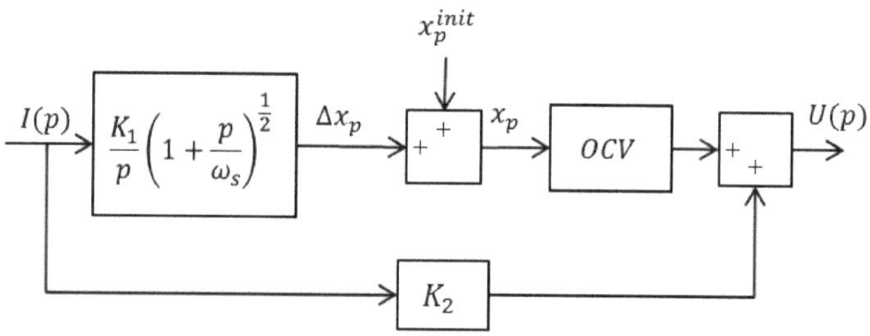

Fig. 15. Single-electrode model

Finally, the solid-phase diffusion impedance is implemented via a recursive distribution of poles and zeros as presented in Oustaloup et al. (2000).

The single electrode model appears to be compact, given its small number of parameters (Guillemard et al., 2012a). In comparison to a cell electric circuit model (Hua et al., 2012), it needs only one parameter, namely ω_C, to take the solid-phase diffusion into account. As shown in Sabatier et al. (2014b) and by Fig. 16 that presents a comparison of the cell voltage and of the model response, this small number of parameters does not detract from the accuracy of such a model.

Li-ion battery packs have to work together with reliable Battery Management Systems (BMS) to ensure their optimal and safe use. Among the tasks ensured by the BMS, State-Of-Charge (SOC) estimation is of crucial importance. To obtain this information that cannot be measured, an output feedback based observer was used in Sabatier et al. (2014), Guillemard et al. (2012b; 2012c) and Nouillant et al. (2012). This observer is represented by Fig. 17 (Francisco et al., 2014).

As shown by Fig. 18, the observer permits an SOC estimation with an error less than 3%.

5 Conclusion

This chapter has shown how fractional order models can be used to capture the dynamical behavior of electrochemical devices such as batteries or supercapacitors. It has also shown how the models can be used to build state of charge or state of health

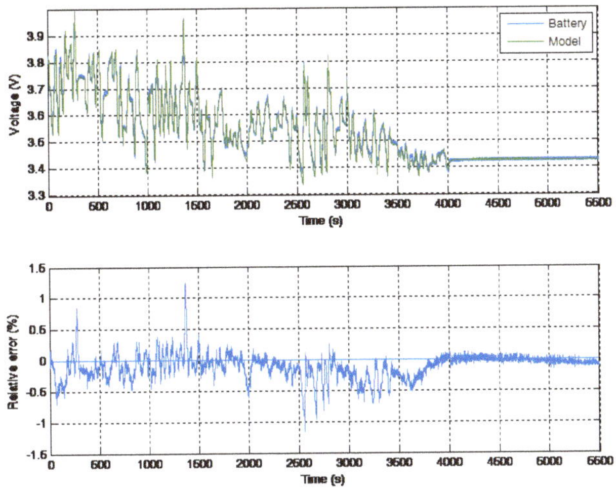

Fig. 16. Cell voltage variations / model response comparison for a current cycle and relative error

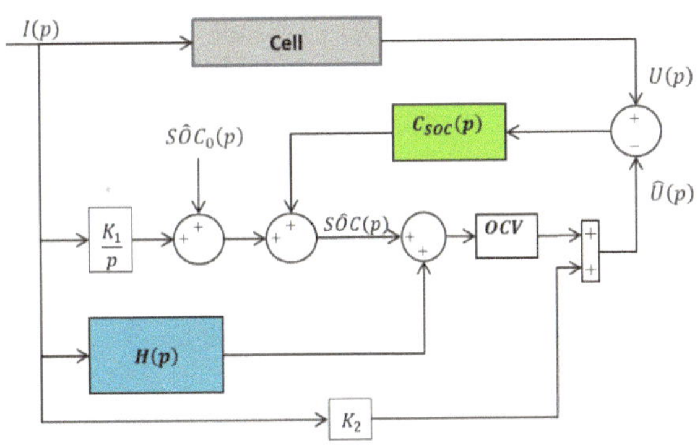

Fig. 17. SOC feedback observer

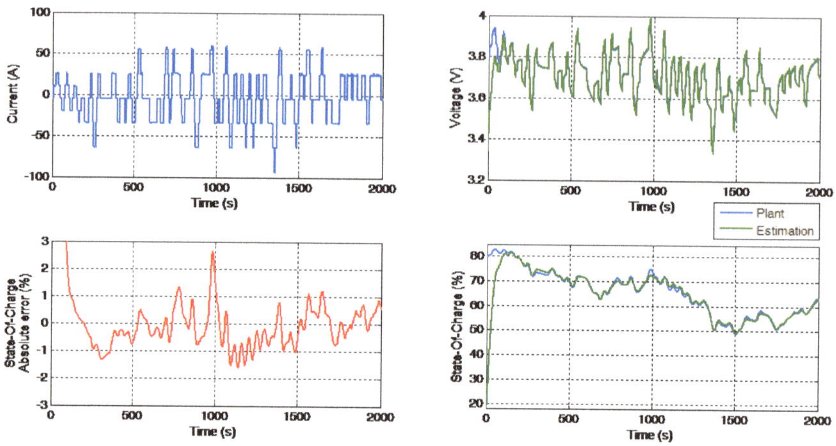

Fig. 18. SOC feedback observer validation

estimators. It is obvious that this work can be extended to other electrochemical devices, but it also raises questions about the nature of non-integer systems and the physical interpretation of the models.

Fractional order models are not quite conventional linear models, and not quite conventional distributed parameter models described by a diffusion equation. They are in fact halfway between these two classes of systems, which explains why they are particularly suited for diffusion phenomena modeling, although there is still a lively debate on the physical interpretation of these models, as now explained.

While integer order differential equations derive from an infinitesimal analysis of phenomena, fractional differential equations are mainly a mathematical extension of their integer counterparts. Despite this, the fractional system or more exactly the fractional model concept, which derives directly from fractional differential equations, has given rise to many results: modeling studies in many domains and the associated system identification methods; approximation, discretization and simulation methods; fractional models properties analysis; fractional models control methods.

One may wonder whether the fractional model concept was created hastily. For instance, a look at the literature indicates that the definition of these models using Caputo's derivative has been questioned (Sabatier et al., 2008; 2010b; 2012; Trigeassou and Maamri, 2009). Until now, there have been few physical interpretations of these models (and of fractional differentiation), some of which are unconvincing. It is therefore questionable whether these models are physically consistent or whether they should only be considered as mathematical tools to be handled with care when they are used to characterize the dynamical behavior of a real system. Indeed, for these models, the properties mentioned in the literature are stability in the Lyapunov sense, initialization, observability, controllability, etc., while their internal state is still poortly understood (Sabatier et al., 2008; 2010b). What is the internal state of a fractional model? Is the internal state proposed in the literature physically consistent with the internal state of the modeled system? Some preliminary results suggest not (Sabatier and Farges, 2014c; Sabatier et al., 2015). Thus, the stability, initialization, observability, and controllability properties of the models may have nothing to do with the properties of the modeled system. This means that these models may only be good for capturing the input-output behavior of the system in a given range of time and provide no information about its internal structure. Is the study of the internal properties of these models useful?

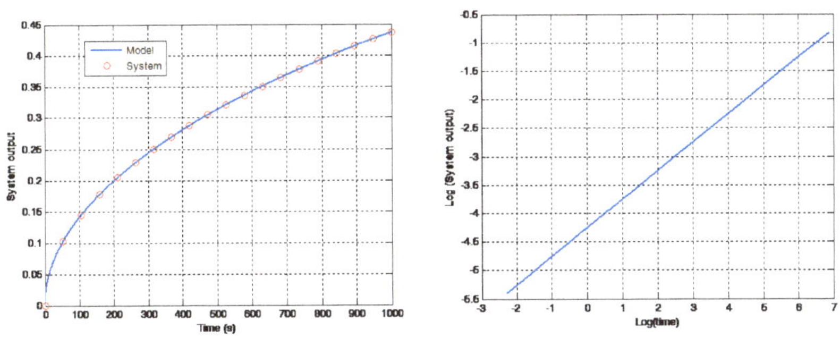

Fig. 19. System time response and its Log-Log representation

An example is now used to illustrate how different the internal behavior of a system and its fractional model may be. A system step response and its Log-Log representation are given in Fig. 19. This system seems to behave like the fractional model:

$$H(s) = \frac{0.014}{s^{0.5}} \quad . \tag{10}$$

Based on this model, the system internal properties could be studied. But are these properties representative of the properties of the modeled system, as it is defined by the following state space representation (integer non-linear)?:

$$\begin{cases} \frac{d}{dt}x(t) = 10^{-4}x(t) + 2.10^{-4}u(t) \\ y(t) = x(t)^{0.5} \end{cases} \quad . \tag{11}$$

Note that the non-linearity can be very difficult to detect due to physical limitations on the system output or input (as for ultra-capacitors, batteries, lungs, ...).

Understanding the physical nature of fractional models and more broadly of fractional differentiation seems a major challenge. This will give credit (or not) to a series of studies based on a simple extension of their integer counterparts. It should also make it possible to understand the origin(s) of what produces fractional behaviors (or more exactly $t^{\nu-1}$, 0<v<1, dynamics), and thus specify the interest of fractional differentiation for modeling real physical systems.

Whatever the result, and beyond the field of modeling, it remains undeniable that fractional differentiation has led to the emergence of powerful new tools in many areas.

Bibliography

Belhachemi F., Rael S. & Davat B. (2000) A physical based model of power electric double-layer supercapacitors, IEEE Industry Applications Conference, vol. 5, pp. 3069-3076, Oct. 8-12.

Bertrand N., Sabatier J., Briat O. & Vinassa J.M. (2008). An ultracapacitor non-linear fractional model, Vehicle Power and Propulsion Conference VPPC, pp. 1-6

Bertrand N., Sabatier J., Briat O. & Vinassa J. M. (2010a). Fractional Non-Linear modelling of Ultracapacitors, Communications in Nonlinear Science and Numerical Simulation, Volume 15, Issue 5, pp 1327-1337.

Bertrand N., Sabatier J., Vinassa J. M. & Briat O. (2010b). Embedded Fractional Nonlinear Supercapacitor Model and Its Parametric Estimation Method, IEEE Transactions on Industrial Electronics, Vol. 57, No 12, pp 3991-4000.

Botte G. G., Subramanian V. R. & White R. E. (2000). Mathematical modeling of secondary lithium batteries, Electrochimica Acta 45, 2595–2609.

Cugnet M., Laruelle S., Grugeon S., Sahut B., Sabatier J., Tarascon J. M. & Oustaloup A. (2009a). A Mathematical Model for the Simulation of New and Aged Automotive Lead-Acid Batteries, Journal of electrochemical society, Vol. 156, Issue 12, pp A974-A985.

Cugnet M., Sabatier J., Laruelle S., Grugeon S., Chanteur I., Sahut B., Oustaloup A. & Tarascon J. M. (2009b). A Solution for Lead-acid Battery Global State Estimation, ECS transactions, Volume 19, n°25, pp 77-88.

Cugnet M., Sabatier J. & Chanteur I. (2009c). Estimateur d'état batterie par invalidation de modčle– patent FR20070059784 20071212 published on 06/06/2009

Cugnet M., Sabatier J. & Chanteur I. (2009d). Procédé d'estimation de la résistance interne d'une batterie de véhicule automobile, patent EP 2071345 A1published on 17/06/2009

Cugnet M., Sabatier J., Laruelle S., Grugeon S., Sahut B., Oustaloup A. & Tarascon J. M. (2010). Lead-acid battery fractional modeling associated to a model validation method for resistance and cranking capability estimation, IEEE Transactions on Industrial Electronics, Vol. 57, No 3, pp 909-917.

Francisco J. M., Sabatier J., Lavigne L., Guillemard F., Moze M., Tari M., Merveillaut M., Noury A. (2014). Lithium-ion battery state of charge estimation using a fractional battery model, IEEE International Conference on Fractional Differentiation and its Applications, Catania, Italy

Gregoire G., Ragot F., Sabatier J., Oustaloup A. & Malti R. (2006). Systčme d'estimation de l'état de charge de moyens de stockage d'énergie électrique de véhicule et procédé de calibrage, patent FR2877097 published on 28/04/2006.

Hashim Ali S. A. (2005). Thermodynamic Analysis of Lithium Ion Cells, Ionics 11.

Hua 8., Lia S. & Penga H. (2012). A comparative study of equivalent circuit models for Li-ion batteries, Journal of Power Sources 198, 359–367.

Karthikeyan D. K., Sikha G. & White R. E. (2008). Thermodynamic model development for lithium intercalation electrodes, Journal of Power Sources 185, 1398–1407.

Quintana J. J., Ramos A & Nuez I. (2006). Identification of the fractional impedance of ultracapacitors, IFAC Workshop FDA, IFAC.
Newman J. & Thomas-Alyea K. E (2004). Electrochemical Systems, 3rd Edition, Wiley.
Newman H. (1997). Impedance Modelling of Porous Electrode, Electrochemical Applications, 1/97, p.2.
Guillemard F., Nouillant C., Ouaksel K., Sabatier J., Merveillaut M. (2012a). Dispositif de détermination de grandeur(s) d'une batterie ŕ cellules en serie, patent 2964196 (A1) published on 02/03/2012
Guillemard F., Porcellato D., Sabatier J. & Merveillaut M. (2012b). Dispositif d'estimation de variables d'état d'une cellule d'une batterie multicellulaire en fonction au moins du courant qui la traverse, patent 2973517 (A1) published on 05/10/2012
Guillemard F., Porcellato D., Ouaksel K., Nouillant C., Sabatier J. & Merveillaut M. (2012c). Dispositif de détermination prédictive d'information d'au moins une cellule d'une batterie multicellulaire, en présence d'une sollicitation en puissance, patent 2973516 (A1) published on 05/10/2012
Maxwell Technologies Inc (2004). Multi-electrode double layer capacitor, paptent W0 98/15962
Mouyon P. H. (1986). Pseudolinearisation et Pseudocompensation des systčmes non linéaires, l'ENSAE PhD thesis.
Nouillant C., Ouaksel K., Guillemard F., Sabatier J., Merveillaut M. (2012). Procédé d'estimation amélioré des SOC et SOH d'une batterie d'un véhicule, estimateur et véhicule comportant un tel procédé, patent FR2963434 (A1) published on 3/02/2012
Oustaloup A., Levron F., Mathieu B., Nanot F (2000). Frequency-band complex noninteger differentiator: characterization and synthesis". IEEE Transactions on Circuits and Systems I: Fundamental Theory and Applications, I , Volume: 47 Issue: 1, Page(s): 25 -39.
Ragot F., Gregoire G., Sabatier J., Oustaloup A. & Malti R. (2006a). Systčme de détermination de l'état de charge de moyens de stockage d'énergie électrique de véhicule, patent FR2877095 published on le 28/04/2006.
Ragot F., Gregoire G., Sabatier J., Oustaloup A. & Malti R. (2006b). System for determining the state of charge of electricity storage units of vehicules, patent EP1653241 published on 05/03/2006.
Ragot F., Sabatier J. & Oustaloup A. (2007). Procédé et systčme de l'estimation d'état de charge de moyens de stockage d'énergie électrique, patent FR2889314 published on 02/02/2007.
Riu D., Reitere N. & Linzen D., (2004). Half order modelling of supercapacitor, Industry Applications Conference IAS, Vol.4, p.2550.
Sabatier J., Aoun M., Oustaloup A., Gregoire G., Ragot F. & Roy P. (2006). Fractional system identification for lead acid battery sate charge estimation - Signal Processing Journal, Vol. 86, N°10, pp 2645-2657.
Sabatier J, Merveillaut M., Malti R., Oustaloup A. (2008). On a Representation of Fractional Order Systems: Interests for the Initial Condition Problem – 3rd IFAC Workshop on "Fractional Differentiation and its Applications" (FDA'08) - Ankara, Turkey, November 5-7.
Sabatier J., Cugnet M., Laruelle S., Grugeon S., Sahut B., Oustaloup A. & Tarascon J. M. (2010a). A Fractional Order Model for Lead-Acid Battery Crankability Estimation – Communications in Nonlinear Science and Numerical Simulation, Volume 15, Issue 5, pp 1308-1317.
Sabatier J., Merveillaut M., Malti R., Oustaloup A. (2010b) How to Impose Physically Coherent Initial Conditions to a Fractional System?, Communications in Nonlinear Science and Numerical Simulation, Volume 15, Issue 5.
Sabatier J., Cugnet M., Laruelle S., Grugeon S., Chanteur I., Sahut B., Oustaloup A. & Tarascon J. M. (2011). Diagnosis of the internal resistance of an automotive lead-acid battery by the implementation of a model invalidation-based approach: application to the crankability estimation, Chapter 6 of Electrical Machines Diagnosis, Edited by J.C. TRIGEASSOU, Ed Wiley.

Sabatier J., Merveillaut M., Francisco J. & Guillemard F. (2014a). Lithium-ion batteries modelling involving fractional differentiation, Journal of power sources, Journal of Power Sources 262C, pp. 36-43.

Sabatier J., Francisco J. M., Guillemard F., Lavigne L. & Moze M. (2014b). Lithium-ion batteries modelling: a simple fractional differentiation based model and its associated parameters estimation method, to appear in Signal Processing. doi:10.1016/j.sigpro.2014.06.008

Sabatier J, Farges C (2014c) Long memory models: a first solution to the infinite energy storage ability of linear time-invariant fractional models, 19th World Congress of the International Federation of Automatic Control, Cape Town, South Africa, 24-29 August 2014

Sabatier J, Farges C (2015) Analysis of fractional models physical consistency, to appear in Journal on vibration and control.

Sathyanarayana S., Venugopalan S. & Gopikanth M. L. (1979). Impedance parameters and the state-of-charge I : Nickel-cadmium battery, Journal of applied electrochemistry, Vol. 9, pp. 125-139.

Trigeassou J.C., N. Maamri N. (2009). State-space modelling of fractional differential equations and the initial condition problem, IEEE SSD'09, Djerba, Tunisia.

Zubieta L. & Bonert R. (2000). Characterization of Double Layer Capacitor (DLCs) for Power Electronics Application, IEEE Transaction on Industry Applications, Vol.36, pp.199.

M. K. Lenzi*, G. Gonçalves, F. R. G. B. Silva, R. S. Zola, H. V. Ribeiro, R. Rossato, and E. K. Lenzi

Results for an Electrolytic Cell Containing Two Groups of Ions: PNP - Model and Fractional Approach

Abstract: An extension of the anomalous Poisson – Nernst – Planck (PNP) diffusional model for an electrolytic cell containing two groups of ions by considering fractional time derivatives of distributed order and nonusual boundary conditions is discussed. The equations are solved exactly in the small AC signal approximation and exhibit a rich variety of behaviors suitable to describe the behavior of the experimental data. In this context, the electrical impedance is obtained and the role of the boundary conditions in the low frequency limit is investigated.

Keywords: anomalous diffusion, electrical response, impedance

1 Introduction

Impedance spectroscopy is an experimental technique for studying the electrical response of a system subjected to an external, usually small amplitude, periodic current signal. It is widely used for investigating systems such as batteries (Chen et al., 2001; Jungst et al., 2003; Itagaki et al., 2005; Andre et al., 2011), fuel cells (Danzer and Hofer, 2008; Danzer and Hofer, 2009; Mamlouk and Scott, 2011; Aaron et al., 2012; Niyan and Hoorfar, 2013), colloidal systems (Lvovich and Smiechowski, 2006; Lvovich and Smiechowski 2008; Grosse and Delgado, 2010; Hollingsworth, 2013), and biological tissues (Krishnan and Garnett, 2006; Naumowicz et al., 2006; Ismail et al., 2011; Hilber et al., 2011). It has also been applied in medicine (Jun et al., 2012; Jun et al., 2013;

***Corresponding Author: M. K. Lenzi:** Departamento de Engenharia Química, Universidade Federal do Paraná - Curitiba, PR 81531 - 990, Brazil
G. Gonçalves: Departamento de Engenharia Química, Universidade Tecnológica Federal do Paraná - Ponta Grossa, 84016-210 PR, Brazil
F. R. G. B. Silva: Departamento de Física, Universidade Estadual de Maringá - Maringá, PR 87020-900, Brazil
R. S. Zola: Departamento de Física, Universidade Tecnológica Federal do Paraná - Apucarana, PR 76812-460, Brazil
H. V. Ribeiro: Departamento de Física, Universidade Tecnológica Federal do Paraná - Apucarana, PR 76812-460, Brazil
R. Rossato: Departamento de Física, Universidade Tecnológica Federal do Paraná - Apucarana, PR 76812-460, Brazil
E. K. Lenzi: Departamento de Física, Universidade Estadual de Ponta Grossa - Ponta Grossa, PR 84030-900, Brazil

McGivney et al., 2012) and the food industry (Yang, 2008a; Yang and Bashir, 2008b; Masot et al., 2010; Karásková et al., 2011; Euring et al., 2011; Pérez-Esteve et al., 2014) as a fast and low cost method for food quality control. The great popularity of this tool lies in the fact that from a single measurement, one may obtain results related to complex variables such as mass transfer, chemical reaction rates, and other processes. In particular, when applied to biological systems, a remarkable advantage of impedance spectroscopy is that it is non-destructive due to the low voltages applied (Scandurra et al., 2013). Furthermore, it is a powerful technique to investigate electrical properties of materials (Naumowicz et al., 2013) as well as the interfacial and bulk properties, obtaining, relevant physicochemical parameters (Barsoukov and Macdonald, 2005). The results obtained via impedance spectroscopy have been investigated, in general, by using the Poisson–Nernst–Planck (PNP) model and/or equivalent circuits with constant phase elements (CPE). The PNP model was initially worked out (see, for example, Ref. (Bisquert and Compte, 2001)) by assuming that the electrodes are completely blocking and that only one group of ions is present in the electrolytic cell. Extensions have been proposed by considering a more complete theoretical model and by using different boundary conditions on the electrodes (Macdonald, 1953; Macdonald, 1973). Such extensions are important because completely blocking electrodes models are suitable only for describing the high frequency range of electrolytic cells, when confronted with experimental data. These approaches have been used to investigate experimental data in several contexts (see, for example, Refs. (Lenzi et al., 2013; Ciuchi et al., 2012)). However, the behavior exhibited experimentally often deviate from what is obtained by the standard formalism, where the impedance assumes the asymptotic behavior $\mathcal{Z} \sim 1/(i\omega)$ in the low frequency limit.

The disagreement between experimental data and usual models is a strong motivation for considering changes in the usual description. One of them is concerned with the diffusive motion of the ions, which can be extended to anomalous diffusion, for instance, via fractional derivatives (Metzler and Klafter, 2000) . Another possibility is to consider the complexity of the surface effects by generalizing the boundary conditions (Lenzi et al.; 2014). A remarkable characteristic of anomalous diffusive processes is the dependence of the mean square displacement with time that, unlike a usual diffusive process, is non-linear, e.g., $\langle (z - \langle z \rangle)^2 \rangle \sim t^\gamma$, with $\gamma \neq 1$. Examples connected to electrolytic cells are encountered in fractal electrodes (Kumar and Kant, 2009) , water (Batalioto et al., 2010), and ionic solutions (Lenzi et al., 2013). Thus, anomalous diffusion models, namely, PNPA models, describe behaviors that are not well suited to the usual PNP model. In Refs. (Bisquert and Compte, 2001; Bisquert, 2003a; Bisquert et al., 2003b; Bisquert, 2005) different situations were investigated with the aim of determining the impedance by employing the fractional calculus. In Ref. (Lenzi et al., 2009), a complete approach was presented to determine the impedance in a electrolytic cell by taking anomalous diffusion into account. In (Lenzi et al., 2009), the fractional drift-diffusion problem is analytically solved and, at the same time, the influence of ions present in the sample is considered by applying Poisson's equation.

The model proposed in the Ref. (Lenzi et al., 2011) was extended by considering that the boundary conditions at the electrodes are governed by integro-differential equations (Santoro et al., 2011). This extension of the PNPA model was applied to investigate the electrical response of water (Lenzi et al., 2011) and nematic liquid-crystal cells (Ciuchi et al., 2012). In the latter, the integro-differential equation embodies the usual kinetic equation for describing the adsorption-desorption process at the surfaces of electrodes expressed in terms of a temporal kernel that can be chosen to cover different scenarios. Due to the good agreement between the PNPA model and experimental data of impedance presented in Refs. (Lenzi et al., 2011) and (Ciuchi et al., 2012), showing that anomalous diffusion can provide a new perspective to the investigation of electrical response, the connection between the PNPA model presented in Refs. (Santoro et al., 2011; Lenzi et al., 2011; Ciuchi et al., 2012) and equivalent circuits with constant phase elements was presented in the Ref. (Lenzi et al., 2013a). The presence of CPE in equivalent circuits is usually applied to describe frequency dispersion found in solid electrode/electrolyte interfaces (Jorcin et al., 2006), which can be related to surface disorder and roughness (Liu, 1985; Córdoba-Torres et al.; 2012; Rammelt and Reinhard, 1990; Sapoval, 2005; Hitz and Lasia, 2001), electrode porosity (Hitz and Lasia, 2001) and to electrode geometry (Huang et al., 2007). It is interesting to note that the results presented in Ref. (Lenzi et al., 2013) show that, depending on the choice of the equivalent circuit, the effect of CPE's can be the same as the one obtained using integro-differential boundary conditions to describe anomalous diffusive process.

In this chapter, our goal is to investigate a anomalous Poisson–Nernst–Planck (PNPA) model by considering that two groups of ions are present in the electrolytic cell. Generalizations of the PNP model to take into account the presence of different groups of ions in the sample have been analyzed before (Macdonald and Franceschetti, 1978; Barbero et al., 2008; Alexe-Ionescu et al., 2009). Nonetheless, as discussed above, this approach often results in a poor agreement with experimental data in the low frequency regime and more sophisticated descriptions considering the anomalous behavior of the diffusing elements must be accounted for. Here, the electrical response analysis is carried out in the PNPA approach with integro - differential boundary conditions and the presence of a fractional time derivative of distributed order in the bulk equations by supposing that the positive and negative ions of the same group have the same mobility; however, the mobilities of the positive and negative ions of the different groups are not the same. This extension is presented and discussed in the Sec. II. In this section, the analytical solution of the model, in the linear approximation, is also presented and, in Sec. III, the conclusions are reported.

2 Fractional Diffusion and Impedance

Let us introduce an extension of the PNPA model which was proposed in Refs. (Santoro et al., 2011; Lenzi et al., 2011; Ciuchi et al., 2012) for investigating an electrolytic cell, which takes into account Poisson's equation and a fractional diffusion equation of distributed order with boundary conditions expressed in terms of integro – differential equations. It considers the presence of two groups of ions with different mobilities. For simplicity, we assume that the positive ($\alpha = +$) and negative ($\alpha = -$) ions of the same group have the same mobility, i.e., the diffusion coefficients for positive and negative ions are equal. In addition, for ions of different the groups they are different. Thus, we consider the following fractional diffusion equation of distributed order for the bulk densities of ions

$$\int_0^1 d\gamma' \tau(\gamma') \frac{\partial^{\gamma'}}{\partial t^{\gamma'}} n_{\alpha l}(z,t) = -\frac{\partial}{\partial z} \mathcal{J}_{\alpha l}(z,t), \tag{1}$$

where $\tau(\gamma')$ is a distribution of γ' and the index l ($l = 1$ or $l = 2$) represents the group of ions present in the systems. The fractional operator considered in Eq. (1) is Caputo's (Podlubny, 1999), i.e.,

$$\frac{\partial^{\gamma'}}{\partial t^{\gamma'}} n_{\alpha l}(z,t) = \frac{1}{\Gamma(k-\gamma')} \int_{t_0}^{t} dt' \frac{n_{\alpha l}^{(k)}(z,t)}{(t-t')^{\gamma'-n+1}}, \tag{2}$$

with $k-1 < \gamma' < k$ and $n_{\alpha l}^{(k)}(z,t) \overset{\text{def}}{=} \partial_t^k n_{\alpha l}(z,t)$. In particular, we consider $t_0 \to -\infty$ in order to analyze the response of the system driven by a periodic potential defined later on, as indicated in Ref. (Podlubny, 1999). The current density is given by

$$\mathcal{J}_{\alpha l}(z,t) = -\mathcal{D}_{\alpha l} \frac{\partial}{\partial z} n_{\alpha l}(z,t) \mp \frac{q\mathcal{D}_{\alpha l}}{k_B T} n_{\alpha l}(z,t) \frac{\partial}{\partial z} V(z,t). \tag{3}$$

In Eq. (3), $\mathcal{D}_{\alpha l}$ is the diffusion coefficient for ions of the group l (previously assumed to be equal for the positive and negative ions within the same group, i.e., $\mathcal{D}_{-1} = \mathcal{D}_{+1} = \mathcal{D}_1$ and $\mathcal{D}_{-2} = \mathcal{D}_{+2} = \mathcal{D}_2$, but $\mathcal{D}_1 \neq \mathcal{D}_2$), q is the ion's charge (for simplicity, assumed the same for both groups), $V(z,t)$ is the actual electric potential across a sample of thickness d, with electrodes placed at $z = \pm d/2$ of a Cartesian reference frame where z is the axis normal to them, k_B is the Boltzmann constant, and T is the absolute temperature. The effective potential across the sample, which depends on the difference between the densities of charged particles, i.e., the net charge density of the bulk system whose dielectric coefficient is ε (measured in ε_0 units), is determined by Poisson's equation

$$\frac{\partial^2}{\partial z^2} V(z,t) = -\frac{q}{\varepsilon} \left(n_{+1}(z,t) + n_{+2}(z,t) - n_{-1}(z,t) - n_{-2}(z,t) \right) . \tag{4}$$

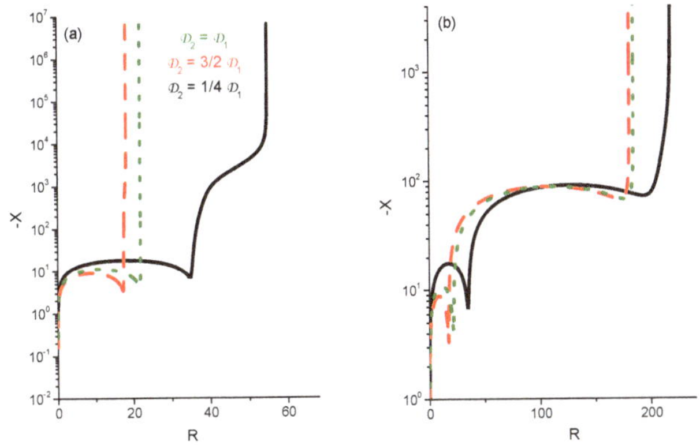

Fig. 1. Nyquist plot ($X = -\mathrm{Im}\mathcal{Z}$ and $R = \mathrm{Re}\mathcal{Z}$) for different diffusion coefficients by considering perfect blocking boundary conditions (Fig. 1a), i.e., $\Phi_{\alpha l}(i\omega) = 0$, and processes governed by Eq.(31) which are connected to a Debye relaxation (Fig. 1b). We consider, for simplicity, $d = 10^{-5}\ m$, $\mathcal{S} = 3.14 \times 10^{-4}\ m^2$, $\kappa = 10^{-4} m/s$, $\tau = 0.01s$, $\lambda_1 = \lambda_2 = 6.28 \times 10^{-8}\ m$, $Y_l(i\omega) = i\omega/\mathcal{D}_l$, $\mathcal{D}_1 = 4 \times 10^{-9}\ m/s^2$, and $\epsilon = 80\epsilon_0$.

We consider that Eq. (1) is subjected to the boundary condition

$$\mathcal{J}_{\alpha l}(z,t)\Big|_{z=\pm\frac{d}{2}} = \pm \int_0^1 d\overline{\vartheta} \int_{-\infty}^{t} d\overline{t}\,\overline{\kappa}_{\alpha l}(t-\overline{t},\overline{\vartheta}) \frac{\partial^{\overline{\vartheta}}}{\partial \overline{t}^{\overline{\vartheta}}} n_{\alpha l}(z,\overline{t})\Bigg|_{z=\pm\frac{d}{2}} . \tag{5}$$

Equation (5) recovers several situations such as the blocking electrodes for $\overline{\kappa}(t,\overline{\vartheta}) = 0$, adsorption-desorption process at the surfaces corresponding to the Henry approximation when $\overline{\kappa}(t,\overline{\vartheta}) \propto e^{-t/\tau}\delta(\overline{\vartheta}-1)$, and a Chang-Jaffe like boundary condition for $\overline{\kappa}(t,\overline{\vartheta}) \propto \delta(t)\delta(\overline{\vartheta})$. In this manner, Eq. (5) provides a unified framework for dealing with several boundary conditions and also the non-usual relaxations (Lenzi et al., 2010) which emerge when $\overline{\kappa}(t,\overline{\vartheta}) = \kappa(t)\delta(\overline{\vartheta})$ for $\kappa(t)$ chosen arbitrarily. For $\overline{\kappa}(t) \propto \delta(t)\widetilde{\tau}(\vartheta)$ with $\widetilde{\tau}(\vartheta)$ arbitrary, we can related the processes at the surface with fractional kinetic equations.

The previous set of equations represent a general PNPA model for an electrolytic cell containing two groups of ions from which is possible to obtain an analytical solution in the linear approximation for the stationary state. In this regime, one can consider that $n_{\alpha l}(z,t) = \mathcal{N}_l + \delta n_{\alpha l}(z,t)$, with $\mathcal{N}_l \gg |\delta n_{\alpha l}(z,t)|$ where $\mathcal{N}_l$ represents the number of ions. In addition, we also consider $\delta n_{\alpha l}(z,t) = \eta_{\alpha l}(z)e^{i\omega t}$ to analyze the impedance when the electrolytic cell is subject to the time dependent potential $V(z,t) = \phi(z)e^{i\omega t}$, with $V(\pm d/2,t) = \pm V_0 e^{i\omega t}/2$. Thus, by substituting these expressions into Eqs. (1), (5), and (4) yields a set of coupled equations which may be simplified by using the auxiliary functions $\psi_{\pm 1}(z) = \eta_{+1} \pm \eta_{-1}$ and $\psi_{\pm 2}(z) = \eta_{+2} \pm \eta_{-2}$, where

$\psi_{\pm 1}(z)$ and $\psi_{\pm 2}(z)$ are solutions of the differential equations

$$\frac{d^2}{dz^2}\psi_{+1} - Y_1(i\omega)\psi_{+1} = 0 \quad (6)$$

$$\frac{d^2}{dz^2}\psi_{-1} - \left(Y_1(i\omega) + \frac{1}{\lambda_1^2}\right)\psi_{-1} - \frac{1}{\lambda_1^2}\psi_{-2} = 0 \quad (7)$$

$$\frac{d^2}{dz^2}\psi_{+2} - Y_2(i\omega)\psi_{+2} = 0 \quad (8)$$

$$\frac{d^2}{dz^2}\psi_{-2} - \left(Y_2(i\omega) + \frac{1}{\lambda_2^2}\right)\psi_{-1} - \frac{1}{\lambda_2^2}\psi_{-1} = 0, \quad (9)$$

with $Y_l(i\omega) = (1/\mathcal{D}_l)\int_0^1 d\gamma' \tau(\gamma')(i\omega)^{\gamma'}$, where

$$\lambda_1 = \sqrt{\frac{\varepsilon k_B T}{2\mathcal{N}_1 q^2}} \quad \text{and} \quad \lambda_2 = \sqrt{\frac{\varepsilon k_B T}{2\mathcal{N}_2 q^2}}, \quad (10)$$

are the Debye's screening lengths for the groups 1 and 2. By solving this system of equations, we obtain complete information about the predictions of the model discussed here for the density of ions and, consequently, the impedance of the system. However, as we are interested in the impedance which is the quantity obtained from the experiment, we focus our attention on the solutions of ψ_{-1} and ψ_{-2}. In fact, potential $\phi(z)$ used to obtain the electric field and, consequently, the impedance is determined by the equation

$$\frac{d^2}{dz^2}\phi = \frac{1}{\varepsilon}\left(\psi_{-1} + \psi_{-2}\right) . \quad (11)$$

Performing some calculation, it is possible to show that

$$\psi_{-1}(z) = \mathcal{A}^{(1)}_{-,1}e^{\delta_1 z} + \mathcal{A}^{(1)}_{-,2}e^{-\delta_1 z} + \mathcal{A}^{(1)}_{-,3}e^{\delta_2 z} + \mathcal{A}^{(1)}_{-,4}e^{-\delta_2 z} \quad (12)$$

$$\psi_{-2}(z) = \mathcal{A}^{(2)}_{-,1}e^{\delta_1 z} + \mathcal{A}^{(2)}_{-,2}e^{-\delta_1 z} + \mathcal{A}^{(2)}_{-,3}e^{\delta_2 z} + \mathcal{A}^{(2)}_{-,4}e^{-\delta_2 z}, \quad (13)$$

with

$$\frac{\mathcal{A}^{(2)}_{-,1}}{\mathcal{A}^{(1)}_{-,1}} = \frac{\mathcal{A}^{(2)}_{-,2}}{\mathcal{A}^{(1)}_{-,2}} = \lambda_1^2\left(\delta_1^2 - \xi_1^2\right) = k_1 \quad (14)$$

$$\frac{\mathcal{A}^{(2)}_{-,3}}{\mathcal{A}^{(1)}_{-,3}} = \frac{\mathcal{A}^{(2)}_{-,4}}{\mathcal{A}^{(1)}_{-,4}} = \lambda_1^2\left(\delta_2^2 - \xi_1^2\right) = k_2, \quad (15)$$

where

$$\delta_{1,2} = \pm\sqrt{\frac{1}{2}\left(\xi_1^2 + \xi_2^2\right) \pm \frac{1}{2}\sqrt{\left(\xi_1^2 - \xi_2^2\right)^2 + 4\beta^2}}, \quad (16)$$

with $\xi_l^2 = 1/\lambda_l^2 + Y_l(i\omega)$ and $\beta^2 = 1/\left(\lambda_1^2\lambda_2^2\right)$. By using the previous equations, it is possible to show that the potential is given by

$$\phi(z) = -\frac{q}{\varepsilon}\left\{\frac{1+k_1}{\delta_1^2}\left(\mathcal{A}^{(1)}_{-,1}e^{\delta_1 z} + \mathcal{A}^{(1)}_{-,2}e^{-\delta_1 z}\right) + \frac{1+k_2}{\delta_2^2}\left(\mathcal{A}^{(1)}_{-,3}e^{\delta_2 z} + \mathcal{A}^{(1)}_{-,4}e^{-\delta_2 z}\right)\right\} + \mathcal{B}_1 z + \mathcal{B}_2 . \quad (17)$$

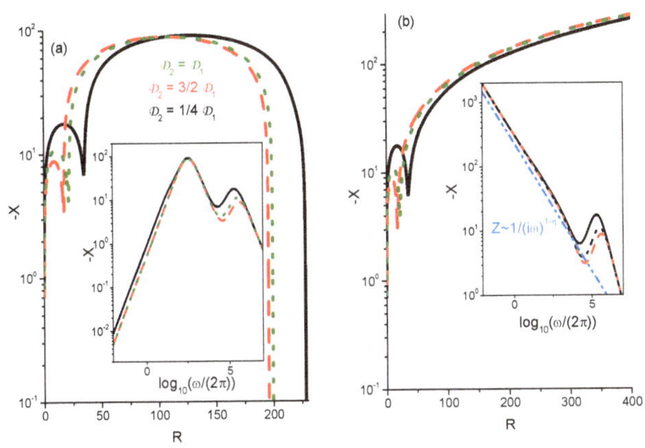

Fig. 2. Figure 2a considers different diffusion coefficients and the boundary condition $\Phi_{\alpha l}(i\omega) = \kappa$. The processes governed by Eq.(32) which are connected to a non–Debye relaxation are illustrate in Fig. 2b. The insets present in Figs. 2a and 2b show the behavior of the imaginary part of the impedance which is connected to the surface effects in the low frequency limit. We consider, for simplicity, $d = 10^{-5}\,m$, $\mathcal{S} = 3.14 \times 10^{-4}\,m^2$, $\kappa = 10^{-4} m/s$, $\eta = 0.6$, $Y_l(i\omega) = i\omega/\mathcal{D}_l$, $\tau = 0.01s$, $\lambda_1 = \lambda_2 = 6.28 \times 10^{-8}\,m$, $\mathcal{D}_1 = 4 \times 10^{-9}\,m/s^2$, and $\epsilon = 80\epsilon_0$.

Equations (12), (13), and (17) can be simplified by applying the condition $\phi(z) = -\phi(-z)$,

$$\psi_1(z) = 2\mathcal{A}_{-,1}^{(1)} \sinh(\delta_1 z) + 2\mathcal{A}_{-,3}^{(1)} \sinh(\delta_2 z) \tag{18}$$

$$\psi_2(z) = 2k_1\mathcal{A}_{-,1}^{(1)} \sinh(\delta_1 z) + 2k_2\mathcal{A}_{-,3}^{(1)} \sinh(\delta_2 z) \tag{19}$$

$$\phi(z) = -\frac{q}{\varepsilon}2\left\{\frac{1+k_1}{\delta_1^2}\mathcal{A}_{-,1}^{(1)} \sinh(\delta_1 z) + \frac{1+k_2}{\delta_2^2}\mathcal{A}_{-,3}^{(1)} \sinh(\delta_2 z)\right\} + \mathcal{B}_1 z. \tag{20}$$

In order to obtain a relation between $\mathcal{A}_{+,1}$ and $\mathcal{A}_{+,3}$, Eq. (5) can be used, yielding

$$\mathcal{A}_{-,3}^{(1)} = -\Lambda(i\omega)\mathcal{A}_{-,1}^{(1)} \tag{21}$$

$$\Lambda(i\omega) = \frac{\lambda_1^2\mathcal{G}_1(i\omega) - \lambda_2^2\mathcal{G}_3(i\omega)}{\lambda_1^2\mathcal{G}_2(i\omega) - \lambda_2^2\mathcal{G}_4(i\omega)}, \tag{22}$$

with

$$\mathcal{G}_1(i\omega)=\left(\delta_1 - \frac{1}{\lambda_1^2\delta_1}(1+k_1)\right)\cosh\left(\delta_1\frac{d}{2}\right) + \frac{1}{\mathcal{D}_1}\Phi_{-,1}(i\omega)\sinh\left(\delta_1\frac{d}{2}\right), \tag{23}$$

$$\mathcal{G}_2(i\omega)=\left(\delta_1 - \frac{1}{\lambda_1^2\delta_2}(1+k_2)\right)\cosh\left(\delta_2\frac{d}{2}\right) + \frac{1}{\mathcal{D}_1}\Phi_{-,1}(i\omega)\sinh\left(\delta_2\frac{d}{2}\right), \tag{24}$$

$$\mathcal{G}_3(i\omega)=\left(\delta_1 k_1 - \frac{1}{\lambda_2^2\delta_1}(1+k_1)\right)\cosh\left(\delta_1\frac{d}{2}\right) + \frac{k_1}{\mathcal{D}_2}\Phi_{-,2}(i\omega)\sinh\left(\delta_1\frac{d}{2}\right), \tag{25}$$

$$\mathcal{G}_4(i\omega) = \left(\delta_2 k_2 - \frac{1}{\lambda_2^2 \delta_2}(1+k_2)\right)\cosh\left(\delta_2 \frac{d}{2}\right) + \frac{k_2}{\mathcal{D}_2}\Phi_{-,2}(i\omega)\sinh\left(\delta_2 \frac{d}{2}\right), \quad (26)$$

where $\Phi_{\alpha l}(i\omega) = \int_0^1 d\overline{\vartheta} \int_0^\infty dv (i\omega)^{\overline{\vartheta}} \overline{\kappa}_{\alpha l}(v, \overline{\vartheta}) e^{i\omega v}$. By using Eq.(21) and the conditions on the potential, it is possible to show that

$$\mathcal{B}_1 = -\frac{\lambda_1^2 q}{\epsilon}\mathcal{C}(i\omega) 2\mathcal{A}_{-,1}^{(1)}, \quad (27)$$

$$\mathcal{A}_{-,1}^{(1)} = -\frac{\epsilon V_0}{2\lambda_1^2 q d}\frac{1}{2\Delta(i\omega) + \mathcal{C}(i\omega)}, \quad (28)$$

where $\mathcal{C}(i\omega) = \mathcal{G}_1(i\omega) - \Lambda(i\omega)\mathcal{G}_2(i\omega)$ and

$$\Delta(i\omega) = \frac{1+k_1}{\delta_1^2 \lambda_1^2 d}\sinh\left(\delta_1 \frac{d}{2}\right) - \frac{1+k_2}{\delta_2^2 \lambda_1^2 d}\Lambda(i\omega)\sinh\left(\delta_2 \frac{d}{2}\right). \quad (29)$$

From the previous results, it is possible to obtain the electric field $E(z,t)$ and, consequently, the surface density (Σ) from Coulomb's theorem $E(d/2,t) = -\Sigma(t)/\epsilon$. These quantities are relevant to obtain the admittance $\mathcal{Y}$ and, consequently, the impedance $\mathcal{Z} = 1/\mathcal{Y}$ of the system. The current at the electrode is determined by the equation

$$I = \mathcal{S}q\left[(\mathcal{J}_{+1} + \mathcal{J}_{+2}) - (\mathcal{J}_{-1} + \mathcal{J}_{-2})\right]\Big|_{z=\frac{d}{2}} + \mathcal{S}\frac{d}{dt}\Sigma \quad (30)$$

from which the admittance, $\mathcal{Y} = I/V$, of the sample (cell) can determined.

Figure 1a shows the Nyquist plot for the impedance obtained from the previous development when different diffusion coefficients are considered for perfect blocking boundary conditions, i.e., $\Phi_{\alpha l}(i\omega) = 0$. Note that the presence of two plateaus depend on the choice of the diffusion coefficients and the behavior of the double layer is essentially the same as the one presented in Ref. (de Paula et al., 2012) for one group of ions. This feature implies that the behavior of the imaginary part of the impedance for this cases is $Z \sim 1/(i\omega)$ which is connected to the boundary condition used in Fig. 1a. Similar behavior in the low frequency limit is obtained for the processes governed by an usual kinetic equation, as illustrated in Fig. 1b. This choice lead us to

$$\Phi_{\alpha l}(i\omega) = \kappa \frac{i\omega\tau}{1 + i\omega\tau} \quad (31)$$

connected to the kernel $\overline{\kappa}_{\alpha l}(v, \vartheta) \sim (\kappa/\tau)\, e^{-v/\tau}\delta(\vartheta - 1)$ which is typical of a relaxation governed by a Poisson distribution. A different behavior is obtained when the boundary condition $\Phi_{\alpha l}(i\omega) = \kappa$ (Chang-Jaffe) is considered, however as illustrated in Fig. 2a the behavior of the imaginary part of the impedance in the limit $\omega \to 0$ lead us to $Z \to 0$ (see the inset) which is very different from the experimental behavior exhibited in Refs. (Lenzi et al., 2011; Ciuchi et al., 2012; Lenzi et al., 2013b) . The case

$$\Phi_{\alpha l}(i\omega) = \kappa\,(i\omega\tau)^{1-\eta} \quad (32)$$

related to the choice $\overline{\kappa}_{\alpha l}(v, \vartheta) \sim \kappa\tau^{1-\eta}\delta(\vartheta + \eta - 1)\delta(v)$ with $0 < \eta < 1$ is illustrated in Fig. 2b. In this case, the behavior of the impedance in the low frequency limit is

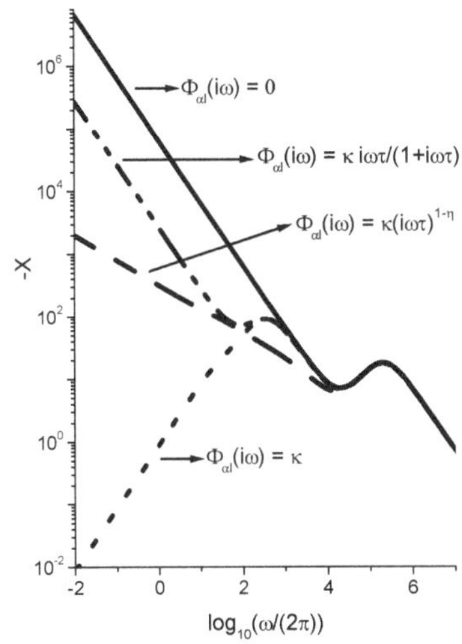

Fig. 3. This figure illustrates the imaginary part of the impedance for the boundary conditions considered in Figs. 1 and 2. The parameters are the same as for Figs. 1 and 2.

different from the one shown in Fig. 1 and may be connected to surface effects whose kinetics are not governed by a first order equation which has the Henry isotherm as stationary state. In addition, the behavior $\mathcal{Z} \sim 1/(i\omega)^{1-\eta}$ for the impedance manifested in many experimental contexts, e.g., water, liquid crystal (Ciuchi et al., 2012) and ionic solutions (Lenzi et al., 2013b), is contained in this formalism. Figure 3 illustrates this point concerning the behavior of the impedance in the low frequency limit where the processes present in the surface play an important role on the electrical response. These features show that depending on the choice of the boundary conditions, different behaviors for the impedance in the low frequency limit may be obtained. In this sense, a suitable choice for the boundary conditions may reproduce the surface effect manifested in this limit. As we are using a phenomenological approach, the answer to this question can be found by analyzing the behavior of the experimental data. In this manner, we may establish a bridge between the results presented here and the experimental behavior manifested by the electrolytical cell.

3 Conclusions

We have investigated an extension of the PNPA model by considering the presence of two group of ions. We have obtained an analytical solution in the linear approximation and, consequently, the behavior of the impedance by considering that the ions of the same group have equal diffusion coefficients and ions of different groups have different diffusion coefficients. The results show that the difference between the diffusion coefficients produces another plateau in the low frequency limit when the boundary conditions are characterized by perfect blocking electrodes. Similar behavior is obtained if a linear kinetic equation of first order (Henry isotherm) is considered to describe the adsorption–desorption on the interfaces between electrode and electrolyte. In this scenario, the behavior of the imaginary part of the impedance in the low frequency limit is asymptotically given by $\mathcal{Z} \sim 1/(i\omega)$ (see. Fig. 1). This behavior, in the low frequency limit, is different from the one presented in Fig. 2b, when Eq.(32) is considered as boundary condition leading us to obtain $\mathcal{Z} \sim 1/(i\omega)^{1-\eta}$ which reminds us of the ones observed, for example, in the experimental data of water (Batalioto et al., 2010; Lenzi et al., 2011) and liquid crystal (Ciuchi et al., 2012). These features suggest that the existing effects in the low frequency limit, for the case worked out here, are governed by the interaction between electrode and electrolyte and not by the bulk effects. Finally, we hope that the results presented here can be useful to investigate situations which are not suitably described in terms of the usual approach.

Acknowledgment: The authors are thankful to CNPq and CAPES (Brazilian agencies) for financial support.

Bibliography

Aaron D., Borole A. P., Yiacoumi S. & Tsouris C. (2012). Effects of operating conditions on internal resistances in enzyme fuel cells studied via electrochemical impedance spectroscopy. Journal of Power Sources, 201, 59 – 65.

Alexe-Ionescu A. L.,Barbero G. & Lelidis I. (2009). Complex dielectric constant of a nematic liquid crystal containing two types of ions: limit of validity of the superposition principle. Journal of Physical Chemistry B, 113, 14747.

Andre D., Meiler M., Steiner K., Wimmer Ch., Soczka-Guth T. & Sauer D. U. (2011). Characterization of high-power lithium-ion batteries by electrochemical impedance spectroscopy. I. Experimental investigation. Journal of Power Sources, 196, 5334 – 5341.

Barbero G., Batalioto F. & Figueiredo Neto A. M. (2008). Theory of small-signal ac response of a dielectric liquid containing two groups of ions. Applied Physics Letters, 92, 172908.

Barsoukov E. & Macdonald J. R. (2005). Impedance Spectroscopy: Theory, Experiment and Applications. New Jersey: Wiley-Interscience.

Batalioto F., Duarte A. R. & Barbero G. , A. M. F. Neto (2010). Dielectric Dispersion of Water in the Frequency Range from 10 mHz to 30 MHz. Journal of Physical Chemistry B, 114, 3467 – 3471.

Bisquert J. & Compte A. (2001). Theory of the electrochemical impedance of anomalous diffusion. Journal of Electroanalytical Chemistry, 499, 112 – 120.

Bisquert J. (2003a). Fractional diffusion in the multiple-trapping regime and revision of the equivalence with the continuous-time random walk. Physical Review Letters, 91, 010602.

Bisquert J.,Garcia-Belmonte G.& Pitarch A. (2003b). An Explanation of Anomalous Diffusion Patterns Observed in Electroactive Materials by Impedance Methods. ChemPhysChem, 4, 287 – 292.

Bisquert J. (2005). Interpretation of a fractional diffusion equation with nonconserved probability density in terms of experimental systems with trapping or recombination. Physical Review E, 72, 011109.

Chen C. H., Liu J. & Amine K. (2001). Symmetric cell approach and impedance spectroscopy of high power lithium-ion batteries. Journal of Power Sources, 96, 321 – 328.

Ciuchi F., Mazzulla A., Scaramuzza N., Lenzi E. K.& Evangelista L. R. (2012). Fractional Diffusion Equation and the Electrical Impedance: Experimental Evidence in Liquid-Crystalline Cells. Journal of Physical Chemistry C, 116, 8773 – 8777.

Córdoba-Torres P., Mesquita T. J., Devos O. ,Tribollet B. , Roche V.& Nogueira R. P. (2012). On the intrinsic coupling between constant-phase element parameters α and Q in electrochemical impedance spectroscopy. Electrochimica Acta, 72, 172 – 178.

Danzer M. A. & Hofer E. P. (2008). Electrochemical parameter identification-An efficient method for fuel cell impedance characterisation. Journal of Power Sources, 183, 55 – 61.

Danzer M. A. & Hofer E. P., (2009). Analysis of the electrochemical behaviour of polymer electrolyte fuel cells using simple impedance models. Journal of Power Sources, 190, 25 – 33.

de Paula J. L., da Cruz J. A., Lenzi E. K. & Evangelista L. R. (2012).Immittance response of an electrolytic cell in the presence of adsorption, generation, and recombination of ions. Journal of Electroanalytical Chemistry, 682, 116 – 120.

Euring F., Russ W., Wilke W. & Grupa U. (2011). Development of an impedance measurement system for the detection of decay of apples. Procedia Food Science, 1, 1188 – 1194.

Grosse C. & Delgado A. V. (2010). Dielectric dispersion in aqueous colloidal systems. Current Opinion in Colloid & Interface Science, 15, 145 – 159.

Hilber W., Lornejad-Schäfer M. R., Schäfer C., Lederer T., Schröder K. & Jakoby B. (2011). Impedance Spectroscopy of a Human Hepatic $3D$ Cell Model in-Vitro: A Comparative Study with Differently Shaped Electrodes. Procedia Engineering, 25, 1000 – 1003.

Hitz C. & Lasia A. (2001). Experimental study and modeling of impedance of the her on porous Ni electrodes. Journal of Electroanalytical Chemistry, 500, 213 – 222.

Hollingsworth A. D. (2013). Remarks on the determination of low-frequency measurements of the dielectric response of colloidal suspensions. Current Opinion in Colloid & Interface Science, 18, 157 – 159.

Huang V. M., Vivier V. & Orazem M. E., N. Pébère, B. Tribollet (2007). The Apparent Constant-Phase-Element Behavior of an Ideally Polarized Blocking Electrode: A Global and Local Impedance Analysi. Journal of The Electrochemical Society, 152, C81 – C88.

Ismail A. H., Schäfer C., Heiss A., Walter M., Jahnen-Dechent W. & Leonhardt S. (2011). An electrochemical impedance spectroscopy (EIS) assay measuring the calcification inhibition capacity in biological fluids. Biosensors and Bioelectronics, 26, 4702 – 4707.

Itagaki M., Kobari N., Yotsuda S., Watanabe K., Kinoshita S. & Ueb M. (2005). $LiCoO_2$ electrode/electrolyte interface of Li-ion rechargeable batteries investigated by in situ electrochemical impedance spectroscopy. Journal of Power Sources, 148, 78 – 84.

Jorcin J., Orazem M. E. & Pébére N., Tribollet B. (2006). CPE analysis by local electrochemical impedance spectroscopy. Electrochimica Acta, 51, 1473 – 1479.

Jun H., Dao L. T. M. & Cho S. (2012). Electrical Impedance Detection of Senescence in Adipose Tissue-derived Stem Cells. Procedia Engineering, 47, 1025 – 1028.

Jun H., DaoL. T. M. & Pyun J., Cho S. (2013). Effect of cell senescence on the impedance measurement of adipose tissue-derived stem cells. Enzyme and Microbial Technology, **53**, 302 – 306.

Jungst R. G., Nagasubraamanian G., Case H. L., Liaw B. Y., Urbina A., Paez T. L. & Doughty D. H. (2003). Accelerated calendar and pulse life analysis of lithium-ion cells. Journal of Power Sources, 119, 870 – 873.

Karásková P., Fuentes A., Fernández-Segovia I., Alcañiz M., Masot R. & Barat J. M. (2011). Development of a low-cost non-destructive system for measuring moisture and salt content in smoked fish products. Procedia Food Science, 1, 1195 – 1201.

Krishnan C. V. & Garnett M. (2006). New insights into the double layer structure from impedance measurements: Implications for biological systems. Electrochimica Acta, 51, 1541 – 1549.

Kumar R. & Kant R. (2009). Theory of Generalized Gerischer Admittance of Realistic Fractal Electrode. Journal of Physical Chemistry C, 113, 19558 – 19567.

Lenzi E. K., Evangelista L. R. & Barbero G. (2009)., Fractional Diffusion Equation and Impedance Spectroscopy of Electrolytic Cells. Journal of Physical Chemical Letters B, 113, 11371 – 11374.

Lenzi E. K., Yednak C. A. R. & Evangelista L. R. (2010). Non-Markovian diffusion and the adsorption-desorption process. Physical Review E, 81, 011116.

Lenzi E. K., Fernandes P. R. G., Petrucci T., Mukai H. & Ribeiro H. V. (2011). Anomalous-diffusion approach applied to the electrical response of water. Physical Review E, 84, 041128.

Lenzi E. K., de Paula J. L., Silva F. R. G. B. & Evangelista L. R. (2013a). A Connection Between Anomalous Poisson-Nernst-Planck Model and Equivalent Circuits with Constant Phase Elements. Journal of Physical Chemistry C, 117, 23685 – 23690.

Lenzi E. K., Fernandes P. R.G., Petrucci T., Mukai H., Ribeiro H. V., Lenzi M. K.& Gonçalves G. (2013b). Anomalous Diffusion and Electrical Response of Ionic Solutions. International Journal of Electrochemical Science, 8, 2849 – 2862.

Lenzi E. K., Lenzi M. K., Silva F.R.G.B., Gonçalves G., Rossato R., Zola R.S.& Evangelista L.R. (2014). A framework to investigate the immittance responses for finite length-situations: Fractional diffusion equation, reaction term, and boundary conditions. Journal of Electroanalytical Chemistry, 712, 82 – 88.

Liu S. H. (1985). Fractal model for the ac response of a rough interface. Physical Review Letters, 55, 529 – 532.

Lvovich V. F. & Smiechowski M. F. (2006). Impedance characterization of industrial lubricants. Electrochimica Acta, 51, 1487 – 1496.

Lvovich V. F. & Smiechowski M. F. (2008). Non-linear impedance analysis of industrial lubricants Electrochimica Acta, 53, 7375 – 7385.

Macdonald J. R. (1953). Theory of ac Space-Charge Polarization Effects in Photoconductors, Semiconductors, and Electrolytes. Physical Review, 92, 4 – 17.

Macdonald J. R. (1973). Theory of Space-charge Polarization and Electrode-discharge Effects. Journal of Chemical Physics, 58, 4982 – 5001.

Macdonald J. R. & Franceschetti D. R. (1978). Theory of Small-signal AC Response of Solids and Liquids with Recombining Mobile Charge. Journal of Chemical Physics, 68, 1614 – 1637.

Mamlouk M. & Scott K. (2011). Analysis of high temperature polymer electrolyte membrane fuel cell electrodes using electrochemical impedance spectroscopy. Electrochimica Acta, 56, 5493 - 5512.

Masot R., Alcañiz M., Fuentes A., Schmidt F. C., Barat J. M., Gil L., Baigts D., Martínez-Máñez R. & Soto J. (2010). Design of a low-cost non-destructive system for punctual measurements of salt levels in food products using impedance spectroscopy. Sensors and Actuators A: Physical, 158, 217 – 223.

McGivney D., Calvetti D. & Somersalo E. (2012). Quantitative imaging with electrical impedance spectroscopy. Physics in Medicine and Biolog, 57, 7289.

Metzler R. & Klafter J. (2000). The random walk's guide to anomalous diffusion: a fractional dynamics approach. Physics Reports, 339, 1 – 77.

Naumowicz M., Petelska A. D. & Figaszewski Z. A. (2006). Impedance analysis of a phosphatidylcholine-phosphatidylethanolamine system in bilayer lipid membranes. Electrochimica Acta, 51, 5024 – 5028.

Naumowicz M., Petelska A. D. & Figaszewski Z. A. (2013). Electrochemical impedance spectroscopy as a method for electrical characterization of the bilayers formed from lipid-amino acid systems. Chemistry and Physics of Lipids, 175, 116 – 122.
Niyan S. M. R. & Hoorfar M. (2013). Study of proton exchange membrane fuel cells using electrochemical impedance spectroscopy technique - A review. Journal of Power Sources, 240, 281 – 293.
Pérez-Esteve E., Fuentes A., Grau R., Fernández-Segovia I., Masot R., Alcañiz M. & Barat J. M. (2014). Use of impedance spectroscopy for predicting freshness of sea bream (Sparus aurata). Food Control, 35, 360
Podlubny I. (1999). Fractional Differential Equations. San Diego (CA): Academic Press.
Rammelt U. & Reinhard, G. (1990).On the applicability of a constant phase element (CPE) to the estimation of roughness of solid metal electrodes. Electrochimica Acta, 35, 1045 – 1049.
Santoro P. A., de Paula J. L., Lenzi E. K. & Evangelista L. R. (2011). Anomalous diffusion governed by a fractional diffusion equation and the electrical response of an electrolytic cell, Journal Chemical Physics, 135, 114704.
Sapoval B. (2005). Transport across irregular interfaces: Fractals Electrodes, Membranes and Catalysts. In Bunde, A., Havlin, S. (Eds), Fractals and Disordered Systems (pp. 233 – 261). Heidelberg: Springer- Verlag.
Scandurra G., Tripodi G. & Verzera A. (2013). Impedance spectroscopy for rapid determination of honey floral origin. Journal of Food Engineering, 119, 738 – 743.
Yang L. (2008a). Electrical impedance spectroscopy for detection of bacterial cells in suspensions using interdigitated microelectrodes. Talanta, 74, 1621 – 1629.
Yang L. & Bashir R. (2008b). Electrical/electrochemical impedance for rapid detection of foodborne pathogenic bacteria. Biotechnology Advances, 26, 135 – 150.

Abdon Atangana*

Application of Fractional Calculus to Epidemiology

Abstract: The aim of this chapter is to show how we can use fractional calculus to model a problem in epidemiology. We start our chapter by presenting the first use of mathematical tools to solve some physical problems. We give a reason to the use of fractional calculus for modeling real world problems rather than the local derivative. The model of the whooping cough was extended within the scope of fractional derivative. A detailed analysis of the stability of the disease free equilibrium points was presented. An iterative method based on the Sumudu transform operator is used to derive a special solution of the generalized system. An analysis of the stability of the method is presented together with the uniqueness of the special solution. An algorithm is proposed and used to simulate numerical results.

1 Introduction

The use of mathematical tools for solving real world problems is as old as the creation of the world. A mathematical model is a depiction of a system by means of mathematical concepts and language. The procedure of developing a mathematical model is termed mathematical modeling. It is worth noting that mathematical models are used not only in natural sciences such as physics, biology, earth science, meteorology and engineering, but also in the social sciences such as economic, psychology, sociology and political sciences. These models help to explain a system and study the effects of its components, and make predictions about behaviours. One of the most used mathematical concepts in modeling is perhaps "the concept of derivative". Due to the complexity of the physical problems encountered in our daily lives, this concept has been correctly extended. For instance, in order to precisely reproduce the non-local, frequency- and history-dependent properties of power law phenomena, some different modelling tools based on fractional operators have to be introduced. In particular, the advantages of fractional calculus and fractional order models mean differential systems involving fractional order integro-differential operators and their applications have already been intensively studied during the last few decades with excellent results. The long-range temporal or spatial dependence phenomena inherent to the fractional order systems presents unique peculiarities not supported by their integer order counterpart which permit better models of the dynamics of complex processes. Although non-integer differentiation has become a popular tool for modelling

***Corresponding Author: Abdon Atangana:** Institute for groundwater study, University of the Free State, 9300, Bloemfontein, South Africa, E-mail: abdonatangana@yahoo.fr and controlling the behaviours of physical systems from diverse branches of applied science, many problems remain to be explored and solved (Baleanu *et al.*, 2012; Atangana and Secer, 2013; Samko *et al.*, 1993; Atangana and Necdet, 2013; Atangana and Oukouomi, 2013; Kilbas *et al.*, 2006; Luchko and Gorenklo, 1998). We have devoted this chapter to the possible application of fractional order derivative to epidemiologics' problems. We shall recall that epidemiology is the science that studies the patterns, causes, and effect of health and disease conditions in defined populations. Experiments with infectious disease spread in human populations are often impossible, unprincipled or costly, thought data is sometimes available from naturally occurring epidemics or from the natural incidence of endemic diseases; nevertheless, data is often incompleted because of underreporting. This lack of reliable data makes accurate parameter estimation difficult so that it may only be possible to estimate a range of values for some parameters. Since repeatable experiments and accurate data are usually not available in epidemiology, mathematical models and computer simulations can be used to perform needed theoretical experiments. Calculations can easily be done for variety of parameter values and data sets. Nevertheless, it is worth noting that mathematical models have both limitations and capabilities. In this Chapter, we analyse disease within the scope of fractional derivatives: the well-known epidemic of whooping cough.

1.1 Modelling Epidemic of Whooping Cough with Concept of Fractional Order Derivative

Pertussis known as whooping cough is a highly contagious bacterial disease caused by Bordetella pertussis. The disease annually affects 48.5 million people worldwide, with about 295,000 deaths. The classical symptoms of perturssis are a paroxysmal cough, respiratory whoop and fainting or vomiting after coughing. The cough from pertussis has been documented to cause ubconjunctival haemorrhages, rib fractures, urinary incontinence, hernias, post-cough fainting, and vertebral artery dissection. Violent coughing can cause the pleura to rupture, leading to a pneumothorax (Carbonetti, 2007; Bettiol *et al.*, 2012; Cornia *et al.*, 2010; Zhang *et al.*, 2014). If there is vomiting after a coughing spell or an inspiratory whooping sound and coughing, the likelihood almost doubles that the illness is pertussis.

A mathematical model underpinning the description of this deadly disease starts with the classic SEIR framework in which individuals are grouped into four epidemiological classes according to their infection status: susceptible; exposed (latent); infectious; and recovered. For a population of size **N**, disease dynamics are given by

$$\begin{cases} \dfrac{dS(t)}{dt} = \mu N(1-p) - \left(\dfrac{\beta(t)I(t)}{N} + \mu\right) S \\ \dfrac{dE(t)}{dt} = \dfrac{\beta(t)I(t)}{N} S(t) - (\sigma + \mu)E(t) \\ \dfrac{dI(t)}{dt} = \sigma E(t) - (\gamma + \mu)I(t) \\ \dfrac{dS(t)}{dt} = \gamma I(t) + \mu Np - \mu R(t) \end{cases} \tag{1}$$

(Domenech, 2014) although the above equations have been used for the purpose of prediction, it is important to note that, this system of equations does not take into account the time scale and also does not precisely reproduce the nonlocal, frequency- and history-dependent properties of power law phenomena. Therefore, the above equation cannot predict exactly the physical problem under consideration. It is also important to mention that the model will be more accurate if more unknowns are incorporated into the input, therefore in order to extend the limitations of this model, we replace the local derivative with the Caputo fractional derivative to obtain the follows:

$$\begin{cases} \dfrac{d^\alpha S(t)}{dt^\alpha} = \mu N(1-p) - \left(\dfrac{\beta(t)I(t)}{N} + \mu\right) S \\ \dfrac{d^\alpha E(t)}{dt^\alpha} = \dfrac{\beta(t)I(t)}{N} S(t) - (\sigma + \mu)E(t) \\ \dfrac{d^\alpha I(t)}{dt^\alpha} = \sigma E(t) - (\gamma + \mu)I(t) \\ \dfrac{d^\alpha R(t)}{dt^\alpha} = \gamma I(t) + \mu Np - \mu R(t) \end{cases} \tag{2}$$

It is important to note that the above set of equation takes into account the time scale. One can find nowadays in the literature different definitions of fractional derivatives. However, the most well-known are the Riemann–Liouville and the Caputo derivatives. For Caputo we have

$${}_0^C D_x^\alpha (f(x)) = \frac{1}{\Gamma(n-\alpha)} \int_0^x (x-t)^{n-\alpha-1} \frac{d^n f(t)}{dt^n} dt \tag{3}$$

For the case of Riemann-Liouville we have the following definition

$$D_x^\alpha (f(x)) = \frac{1}{\Gamma(n-\alpha)} \frac{d^n}{dx^n} \int_0^x (x-t)^{n-\alpha-1} f(t) dt \tag{4}$$

Each one of these fractional derivatives presents some compensations and weakness (Atangana and Doungmo, 2014; Jackson and Rohani; 2013; Domenech, 2014).

Definition 1: Let $\Omega = [a\ b]$ $(-\infty \le a < b \le \infty)$ be a finite or infinite interval of real axis $R = (-\infty, \infty)$. We denote by $L_p(a, b)$ $(1 \le p \le \infty)$ the set of those Lebesgue complex-valued measurable functions f on Ω for which $\|f\|_p < \infty$, where

$$\|f\|_p = \left(\int_a^b |f(t)|^p \, dt\right)^{\frac{1}{p}} \quad (1 \le p \le \infty) \tag{5}$$

We shall additionally present the following useful theorem

Theorem 1: *if $h(t) \in L_1(R)$ and $h_1(t) \in L_p(R)$, then their convolution $(h * h_1)(x) \in L_p(R)\,(1 \le p \le \infty)$, the following inequality holds* (Atangana and Doungmo, 2014)

$$\|f\,(h * h_1)(x)\|_p < \|h\|_1 \|h_1\|_p \tag{6}$$

In particular, if $h(t) \in L_1(R)$ and $h_1(t) \in L_2(R)$, then their convolution $(h * h_1)(x) \in L_2(R)$ then,

$$\|f\,(h * h_1)(x)\|_2 < \|h\|_1 \|h_1\|_2$$

Lemma 1: (Atangana and Doungmo, 2014) the fractional integration operators with $R(\alpha) > 0$ are bounded in $L_p(a,\ b)\,(1 \le p \le \infty)$

$$\left\|I_a^{\alpha} f\right\|_p \le K \|f\|_p, \quad K = \frac{(b-a)^{R(\alpha)}}{R(\alpha)\,|\Gamma(\alpha)|} \tag{7}$$

The literature on Hilbert spaces entertains numerous illustrations of inner product spaces wherein the metric produced by the inner product produces a complete metric space. Inner product spaces have an intuitively defined norm grounded in the inner product of the space itself by the parallelogram equality:

$$\|x\| = \sqrt{(x, x)} \tag{8}$$

It is a well defined by the no negativity axiom of the definition of the inner product space that the following properties can be observed:

$$|(x, y)| \le \|x\| \, \|y\| \tag{9}$$

The above is the well-known Cauchy-Schwarz inequality. Also the following can be obtained

$$\|a.x\| = |a| \, \|x\|$$

The above property is called homogeneity. The last interesting one for this paper will be given as:

$$\|x + y\| \le \|x\| + \|y\|$$

The above relation is called triangular inequality

We shall then provide a detailed analysis of the above set of equations and also the semi-analytical method to solve it.

1.1.1 Steady state solutions, Eigen-Values and Stability analysis

The obtaining of the steady state solutions are achieved via the assumption stating that, the system does not depend on time, for that reason $S(t)$, $E(t)$, $I(t)\, and\, R(t)$ are constants. Due to the benefits provided by the Caputo fractional derivative, we can have that $\frac{d^{\alpha}S(t)}{dt^{\alpha}}, \frac{d^{\alpha}E(t)}{dt^{\alpha}}, \frac{d^{\alpha}I(t)}{dt^{\alpha}} \frac{d^{\alpha}R(t)}{dt^{\alpha}}$ are constants, for simplicity, we chose those constants to be zero each so that,

$$\begin{cases} 0 = \mu N(1-p) - \left(\frac{\beta \overline{I}}{N} + \mu\right)\overline{S} \\ 0 = \frac{\beta \overline{I}}{N}\overline{S} - (\sigma+\mu)\overline{E} \\ 0 = \sigma\overline{E} - (\gamma+\mu)\overline{I} \\ 0 = \gamma\overline{I} + \mu Np - \mu\overline{R} \end{cases} \tag{10}$$

After some manipulations, we obtain the following solutions

$$\overline{I} = \frac{N\mu}{\beta}\left(\frac{(1-p)\beta\sigma}{(\sigma+\mu)(\gamma+\mu)} - 1\right), \ \overline{S} = \frac{N(\sigma+\mu)(\gamma+\mu)}{\beta\sigma} \tag{11}$$

$$\overline{E} = \frac{(\gamma+\mu)}{\sigma}\frac{N\mu}{\beta}\left(\frac{(1-p)\beta\sigma}{(\sigma+\mu)(\gamma+\mu)} - 1\right),$$

$$\overline{R} = \frac{\gamma\frac{N\mu}{\beta}\left(\frac{(1-p)\beta\sigma}{(\sigma+\mu)(\gamma+\mu)} - 1\right) + \mu Np}{\mu}$$

We shall now find the reproductive number, from the infectious solution, given the following information

$$\overline{I} = \frac{N\mu}{\beta}\left(\frac{(1-p)\beta\sigma}{(\sigma+\mu)(\gamma+\mu)} - 1\right) = \frac{N\mu}{\beta}(R_0 - 1)$$

So that, the reproductive number can be obtained as

$$R_0 = \frac{(1-p)\beta\sigma}{(\sigma+\mu)(\gamma+\mu)} \tag{12}$$

If $R_0 = 1$, we will have a disease-free system which is not the purpose of our study. If $R_0 < 1$, the infectious will die out with time and no further infection will be observed. On the other hand if $R_0 > 1$, then the spread of the disease will take place and will need to be monitored or control. We skin our interest in the last event, we shall now study the stability of the disease free equilibrium. The disease free equilibrium will be obtained by assuming that there is no disease or infection amount the population, that will imply $\overline{I} = 0$, such that

$$(N(1-p), 0, 0, 0)$$

The stability can be analysed via the method of Jacobian matrix. And will be presented as follows:

$$J(S,I,E,R) = \begin{pmatrix} -\frac{\beta I}{N} - \mu & -\frac{\beta S}{N} & 0 & 0 \\ \frac{\beta I}{N} & \frac{\beta S}{N} & -\sigma - \mu & 0 \\ 0 & \gamma + \mu & \sigma & 0 \\ 0 & \gamma & 0 & -\mu \end{pmatrix} \tag{13}$$

Thus for the disease free equilibrium, we have the following

$$J(N(1-p),\, 0,0,0) = \begin{pmatrix} -\mu & -\beta(1-p) & 0 & 0 \\ 0 & \beta(1-p) & -\sigma-\mu & 0 \\ 0 & \gamma+\mu & \sigma & 0 \\ 0 & \gamma & 0 & -\mu \end{pmatrix} \tag{14}$$

Now to find the Eigen-values, we solve the following equation

$$Det\left(\begin{pmatrix} -\mu-\lambda & -\beta(1-p) & 0 & 0 \\ 0 & \beta(1-p)-\lambda & -\sigma-\mu & 0 \\ 0 & \gamma+\mu & \sigma-\lambda & 0 \\ 0 & \gamma & 0 & -\mu-\lambda \end{pmatrix}\right) = 0 \tag{15}$$

where Det is the determinant and λ is the Eigen-value to be found. The above equation is equivalent to

$$(\mu+\lambda)^2\left((\beta(1-p)-\lambda)(\sigma-\lambda)+(\mu+\gamma)(\sigma+\mu)\right)=0 \tag{16}$$

then, we have the first Eigen-value to be

$$-\mu = \lambda_{1,2} \tag{17}$$

$$Let\Delta = (\beta(1-p)+\sigma)^2 - 4\{\beta(1-p)\sigma + (\mu+\gamma)(\sigma+\mu)\} \tag{18}$$

If $\Delta < 0$, we have no real solution for the below equation

$$(\beta(1-p)-\lambda)(\sigma-\lambda)+(\mu+\gamma)(\sigma+\mu)=0$$

For this case, we only have one double Eigen-value

$$-\mu = \lambda_{1,2} < 0$$

Then the system will be stable.

If $\Delta > 0$, we have two real solutions for the below equation

$$(\beta(1-p)-\lambda)(\sigma-\lambda)+(\mu+\gamma)(\sigma+\mu)=0.$$

And they are given as

$$\lambda_{3,4} = \frac{\beta(1-p)+\sigma \mp \sqrt{\Delta}}{2}$$

In the event that, $\lambda_{3,4}$ are both negative, the disease free steady state solutions are stable, however in the event that only one is negative further study needs to be done, to have the full stability analysis of the system. We shall present the derivation of special solution of the system 2 by the mean of iteration in the next section.

1.1.2 Iterative methods

The simulation of the behaviour of the physical problem can only be achieved by solving the system of equation describing the system. The solvability can be achieved numerically or analytically. In this chapter, we shall solve the system analytically using iterative methods because of the nonlinearity of the system. The method used here is called the homotopy Sumudu decomposition method proposed in (Weerakoon, 1994; Weerakoon, 1997; Watugala, 1993; Eltayeb and Kilicman, 2010; Singh *et al.*, 2011). We illustrate the basic idea of this method, by considering a general fractional nonlinear non-homogeneous partial differential equation with the initial condition of the form:

$$D_t^{\alpha} U(x,t) = L(U(x,t)) + N(U(x,t)) + f(x,t), \quad \alpha > 0 \tag{19}$$

subjected to the initial conditions

$$D_0^k U(x,0) = g_k, \ (k = 0, \ldots\ldots n-1), \ D_0^n U(x,0) = 0 \; and \; n = [\alpha]$$

where, D_t^{α} (denotes without loss of generality) the Caputo fraction derivative operator, f is a known function, N is the general nonlinear fractional differential operator and L represents a linear fractional differential operator.

Applying the Sumudu Transform on Both sides of Eq. 19, we obtain:

$$S[D_t^{\alpha} U(x,t]) = S[L(U(x,t))] + S[N(U(x,t))] + S[f(x,t)] \tag{20}$$

Where the Sumudu transform of a function f is given as:

$$S(f(t)) = F_S(u) = \int_0^{\infty} \frac{1}{u} exp\left[-\frac{t}{u}\right] f(t)dt \tag{21}$$

And

$$S\left[D_t^{\alpha} f(t)\right] = u^{-\alpha} S\left[f(t)\right] - \sum_{k=0}^{m-1} u^{-\alpha+k} f^{(k)}\left(0^+\right), \ (m-1 < \alpha \le m) \tag{22}$$

And the following are some useful properties of the Sumudu operator 30-33.

- The transform of a Heaviside unit ramp function is a Heaviside unit ramp function in the transformed domain.

The transform of a monomial t^n is the called monomial $S\left(t^n\right) = n!u^n$.

If $f(t)$ is a monotonically increasing function, so is $F(u)$ and the converse is true for decreasing functions.

The Sumudu transform can be defined for functions which are discontinuous at the origin. In this case, the two branches of the function should be transformed separately.

If $f(t)$ is C^n continuous at the origin, so is the transformation $F(u)$.

The limit of $f(t)$ as t tends to zero is equal to the limit of $F(u)$ as u tends to zero provided that both limits exist.

The limit of $f(t)$ as t tends to infinity is equal to the limit of $F(u)$ as u tends to infinity provided both limits exist.

Scaling of the function by a factor $c > 0$ to form the function $f(ct)$ gives a transform $F(cu)$ which is the result of scaling by the same factor.

Using the property of the Sumudu transform, we have

$$S\left[U(x,t)\right] = u^{\alpha} S\left[L\left(U(x,t)\right)\right] + u^{\alpha} S\left[N\left(U(x,t)\right)\right] + u^{\alpha} S\left[f(x,t)\right] + g(x,t) \tag{23}$$

Now applying the Sumudu inverse on both sides of **??** we obtain:

$$U(x,t) = S^{-1}\left[u^{\alpha} S\left[L\left(U(x,t)\right)\right] + u^{\alpha} S\left[N\left(U(x,t)\right)\right]\right] + G(x,t) \tag{24}$$

$G(x,t)$ represents the term arising from the known function $f(x,t)$ and the initial conditions.

Now we apply the HPM:

$$U(x,t) = \sum_{n=0}^{\infty} p^n U_n(x,t) \tag{25}$$

The nonlinear term can be decomposed

$$NU(x,t) = \sum_{n=0}^{\infty} p^n H_n(U) \tag{26}$$

using the He's polynomial $H_n(U)$ [11] given as:

$$H_n(U_0, \cdots\cdots\cdots, U_n) = \frac{1}{n!}\frac{\partial^n}{\partial p^n}\left[N\left(\sum_{j=0}^{\infty} p^j U_j(x,t)\right)\right] \quad n = 0, 1, 2 \cdots\cdots\cdots \tag{27}$$

Substituting 25 and 26

$$\sum_{n=0}^{\infty} p^n U_n(x,t) =$$

$$G(x,t) + p\left[S^{-1}\left[u^{\alpha} S\left[L\left(\sum_{n=0}^{\infty} p^n U_n(x,t)\right)\right] + u^{\alpha} S\left[N\left(\sum_{n=0}^{\infty} p^n U_n(x,t)\right)\right]\right]\right]$$

which is the coupling of the Sumudu transform and the HPM using He's polynomials. Comparing the coefficients of like powers of p, the following approximations are obtained.

$$p^0 : U_0(x,t) = G(xt), \tag{28}$$

$$p^1 : U_1(x,t) = S^{-1}\left[u^{\alpha} S\left[L\left(U_0(x,t)\right) + H_0(U)\right]\right],$$

$$p^2 : U_2(x,t) = S^{-1}\left[u^\alpha S\left[L\left(U_1(x,t)\right) + H_1(U)\right]\right],$$

$$p^3 : U_3(x,t) = S^{-1}\left[u^\alpha S\left[L\left(U_2(x,t)\right) + H_2(U)\right]\right],$$

$$p^n : U_n(x,t) = S^{-1}\left[u^\alpha S\left[L\left(U_{n-1}(x,t)\right) + H_{n-1}(U)\right]\right],$$

Finally, we approximate the analytical solution $U(x, t)$ by the truncated series:

$$U(x,t) = \sum_{n=0}^{N} U_n(x,t) \tag{29}$$

The above series solution generally converges very rapidly [14]. We shall present the application for system 2

Now making use of the above operator 21, on both sides of Eq. 2, we obtain

$$\begin{cases} S(u) = u^\alpha S(0) + u^\alpha S\left\{N(1-p) - \left(\frac{\beta(t)I(t)}{N} + \mu\right)S\right\}(u) \\ E(u) = u^\alpha E(0) + u^\alpha S\left\{\frac{\beta(t)I(t)}{N}S(t) - (\sigma+\mu)E(t)\right\}(u) \\ I(u) = u^\alpha I(0) + u^\alpha S\left\{\sigma E(t) - (\gamma+\mu)I(t)\right\}(u) \\ R(u) = u^\alpha R(0) + u^\alpha S\left\{\gamma I(t) + \mu N p - \mu R(t)\right\}(u) \end{cases} \tag{30}$$

Thus applying the inverse Sumudu operator on both sides of the above, we obtain

$$\begin{cases} S(t) = S(0) + u^\alpha S\left\{N(1-p) - \left(\frac{\beta(t)I(t)}{N} + \mu\right)S\right\}(u) \\ E(t) = E(0) + u^\alpha S\left\{\frac{\beta(t)I(t)}{N}S(t) - (\sigma+\mu)E(t)\right\}(u) \\ I(t) = I(0) + u^\alpha S\left\{\sigma E(t) - (\gamma+\mu)I(t)\right\}(u) \\ R(t) = R(0) + u^\alpha S\left\{\gamma I(t) + \mu N p - \mu R(t)\right\}(u) \end{cases} \tag{31}$$

The iterative method of 28 can now be employed to put frontward the main recursive formula connecting the Lagrange multiplier as

$$\begin{cases} S_0(t) = S(0) \\ I_0(t) = I(0) \\ R_0(t) = R(0) \\ D_0(t) = D(0) \end{cases} \quad \begin{cases} S_{n+1}(t) = S_n(t) + S^{-1}\left\{u^\alpha S\left\{N(1-p) - \left(\frac{\beta(t)I_n(t)}{N} + \mu\right)S_n\right\}(u)\right\}(t) \\ E_{n+1}(t) = E_n(t) + S^{-1}\left\{u^\alpha S\left\{\frac{\beta(t)I_n(t)}{N}S_n(t) - (\sigma+\mu)E_n(t)\right\}(u)\right\}(t) \\ I_{n+1}(t) = I_n(t) + S^{-1}\left\{u^\alpha S\left\{\sigma E_n(t) - (\gamma+\mu)I_n(t)\right\}(s)\right\}(t) \\ D_{n+1}(t) = D_n(t) + S^{-1}\left\{u^\alpha S\left\{\gamma I_n(t) + \mu N p - \mu R_n(t)\right\}(u)\right\}(t) \end{cases} \tag{32}$$

The special solution of this equation is therefore given as

$$\begin{cases} S_n(t) = S(t) \\ I_n(t) = I(t) \\ R_n(t) = R(t) \\ E_n(t) = D(t) \end{cases} \tag{33}$$

1.1.3 Stability of iterative methods

The effectiveness of this used technique can be described by examining the stability and the convergence analysis. Consequently, we present in the next section the stability analysis of the used method for solving the novel system equation 8. To achieve this we consider the following operator

$$E(S,I,R,D)=\begin{cases} {}_0^C D_t^\beta S(t) \\ {}_0^C D_t^\beta E(t) \\ {}_0^C D_t^\beta I(t) \\ {}_0^C D_t^\beta R(t) \end{cases} = \begin{cases} N(1-p)-\left(\frac{\beta(t)I(t)}{N}+\mu\right)S \\ \frac{\beta(t)I(t)}{N}S(t)-(\sigma+\mu)E(t) \\ \sigma E(t)-(\gamma+\mu)I(t) \\ \gamma I(t)+\mu Np-\mu R(t) \end{cases} \tag{34}$$

Theorem 1: *Let us think about the above operator E* and *think about the initial condition for the system of equations 8, then the method used leads to a special solution of system equation 8*

Proof. To achieve this we shall think about the following $Z-$ sub-Hilbert space of the Hilbert space $H = L^2((0,T))$ [16] that can be defined as the set of those functions the following space

$$P:(0,T)\to R,\ Z=\left\{u,v\,\middle|\,\int_0^t uv d\tau <\infty\right\}$$

We require that the differential operators are limited under the L^2 norms. Exploiting the description of the operator, E we ensure the succeeding

$$E(S,E,I,R)-E(S_1,E_1,I_1,R_1)=$$
$$\begin{cases} -\left(\frac{\beta(t)I(t)}{N}+\mu\right)S+\left(\frac{\beta(t)I_1(t)}{N}+\mu\right)S_1 \\ \frac{\beta(t)I(t)}{N}S(t)-(\sigma+\mu)E(t)-\frac{\beta(t)I_1(t)}{N}S_1(t)+(\sigma+\mu)E_1(t) \\ \sigma(E(t)-E_1(t))-(\gamma+\mu)(I(t)-I_1(t)) \\ \gamma(I(t)-I_1(t))-\mu(R(t)-R_1(t)) \end{cases} \tag{35}$$

We shall evaluate next the inner product of $G = A(E(S,E,I,R)-E(S_1,E_1,I_1,R_1),$ $(S-S_1,E-E_1,I-I_1,R-R_1))$ where the fractional-inner product is defined as [7]

$$\begin{aligned} G &= A(E(S,E,I,R)-E(S_1,E_1,I_1,R_1),\ (S-S_1,E-E_1,I-I_1,R-R_1)) \\ &\le (E(S,E,I,R)-E(S_1,E_1,I_1,R_1),\ (S-S_1,E-E_1,I-I_1,R-R_1)) \end{aligned}$$

Our next concern now is to evaluate the inner product

$$(E(S,E,I,R)-E(S_1,E_1,I_1,R_1),\ (S-S_1,E-E_1,I-I_1,R-R_1))=$$

$$\begin{cases} \left(\mu(S-S_1)+\left(-\frac{\beta(t)I}{N}+\frac{\beta(t)I_1}{N}\right),(S-S_1)\right) \\ \left(\frac{\beta(t)I}{N}S-\frac{\beta(t)I_1}{N}S_1+(\sigma+\mu)(E-E_1),(E-E_1)\right) \\ (\sigma(E-E_1)-(\gamma+\mu)(I-I_1),(I-I_1)) \\ (\gamma(I-I_1)+\mu(-R+R_1),(R-R_1)) \end{cases} \tag{36}$$

We shall examine case by case

$$\left(\mu(S-S_1)+\left(-\frac{\beta(t)I}{N}+\frac{\beta(t)I_1}{N}\right),(S-S_1)\right) \le \|S-S_1\|\left\{\mu\|S-S_1\|+\|\beta\|\left\|\frac{S_1I_1}{N}-\frac{cSI}{N}\right\|\right\} \tag{37}$$

However due to the physical problem under investigation, all the parameters are bounded, implied, we can find some positive parameters O_1, O_2, such that

$$\|S_1-S\| \le O_1 \text{ and } \|S_1I_1-cSI\| \le O_2$$

Thus replacing the above inequalities in 37 to obtain

$$\left(\mu(S-S_1)+\left(-\frac{\beta(t)I}{N}+\frac{\beta(t)I_1}{N}\right),(S-S_1)\right) \le \|S-S_1\|\left\{\mu O_1+\frac{1}{N}O_2\right\} = p_1\|S-S_1\| \tag{38}$$

Using the same foot-steps we obtain the following

$$\begin{aligned} \left(\frac{\beta(t)I}{N}S-\frac{\beta(t)I_1}{N}S_1+(\sigma+\mu)(E-E_1),(E-E_1)\right) &\le p_2\|(E-E_1)\| \\ (\sigma(E-E_1)-(\gamma+\mu)(I-I_1),(I-I_1)) &\le p_3\|I-I_1\| \\ (\gamma(I-I_1)+\mu(-R+R_1),(R-R_1)) &\le p_4\|R-R_1\| \end{aligned} \tag{39}$$

Thus replacing the above inequalities 39, 38 and 37 into 36 to obtain

$$(E(S,I,R,D)-E(S_1,I_1,R_1,D_1),\ (S-S_1,I-I_1,R-R_1,D-D_1)) \le \begin{cases} p_1\|S-S_1\| \\ p_2\|(E-E_1)\| \\ p_3\|I-I_1\| \\ p_4\|R-R_1\| \end{cases} \tag{40}$$

Thus, Eq. 34 becomes

$$E \le \begin{cases} p_1\|S-S_1\| \\ p_2\|(E-E_1)\| \\ p_3\|I-I_1\| \\ p_4\|R-R_1\| \end{cases} \tag{41}$$

Following the same line of thinking, we can find a positive vector $F(f_1f_2f_3f_4)$ such that, for all vector $V(V_1,V_2,V_3,V_4) \in H$

$$E \le \begin{cases} f_1\|S-S_1\|\,\|V_1\| \\ f_2\|(E-E_1)\|\,\|V_2\| \\ f_3\|I-I_1\|\,\|V_3\| \\ f_4\|R-R_1\|\,\|V_4\| \end{cases} \tag{42}$$

Putting together 42 and 41 completes the proof of theorem 1.

1.1.4 Uniqueness

Theorem 2: *Taking into account the initial conditions of Eq. 8, there is only one unique special solution for Eq. 8 while using the new variational iteration method.*
Proof: Assuming that I is the exact solution of system 8, let T and T_1 be two difference special solutions of system and converge to $I \neq 0$ for some large number n and m 2 while using the homotopy method, then using theorem 1, we have the following inequality

$$(E(u, v, w, s, z) - E(u_1, v_1, w_1, s_1, z_1), (w_1, w_2, w_3, w_4, w_5)) \leq F \|T - T_1\| \|I\|$$

$$F \|T - T_1\| \|I\| \leq F \|T - I + I - T_1\| \|I\|$$

Employing the triangular inequality, we arrive at the succeeding

$$F \|T - T_1\| \|I\| \leq F \{\|I - T_1\| + \|T - I\|\} \|I\| \tag{43}$$

Nevertheless, subsequently T and T_1 convergence to W for large number n and m, then we can find a small positive parameter ε, such that:

$$\|I - T_1\| < \frac{\varepsilon}{2P \|I\|}, \text{ for } n$$

$$\|T - I\| < \frac{\varepsilon}{2P \|I\|}, \text{ for } m$$

Now consider $M = max(n, m)$, then

$$F \|T - T_1\| \|I\| \leq F \{\|I - T_1\| + \|T - I\|\} \|I\| < \frac{\varepsilon}{2F \|I\|} + \frac{\varepsilon}{2F \|I\|} = \varepsilon \, for \, M$$

Nonetheless using the topological knowledge, we have that

$$F \|T - T_1\| \|I\| = 0$$

Since $I \neq 0$ and $F \neq 0$, then $\|T - T_1\| = 0$ implying $T = T_1$. This shows the uniqueness of the special solution.

1.1.5 Numerical simulations

It is important to give an instruction to the computer to generate the iteratively the special solution. To achieve this, we shall give the following code that will be used to derive the special solution of system 2

$$\text{Input:} \begin{cases} S_0(t) = S(0) \\ I_0(t) = I(0) \\ R_0(t) = R(0) \\ E_0(t) = E(0) \end{cases} \text{ as preliminary input,}$$

- i–number terms in the rough calculation
- Output: $\begin{cases} S_{Ap}(t) \\ E_{Ap}(t) \\ I_{Ap}(t) \\ R_{Ap}(t) \end{cases}$, the approximate solution

Step 1: Put $\begin{cases} S_0(t) = S(0) \\ \\ I_0(t) = I(0) \\ R_0(t) = R(0) \\ E_0(t) = E(0) \end{cases}$ and $\begin{cases} S_{Ap}(t) \\ \\ I_{Ap}(t) \\ R_{Ap}(t) \\ E_{Ap}(t) \end{cases} = \begin{cases} S_{Ap}(t) \\ \\ I_{Ap}(t) \\ E_{Ap}(t) \\ E_{Ap}(t) \end{cases}$

Step 2: for $i = 1$ to $n-1$ do step 3, step 4 and step 5

$$\begin{cases} S_{n+1}(t) = S_n(t) + S^{-1}\left\{u^\alpha S\left\{N(1-p) - \left(\dfrac{\beta(t)I_n(t)}{N} + \mu\right) S_n\right\}(u)\right\}(t) \\ E_{n+1}(t) = E_n(t) + S^{-1}\left\{u^\alpha S\left\{\dfrac{\beta(t)I_n(t)}{N} S_n(t) - (\sigma+\mu)E_n(t)\right\}(u)\right\}(t) \\ I_{n+1}(t) = I_n(t) + S^{-1}\left\{u^\alpha S\{\sigma E_n(t) - (\gamma+\mu)I_n(t)\ \}(s)\right\}(t) \\ D_{n+1}(t) = D_n(t) + S^{-1}\left\{u^\alpha S\{\gamma I_n(t) + \mu Np - \mu R_n(t)\}(u)\right\}(t) \end{cases}$$

Step 3: compute

$$\begin{cases} \beta_{1(n+1)}(t) = \beta_{1(n)}(t) + S_{Ap}(t) \\ \beta_{2(n+1)}(t) = \beta_{2(n)}(t) + I_{Ap}(t) \\ \beta_{3(n+1)}(t) = \beta_{3(n)}(t) + R_{Ap}(t) \\ \beta_{4(n+1)}(t) = \beta_{4(n)}(t) + E_{Ap}(t) \end{cases}$$

Step 4: Compute:

$$\begin{cases} S_{Ap}(t) \\ I_{Ap}(t) \\ R_{Ap}(t) \\ \\ E_{Ap}(t) \end{cases} = \begin{cases} S_{Ap}(t) + \beta_{1(n+1)}(t) \\ I_{Ap}(t) + \beta_{2(n+1)}(t) \\ R_{Ap}(t) + \beta_{3(n+1)}(t) \\ \\ E_{Ap}(t) + \beta_{4(n+1)}(t) \end{cases}$$

Stop.

The above procedure shall be employed to yield the numerical replication of the physical problem under investigation. The numerical simulations are achieved here using the following theoretical parameters: N= 1000, p = 0.3, $\mu = 0.4$, $\gamma = 0.4$, $\sigma = 0.3$. We shall mention that, S(t), E(t), I(t) and R(t) represent: the susceptible, latent, infected and recovery populations respectively. N is the initial population and increase as time goes. $\beta(t) = 0.5$ is the infected rate, μ, $\sigma\gamma$ are susceptible rate, latent rate and recovery rate respectively.

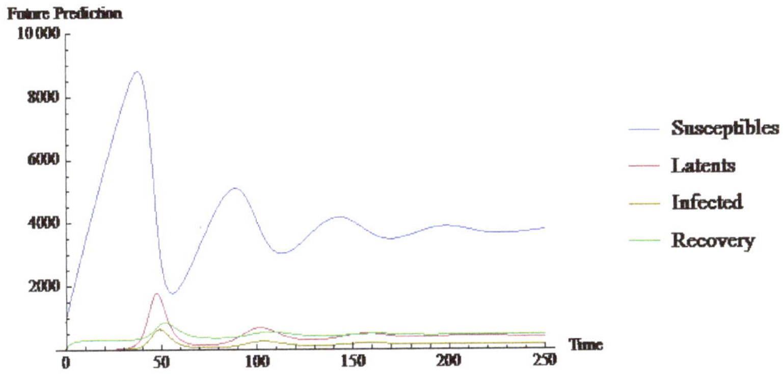

Fig. 1. Prediction for beta = 0.925

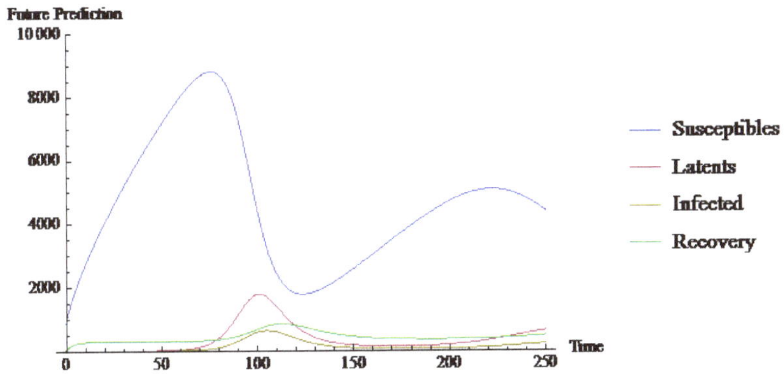

Fig. 2. Prediction for beta = 0.725

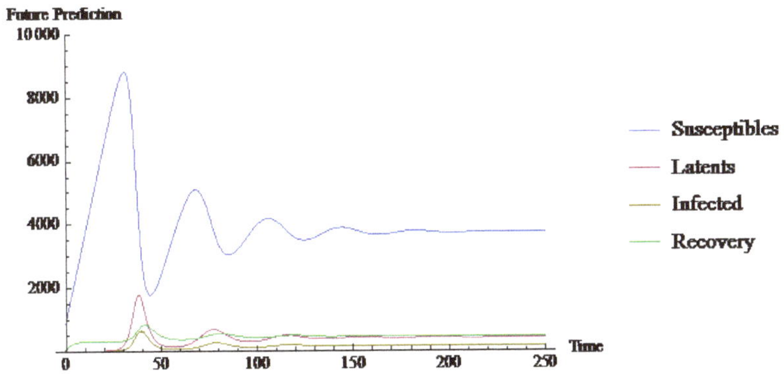

Fig. 3. Prediction for Beta =1

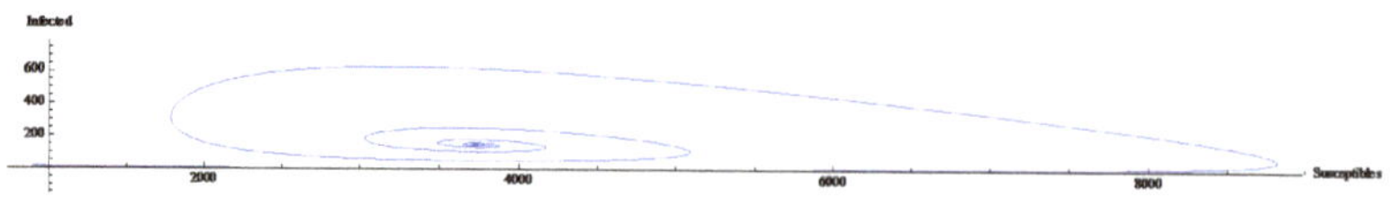

Fig. 4. Parametric representation of Susceptible against Infected populations for Beta = 0.995

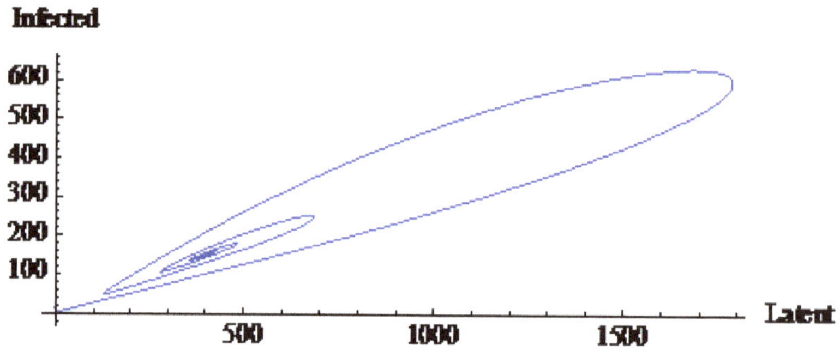

Fig. 5. Parametric representation of Latent against Infected populations for Beta = 0.995

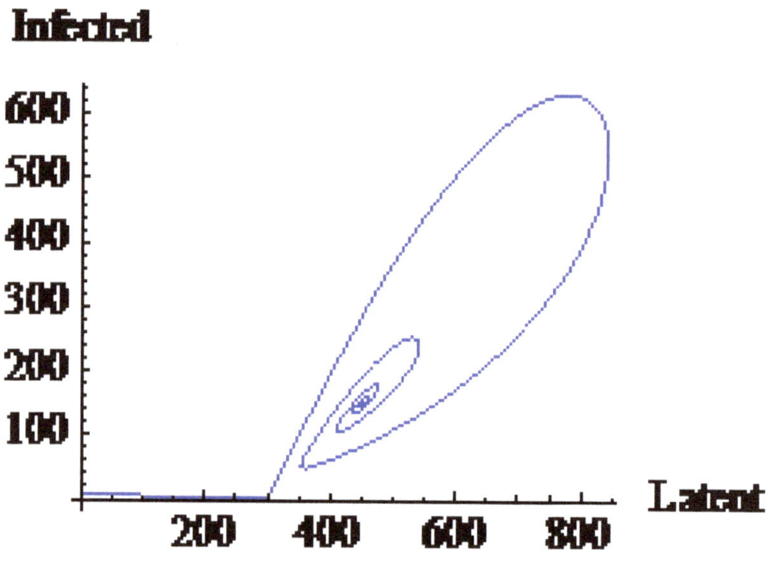

Fig. 6. Parametric representation of Latent against Infected populations for Beta = 0.995

2 Conclusion

The aim of this chapter was to show the possibility of extending the application of the fractional calculus to model some epidemiological problems. Since, it has been recognized in the recent decade that, it is rather better to use derivative that has a new parameter that the classical version of Newtonian derivative. These derivatives with new parameters are referred to fractional order derivatives. As epidemiologist, the cornerstone of public health, and informs policy decisions and evidence-based practice by identifying risk factors for disease and targets for preventive healthcare. Since most real world problems can be modelled via mathematical equations, in this chapter we modelled the spread of the deadly disease called whooping cough within the scope of Caputo derivative. We presented a detailed analysis of stability of disease free equilibrium points. We presented the derivation of a special solution of this set of equation using the Sumudu homotopy perturbation method. We presented in detail the stability of the used method for solving this problem. We further our investigation by presenting the uniqueness of the special solution under some conditions.

Bibliography

Apostol T.M., Calculus, One-Variable Calculus with an Introduction to Linear Algebra. (1967), Vol. 1., Wiley.

Atangana A., Bildik N., The Use of Fractional Order Derivative to Predict the Groundwater Flow. Mathematical Problems in Engineering, (2013), 9.

Atangana A., Doungmo Goufo E.F., On the Mathematical Analysis of Ebola Hemorrhagic Fever: Deathly Infection Disease in West African Countries. BioMed Research International, (2014), 7.

Atangana A., Oukouomi Noutchie S.C., Stability and convergence of a time-fractional variable order Hantush equation for a deformable aquifer. Abstract and Applied Analysis, (2014), 8.

Atangana A., Secer A., A note on fractional order derivatives and table of fractional derivatives of some special functions. Abstract and Applied Analysis, (2013), 8.

Baleanu D., Diethelm K., Scalas E., Trujillo J.J., Fractional Calculus Models and Numerical Methods. Series on Complexity, Nonlinearity and Chaos, (2012), World Scientific, Singapore.

Bettiol S., Wang K., Thompson M.J., Roberts N.W., Perera R., Heneghan C.J., Harnden A., Symptomatic treatment of the cough in whooping cough. Cochrane Database Syst Rev, (2012), 5 (5): CD003257.

Carbonetti N.H., Immunomodulation in the pathogenesis of Bordetella pertussis infection and disease. Curr Opin Pharmacol, (2007), 7 (3): 272–278.

Cornia P.B., Hersh A.L., Lipsky B.A., Newman T.B., Gonzales R., et al, Does this coughing adolescent or adult patient have pertussis?. JAMA, (2010), 304 (8): 890–896.

Domenech de Celles M., Riolo M.A., Magpantay F.M.G., Rohani P., King A.A., et al, Epidemiological evidence for herd immunity induced by acellular pertussis vaccines. Proc Nat Acad Sci USA, (2014).

Eltayeb H. Kılıçman A., A note on the Sumudu transforms and differential equations. Applied Mathematical Sciences, (2010), 1089–1098.

Jackson D.W., Rohani P., Perplexities of pertussis: recent global epidemiological trends and their potential causes Epidemiology and Infection, (2013), 1-13.

Kilbas A.A., Srivastava H.M., Trujillo J.J., Theory and Applications of Fractional Differential Equations. Vol. 204 of North-Holland Mathematics Studies, (2006), Elsevier Science B.V., Amsterdam, The Netherlands.

Luchko Y., Gorenflo R., The initial value problem for some fractional differential equations with the Caputo derivative. Preprint Series A08–98, Freie Universitat Berlin, (1998), Fachbreich Mathematik and Informatik.

Samko S.G., Kilbas A.A., Marichev O.I., et al., Fractional Integrals and Derivatives. Translated from the 1987 Russian Original, (1993), Gordon and Breach, Yverdon, Switzerland.

Singh J., Kumar D., Sushila, Homotopy perturbation Sumudu transform method for non-linear equations. Advances in Applied Mathematics and Mechanics, (2011), 165–175.

Watugala G.K., Sumudu transform: a new integral transform to solve differential equations and control engineering problems. International Journal of Mathematical Education in Science and Technology, (1993), 35–43.

Weerakoon S., Application of Sumudu transform to partial differential equations. International Journal of Mathematical Education in Science and Technology,(1994), 277–283.

Weerakoon S., The Sumudu transform and the Laplace transform: reply. International Journal of Mathematical Education in Science and Technology, (1997), 160.

Zhang L., Prietsch S.O., Axelsson, I; Halperin S.A., et al., Acellular vaccines for preventing whooping cough in children. The Cochrane database of systematic reviews, (2014), 9: CD001478.

A.H. Bhrawy, T.M. Taha, M.A. Abdelkawy, and R.M. Hafez

On Numerical Methods for Fractional Differential Equation on a Semi-infinite Interval

Abstract: Chapter 11 is devoted to numerical solutions of fractional differential equations (FDEs) on a semi-infinite interval. This chapter presents a broad discussion of spectral techniques based on operational matrices of fractional derivatives and integration methods for solving several kinds of linear and nonlinear FDEs. We present the operational matrices of fractional derivatives and integrals for some orthogonal polynomials/functions on a semi-infinite interval, and use them together with different spectral techniques for solving the aforementioned equations on a semi-infinite interval. Numerous examples are presented to illustrate the numerical and theoretical properties of various spectral techniques for solving FDEs on a semi-infinite interval.

Keywords: Multi-term FDEs; fractional-order generalized Laguerre orthogonal functions, operational matrices; Generalized Laguerre polynomials, Tau method, collocation method

1 Introduction

In recent years there has been a high level of interest in employing spectral methods for numerically solving many types of integral and differential equations, due to their ease in implementation over finite and infinite domains (Canuto et al., 2006; Bhrawy & Zaky, 2015a; Zayernouri & Karniadakis, 2014; Parand & Nikarya, 2014; Bhrawy & Zaky, 2015b; Keshavarz et al., 2014; Bhrawy & Abdelkawy, 2015; Tatari & Haghighi, 2014; Doha et al., 2009). Their speed of convergence is one of the great advantages of spectral methods. Spectral methods have exponential rates of convergence; they also have high level of accuracy. They have been classified into three types namely, collocation (Ma et al. 2014; Zhang et al. 2014; Ma & Huang 2014), tau (Doha et al., 2011; Lau & Price, 2012) and Galerkin (Zayernouri et al., 2014; Yang, 2014) methods.

FDEs have drawn the interest of many researchers due to their important applications in science and engineering. Consequentially, creating and developing numerical techniques for solving FDEs are gaining much attention. In recent years, there

A.H. Bhrawy: Department of Mathematics, Faculty of Science, King Abdulaziz University, Jeddah, Saudi Arabia, E-mail: alibhrawy@yahoo.co.uk
T.M. Taha, M.A. Abdelkawy: Department of Mathematics, Faculty of Science, Beni-Suef University, Beni-Suef, Egypt, E-mail: taha_m_taha@yahoo.com, melkawy@yahoo.com
R.M. Hafez: Department of Basic Science, Institute of Information Technology, Modern Academy, Cairo, Egypt, E-mail: r_mhafez@yahoo.com

has been a great interest in efficient numerical methods to find more accurate approximate solution of FDEs (see, for instance, (Katsikadelis, 2014; Daftardar-Gejji et al., 2014; Gaur & Singh, 2014; Chen et al., 2014; Dubey & Sharma, 2014; Liu & Lu, 2014; Bhrawy et al., 2015; Yang et al., 2014; Huang & Zhdanov, 2014; Liu et al., 2014; Takači et al., 2014)). It is well known that one of the most accurate methods of discretization for solving numerous differential equations is the spectral method. Spectral methods employ linear combinations of orthogonal polynomials as basis functions and provide accurate approximate solutions (Bhrawy, 2014; Doha et al., 2014). The spectral methods based on orthogonal systems like Jacobi polynomials and their special cases are only available for bounded domains for approximation of FDEs; see (Doha & Bhrawy, 2012; Doha et al., 2012b). Indeed, several problems in finance, plasma physics, porous media, dynamical processes and engineering are set on unbounded domains.

In the last few years, there has been a growing interest in the use of spectral method for the numerical treatments of FDEs in bounded domains. Doha et al. (Doha et al., 2012a) introduced the fractional derivatives of the Jacobi operational matrix and applied them in combination with the Jacobi tau scheme for solving linear multi-term FDEs. The authors of (Saadatmandi & Dehghan, 2010) presented a Legendre tau scheme combined with the operational matrix of Legendre polynomials for the numerical solution of multi-term FDEs. Recently, Kazem et al. (Kazem et al. 2013) defined new orthogonal functions based on Legendre polynomials to obtain an efficient spectral technique for multi-term FDEs, the authors of (Yin et al., 2013) extended this definition and presented the operational matrix of fractional derivatives and integration for such functions to construct a new tau technique for solving two-dimensional FDEs. Moreover, the authors of (Ahmadian et al., 2013) adopted the operational matrix of fractional derivatives for Legendre polynomials which is applied with a tau method for solving a class of fuzzy FDEs. Indeed, with a few noticeable exceptions, little work was done to use spectral methods in unbounded domains to solve such important classes of FDEs.

For FDEs in unbounded domains, the operational matrices of fractional derivatives and fractional integrals of generalized Laguerre polynomials were investigated for solving multi-term FDEs on a semi-infinite interval, see (Bhrawy et al., 2014; Baleanu et al., 2013). The generalized Laguerre spectral tau and collocation techniques were given in (Baleanu et al., 2013) to solve linear and nonlinear FDEs on the half line. These spectral techniques were developed and generalized by using modified generalized Laguerre polynomials in (Bhrawy et al., 2012). Indeed, the authors of (Ishteva & Boyadjiev, 2005; Ishteva et al., 2005) presented a Caputo fractional extension of the classical Laguerre polynomials and proposed new C-Laguerre functions.

The objective of this chapter is to present a broad survey of recent and new spectral methods for solving FDEs on a semi-infinite interval. The operational matrices of fractional derivatives and integrals for some orthogonal polynomials/functions on a semi-infinite interval are presented. These operational matrices are employed in combination with spectral tau and collocation schemes for solving several kinds of linear and nonlinear FDEs. The fractional-order generalized Laguerre functions (FGLFs) are presented. Moreover, we present the construction of generalized Laguerre operational matrix (GLOM) and fractional-order generalized Laguerre operational matrix (FGLOM) of fractional derivatives and integrals which are employed with tau methods to provide two efficient numerical schemes for solving linear FDEs. We also aim to propose a fractional-order generalized Laguerre collocation (FGLC) method, for solving fractional initial value problems of fractional order ν $(0 < \nu < 1)$ with nonlinear terms, in which the the nonlinear FDE is collocated at the N zeros of the new function which is defined on the interval $\Lambda = (0, \infty)$. We extend the application of FGLC method based on these functions to solve a system of FDEs with fractional orders less than 1. Several illustrative examples are implemented to confirm the high accuracy and effectiveness of the spectral techinques for solving FDES of fractional order ν $(0 < \nu < 1)$ on a semi-infinite interval.

The remainder of this chapter is organized as follows: In the next section we start by presenting some necessary definitions from fractional calculus theory. Section 3 is devoted to generalized Laguerre polynomials/functions and the quadrature rule of such functions. In Section 4, we present the GLOM and FGLOM of fractional derivatives in the Caputo sense. The GLOM and FGLOM of Riemann-Liouville fractional integrals are presented in Section 5. Section 6 is devoted to empoly the spectral methods based on generalized Laguerre polynomials and FGLFs in combination with the GLOM and FGLOM for solving FDEs and system of FDEs including linear and nonlinear terms. Several examples to illustrate the main ideas of this chapter are presented in Section 7.

2 Preliminaries and Notations

In this section, we give some basic definitions and properties of fractional calculus theory which are further used in the subsequent sections.

Definition 16. *The Riemann-Liouville integral $J^\nu f(x)$ and the Riemann-Liouville fractional derivative $D^\nu f(x)$ of order $\nu > 0$ are defined by*

$$\begin{aligned} J^\nu f(x) &= \frac{1}{\Gamma(\nu)} \int_0^x (x-t)^{\nu-1} f(t)dt, \quad x > 0, \\ J^0 f(x) &= f(x), \end{aligned} \tag{1}$$

and

$$D^\nu f(x) = J^{m-\nu} D^m f(x) = \frac{1}{\Gamma(m-\nu)} \int_0^x (x-t)^{m-\nu-1} \frac{d^m}{dt^m} f(t)dt, \quad x > 0, \tag{2}$$

respectively, where $m - 1 < \nu \leq m$, $m \in N^+$ and $\Gamma(.)$ denotes the Gamma function.

Definition 17. *The Caputo fractional integral and derivative operator satisfies*

$$J^\nu x^\beta = \frac{\Gamma(\beta+1)}{\Gamma(\beta+1+\nu)} x^{\beta+\nu}, \tag{3}$$

$$D^\nu x^\beta = \begin{cases} 0, & \text{for } \beta \in N_0 \text{ and } \beta < \lceil \nu \rceil, \\ \dfrac{\Gamma(\beta+1)}{\Gamma(\beta+1-\nu)} x^{\beta-\nu}, & \text{for } \beta \in N_0 \text{ and } \beta \geq \lceil \nu \rceil \text{ or } \beta \notin N \text{ and } \beta > \lfloor \nu \rfloor, \end{cases} \tag{4}$$

where $\lfloor \nu \rfloor$ *and* $\lceil \nu \rceil$ *are the floor and ceiling functions respectively, while* $N = \{1, 2, \dots\}$ *and* $N_0 = \{0, 1, 2, \dots\}$.

Caputo's fractional differentiation is a linear operator,

$$D^\nu(\lambda f(x) + \mu g(x)) = \lambda D^\nu f(x) + \mu D^\nu g(x), \tag{5}$$

where λ and μ are constants.

Lemma 4. *If* $m - 1 < \nu \leq m$, $m \in N$, *then*

$$D^\nu J^\nu f(x) = f(x), \qquad J^\nu D^\nu f(x) = f(x) - \sum_{i=0}^{m-1} f^{(i)}(0^+)\frac{x^i}{i!}, \quad x > 0. \tag{6}$$

3 Generalized Laguerre Polynomials/Functions

We introduce an orthogonal polynomial and another orthogonal function on a semi-infinite interval. Also the quadrature rules of these orthogonal functions are investigated.

3.1 Generalized Laguerre Polynomials

We recall below some relevant properties of the generalized Laguerre polynomials (Szegö, 1985; Funaro, 1992). Let $\Lambda = (0, \infty)$ and $w^{(\alpha)}(x) = x^\alpha e^{-x}$ be a weight function on Λ in the usual sense. Define

$$L^2_{w^{(\alpha)}}(\Lambda) = \{v \mid v \text{ is measurable on } \Lambda \text{ and } ||v||_{w^{(\alpha)}} < \infty\},$$

equipped with the following inner product and norm

$$(u, v)_{w^{(\alpha)}} = \int_\Lambda u(x)\, v(x)\, w^{(\alpha)}(x)\, dx, \qquad ||v||_{w^{(\alpha)}} = (u, v)^{\frac{1}{2}}_{w^{(\alpha)}}.$$

Next, let $L_i^{(\alpha)}(x)$ be the generalized Laguerre polynomials of degree i. We know from (Szegö, 1985) that for $\alpha > -1$,

$$L_{i+1}^{(\alpha)}(x) = \frac{1}{i+1}[(2i+\alpha+1-x)L_i^{(\alpha)}(x) - (i+\alpha)L_{i-1}^{(\alpha)}(x)], \quad i = 1, 2, \ldots, \tag{7}$$

where $L_0^{(\alpha)}(x) = 1$ and $L_1^{(\alpha)}(x) = 1 + \alpha - x$.

The set of generalized Laguerre polynomials is the $L^2_{w^{(\alpha)}}(\Lambda)$-orthogonal system, namely

$$\int_0^\infty L_j^{(\alpha)}(x)L_k^{(\alpha)}(x)w^{(\alpha)}(x)dx = h_k\delta_{jk}, \tag{8}$$

where δ_{jk} is the Kronecker function and $h_k = \frac{\Gamma(k+\alpha+1)}{k!}$.
The generalized Laguerre polynomial of degree i on the interval Λ, is given by (Baleanu et al., 2013)

$$L_i^{(\alpha)}(x) = \sum_{k=0}^{i} (-1)^k \frac{\Gamma(i+\alpha+1)}{\Gamma(k+\alpha+1)\,(i-k)!\,k!} x^k, \quad i = 0, 1, \ldots . \tag{9}$$

The special value

$$D^q L_i^{(\alpha)}(0) = (-1)^q \sum_{j=0}^{i-q} \frac{(i-j-1)!}{(q-1)!(i-j-q)!} L_j^{(\alpha)}(0), \quad i \geqslant q, \tag{10}$$

where $L_j^{(\alpha)}(0) = \frac{\Gamma(j+\alpha+1)}{\Gamma(\alpha+1)j!}$, will becom important later.

3.2 Fractional-order Generalized Laguerre Functions

We define new fractional orthogonal functions based on generalized Laguerre polynomials to obtain the solution of FDEs more simply and efficiently. The FGLFs can be defined by introducing the change of variable $t = x^\lambda$ and $\lambda > 0$ on generalized Laguerre polynomials. Let FGLFs $L_i^{(\alpha)}(x^\lambda)$ be denoted by $L_i^{(\alpha,\lambda)}(x)$; by using (7) $L_i^{(\alpha,\lambda)}(x)$ may be obtained from the recurrence relation

$$L_{i+1}^{(\alpha,\lambda)}(x) = \frac{1}{i+1}[(2i+\alpha+1-x^\lambda)L_i^{(\alpha,\lambda)}(x) - (i+\alpha)L_{i-1}^{(\alpha,\lambda)}(x)], \quad i = 1, 2, \ldots, \tag{11}$$

with $L_0^{(\alpha,\lambda)}(x) = 1$ and $L_1^{(\alpha,\lambda)}(x) = 1 + \alpha - x^\lambda$.

It is clear that the analytic form of $L_i^{(\alpha,\lambda)}(x)$ of fractional degree $i\lambda$ is

$$L_i^{(\alpha,\lambda)}(x) = \sum_{k=0}^{i} (-1)^k \frac{\Gamma(i+\alpha+1)}{\Gamma(k+\alpha+1)\,(i-k)!\,k!} x^{\lambda k}, \quad i = 0, 1, \ldots . \tag{12}$$

Lemma 5. *The set of FGLFs is a $L^2_{w^{(\alpha,\lambda)}}(\Lambda)$-orthogonal system,*

$$\int_0^\infty L_j^{(\alpha,\lambda)}(x)L_k^{(\alpha,\lambda)}(x)w^{(\alpha,\lambda)}(x)dx = h_k, \tag{13}$$

where $w^{(\alpha,\lambda)}(x) = \lambda\, x^{(\alpha+1)\lambda-1} e^{-x^\lambda}$ *and* $h_k = \begin{cases} \frac{\Gamma(k+\alpha+1)}{k!}, & j = k, \\ 0, & j \neq k. \end{cases}$

Proof. The proof of this lemma can be accomplished directly by using the definition of FGLFs and the orthogonality condition of generalized Laguerre polynomials.

3.3 Fractional-order Generalized Laguerre-Gauss-type Quadratures

We now construct the fractional-order generalized Laguerre-Gauss quadrature rules. We have the advantage of building on the generalized Laguerre-Gauss quadrature rule. We denote by $x_{N,j}^{(\alpha)}$, $0 \leqslant j \leqslant N$, the nodes of the generalized Laguerre-Gauss interpolation on the interval Λ. Their corresponding Christoffel numbers are $\omega_{N,j}^{(\alpha)}$, $0 \leqslant j \leqslant N$. The nodes of the fractional-order generalized Laguerre-Gauss interpolation on the interval Λ are the zeros of $L_{N+1}^{(\alpha,\lambda)}(x)$, which we denote by $x_{N,j}^{(\alpha,\lambda)}$, $0 \leqslant j \leqslant N$. Clearly $x_{N,j}^{(\alpha,\lambda)} = (x_{N,j}^{(\alpha)})^{\frac{1}{\lambda}}$, and their corresponding Christoffel numbers are $\omega_{N,j}^{(\alpha,\lambda)}$, $0 \leqslant j \leqslant N$.

$$\begin{aligned}\int_\Lambda \phi(x)w^{(\alpha,\lambda)}(x)dx &= \int_\Lambda \phi(x^{\frac{1}{\lambda}})w^{(\alpha)}(x)dx = \sum_{j=0}^{N} \omega_{N,j}^{(\alpha)}\phi((x_{N,j}^{(\alpha)})^{\frac{1}{\lambda}}) \\ &= \sum_{j=0}^{N} \omega_{N,j}^{(\alpha,\lambda)}\phi(x_{N,j}^{(\alpha,\lambda)}),\end{aligned} \tag{14}$$

for the previous relation, the Christoffel numbers are obtained from

$$\begin{aligned}\omega_{N,j}^{(\alpha,\lambda)} &= -\frac{\Gamma(i+\alpha+1)}{(i+1)!L_i^{(\alpha,\lambda)}(x_{N,j}^{(\alpha,\lambda)})\partial_x L_{i+1}^{(\alpha,\lambda)}(x_{N,j}^{(\alpha,\lambda)})} \\ &= \frac{\Gamma(i+\alpha+1)\, x_{N,j}^{(\alpha,\lambda)}}{(i+\alpha+1)(i+1)![L_i^{(\alpha,\lambda)}(x_{N,j}^{(\alpha,\lambda)})]^2}, \quad 0 \le j \le i.\end{aligned} \tag{15}$$

Corollary 4. *In particular, the special case for generalized Laguerre polynomials may be obtained directly by taking $\lambda = 1$ in the FGLFs, which are denoted by $L_i^{(\alpha)}(x)$, see, (Bhrawy et al., 2014).*

4 Operational Matrices of Caputo Fractional Derivatives

In this section, we present the operational matrices of Caputo fractional derivatives for generalized Laguerre polynomials and FGLFs on the half line.

4.1 GLOM of Fractional Derivatives

Let $u(x) \in L^2_{w^{(\alpha)}}(\Lambda)$, then $u(x)$ may be expressed in terms of generalized Laguerre polynomials as

$$u(x) = \sum_{j=0}^{\infty} a_j L_j^{(\alpha)}(x), \quad a_j = \frac{1}{h_k} \int_0^{\infty} u(x) L_j^{(\alpha)}(x) w^{(\alpha)}(x) dx, \quad j = 0, 1, 2, \cdots. \tag{16}$$

In practice, only the first $(N + 1)$-terms generalized Laguerre polynomials are considered. Then we obtain

$$u_N(x) = \sum_{j=0}^{N} a_j L_j^{(\alpha)}(x) = C^T \phi(x). \tag{17}$$

where the generalized Laguerre coefficient vector C and the generalized Laguerre vector $\phi(x)$ are given by

$$C^T = [c_0, c_1, \ldots, c_N], \quad \phi(x) = [L_0^{(\alpha)}(x), L_1^{(\alpha)}(x), \ldots, L_N^{(\alpha)}(x)]^T. \tag{18}$$

Lemma 6. *Let $L_i^{(\alpha)}(x)$ be a generalized Laguerre polynomial then*

$$D^\nu L_i^{(\alpha)}(x) = 0, \quad i = 0, 1, \cdots, \lceil \nu \rceil - 1, \quad \nu > 0. \tag{19}$$

In the following theorem we prove that this is the correct form of the operational matrix of Caputo fractional derivative for the generalized Laguerre vector (Bhrawy et al., 2014).

Theorem 18. *Suppose $\nu > 0$, the fractional derivative of order ν of $\phi(x)$ is given by*

$$D^\nu \phi(x) \simeq \mathbf{D}^{(\nu)} \phi(x), \tag{20}$$

where $\boldsymbol{D}^{(v)}$ is the $(N+1)\times(N+1)$ operational matrix of fractional derivative of order v in the Caputo sense and is defined as follows:

$$\boldsymbol{D}^{(v)} = \begin{pmatrix} 0 & 0 & 0 & \cdots & 0 \\ \vdots & \vdots & \vdots & \cdots & \vdots \\ 0 & 0 & 0 & \cdots & 0 \\ \Psi v(\lceil v\rceil,0) & \Psi v(\lceil v\rceil,1) & \Psi v(\lceil v\rceil,2) & \cdots & \Psi v(\lceil v\rceil,N) \\ \vdots & \vdots & \vdots & \cdots & \vdots \\ \Psi v(i,0) & \Psi v(i,1) & \Psi v(i,2) & \cdots & \Psi v(i,N) \\ \vdots & \vdots & \vdots & \cdots & \vdots \\ \Psi v(N,0) & \Psi v(N,1) & \Psi v(N,2) & \cdots & \Psi v(N,N) \end{pmatrix}, \tag{21}$$

where

$$\Psi v(i,j) = \sum_{k=\lceil v\rceil}^{i}\sum_{\ell=0}^{j} \frac{(-1)^{k+\ell}\, j!\,\Gamma(i+\alpha+1)\,\Gamma(k-v+\alpha+\ell+1)}{(i-k)!\,(j-\ell)!\,\ell!\,\Gamma(k-v+1)\,\Gamma(k+\alpha+1)\,\Gamma(\alpha+\ell+1)}.$$

(For the proof, see (Baleanu et al., 2013)).

4.2 FGLOM of Fractional Derivatives

Let $u(x) \in L^2_{w^{(\alpha,\lambda)}}(\Lambda)$, then $u(x)$ may be expressed in terms of FGLFs as

$$u(x) = \sum_{j=0}^{\infty} a_j L_j^{(\alpha,\lambda)}(x), \quad a_j = \frac{1}{h_k}\int_0^{\infty} u(x) L_j^{(\alpha,\lambda)}(x) w^{(\alpha,\lambda)}(x)dx, \quad j = 0,1,2,\cdots. \tag{22}$$

In practice, only the first $(N+1)$-terms FGLFs are considered. We have that:

$$u_N(x) = \sum_{j=0}^{N} a_j L_j^{(\alpha,\lambda)}(x) = C^T \Phi(x), \tag{23}$$

where the fractional-order generalized Laguerre coefficient vector C and the fractional-order generalized Laguerre vector $\Phi(x)$ are given respectively by

$$C^T = [c_0, c_1, \ldots, c_N], \quad \Phi(x) = [L_0^{(\alpha,\lambda)}(x), L_1^{(\alpha,\lambda)}(x), \ldots, L_N^{(\alpha,\lambda)}(x)]^T. \tag{24}$$

Lemma 7. *Let $L_i^{(\alpha,\lambda)}(x)$ be a FGLFs then*

$$D^v L_i^{(\alpha,\lambda)}(x) = 0, \quad \lambda i \le \lceil v\rceil - 1, \quad v > 0. \tag{25}$$

Theorem 19. *Let $\Phi(x)$ be fractional-order generalized Laguerre vector defined in Eq. (24) and also suppose $0 < v < 1$ then*

$$D^v \Phi(x) \simeq \boldsymbol{D}^{(v)} \Phi(x), \tag{26}$$

where $\boldsymbol{D}^{(\nu)}$ is the $(N+1)\times(N+1)$ operational matrix of fractional derivative of order ν in the Caputo sense and is defined as follows:

$$\boldsymbol{D}^{(\nu)} = \begin{pmatrix} 0 & 0 & 0 & \cdots & 0 \\ S_\nu(1,0,\lambda) & S_\nu(1,1,\lambda) & S_\nu(1,2,\lambda) & \cdots & S_\nu(1,N,\lambda) \\ \vdots & \vdots & \vdots & \cdots & \vdots \\ S_\nu(i,0,\lambda) & S_\nu(i,1,\lambda) & S_\nu(i,2,\lambda) & \cdots & S_\nu(i,N,\lambda) \\ \vdots & \vdots & \vdots & \cdots & \vdots \\ S_\nu(N,0,\lambda) & S_\nu(N,1,\lambda) & S_\nu(N,2,\lambda) & \cdots & S_\nu(N,N,\lambda) \end{pmatrix}, \tag{27}$$

where

$$S_\nu(i,j,\lambda) = \sum_{k=1}^{i}\sum_{s=0}^{j} \frac{(-1)^{k+s}\, j!\, \Gamma(i+\alpha+1)\, \Gamma(\lambda k+1)\, \Gamma(k-\frac{\nu}{\lambda}+\alpha+s+1)}{s!\, k!\, (i-k)!\, (j-s)!\, \Gamma(\lambda k-\nu+1)\, \Gamma(k+\alpha+1)\, \Gamma(\alpha+s+1)}.$$

Proof. The analytic form of the FGLFs $L_i^{(\alpha,\lambda)}(x)$ of degree i is given by (12). Using Eqs. (4), (5) and (12) we have

$$\begin{aligned} D^\nu L_i^{(\alpha,\lambda)}(x) &= \sum_{k=0}^{i} (-1)^k \frac{\Gamma(i+\alpha+1)}{(i-k)!\, k!\, \Gamma(k+\alpha+1)} D^\nu x^{\lambda k} \\ &= \sum_{k=1}^{i} (-1)^k \frac{\Gamma(i+\alpha+1)\Gamma(\lambda k+1)}{(i-k)!\, k!\, \Gamma(\lambda k-\nu+1)\, \Gamma(k+\alpha+1)} x^{\lambda k-\nu}, \quad i=1,\ldots,N. \end{aligned} \tag{28}$$

Now, approximating $x^{\lambda k-\nu}$ by $N+1$ terms of fractional generalized Laguerre series yields

$$x^{\lambda k-\nu} = \sum_{j=0}^{N} b_j L_j^{(\alpha,\lambda)}(x), \tag{29}$$

where b_j is given from (22) with $u(x) = x^{\lambda k-\nu}$, and

$$b_j = \sum_{s=0}^{j} (-1)^s \frac{j!\, \Gamma(k-\frac{\nu}{\lambda}+\alpha+s+1)}{(j-s)!\, (s)!\, \Gamma(s+\alpha+1)}. \tag{30}$$

Employing Eqs. (28)-(30) we get

$$D^\nu L_i^{(\alpha,\lambda)}(x) = \sum_{j=0}^{N} S_\nu(i,j,\lambda) L_j^{(\alpha,\lambda)}(x), \quad i=1,\cdots,N, \tag{31}$$

where

$$S_\nu(i,j,\lambda) = \sum_{k=1}^{i}\sum_{s=0}^{j} \frac{(-1)^{k+s}\, j!\, \Gamma(i+\alpha+1)\, \Gamma(\lambda k+1)\, \Gamma(k-\frac{\nu}{\lambda}+\alpha+s+1)}{s!\, k!\, (i-k)!\, (j-s)!\, \Gamma(\lambda k-\nu+1)\, \Gamma(k+\alpha+1)\, \Gamma(\alpha+s+1)}.$$

Accordingly, Eq. (31) can be written in a vector form as follows:

$$D^{\nu} L_i^{(\alpha,\lambda)}(x) \simeq \Big[S_\nu(i,0,\lambda), S_\nu(i,1,\lambda), S_\nu(i,2,\lambda), \ldots, S_\nu(i,N,\lambda)\Big]\,\Phi(x), \quad i = 1,\ldots,N. \tag{32}$$

Eq. (32) leads to the desired result.

Remark 20. *In the case of $\lambda = 1$, Theorem 4.4 gives the same result as Theorem 4.2 see (Baleanu et al., 2013).*

5 Operational Matrices of Riemann-Liouville Fractional Integrals

In this section, we present the operational matrices of Riemann-Liouville fractional integrals for generalized Laguerre polynomials and FGLFs on the half line.

5.1 GLOM of Fractional Integration

The main objective of this subsection is to derive an operational matrix of fractional integration for generalized Laguerre vector $\phi(x)$.

Theorem 21. *Let $\phi(x)$ be the generalized Laguerre vector and $\nu > 0$ then*

$$J^{\nu}\phi(x) \simeq \mathbf{P}^{(\nu)}\phi(x), \tag{33}$$

where $\mathbf{P}^{(\nu)}$ is the $(N+1)\times(N+1)$ operational matrix of fractional integration of order ν in the Riemann-Liouville sense and is defined as follows:

$$\mathbf{P}^{(\nu)} = \begin{pmatrix} \Theta_\nu(0,0) & \Theta_\nu(0,1) & \Theta_\nu(0,2) & \cdots & \Theta_\nu(0,N) \\ \Theta_\nu(1,0) & \Theta_\nu(1,1) & \Theta_\nu(1,2) & \cdots & \Theta_\nu(1,N) \\ \vdots & \vdots & \vdots & \cdots & \vdots \\ \Theta_\nu(i,0) & \Theta_\nu(i,1) & \Theta_\nu(i,2) & \cdots & \Theta_\nu(i,N) \\ \vdots & \vdots & \vdots & \cdots & \vdots \\ \Theta_\nu(N,0) & \Theta_\nu(N,1) & \Theta_\nu(N,2) & \cdots & \Theta_\nu(N,N) \end{pmatrix}, \tag{34}$$

where

$$\Theta_\nu(i,j) = \sum_{k=0}^{i}\sum_{r=0}^{j} \frac{(-1)^{k+r}\, j!\, \Gamma(i+\alpha+1)\, \Gamma(k+\nu+\alpha+r+1)}{(i-k)!\,(j-r)!\,r!\,\Gamma(k+\nu+1)\,\Gamma(k+\alpha+1)\,\Gamma(\alpha+r+1)}.$$

5.2 FGLOM of Fractional Integration

We aim to construct an operational matrix of fractional integration for fractional-order generalized Laguerre vector $\Phi(x)$.

Theorem 22. *Let $\Phi(x)$ be the fractional-order generalized Laguerre vector and $0 < \nu < 1$ then*

$$J^{\nu}\Phi(x) \simeq \mathbf{P}^{(\nu)}\Phi(x), \tag{35}$$

where $\mathbf{P}^{(\nu)}$ is the $(N+1)\times(N+1)$ operational matrix of fractional integration of order ν and $0 < \nu < 1$ in the Riemann-Liouville sense and is defined as follows:

$$\mathbf{P}^{(\nu)} = \begin{pmatrix} \Omega_\nu(0,0,\lambda) & \Omega_\nu(0,1,\lambda) & \Omega_\nu(0,2,\lambda) & \cdots & \Omega_\nu(0,N,\lambda) \\ \Omega_\nu(1,0,\lambda) & \Omega_\nu(1,1,\lambda) & \Omega_\nu(1,2,\lambda) & \cdots & \Omega_\nu(1,N,\lambda) \\ \vdots & \vdots & \vdots & \cdots & \vdots \\ \Omega_\nu(i,0,\lambda) & \Omega_\nu(i,1,\lambda) & \Omega_\nu(i,2,\lambda) & \cdots & \Omega_\nu(i,N,\lambda) \\ \vdots & \vdots & \vdots & \cdots & \vdots \\ \Omega_\nu(N,0,\lambda) & \Omega_\nu(N,1,\lambda) & \Omega_\nu(N,2,\lambda) & \cdots & \Omega_\nu(N,N,\lambda) \end{pmatrix}, \tag{36}$$

and

$$\Omega_\nu(i,j,\lambda) = \sum_{k=0}^{i} \frac{(-1)^k\,\Gamma(i+\alpha+1)\,j!\,\Gamma(k\lambda+1)}{\Gamma(k+\alpha+1)\,(i-k)!\,k!\,\Gamma(k\lambda+\nu+1)} \times \sum_{r=0}^{j} \frac{(-1)^r\,\Gamma(r+k+\frac{\nu}{\lambda}+\alpha+1)}{(j-r)!\,r!\,\Gamma(r+\alpha+1)}. \tag{37}$$

Proof. From (12) and (3), we have

$$\begin{aligned} J^{\nu}L_i^{(\alpha,\lambda)}(x) &= \sum_{k=0}^{i}(-1)^k \frac{\Gamma(i+\alpha+1)}{(i-k)!\,k!\,\Gamma(k+\alpha+1)}\,J^{\nu}x^{k\lambda} \\ &= \sum_{k=0}^{i}(-1)^k \frac{\Gamma(i+\alpha+1)\,\Gamma(k\lambda+1)}{(i-k)!\,k!\,\Gamma(k\lambda+\nu+1)\,\Gamma(k+\alpha+1)}\,x^{k\lambda+\nu}, \quad i=0,1,\cdots,N. \end{aligned} \tag{38}$$

The approximation of $x^{k\lambda+\nu}$ using $N+1$ terms of FGLFs, yields

$$x^{k\lambda+\nu} = \sum_{j=0}^{N} c_j L_j^{(\alpha,\lambda)}(x), \tag{39}$$

where c_j is given from (22) with $u(x) = x^{k\lambda+\nu}$, that is

$$c_j = \sum_{r=0}^{j}(-1)^r \frac{j!\,\Gamma(r+k+\frac{\nu}{\lambda}+\alpha+1)}{(j-r)!\,r!\,\Gamma(r+\alpha+1)}, \quad j=1,2,\cdots,N. \tag{40}$$

By virtue of (38) and (39), we obtain

$$J^{\nu}L_i^{(\alpha,\lambda)}(x) = \sum_{j=0}^{N} \Omega_{\nu}(i,j)L_j^{(\alpha,\lambda)}(x), \quad i = 0, 1, \cdots, N, \tag{41}$$

where

$$\Omega_{\nu}(i,j,\lambda) = \sum_{k=0}^{i} \frac{(-1)^k\, \Gamma(i+\alpha+1)\, j!\, \Gamma(k\lambda+1)}{\Gamma(k+\alpha+1)\,(i-k)!\,k!\,\Gamma(k\lambda+\nu+1)}$$
$$\times \sum_{r=0}^{j} \frac{(-1)^r\, \Gamma(r+k+\frac{\nu}{\lambda}+\alpha+1)}{(j-r)!\,r!\,\Gamma(r+\alpha+1)}, \quad j = 1, 2, \cdots N.$$

The vector form of Eq. (41) is

$$J^{\nu}L_i^{(\alpha,\lambda)}(x) \simeq \left[\Omega_{\nu}(i,0,\lambda), \Omega_{\nu}(i,1,\lambda), \Omega_{\nu}(i,2,\lambda), \cdots, \Omega_{\nu}(i,N,\lambda)\right]\Phi(x), \quad i = 0, 1, \cdots, N. \tag{42}$$

Eq. (42) leads to the desired result.

6 Spectral Methods for FDEs

The main aim of this section is to propose different ways to approximate linear FDEs using the generalized Laguerre tau method based on fractional-order Laguerre operational matrices of differentiation and integration such that it can be implemented efficiently. Also, we propose a new collocation method for solve nonlinear FDEs and systems of FDEs based on the FGLFs.

6.1 Generalized Laguerre Tau Operational Matrix Formulation Method

Here, we propose a direct solution technique to approximate linear multi-term FDEs with constant coefficients using the generalized Laguerre tau method in combination with GLOM. Consider the linear multi-order FDE

$$D^{\nu}u(x) = \sum_{j=1}^{k} \gamma_j D^{\beta_j}u(x) + \gamma_{k+1}u(x) + g(x), \quad \text{in } \Lambda, \tag{43}$$

with initial conditions

$$u^{(i)}(0) = d_i, \qquad i = 0, \ldots, m-1, \tag{44}$$

where γ_j $(j = 1, \dots, k+1)$ are real constant coefficients and also $m-1 < \nu \le m$, $0 < \beta_1 < \beta_2 < \dots < \beta_k < \nu$. Moreover $D^{\nu}u(x) \stackrel{\text{def}}{=} u^{(\nu)}(x)$ denotes the Caputo fractional derivative of order ν for $u(x)$. The values of d_i $(i = 0, \dots, m-1)$ describe the initial state of $u(x)$, and $g(x)$ is a given source function (for details, see (Baleanu et al., 2013)).
To solve the fractional initial value problem; (43)-(44), we approximate $u(x)$ and $g(x)$ by generalized Laguerre polynomials as

$$u(x) \simeq \sum_{i=0}^{N} c_i L_i^{(\alpha)}(x) = C^T \phi(x), \tag{45}$$

$$g(x) \simeq \sum_{i=0}^{N} g_i L_i^{(\alpha)}(x) = G^T \phi(x), \tag{46}$$

where vector $G = [g_0, \dots, g_N]^T$ is known and $C = [c_0, \dots, c_N]^T$ is an unknown vector. By using Theorem 4.2 (relation Eqs. (20) and (45)) we have

$$D^{\nu}u(x) \simeq C^T D^{\nu}\phi(x) = C^T D^{(\nu)}\phi(x), \tag{47}$$

$$D^{\beta_j}u(x) \simeq C^T D^{\beta_j}\phi(x) = C^T D^{(\beta_j)}\phi(x), \quad j = 1, \dots, k. \tag{48}$$

Employing Eqs (45)-(48), the residual $R_N(x)$ for Eq. (43) can be written as

$$R_N(x) = (C^T D^{(\nu)} - C^T \sum_{j=1}^{k} \gamma_j \mathbf{D}^{(\beta_j)} - \gamma_{k+1} C^T - G^T)\phi(x). \tag{49}$$

As in a typical tau method, see (Canuto & Hussaini 2006; Saadatmandi & Dehghan, 2010; Doha et al., 2011), we generate $(N - m + 1)$ linear equations by applying

$$\langle R_N(x), L_j^{(\alpha)}(x) \rangle = \int_0^{\infty} w(x) R_N(x) L_j^{(\alpha)}(x) dx = 0, \qquad j = 0, 1, \dots, N - m. \tag{50}$$

Also by substituting Eqs. (20) and (45) into Eq (44) we get

$$u^{(i)}(0) = C^T \mathbf{D}^{(i)} \phi(0) = d_i, \qquad i = 0, 1, \dots, m-1, \tag{51}$$

The combination of Eq. (50) and Eq. (51) reduces the solution of (43)-(44) to a linear system of algebraic equations, which can be solved for unknown coefficients of the vector C by any direct solver technique to find the spectral solution $u(x)$.

6.2 FGLFs Tau Operational Matrix Formulation Method

A direct solution technique is proposed here, to solve linear FDEs using the fractional-order generalized Laguerre tau method in combination with FGLOM.

Let us consider the linear FDE

$$D^{\nu}u(x) + \gamma u(x) = g(x), \quad \text{in } \Lambda = (0, \infty), \tag{52}$$

subject to

$$u(0) = u_0, \tag{53}$$

where γ is a real constant coefficient and also $0 < \nu \leq 1$, while $D^{\nu}u(x) \stackrel{\text{def}}{=} u^{(\nu)}(x)$ is the Caputo fractional derivative of order ν.
We implement an efficient algorithm to solve the fractional initial value problem; (52)-(53), and approximate $u(x)$ and $g(x)$ by FGLFs as

$$u_N(x) \simeq \sum_{i=0}^{N} c_i L_i^{(\alpha,\lambda)}(x) = C^T \Phi(x), \tag{54}$$

$$g_N(x) \simeq \sum_{i=0}^{N} g_i L_i^{(\alpha,\lambda)}(x) = G^T \Phi(x), \tag{55}$$

where vector $G = [g_0, \ldots, g_N]^T$ is known and $C = [c_0, \ldots, c_N]^T$ is an unknown vector. By using Theorem 4.4 (relations (26) and (54)) we have

$$D^{\nu}u(x) \simeq C^T D^{\nu}\Phi(x) = C^T D^{(\nu)}\Phi(x). \tag{56}$$

Employing Eqs (54)-(56), the residual $R_N(x)$ for Eq. (52) can be written as

$$R_N(x) = (C^T D^{(\nu)} + \gamma C^T - G^T)\Phi(x). \tag{57}$$

The application of a spectral tau scheme, see (Doha et al., 2012a), provides a system of (N) linear equations,

$$\langle R_N(x), L_j^{(\alpha,\lambda)}(x)\rangle = \int_0^{\infty} w^{(\alpha,\lambda)}(x) R_N(x) L_j^{(\alpha,\lambda)}(x)dx = 0, \qquad j = 0, 1, \ldots, N. \tag{58}$$

Substituting (54) in (53) yields

$$u(0) = C^T \mathbf{D}^{(0)}\Phi(0) = u_0, \tag{59}$$

The combination of (58) and (59) gives a system of algebraic equations, which may be solved by any direct solver technique to obtain the spectral solution $u_N(x)$.

6.3 Tau Method Based on FGLOM of Fractional Integration

The fractional-order generalized Laguerre tau scheme in conjunction with the derived operational matrix is now proposed as a method for solving the linear FDEs. The basic steps of such a scheme are: (i) The aforementioned fractional differential equation is converted into a fractional integrated form equation by making use of fractional integration for this equation. (ii) This integrated form equation is approximated by expressing the numerical solution as a linear combination of FGLFs. (iii) The problem is transformed into a system of algebraic equations by using the operational matrix of fractional integration of FGLFs.

In order to show the importance of FGLOM of fractional integration, we apply it to solve the following FDE:

$$D^{\nu}u(x) + \gamma u(x) = f(x), \quad \text{in } \Lambda = (0, \infty), \tag{60}$$

with initial condition

$$u(0) = u_0, \tag{61}$$

where γ is a real constant coefficient and also $0 < \nu \leq 1$. Moreover, $D^{\nu}u(x)$ denotes the Caputo fractional derivative of order ν for $u(x)$ and the value u_0 describes the initial condition of $u(x)$. If we apply the Riemann-Liouville integral of order ν on (60) and after making use of (6), we get the integrated form of (60), namely

$$u(x) - \sum_{j=0}^{m-1} u(0^+)\frac{x^j}{j!} + \gamma J^{\nu}u(x) = J^{\nu}f(x), \tag{62}$$

this implies that

$$u(x) + \gamma J^{\nu}u(x) = g(x), \tag{63}$$

where

$$g(x) = J^{\nu}f(x) + \sum_{j=0}^{m-1} u_0\frac{x^j}{j!}.$$

Approximating $u(x)$ and $g(x)$ by employing the FGLFs gives

$$u_N(x) \simeq \sum_{i=0}^{N} c_i L_i^{(\alpha,\lambda)}(x) = C^T \Phi(x), \tag{64}$$

$$g(x) \simeq \sum_{i=0}^{N} g_i L_i^{(\alpha,\lambda)}(x) = G^T \Phi(x). \tag{65}$$

In virtue of Theorem 5.2 (relation (35)), the Riemann-Liouville integral of order ν of (64), can be obtained from

$$J^{\nu}u_N(x) \simeq C^T J^{\nu}\Phi(x) \simeq C^T \mathbf{P}^{(\nu)}\Phi(x). \tag{66}$$

Employing Eq. (64) the residual $R_N(x)$ for Eq. (63) can be written as

$$R_N(x) = (C^T + \gamma C^T \mathbf{P}^{(\nu)} - G^T)\Phi(x). \tag{67}$$

Finally, applying the spectral tau method to the residual gives

$$(R_N(x), L_j^{(\alpha,\lambda)}(x))_{w^{(\alpha,\lambda)}(x)} = \int_0^\infty R_N(x)\, w^{(\alpha,\lambda)}(x)\, L_j^{(\alpha,\lambda)}(x)dx = 0, \qquad j = 0, 1, \cdots, N. \tag{68}$$

Also from (64) into (61) yields

$$u(0) = C^T \Phi(0) = u_0. \tag{69}$$

Eqs. (68) and (69) generate $N+1$ linear algebraic equations which may be solved using an LU factorization scheme.

6.4 Collocation Method for Nonlinear FDEs

We use the FGLC method to numerically solve the nonlinear FDE, namely

$$D^\nu u(x) = f\big(x, u(x)\big), \quad x \in \Lambda, \tag{70}$$

with initial conditions

$$u(0) = u_0, \tag{71}$$

where $0 < \nu \leq 1$.

Let

$$u_N(x) = \sum_{j=0}^{N} a_j L_j^{(\alpha,\lambda)}(x), \tag{72}$$

then, making use of formula (31) enables one to express explicitly the derivatives $D^\nu u(x)$, in terms of the expansion coefficients a_j. The criterion of spectral FGLC method for solving approximately (70)-(71) is to find $u_N(x)$ such that

$$D^\nu u_N(x) = F\big(x, u_N(x)\big), \tag{73}$$

is satisfied exactly at the collocation points $x_{N,k}^{(\alpha,\lambda)}$, $k = 0, 1, \cdots, N-1$. In other words, we have to collocate Eq. (73) at the N fractional-order generalized Laguerre roots $x_{N,k}^{(\alpha,\lambda)}$, which immediately yields

$$\sum_{j=0}^{N} a_j D^\nu L_j^{(\alpha,\lambda)}(x_{N,k}^{(\alpha,\lambda)}) = P\big(x_{N,k}^{(\alpha,\lambda)}, \sum_{j=0}^{N} a_j L_j^{(\alpha,\lambda)}(x_{N,k}^{(\alpha,\lambda)})\big), \tag{74}$$

with (71) written in the form

$$\sum_{j=0}^{N} a_j L_j^{(\alpha,\lambda)}(0) = u_0. \tag{75}$$

This constitutes a system of $(N+1)$ nonlinear algebraic equations in the unknown expansion coefficients a_j $(j = 0, 1, \cdots, N)$, which can be solved by using any standard iteration technique, such as Newton's iteration method.

6.5 Collocation Method for System of FDEs

We use the FGLC method to numerically solve the general form of system of nonlinear FDEs, namely

$$D^{\nu_i}u_i(x) = f_i\big(x, u_1(x), u_2(x), \ldots, u_n(x)\big), \quad x \in \Lambda, \quad i = 1, \ldots, n, \tag{76}$$

with initial conditions

$$u_i(0) = u_{i0}, \qquad i = 1, \ldots, n, \tag{77}$$

where $0 < \nu_i \le 1$. The approximate solutions of (76) can be expanded in the form:

$$u_{iN}(x) = \sum_{j=0}^{N} a_{ij} L_j^{(\alpha,\lambda)}(x). \tag{78}$$

The fractional derivatives $D^{\nu_i}u(x)$, can be expressed in terms of the expansion coefficients a_{ij} using (31). The implementation of FGLC method to solve (76)-(77) is to find $u_{iN}(x)$ such that

$$D^{\nu_i}u_{iN}(x) = F_i\big(x, u_{1N}(x), u_{2N}(x), \ldots, u_{nN}(x)\big), \quad x \in \Lambda, \tag{79}$$

is satisfied exactly at the collocation points $x_{i,N,k}^{(\alpha,\lambda)}$, $k = 0, 1, \cdots, N-1, i = 1, \cdots, n$, which immediately yields

$$\sum_{j=0}^{N} a_{ij} D^{\nu_i} L_j^{(\alpha,\lambda)}(x_{i,N,k}^{(\alpha,\lambda)}) = P_i\Big(x_{i,N,k}^{(\alpha,\lambda)}, \sum_{j=0}^{N} a_{1j} L_j^{(\alpha,\lambda)}(x_{1,N,k}^{(\alpha,\lambda)}), \sum_{j=0}^{N} a_{2j} L_j^{(\alpha,\lambda)}(x_{2,N,k}^{(\alpha,\lambda)}),$$
$$\ldots, \sum_{j=0}^{N} a_{nj} L_j^{(\alpha,\lambda)}(x_{n,N,k}^{(\alpha,\lambda)})\Big), \tag{80}$$

with (77) written in the form

$$\sum_{j=0}^{N} a_{ij} L_j^{(\alpha,\lambda)}(0) = u_{i0}, \qquad i = 1, \cdots, n. \tag{81}$$

This means that the system (76) with its initial conditions has been reduced to a system of $n(N+1)$ nonlinear algebraic equations (80)-(81), which may be solved by using any standard iteration technique.

Corollary 5. *In particular, the special case for fractional Laguerre functions may be obtained directly by taking $\alpha = 0$. However, the generalized Laguerre polynomials may be achieved by replacing $\lambda = 1$, which are used in (Baleanu et al., 2013) for solving FDEs.*

Table 1. MAE with various choices of η, ζ, v and N in $x \in [0, 1]$

N	η	$\zeta = \lambda$	$\alpha = 0$	$\alpha = 2$	$\eta = \zeta = \lambda$	$\alpha = 0$	$\alpha = 2$
8	0.9	0.5	$3.56.10^{-3}$	$9.34.10^{-3}$	0.75	$5.55.10^{-17}$	$4.37.10^{-16}$
16			$9.38.10^{-4}$	$2.42.10^{-3}$		$5.55.10^{-17}$	$4.44.10^{-16}$
24			$3.98.10^{-4}$	$1.31.10^{-3}$		$5.55.10^{-17}$	$4.51.10^{-16}$
32			$2.31.10^{-4}$	$7.35.10^{-4}$		$5.55.10^{-17}$	$4.44.10^{-16}$
40			$1.49.10^{-4}$	$4.38.10^{-4}$		$5.55.10^{-17}$	$4.44.10^{-16}$
48			$1.12.10^{-4}$	$2.69.10^{-4}$		$5.55.10^{-17}$	$4.51.10^{-16}$
8	0.8	0.6	$8.69.10^{-3}$	$2.30.10^{-2}$	0.999	$5.55.10^{-17}$	$3.33.10^{-16}$
16			$2.90.10^{-3}$	$9.69.10^{-3}$		$5.55.10^{-17}$	$3.33.10^{-16}$
24			$1.61.10^{-3}$	$5.85.10^{-3}$		$5.55.10^{-17}$	$3.12.10^{-16}$
32			$1.08.10^{-3}$	$4.11.10^{-3}$		$5.55.10^{-17}$	$3.33.10^{-16}$
40			$7.90.10^{-4}$	$3.09.10^{-3}$		$5.55.10^{-17}$	$3.33.10^{-16}$
48			$6.16.10^{-4}$	$2.37.10^{-3}$		$5.55.10^{-17}$	$3.12.10^{-16}$

7 Applications and Numerical Results

In this section we give some numerical results obtained by using the algorithms presented in the previous sections. Comparisons of the spectral methods with those obtained by other methods reveal that spectral methods are very effective and convenient.

Example 23. *Consider the equation, see (Bhrawy et al., 2014)*

$$D^{\zeta} u(x) + u(x) = \frac{\Gamma(\eta + 1)}{\Gamma(\eta - \zeta + 1)} x^{\eta-\zeta} + x^{\eta}, \qquad 0 < \zeta \le \eta < 1, \quad x \in \Lambda,$$

the exact solution is given by $u(x) = x^{\eta}$.

The solution of this problem is obtained by applying the FGLT method. In Table 1, the maximum absolute errors (MAEs) of $u(x) - u_N(x)$ using the FGLT method based on treating the right hand side of this problem by Gauss quadrature of fractional order generalized Laguerre functions, with various choices of η, ζ, α and N are compared with the results of the improved generalized Laguerre tau (GLT) method (see (Bhrawy et al., 2014)) with various choices of η, ζ, α and N. From Table 1, we achieve an accurate approximation with the exact solution by using FGLFs and that our method is more accurate than GLT (Bhrawy et al., 2014). Fig. 1 displays comparison between the curves of exact solutions and approximate solutions at $N = 10$, $\alpha = 3$ and variable choices of η, ζ, and λ. Meanwhile, MAEs for $N = 10$ and different values of $\eta = \zeta = \lambda$ and α are shown in Fig. 2 and Fig. 3.

Example 24. *Consider the equation, see (Ford & Connolly, 2009)*

$$D^2 u(x) + D^{\frac{1}{2}} u(x) + u(x) = x^2 + 2 + \frac{2.6666666667}{\Gamma(0.5)} x^{1.5}, \quad u(0) = 0, \quad u'(0) = 0, \quad x \in \Lambda,$$

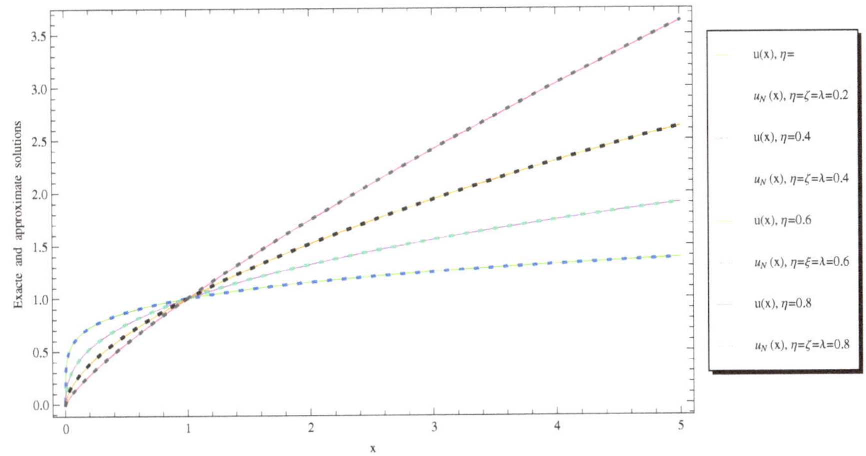

Fig. 1. Comparing the exact and approximate solutions at $N = 10, \alpha = 3$ and $\eta = \zeta = \lambda = \{0.2, 0.4, 0.6, 0.8\}$.

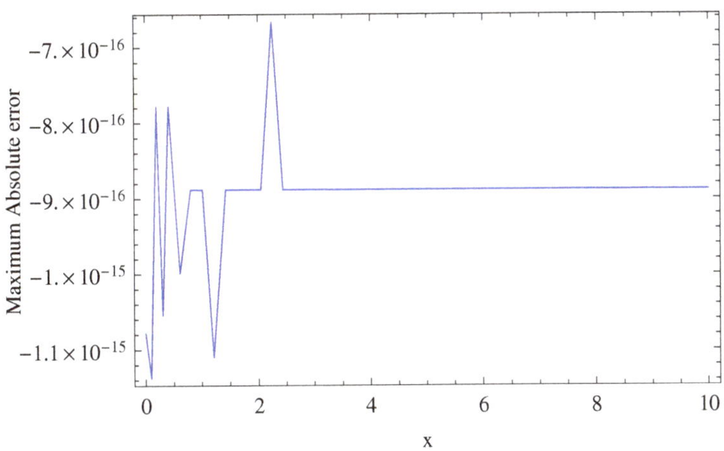

Fig. 2. Graph of maximum absolute error functions at $N = 10, \alpha = 3$ and $\eta = \zeta = \lambda = 0.85$ for Example 1.

whose exact solution is given by $u(x) = x^2$.

The use of the technique described in Section 6.1 with $N = 2$ enables one to approximate the solution as

$$u(x) = c_0 L_0^{(\alpha)}(x) + c_1 L_0^{(\alpha)}(x) + c_2 L_2^{(\alpha)}(x) = C^T \phi(x).$$

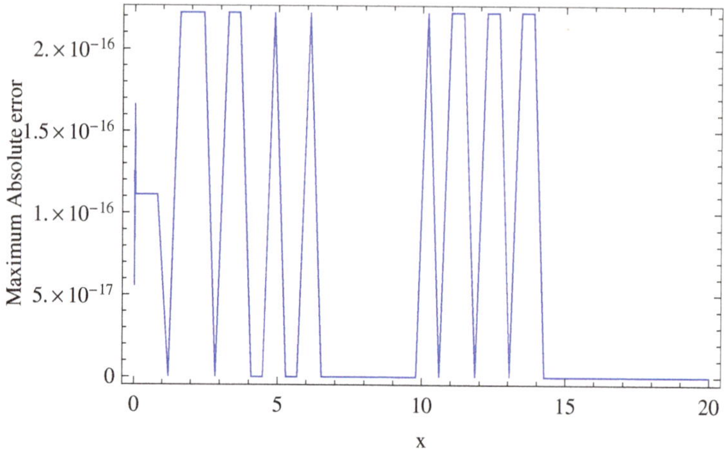

Fig. 3. Graph of maximum absolute error functions at $N = 10$, $\alpha = 0$ and $\eta = \zeta = \lambda = 0.25$ for Example 23.

Here, we have

$$\mathbf{D}^{(2)} = \begin{pmatrix} 0 & 0 & 0 \\ 0 & 0 & 0 \\ 1 & 0 & 0 \end{pmatrix}, \ \mathbf{D}^{(\frac{1}{2})} = \begin{pmatrix} 0 & 0 & 0 \\ \Psi_{1/2}(1,0) & \Psi_{1/2}(1,1) & \Psi_{1/2}(1,2) \\ \Psi_{1/2}(2,0) & \Psi_{1/2}(2,0) & \Psi_{1/2}(2,0) \end{pmatrix},$$

$$G = \begin{pmatrix} g_0 \\ g_1 \\ g_2 \end{pmatrix}.$$

Now, by applying the technique described in Section 6.1. we have

$$c_0 + \Psi_{1/2}(1,0)c_1 + [\Psi_{1/2}(2,0) + 1]c_2 = 6, \tag{82}$$

$$c_0 + (\alpha+1)c_1 + \frac{(\alpha+1)(\alpha+2)}{2}c_2 = 0, \tag{83}$$

$$-c_1 - (\alpha+2)c_2 = 0. \tag{84}$$

Finally by solving Eqs. (82)-(84), we have the 3 unknown coefficients with various choices of α given in Table 2. Then, we get

$$c_0 = \alpha^2 + 3\alpha + 2, \qquad c_1 = -2\alpha - 4, \qquad c_2 = 2.$$

Thus we can write

$$u(x) = \begin{pmatrix} c_0, & c_1, & c_2 \end{pmatrix} \begin{pmatrix} L_0^{(\alpha)}(x) \\ L_1^{(\alpha)}(x) \\ L_2^{(\alpha)}(x) \end{pmatrix} = x^2,$$

which is the exact solution.

Table 2. c_0, c_1 and c_2 for different values of α for Example 2

α	c_0	c_1	c_2
-0.5	0.75	-3	2
0	2	-4	2
0.5	3.75	-5	2
1	6	-6	2
2	12	-8	2
3	20	-10	2

Example 25. *Consider the equation*

$$D^{\nu}u(x) + u(x) = \frac{\Gamma(3)}{\Gamma(3-\nu)} x^{2-\nu} + x^2, \qquad 0 < \nu < 1, \quad x \in \Lambda,$$

the exact solution is given by $u(x) = x^2$.

Now, we implement the spectral tau scheme based on the FGLOM of fractional derivative (Section 6.2) with $N = 6$, then the approximate solution can be expanded as

$$u_N(x) = \sum_{i=0}^{N} c_i L_i^{(\alpha,\lambda)}(x) = C^T \phi(x).$$

If we choose $\lambda = \frac{1}{3}$ and $\nu = \frac{1}{3}$, then

$$\mathbf{D}^{(\nu)} = \begin{pmatrix} 0 & 0 & 0 & \dots & 0 \\ S_\nu(1,0,\frac{1}{3}) & S_\nu(1,1,\frac{1}{3}) & S_\nu(1,2,\frac{1}{3}) & \dots & S_\nu(1,6,\frac{1}{3}) \\ \vdots & \vdots & \vdots & \dots & \vdots \\ S_\nu(i,0,\frac{1}{3}) & S_\nu(i,1,\frac{1}{3}) & S_\nu(i,2,\frac{1}{3}) & \dots & S_\nu(i,6,\frac{1}{3}) \\ \vdots & \vdots & \vdots & \dots & \vdots \\ S_\nu(6,0,\frac{1}{3}) & S_\nu(6,1,\frac{1}{3}) & S_\nu(6,2,\frac{1}{3}) & \dots & S_\nu(6,6,\frac{1}{3}) \end{pmatrix}, \quad G = \begin{pmatrix} g_0 \\ g_1 \\ g_2 \\ \vdots \\ g_6 \end{pmatrix}.$$

Using Eq. (58), we obtain

$$\begin{aligned}
&c_0 + S_\nu(1,0,\tfrac{1}{3})c_1 + S_\nu(2,0,\tfrac{1}{3})c_2 + S_\nu(3,0,\tfrac{1}{3})c_3 + S_\nu(4,0,\tfrac{1}{3})c_4 + S_\nu(5,0,\tfrac{1}{3})c_5 + S_\nu(6,0,\tfrac{1}{3})c_6 = g_0,\\
&c_1 + S_\nu(1,1,\tfrac{1}{3})c_1 + S_\nu(2,1,\tfrac{1}{3})c_2 + S_\nu(3,1,\tfrac{1}{3})c_3 + S_\nu(4,1,\tfrac{1}{3})c_4 + S_\nu(5,1,\tfrac{1}{3})c_5 + S_\nu(6,1,\tfrac{1}{3})c_6 = g_1,\\
&c_2 + S_\nu(1,2,\tfrac{1}{3})c_1 + S_\nu(2,2,\tfrac{1}{3})c_2 + S_\nu(3,2,\tfrac{1}{3})c_3 + S_\nu(4,2,\tfrac{1}{3})c_4 + S_\nu(5,2,\tfrac{1}{3})c_5 + S_\nu(6,2,\tfrac{1}{3})c_6 = g_2,\\
&c_3 + S_\nu(1,3,\tfrac{1}{3})c_1 + S_\nu(2,3,\tfrac{1}{3})c_2 + S_\nu(3,3,\tfrac{1}{3})c_3 + S_\nu(4,3,\tfrac{1}{3})c_4 + S_\nu(5,3,\tfrac{1}{3})c_5 + S_\nu(6,3,\tfrac{1}{3})c_6 = g_3,\\
&c_4 + S_\nu(1,4,\tfrac{1}{3})c_1 + S_\nu(2,4,\tfrac{1}{3})c_2 + S_\nu(3,4,\tfrac{1}{3})c_3 + S_\nu(4,4,\tfrac{1}{3})c_4 + S_\nu(5,4,\tfrac{1}{3})c_5 + S_\nu(6,4,\tfrac{1}{3})c_6 = g_4,\\
&c_5 + S_\nu(1,5,\tfrac{1}{3})c_1 + S_\nu(2,5,\tfrac{1}{3})c_2 + S_\nu(3,5,\tfrac{1}{3})c_3 + S_\nu(4,5,\tfrac{1}{3})c_4 + S_\nu(5,5,\tfrac{1}{3})c_5 + S_\nu(6,5,\tfrac{1}{3})c_6 = g_5.
\end{aligned} \tag{85}$$

Table 3. The values $c_0, c_1, c_2, \ldots$ and c_6 for different values of α at $v = \frac{1}{3}$ for Example 25

α	c_0	c_1	c_2	c_3	c_4	c_5	c_6
0	720	-4320	10800	-14400	10800	-4320	720
1	5040	-15120	25200	-25200	15120	-5040	720
2	20160	-40320	50400	-40320	20160	-5760	720
3	60480	-90720	90720	-60480	25920	-6480	720

Table 4. The values $c_0, c_1, c_2, \ldots$ and c_6 for different values of α at $v = \frac{1}{4}$ for Example 25

α	c_0	c_1	c_2	c_3	c_4	c_5	c_6
0	720	-4300	10800	-14000	10800	-4300	720
1	5000	-15000	2500	-25000	15000	-5000	720
2	20000	-40000	50000	-40000	20000	-6000	700
3	60000	-90000	90000	-60000	30000	-6000	700

The treatment of the initial condition using Eq. (59), yields

$$c_0 + (\alpha+1)c_1 + \frac{(\alpha+1)(\alpha+2)}{2}c_2 + \frac{(\alpha+1)(\alpha+2)(\alpha+3)}{6}c_3 + \frac{(\alpha+1)(\alpha+2)(\alpha+3)(\alpha+4)}{24}c_4 + \frac{(\alpha+1)(\alpha+2)(\alpha+3)(\alpha+4)(\alpha+5)}{120}c_5 + \frac{(\alpha+1)(\alpha+2)(\alpha+3)(\alpha+4)(\alpha+5)(\alpha+6)}{720}c_6 = 0. \quad (86)$$

Solving the resulting system of algebraic equations (85)-(86) provides the unknown coefficients in terms of α.
Accordingly, the approximate solution can be written as

$$u_N(x) = \sum_{i=0}^{6} c_i L_i^{(\alpha,\frac{1}{3})}(x) = x^2.$$

Tables 3 and 4 list the values of $c_0, c_1, c_2, c_3, c_4, c_5$ and c_6 with different choices of α and two choices of $v = 1/3$ and $v = 1/4$. Indeed, we can achieve the exact solution of this problem with all choices of the fractional-order generalized Laguerre parameter α.

Example 26. *Consider the following fractional initial value problem*

$$D^{\frac{3}{2}}u(x) + 3u(x) = 3x^3 + \frac{8}{\Gamma(0.5)}x^{1.5}, \quad u(0) = 0, \quad u'(0) = 0, \quad x \in \Lambda,$$

whose exact solution is given by $u(x) = x^3$.

If we apply the technique described in Section 6.3 with $\lambda = 1$ (generalized Laguerre tau) and $N = 3$, then the approximate solution can be written as

$$u_N(x) = \sum_{i=0}^{3} c_i L_i^{(\alpha)}(x) = C^T\phi(x),$$

Table 5. c_0, c_1, c_2 and c_3 for different values of α for Example 26

α	c_0	c_1	c_2	c_3
-0.5	$\frac{15}{8}$	$-\frac{45}{4}$	15	-6
0	6	-18	18	-6
0.5	$\frac{105}{8}$	$-\frac{105}{4}$	21	-6
1	24	-36	24	-6
2	60	-60	30	-6
3	120	-90	36	-6

and

$$\mathbf{P}^{(\frac{3}{2})} = \begin{pmatrix} \Theta_{\frac{3}{2}}(0,0) & \Theta_{\frac{3}{2}}(0,1) & \Theta_{\frac{3}{2}}(0,2) & \Theta_{\frac{3}{2}}(0,3) \\ \Theta_{\frac{3}{2}}(1,0) & \Theta_{\frac{3}{2}}(1,1) & \Theta_{\frac{3}{2}}(1,2) & \Theta_{\frac{3}{2}}(1,3) \\ \Theta_{\frac{3}{2}}(2,0) & \Theta_{\frac{3}{2}}(2,1) & \Theta_{\frac{3}{2}}(2,2) & \Theta_{\frac{3}{2}}(2,3) \\ \Theta_{\frac{3}{2}}(3,0) & \Theta_{\frac{3}{2}}(3,1) & \Theta_{\frac{3}{2}}(3,2) & \Theta_{\frac{3}{2}}(3,3) \end{pmatrix}, \quad G = \begin{pmatrix} g_0 \\ g_1 \\ g_2 \\ g_3 \end{pmatrix}.$$

Now, by applying the technique described in Section 6.3. we have

$$\begin{aligned} &3\Theta_{\frac{3}{2}}(0,2)\,c_0 + 3\Theta_{\frac{3}{2}}(1,2)\,c_1 + (1 + 3\Theta_{\frac{3}{2}}(2,2))\,c_2 + 3\Theta_{\frac{3}{2}}(3,2)\,c_3 + g_2 = 0, \\ &3\Theta_{\frac{3}{2}}(0,3)\,c_0 + 3\Theta_{\frac{3}{2}}(1,3)\,c_1 + 3\Theta_{\frac{3}{2}}(2,3)\,c_2 + (1 + 3\Theta_{\frac{3}{2}}(3,3))\,c_3 + g_3 = 0, \end{aligned} \tag{87}$$

$$\begin{aligned} &C^T\phi(0) = c_0 + (\alpha+1)c_1 + \frac{(\alpha+1)(\alpha+2)}{2}c_2 + \frac{(\alpha+1)(\alpha+2)(\alpha+3)}{6}c_3 = 0, \\ &C^T\mathbf{D}^{(1)}\phi(0) = -c_1 - (\alpha+2)c_2 - \frac{(\alpha+3)(\alpha+2)}{2}c_3 = 0. \end{aligned} \tag{88}$$

By solving the linear system (87)-(88), we obtain the 4 unknown coefficients with various choices of α in Table 5, we get

$$c_0 = \alpha^3 + 6\alpha + 11\alpha + 6, \quad c_1 = -3\alpha^2 - 15\alpha - 18, \quad c_2 = 6\alpha + 18, \quad c_3 = -6.$$

Therefore we can write

$$u_N(x) = \sum_{i=0}^{3} c_i L_i^{(\alpha)}(x) = x^3,$$

which is the exact solution.

Example 27. *Consider the following FDE*

$$D^\nu u(x) + u(x) = g(x), \qquad u(0) = 1, \qquad x \in [0, 100],$$

where

$$g(x) = cos(\gamma x) + \frac{1}{\Gamma(-\nu)} \int_0^x (x-t)^{-\nu-1} u(t)dt,$$

and the exact solution is given by $u(x) = cos(\gamma x)$.

Table 6. MAE for $\gamma = 0.01$, $\lambda = \frac{1}{2}$ and different values of N and α in $x \in [0, 100]$ for Example 27

N	α	error	α	error	α	error	α	error
2		$1.46.10^{-2}$		$2.09.10^{-2}$		$2.41.10^{-2}$		$2.18.10^{-2}$
4	0	$3.30.10^{-3}$	1	$6.62.10^{-3}$	2	$1.13.10^{-2}$	3	$2.03.10^{-2}$
6		$8.80.10^{-4}$		$1.90.10^{-3}$		$3.00.10^{-3}$		$4.00.10^{-3}$
8		0		0		0		0

The solution of this problem is obtained by applying the technique described in Section 6.4 based on the FGLOM of fractional integration. The MAE for $\gamma = 0.01, \lambda = \frac{1}{2}$ and various choices of N and α are shown in Table 6. Moreover, the approximate solution obtained by the proposed method for $\alpha = 0,\ \lambda = \frac{3}{4},\ \gamma = 0.1$ and two choices of N is shown in Figure 4 to make it easier to compare with the analytic solution. The MAE for $N = 6,\ \alpha = 0,\ \lambda = \frac{3}{4}$ and $\gamma = 0.1$ is shown in Figure 5.

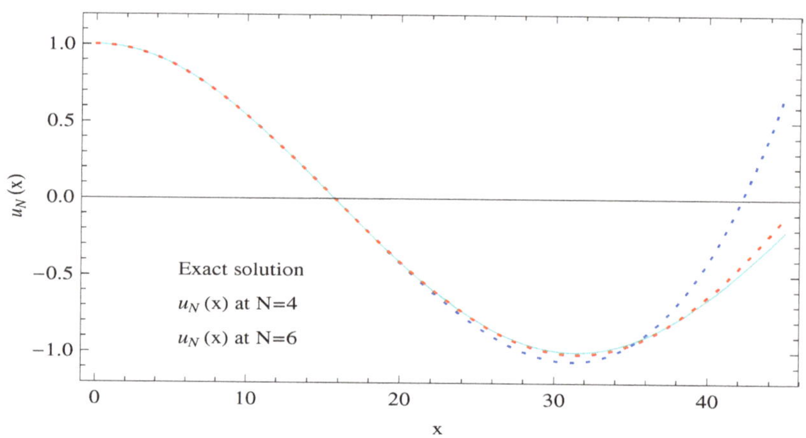

Fig. 4. Comparing the exact solution and the approximate solutions at $N = 4, 6$, where $\alpha = 0, \lambda = \frac{3}{4}$ and $\gamma = 0.1$ for Example 27.

Example 28. *Consider the following nonlinear initial value problem*

$$D^{\nu}u(x) + 2u^2(x) = \Gamma(\nu + 2)x + 2(x^{\nu+1})^2, \qquad 0 < \nu \leq 1,$$

whose exact solution is given by $u(x) = x^{\nu+1}$.

The absolute residual error functions of $u_N(x)$, using the fractional-order Laguerre collocation scheme, with various choices of $\nu,\ \lambda$ and α at $N = 16$ in $x \in [0, 40]$ are listed in Table 7.

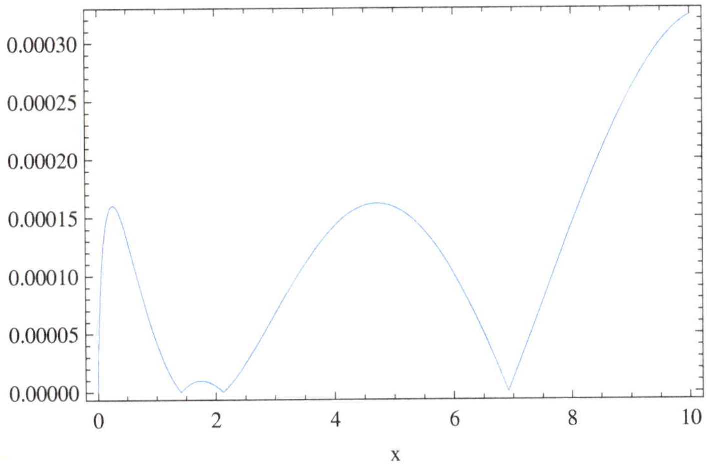

Fig. 5. Graph of the absolute error function for $N = 6$, $\alpha = 0$, $\lambda = \frac{3}{4}$ and $\gamma = 0.1$for Example 27.

Table 7. MAE with various choices of v, λ and α at $N = 16$ in $x \in [0, 40]$, for Example 28.

x	v	λ	$\alpha = 0$	$\alpha = 2$	v	λ	$\alpha = 0$	$\alpha = 2$
0			$1.19.10^{-15}$	$1.80.10^{-14}$			$9.96.10^{-15}$	$4.27.10^{-15}$
10			$1.04.10^{-12}$	$1.10.10^{-10}$			$1.96.10^{-3}$	$3.39.10^{-4}$
20	0.5	0.5	$9.84.10^{-12}$	$1.88.10^{-11}$	0.6	0.6	$4.57.10^{-3}$	$2.14.10^{-4}$
30			$2.80.10^{-11}$	$3.88.10^{-10}$			$1.30.10^{-2}$	$1.05.10^{-3}$
40			$1.97.10^{-11}$	$1.07.10^{-10}$			$1.03.10^{-2}$	$2.33.10^{-3}$

Example 29. *Consider the initial value problem*

$$D^2u(x) - D^{(\frac{3}{2})}u(x) + \frac{6}{5}D^{(1)}u(x) + D^{(\frac{1}{2})}u(x) + \frac{1}{5}u(x) = f(x), \qquad u(0) = 0, \qquad u'(0) = 0, \tag{89}$$

with an exact solution $u(x) = x^{\frac{5}{2}} + x^2$.

We convert the multi-term FDEs (89) into a system of FDE of orders less than or equal 1, by changing variable $u_1(x) = u(x)$ and get

$$\begin{aligned} D^{\frac{1}{2}}u_1(x) &= u_2(x), \\ D^{\frac{1}{2}}u_2(x) &= u_3(x), \\ D^{\frac{1}{2}}u_3(x) &= u_4(x), \\ D^{\frac{1}{2}}u_4(x) &= u_4(x) - \tfrac{6}{5}u_3(x) - u_2(x) - \tfrac{1}{5}u_1(x) + f(x), \end{aligned} \tag{90}$$

with initial conditions

$$u_1(0) = u(0), \qquad u_2(0) = 0, \qquad u_3(0) = u'(0), \qquad u_4(0) = 0. \tag{91}$$

Table 8. Absolute error using FGLC method with various choices of α, $N = 10$ and $\nu = \lambda = 0.5$ for Example 29.

SCT	FGLC method (N=10)					
(N=64) (Bhrawy et al.,2013)	$\alpha = -\frac{1}{2}$	$\alpha = 0$	$\alpha = \frac{1}{2}$	$\alpha = 1$	$\alpha = 2$	$\alpha = 3$
$2.2.10^{-8}$	$2.6.10^{-10}$	$1.0.10^{-13}$	$3.0.10^{-12}$	$3.9.10^{-12}$	$7.3.10^{-12}$	$8.0.10^{-12}$

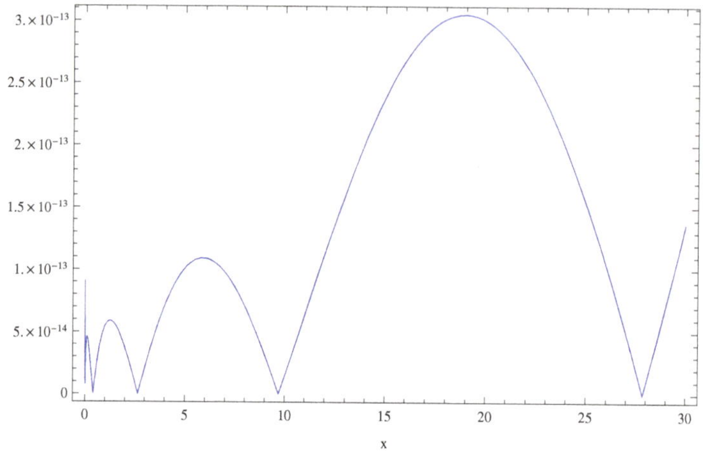

Fig. 6. Graph of the absolute error function for $N = 10$, $\alpha = 0$ and $\nu = \lambda = 0.5$ for Example 29.

In Table 8, we list the results obtained by the FGLC method with various choices of α, $N = 10$, and $\nu = \lambda = 0.5$. The present method is compared with shifted Chebyshev spectral tau (SCT) method given in (Bhrawy et al., 2013). It is observed from this table, that the result obtained by the FGLC method for each choice of the parameter α is superior to that obtained by SCT method. Fig. 6 displays the absolute error function at $N = 10$, $\alpha = 0$ and $\nu = \lambda = 0.5$.

Bibliography

Ahmadian, A., Suleiman, M., & Salahshour, S. (2013), An operational matrix based on Legendre polynomials for solving fuzzy fractional-order differential equations. Abstract and Applied Analysis. 2013, 505903-29.

Baleanu, D., Bhrawy, A.H., & Taha, T.M. (2013), Two efficient generalized Laguerre spectral algorithms for fractional initial value problems. Abstract and Applied Analysis. 2013, 546502-10.

Bhrawy, A.H., (2014), An efficient Jacobi pseudospectral approximation for nonlinear complex generalized Zakharov system. Applied Mathematics and Computation. 247, 30-46.

Bhrawy, A.H., & Abdelkawy, M.A. (2015), A fully spectral collocation approximation for multi-dimensional fractional Schrodinger equations. Journal of Computational Physics. 294, 462-483.

Bhrawy, A.H., Alghamdi, M. M., & Taha, T.M. (2012), A new modified generalized Laguerre operational matrix of fractional integration for solving fractional differential equations on the half line. Advances in Difference Equations. 2012, 0:179.
Bhrawy, A.H., Baleanu, D., L.M. & Assas (2014), Efficient generalized Laguerre-spectral methods for solving multi-term fractional differential equations on the half line. Journal of Vibration and Control. 20, 973-985.
Bhrawy, A.H., Tharwat, M.M., & Yildirim, A. (2013),A new formula for fractional integrals of Chebyshev polynomials: Application for solving multi-term fractional differential equations. Applied Mathematical Modelling. 37, 4245-4252.
Bhrawy, A.H., & Zaky, M.A. (2015). A method based on the Jacobi tau approximation for solving multi-term time-space fractional partial differential equations. Journal of Computational Physics. 281, 876-895.
Bhrawy, A.H., & Zaky, M.A. (2015), Numerical simulation for two-dimensional variable-order fractional nonlinear cable equation. Nonlinear Dynamics. 80(1), 101-116.
Bhrawy, A.H., Zaky, M.A., & Gorder, R.A.V. (2015) , A space-time Legendre spectral tau method for the two-sided space-time Caputo fractional diffusion-wave equation. Numerical Algorithms. DOI 10.1007/s11075-015-9990-9
Canuto C., Hussaini M.Y., Quarteroni A., & Zang T.A. (2006), Spectral Methods: Fundamentals in Single Domains. New York: Springer-Verlag.
Chen, J., Liu, F., Liu, Q., Chen, X., Anh, V., Turner, I., & Burrage, K. (2014), Numerical simulation for the three-dimension fractional sub-diffusion equation. Applied Mathematical Modelling. 38, 3695-3705.
Daftardar-Gejji, V., Sukale, Y., & Bhalekar, S. (2014), A new predictor-corrector method for fractional differential equations. Applied Mathematics and Computation. 244, 158-182.
Doha, E.H., & Bhrawy, A.H. (2012), An efficient direct solver for multidimensional elliptic Robin boundary value problems using a Legendre spectral-Galerkin method. Computers & Mathematics with Applications. 64, 558-571
Doha, E.H., Bhrawy, A.H., & Abd-Elhameed W.M. (2009), Jacobi spectral Galerkin method for elliptic Neumann problems. Numerical Algorithms. 50(1), 67-91.
Doha, E.H., Bhrawy, A.H., Abdelkawy, M.A., & Van Gorder, R.A. (2014), Jacobi-Gauss-Lobatto collocation method for the numerical solution of 1 + 1 nonlinear Schrödinger equations. Journal of Computational Physics. 261, 244-255.
Doha, E.H., Bhrawy, A.H., & Ezz-Eldien, S.S. (2011), A Chebyshev spectral method based on operational matrix for initial and boundary value problems of fractional order. Computers & Mathematics with Applications. 62, 2364-2373.
Doha, E.H., Bhrawy, A.H., & Ezz-Eldien, S.S. (2012), A new Jacobi operational matrix: an application for solving fractional differential equations. Applied Mathematical Modelling. 36, 4931-4943.
Doha, E.H., Bhrawy, A.H., & Hafez, R.M. (2012), On shifted Jacobi spectral method for high-order multi-point boundary value problems. Communications in Nonlinear Science and Numerical Simulation. 17, 3802-3810.
Dubey, S., & Sharma, M. (2014), Solutions to fractional functional differential equations with nonlocal conditions. Fractional Calculus and Applied Analysis. 17 654-673.
Ford, N.J., & Connolly, J.A. (2009), Systems-based decomposition schemes for the approximate solution of multi-term fractional differential equations, Journal of Computational and Applied Mathematics. 229, 382-391.
Funaro, D. (1992), Polynomial Approximations of Differential Equations. Heidelberg: Springer-Verlag.
Gaur, M., & Singh, K. (2014), On group invariant solutions of fractional order Burgers-Poisson equation. Applied Mathematics and Computation. 244, 870-877.

Huang, Q., & Zhdanov, R. (2014), Symmetries and exact solutions of the time fractional Harry-Dym equation with Riemann-Liouville derivative. Physica A. 409, 110-118.

Ishteva, M., & Boyadjiev, L. (2005), On the C-Laguerre functions, Comptes rendus de l'Acade?mie bulgare des sciences. 58, 1019-1024.

Ishteva, M., Boyadjiev, L., & Scherer, R. (2005), On the Caputo operator of fractional calculus and C-Laguerre functions. Math. Sci. Res. 9, 161-170.

Katsikadelis, J. T. (2014), Numerical solution of distributed order fractional differential equations. Journal of Computational Physics. 259, 11-22.

Kazem, S., Abbasbandy, S., & Kumar, S. (2013), Fractional-order Legendre functions for solving fractional-order differential equations. Applied Mathematical Modelling. 37, 5498-5510.

Keshavarz, E., Ordokhani, Y., & Razzaghi M. (2014), Bernoulli wavelet operational matrix of fractional order integration and its applications in solving the fractional order differential equations. Applied Mathematical Modelling. 38, 6038-6051.

Lau S.R., & Price, H. (2012), Sparse spectral-tau method for the three-dimensional helically reduced wave equation on two-center domains. Journal of Computational Physics. 231 7695-7714.

Liu, Z., & Lu, P.(2014), Stability analysis for HIV infection of CD4+ T-cells by a fractional differential time-delay model with cure rate. Advances in Difference Equations. 2014, 298.

Liu, Y., Fang, Z., Li, H., & He, S. (2014), A mixed finite element method for a time-fractional fourth-order partial differential equation. Applied Mathematics and Computation. 243, 703-717

Ma, X., & Huang, C. (2014), Spectral collocation method for linear fractional integro-differential equations. Applied Mathematical Modelling. 38, 1434-1448.

Ma, J., Li, B-W., & Howell, J. R. (2014), Thermal radiation heat transfer in one- and two-dimensional enclosures using the spectral collocation method with full spectrum k-distribution model. International Journal of Heat and Mass Transfer. 71, 35-43.

Parand, K., & Nikarya, M. (2014), Application of Bessel functions for solving differential and integro-differential equations of the fractional order. Applied Mathematical Modelling. 38, 4137-4147.

Takači, D., Takači, A., & Takači, A. (2014), On the operational solutions of fuzzy fractional differential equations. Fractional Calculus and Applied Analysis. 17, 1100-1113.

Tatari, M., & Haghighi, M. (2014), A generalized Laguerre-Legendre spectral collocation method for solving initial-boundary value problems. Applied Mathematical Modelling. 38, 1351-1364.

Saadatmandi, A., & Dehghan, M. (2010),A new operational matrix for solving fractional-order differential equations. Computers & Mathematics with Applications. 59, 1326-1336.

Szegö, G. (1938), Orthogonal Polynomials. Am. Math. Soc. Colloq. Pub.

Yang, Y. (2014), Jacobi spectral Galerkin methods for fractional integro-differential equations. Calcolo. DOI 10.1007/s10092-014-0128-6

Yang, A. M, Zhang, Y. Z., Cattani, C., Xie, G.N., Rashidi, M. M., Zhou, Y.J., & Yang, X-J. (2014), Application of Local Fractional Series Expansion Method to Solve Klein-Gordon Equations on Cantor Sets. Abstract and Applied Analysis, 2014, 372741-6.

Yin, F., Song, J., Wu, Y., & Zhang, L. (2013), Numerical solution of the fractional partial differential equations by the two-dimensional fractional-order Legendre functions. Abstract and Applied Analysis. 2013, 562140-13.

Zayernouri, M., Ainsworth, M., & Em Karniadakis, G. (2014), A unified Petrov-Galerkin spectral method for fractional PDEs. Computer Methods in Applied Mechanics and Engineering. 283(1), 1545-1569.

Zayernouri, M., & Em Karniadakis, G. (2014), Exponentially accurate spectral and spectral element methods for fractional ODEs, Journal of Computational Physics. 257, 460-480.

Zhang, H., Yang, X., & Hanc, X. (2014), Discrete-time orthogonal spline collocation method with application to two-dimensional fractional cable equation. Computers & Mathematics with Applications. 68, 1710-1722.

Hong-Yan Liu and Ji-Huan He

From Leibniz's Notation for Derivative to the Fractal Derivative, Fractional Derivative and Application in Mongolian Yurt

Abstract: The fractal derivative is a natural extension of Leibniz's derivative for discontinuous fractal media. In this chapter three definitions are discussed, one of which is newly introduced and geometrically motivated. The fractal differential equations can be transferred into ordinary differential equations.
Some fractional derivatives are derived from variational iteration algorithms, such as the well-known Riemann-Liouville fractional derivative, Jumarie's modification, and a new one.
A yurt is a portable tent-like dwelling structure favored by Mongolian nomads for more than three millennia, it can be used even in harsh environments as low as negative 50 degrees centigrade. We conclude that the multi-layer structure of the felt cover, which mimics the design of a cocoon, is the key for weatherproofing. A local fractional differential model is established to find an optimal thickness of the fractal hierarchy of the felt cover. A better understanding of the yurt mechanism could help the further design of yurt-like space suits and other protective clothes for special applications.

Keywords: Fractal; fractional differential equation; fractal derivative; the fractional complex transform; variational iteration method; Yurt; cocoon; biomimic design; human selection; local fractional calculus

1 Introduction

The most common definition of a derivative in mathematics textbook is given as follows

$$f'(x) = \lim_{\Delta x \to 0} \frac{\Delta f}{\Delta x}. \tag{1}$$

This definition is useful for any continuum media. For any discontinuous media (fractal media), however, a fractal derivative or a fractional derivative has to be adopted. Fractional calculus began with the seminal work of Leibniz (July 1, 1646 – November 14, 1716), who authorized a very influential essay on monadology as the forerunner for

Hong-Yan Liu: School of fashion technology, Zhongyuan University of Technology, Zhengzhou 450007, China, E-mail: phdliuhongyan@gmail.com
Ji-Huan He: National Engineering Laboratory for Modern Silk, College of Textile and Clothing Engineering, Soochow University, 199 Ren-ai Road, Suzhou 215123, China,
and Nantong Textile Institute, Soochow University, 58 Chong Chuan Road, Nantong, China E-mail: hejihuan@suda.edu.cn

derivatives on Cantor sets and fractals. It is a generalization of ordinary differentiation and integration to non-integer order(s). In a letter to L'Hospital in 1695, Leibniz raised the following question: "Can the meaning of derivatives with integer order be generalized to derivatives with non-integer orders?" To answer this question, we consider a continuous function, $f(x)$, the first-order derivation is

$$Df(x) = \frac{df(x)}{dx} = f'(x) \tag{2}$$

Now we define a new operator H defined as

$$D = H^{1/\alpha}, \quad \alpha < 1 \tag{3}$$

A new problem arises

$$Hf(x) = D^{\alpha}f(x) = ? \tag{4}$$

This is the original idea of the fractional calculus. Leibniz's calculus is quite different from that of Newton because Leibniz did not take the limit in his infinitesimal calculus. The derivative of $f(x)$ with respect to$x$, in the sense of Leibniz's notation, is the standard part of the infinitesimal ratio:

$$f'(x) = st(\frac{\Delta f}{\Delta x}) = st(\frac{f(x_1) - f(x_2)}{x_1 - x_2}) \tag{5}$$

Leibniz's definition is very close to the definition of the fractal derivative (He, 2011a). In a fractal medium, the distance between x_1 and x_2 tends to infinity ($\Delta x \to \infty$) even when $x_1 \to x_2$, and f can be continuous and nondifferentiable.

Although fractional calculus was invented over three centuries ago, it only became a hot topic recently due to the development of the fractal theory, computer science, effective analytical methods (e.g. the variational iteration method, the homotopy perturbation method) and its exact description of many real-life problems (He, 2006; He, 2012).

2 Fractal Derivative

The fractal derivative can be categorized as a special local fractional derivative (Yang, 2012). There are several different definitions.

Definition 1 Chen's definition is as follows (Chen etc., 2010; Chen, 2006)

$$\frac{du(x)}{dx^{\alpha}} = \lim_{s \to x} \frac{u(x) - u(s)}{x^{\alpha} - s^{\alpha}}, \tag{6}$$

where α is the order of the fractal derivative. This definition is much simpler but lacks physical meaning.

Definition 2 Consider a fractal media illustrated in Fig. 1, and assume the smallest measure is L_0 (any discontinuity less than L_0 is ignored). The distance between two points of A and B in Fig. 1 can be expressed using fractal geometry.

This fractal derivative has the form (He, 2011a)

$$\frac{Du}{Dx^\alpha} = \lim_{\Delta x \to L_0} \frac{u(A) - u(B)}{ds} = \lim_{\Delta x \to L_0} \frac{u(A) - u(B)}{kL_0^\alpha} \tag{7}$$

where α is the fractal dimension, the distance (ds) between two points in a discontinuous space can be expressed as

$$ds = kL_0^\alpha \tag{8}$$

where$k$ is a function of fractal dimensions, $k = k(\alpha)$, and it follows that $k(1) = 1$.

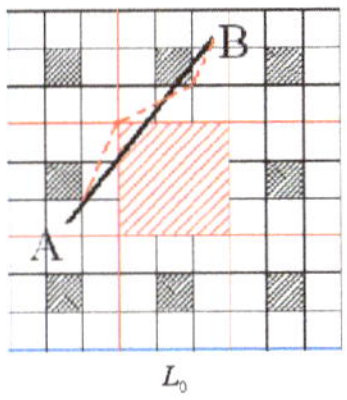

Fig. 1. The distance between two points in a discontinuous space or spacetime (He et al., 2012; He, 2009).

Definition 3 The fractal derivative can alternatively be defined as

$$\frac{Du}{Dx^\alpha} = \Gamma(1 + \alpha) \lim_{\Delta x = x_A - x_B \to L_0} \frac{u(A) - u(B)}{(x_A - x_B)^\alpha} \tag{9}$$

when $\alpha \to 1$ and$L_0 \to 0$, Eq. 8 turns out to be the ordinary differentiation.

Applications of the fractal derivative to fractal media have generated much attention, for example, it can model heat transfer and water permeation in multi-scale fabric and wool fibers (Fan et al., 2012a; Fan et al., 2012b; Fan et al., 2013a; Fan et al., 2013b; Chen et al., 2013; He, 2013a).

Consider a heat conduction in a fractal medium, we have the following equation using the fractal derivative

$$\frac{\partial T}{\partial t} + \frac{D}{D^\alpha x}(k\frac{DT}{D^\alpha x}) = 0 \tag{10}$$

To solve this equation, according to Eq. 8, we introduce a transformation

$$s = x^\alpha \tag{11}$$

This transformation is similar to the fractional complex transform (He, 2006; Li et al., 2010; Wang et al., 2012; Li et al., 2012). As a result, Eq. 10 is converted to a partial differential equation, which is

$$\frac{\partial T}{\partial t} + \frac{\partial}{\partial s}(k\frac{\partial T}{\partial s}) = 0 \tag{12}$$

This equation can be easily solved using a calculus. It should be pointed out that T is locally continuous when the scale is smaller than $\Delta x = L_0$, where L_0 is the smallest measure.

3 On Definitions of Fractional Derivatives

There are many definitions of fractional derivatives. A systematic study of various fractional derivatives is given by Yang in his monograph (Yang, 2012).

The Caputo fractional derivative is defined as (Yang, 2012)

$$D_x^{\alpha}(f(x)) = \frac{1}{\Gamma(n-\alpha)} \int_0^x (x-t)^{n-\alpha-1} \frac{d^n f(t)}{dt^n})dt \tag{13}$$

Riemann-Liouville fractional derivative is defined as: (Yang, 2012)

$$D_x^{\alpha}(f(x)) = \frac{1}{\Gamma(n-\alpha)} \frac{d^n}{dx^n} \int_0^x (x-t)^{n-\alpha-1} f(t)dt \tag{14}$$

Caputo derivatives are defined only for differentiable functions, while f can be a continuous (but not necessarily differentiable) function. The Riemann-Liouville definition can be used for any functions that are continuous but not differentiable anywhere, however, one requires that $D_x^{\alpha}(f(x)) \neq 0$ when f(x) is a constant.

To overcome the shortcomings, Jumarie (Jumarie, 2007a; Jumarie, 2007b; Jumarie, 2006) suggested the following modification of Riemann-Liouville fractional derivative

$$D_x^{\alpha}(f(x)) = \frac{1}{\Gamma(n-\alpha)} \frac{d^n}{dx^n} \int_0^x (x-t)^{n-\alpha-1} [f(t) - f(0)]dt \tag{15}$$

There are alternative definitions, see Ref. (Yang, 2012) .

3.1 Variational Iteration Method

The variational iteration method was first used to solve fractional differential equations in 1998 (He, 1998). Hereby we will introduce the basic properties of fractional derivatives by the variational iteration method (He, 1999; He,2007; He et al., 2007; He, 2011b).

The variational iteration method (He, 1998; He, 1999; He,2007; He et al., 2007; He, 2011b) has been shown to solve a large class of nonlinear differential problems effectively, easily, and accurately with the approximations converging rapidly to accurate solutions, and has now matured into a robust theory for various nonlinear problems, especially for fractional calculus (Wu, 2012a; Ghaneai et al., 2012; Wu, 2012b; He, 2013b; Yang et al., 2013). A complete review on its development and its application is available in Refs. (He, 2006; He, 2012).

We consider the following linear equation of n-th order

$$u^{(n)} = f(t) \tag{16}$$

By the variational iteration method (He, 1999; He,2007; He et al., 2007; He, 2011b), we have the following variational iteration algorithms:

Variational Iteration Algorithm-I:

$$u_{m+1}(t) = u_m(t) + (-1)^n \int_{t_0}^{t} \frac{1}{(n-1)!}(s-t)^{n-1}\left[u_m^{(n)}(s) - f_m(s)\right] ds. \tag{17}$$

Variational Iteration Algorithm-II (He, 2012; He et al., 2012):

$$u_{m+1}(t) = u_0(t) - (-1)^n \int_{t_0}^{t} \frac{1}{(n-1)!}(s-t)^{n-1} f_m(s)ds = u_0(t) - \frac{(-1)^n}{\Gamma(n)} \int_{t_0}^{t} (s-t)^{n-1} f_m(s)ds \tag{18}$$

Note: u_0 must satisfy the initial/boundary conditions.

For a linear equation, we have the following exact solution

$$u(t) = u_0(t) + (-1)^n \int_{t_0}^{t} \frac{1}{(n-1)!}(s-t)^{n-1}\left[u_0^{(n)}(s) - f(s)\right] ds. \tag{19}$$

or

$$u(t) = u_0(t) - \frac{(-1)^n}{\Gamma(n)} \int_{t_0}^{t} (s-t)^{n-1} f(s)ds \tag{20}$$

where $u_0(t)$ satisfies the boundary/initial conditions.

3.2 Definitions on Fractional Derivatives

Corresponding with the Variational Iteration Algorithm-I, we introduce an integration operator I^n defined by

$$I^n f = \int_{t_0}^{t} \frac{1}{(n-1)!}(s-t)^{n-1}\left[u_0^{(n)}(s) - f(s)\right] ds = \frac{1}{\Gamma(n)} \int_{t_0}^{t} (s-t)^{n-1}\left[f_0(s) - f(s)\right] ds \tag{21}$$

where $f_0(t) = u_0^{(n)}(t)$.

Definition 4

We can define a fractional derivative in the form

$$D_t^\alpha f = D_t^\alpha \frac{d^n}{dt^n}(I^n f) = \frac{d^n}{dt^n}(I^{n-\alpha} f) = \frac{1}{\Gamma(n-\alpha)} \frac{d^n}{dt^n} \int_{t_0}^{t} (s-t)^{n-\alpha-1}\left[f_0(s) - f(s)\right] ds \tag{22}$$

For a continuous and differentiable f, we have that:

$$f_0(t) = f(t_0) + (t-t_0)f'(t_0) + \frac{1}{2}(t-t_0)^2 f''(t_0) + \cdots + \frac{1}{(n-1)!}(t-t_0)^{n-1} f^{(n-1)}(t_0) \tag{23}$$

If $f_0(t)$ is continuous but not differentiable anywhere, we have

$$f_0(t) = f(t_0) + \frac{(t-t_0)}{\Gamma(1+\alpha)} f^{(\alpha)}(t_0) + \frac{(t-t_0)^2}{\Gamma(1+2\alpha)} f^{(2\alpha)}(t_0) + \cdots + \frac{(t-t_0)^{n-1}}{\Gamma(1+(n-1)\alpha)} f^{((n-1)\alpha)}(t_0) \tag{24}$$

where $f^{(n\alpha)}(t) = \underbrace{D_t^\alpha D_t^\alpha \cdots D_t^\alpha}_{n \quad times} f(t)$.

Keeping only the first term of $f_0(s)$, we offer another definition of fractional derivative in the form

$$D_t^\alpha f = \frac{1}{\Gamma(n-\alpha)} \frac{d^n}{dt^n} \int_{t_0}^{t} (s-t)^{n-\alpha-1} \left[f(t_0) - f(s)\right] ds \tag{25}$$

In the above f can be continuous and possibly not differentiable anywhere. Eq. 25 is equivalent to Jumarie's derivative when $t_0 = 0$.

According to Variational Iteration Algorithm-II, we can define another fractional derivative in the form

$$D_t^\alpha f(t) = \frac{1}{\Gamma(n-\alpha)} \frac{d^n}{dt^n} \int_{t_0}^{t} (s-t)^{n-\alpha-1} f(s) ds \tag{26}$$

Eq. 26 is equivalent to Riemann-Liouville fractional derivative when $t_0 = 0$.

4 Mongolian Yurt, Biomimic Design of Cocoon and its Evolution

When Genghis Khan (1162? – 1227) founded his Mongol Empire, which became the largest contiguous empire in history after his demise, the Mongolian yurt (see Fig. 2 adopted few modifications in decoration, the traditional yurt is still prevalent even today especially in Mongolia, particularly in the capital city of Ulan Bator where more than half of the population lives in yurts (Guo, 2010). The moving house is very easy to collapse and assemble again, and it is favorable for a nomadic lifestyle.

Yurts have been a distinctive feature of life in Central Asia for at least three thousand years. The first written description of a yurt used as a dwelling was recorded by Herodotus (484BC~424BC) (Grene, 1987), a historical development of various yurts is systematically illustrated in Ref. (Guo, 2010).

Yurts are so efficient to heat and cool, and they offer the sturdiness of a conventional home with fewer structural materials, which means they are more sustainable and can cost a lot less. Contemporary yurts in some scenic spots near Hohhot, however, have only geometrical similarity and many functional differences. The main difference lies in the felt making. A detailed description of the felt making is given in Guo's monograph (Guo, 2010).

The felt cover, which consists of layers of fabric and sheep's wool, mimics the design of a cocoon. If the yurt were not weatherproofed, the people in it might have frozen to death in winter. The felt cover evolved over time.

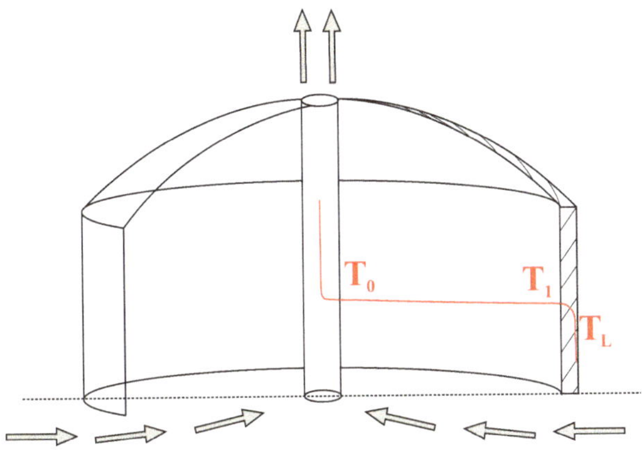

Fig. 2. Mongolian yurt, and temperature drop, the its slope of the temperature at the inner wall tends to zero.

4.1 Pupa-cocoon System

The Mongolian yurt is a biomimic design of the cocoon of the silkworm or tusser, which, while creating a tough barrier offering mechanical protection to the pupa, imposes no barrier to the diffusion of oxygen or water vapor (Chen et al., 2012; Blossman-Myer et al., 2010). The cocoon is also weatherproof. A theoretical analysis was given to explain the fascinating phenomenon by a fractal hydrodynamic model for a discontinuous membrane composed of a hierarchical silk cascade (Chen et al., 2013). It explains the intriguing phenomenon of the pupa's survival at extremes of weather from negative 40 degrees to 50 degrees.

The evolution of the yurt is similar to that of a cocoon, but its evolution accelerated is remarkably quickly. Instead of natural selection, the yurt follows the human selection (Song, 1637), which had long been used by the ancient Chinese for domestication and sericiculture before Charles Darwin introduced the fundamental notion of natural selection and expounded his theory of evolution in his famous treaty On the Origin of Species in 1868. In Ying-Xing Song's encyclopaedia Tiangong Kaiwu (Exploitation of the Works of Nature) published in 1637 (Song, 1637), he highlighted in particular that the Chinese had practiced human selection for over 5 millennia and that it had become an integral part of the Chinese culture. Indeed, the Tiangong Kaiwu gives a clear picture of the state of Chinese technologies during the Ming dynasty and contains (in the second chapter) detailed accounts of the various tools and methods of weaving that existed at that time. These historical accounts bear testimony to the advanced state of China's sericiculture and textile engineering several millennia ago.

4.2 Fractal Hierarchy and Local Fractional Model

Both cocoon(s) and the yurt have special hierarchical structures (Chen et al., 2012; Blossman-Myer et al., 2010) which enable the cocoon/yurt to have excellent protective properties. A cocoon has a multi-layer structure, the fiber diameter reduces greatly from 26 micrometers in the outer layer to 16 micrometers in the inner layer (pelade) (Chen et al., 2012; Blossman-Myer et al., 2010). Similar hierarchical structures are observed in yurts.

To illustrate the heat transfer property in a fractal hierarchy, we establish a one-dimensional model using the local fractional calculus (Yang et al., 2013; Yang, 2012; Yang, 2011).

Fourier's Law of thermal conduction in the yurt and a fractal hierarchy of its wall can be expressed, respectively, as (Yang, 2012; Yang, 2011)

$$\frac{d}{dx}(k_0\frac{dT}{dx}) = 0 \quad 0 < x < x_{L_1} \tag{27}$$

$$\frac{d^\alpha}{dx^\alpha}(k_1\frac{d^\alpha T}{dx^\alpha}) = 0(x_{L_1} \leq x \leq L) \tag{28}$$

with following boundary conditions

$$T(0) = T_0, \tag{29}$$

$$T(L) = T_L \tag{30}$$

$$(k_0\frac{dT}{dx})\bigg|_{x=L_1} = (k_1\frac{d^\alpha T}{dx^\alpha})\bigg|_{x=L_1} \tag{31}$$

where $x = x_{L_1}$ is the inner wall, $x = L$ is the outer wall of the cocoon/yurt, k_0 is thermal conductivity of heat flux in air, k_1 is thermal conductivity of heat flux in the fractal hierarchy, d^α/dx^α is the local fractional derivative defined as (Yang et al., 2013; Yang, 2012; Yang, 2011).

$$\frac{d^\alpha T}{dx^\alpha}|_{x=x_0} = \lim_{x\to x_0}\frac{\Delta^\alpha(T(x)-T(x_0))}{(x-x_0)^\alpha}$$

with $\Delta^\alpha(T(x)-T(x_0)) \cong \Gamma(1+\alpha)\Delta(T(x)-T(x_0))$, α is the fractal dimensions of fractal hierarchy. The local fractional derivative is very much similar to the fractal derivative.

The general solution to Eq. 27 is

$$T = T_0 + ax \tag{32}$$

where$a$ is a constant.

The boundary condition, Eq. 31, becomes

$$\left.\left(k_1 \frac{d^\alpha T}{dx^\alpha}\right)\right|_{x=L_1} = k_0 a \tag{33}$$

To solve the local fractional differential equation, Eq. 28, with boundary conditions, Eqs. 30 and 33, we adopt the following fractional complex transform(Li et al., 2010; Wang et al., 2012; Li et al., 2012).

$$s = \frac{x^\alpha}{\Gamma(1+\alpha)} \tag{34}$$

Eqs. 28 and 31 become, respectively, as

$$\frac{d}{ds}\left(k_1 \frac{dT}{ds}\right) = 0 \tag{35}$$

and

$$\left.\left(k_1 \frac{dT}{ds}\right)\right|_{s=L_1^\alpha/\Gamma(1+\alpha)} = k_0 a \tag{36}$$

The general solution of Eq. 35 is:

$$T = b + cs \tag{37}$$

or equivalently

$$T = b + \frac{c}{\Gamma(1+\alpha)} x^\alpha \tag{38}$$

where b and c are constants.

Incorporating the boundary conditions with Eqs. 30, 33 and 36, we have that:

$$T_L = b + \frac{c}{\Gamma(1+\alpha)} L^\alpha \tag{39}$$

$$k_1 c = k_0 a \tag{40}$$

$$T_0 + \frac{a}{\Gamma(1+\alpha)} L_1^\alpha = b + \frac{c}{\Gamma(1+\alpha)} L_1^\alpha. \tag{41}$$

Solving Eqs. 39-41 simultaneously, we can identify a, b, and c as follows

$$a = \frac{T_L - T_0}{\frac{k_0(L^\alpha - L_1^\alpha)}{k_1 \Gamma(1+\alpha)} + \frac{L_1^\alpha}{\Gamma(1+\alpha)}}, \tag{42}$$

$$b = T_L - \frac{(T_L - T_0)L^\alpha}{L^\alpha - L_1^\alpha + \frac{k_1}{k_0} L_1^\alpha}, \tag{43}$$

$$c = \frac{T_L - T_0}{\frac{L^\alpha - L_1^\alpha}{\Gamma(1+\alpha)} + \frac{k_1}{k_0}\frac{L_1^\alpha}{\Gamma(1+\alpha)}}. \tag{44}$$

We can, therefore, determine the temperature and its slope at the inner wall:

$$T_1 = T(L_1) = T_0 + \frac{aL_1^\alpha}{\Gamma(1+\alpha)} = T_0 + \frac{(T_L - T_0)\frac{L_1^\alpha}{\Gamma(1+\alpha)}}{\frac{k_0(L^\alpha - L_1^\alpha)}{k_1\Gamma(1+\alpha)} + \frac{L_1^\alpha}{\Gamma(1+\alpha)}} = T_0 + \frac{T_L - T_0}{\frac{k_0 L^\alpha(1-(L_1/L)^\alpha)}{k_1 L_1^\alpha} + 1} \tag{45}$$

and

$$T_1^{(\alpha)} = \frac{d^\alpha T}{dx^\alpha}(x = L_1) = \frac{T_L - T_0}{\left(L^\alpha - L_1^\alpha\right) + \frac{k_1 L_1^\alpha}{k_0}} = \frac{T_L - T_0}{L^\alpha + \left(\frac{k_1}{k_0} - 1\right) L_1^\alpha}. \tag{46}$$

It is obvious that T_1 should be closed to the body temperature and its slope ($T_1^{(\alpha)}$) should tend to zero (see Fig. 1), which requires that:

$$T_1^{(\alpha)} = \frac{T_L - T_0}{L^\alpha + \left(\frac{k_1}{k_0} - 1\right) L_1^\alpha} = 0 \tag{47}$$

$$\frac{d^{2\alpha} T_1}{dL_1^{2\alpha}} = 0 \tag{48}$$

T_{L_1} is closed to body temperature

From Eqs. 47-48, L_1, k_1 and α can be optimally determined.

As illustrated in Fig. 1, the slope of the temperature change is almost zero near the inner wall of the yurt, while the temperature changes quickly near the outer wall. This can only be achieved by fractional calculus as studied above. Fractional calculus effectively describes phenomena arising in discontinuous media. In most practical applications, $T_1^{(\alpha)} \neq 0$ or $d^{2\alpha}T_1/dL_1^{2\alpha} \neq 0$, L_1, k_1 and α should be such chosen that $d^{2\alpha}T_1/dL_1^{2\alpha} \neq 0$ or $T_1^{(\alpha)} \neq 0$ is small enough and tends to zero, the smaller, the better.

5 Conclusions

This chapter introduced a new fractal derivative which is especially useful for hierarchical cascades, the last hierarchical size (L_0) will greatly affect the fractal derivative and mechanical properties. The fractal differential equations can be easily converted into ordinary differential equations by the transformation of Eq. 11.

Using the variational iteration method, we can easily derive the Riemann-Liouville fractional derivative and Jumarie fractional derivative, and obtain a more generalized fractional derivative. The fractional differential equations with fractional derivatives are extremely easy to be solved using the variational iteration method.

The paper highlights that the human-yurt system in the central Asia is a biomimic design of pupa-cocoon structure. A local fractional model is established to illustrate its mechanism and to optimally determine the thickness of the felt cover. A better understanding of the yurt mechanism could help the further design of yurt-like space suits and other protective clothes for special applications.

Acknowledgment: The work is supported by Priority Academic Program Development of Jiangsu Higher Education Institutions (PAPD), National Natural Science Foundation of China under grant No. 61303236 and No.11372205 and Project for Six Kinds of Top Talents in Jiangsu Province under grant No. ZBZZ-035, Science & Technology Pillar Program of Jiangsu Province under grant No. BE2013072, Jiangsu Province Key Laboratory No.KJS1314 and Jiangsu Planned Projects for Postdoctoral Research Funds1401076B and China Postdoctoral Science Foundation 2015M571806.

Bibliography

Blossman-Myer, B., & Burggren, W. W. (2010). The silk cocoon of the silkworm, Bombyx mori: Macro structure and its influence on transmural diffusion of oxygen and water vapor. Comparative Biochemistry and Physiology Part A: Molecular & Integrative Physiology, 155(2), 259-263.

Chen, W. (2006). Time–space fabric underlying anomalous diffusion. Chaos, Solitons & Fractals, 28(4), 923-929.

Chen, W., Zhang, X. D., & Korošak, D. (2010). Investigation on fractional and fractal derivative relaxation-oscillation models. International Journal of Nonlinear Sciences and Numerical Simulation, 11(1), 3-10.

Chen, F., Porter, D., & Vollrath, F. (2012). Silk cocoon (Bombyx mori): Multi-layer structure and mechanical properties. Acta biomaterialia, 8(7), 2620-2627.

Chen, R. X., Liu, F. J., He, J. H., & Fan, J. (2013). Silk Cocoon:" Emperor's New Clothes" for Pupa: Fractal Nano-hydrodynamical Approach. Journal of Nano Research, 22, 65-70.

Fan, J., & He, J. (2012a, November). Fractal derivative model for air permeability in hierarchic porous media. In Abstract and Applied Analysis (Vol. 2012). Hindawi Publishing Corporation.

Fan, J., & He, J. H. (2012b). Biomimic design of multi-scale fabric with efficient heat transfer property. Thermal science, 16(5), 1349-1352.

Fan, J., & Shang, X. (2013a). Water permeation in the branching channel net of wool fiber. Heat Transfer Research, 44(5).

Fan, J., & Shang, X. (2013b). Fractal Heat transfer in wool fiber hierarchy. Heat Transfer Research, 44(5).

Ghaneai, H., Hosseini, M. M., & Mohyud-Din, S. T. (2012). Modified variational iteration method for solving a neutral functional-differential equation with proportional delays. International Journal of Numerical Methods for Heat & Fluid Flow, 22(8), 1086-1095.

Guo, Y. Q., (2010). Xishou Menggubao (On Mongolian yurt), The Orient Publisher, Beijing, China, 2010 (in Chinese)

Grene, D. (1987). Herodotus: the history. Chicago: The Univ. of Chicago Press Grene Herodotus: The History 1987.

He, J. H. (1998). Approximate analytical solution for seepage flow with fractional derivatives in porous media. Computer Methods in Applied Mechanics and Engineering, 167(1), 57-68.

He, J. H. (1999). Variational iteration method–a kind of non-linear analytical technique: some examples. International journal of non-linear mechanics, 34(4), 699-708.

He, J. H. (2006). Some asymptotic methods for strongly nonlinear equations. International Journal of Modern Physics B, 20(10), 1141-1199.

He, J. H. (2007). Variational iteration method—some recent results and new interpretations. Journal of computational and applied mathematics, 207(1), 3-17.

He, J. H., & Wu, X. H. (2007). Variational iteration method: new development and applications. Computers & Mathematics with Applications, 54(7), 881-894.

He, J. H. (2009). A generalized Poincaré-invariant action with possible application in strings and E-infinity theory. Chaos, Solitons & Fractals, 39(4), 1667-1670.

He, J. H. (2011a). A new fractal derivation. Thermal science, 15 (suppl. 1), 145-147.

He, J. H. (2011b). A short remark on fractional variational iteration method. Physics Letters A, 375(34), 3362-3364.

He, J. H. (2012, November). Asymptotic methods for solitary solutions and compactons. In Abstract and Applied Analysis (Vol. 2012). Hindawi Publishing Corporation.

He, J. H., Elagan, S. K., & Li, Z. B. (2012). Geometrical explanation of the fractional complex transform and derivative chain rule for fractional calculus. Physics Letters A, 376(4), 257-259.

He, J. H. (2013a). Local fractional variational iteration method for fractal heat transfer in silk cocoon hierarchy. Nonlinear Science Letters A, 4(1), 15-20.

He, J. H. (2013b). Exp-function method for fractional differential equations. International Journal of Nonlinear Sciences and Numerical Simulation, 14(6), 363-366.

Li, Z. B., & He, J. H. (2010). Fractional complex transform for fractional differential equations. Mathematical and Computational Applications, 15(5), 970-973.

Li, Z. B., Zhu, W. H., & He, J. H. (2012). Exact solutions of time-fractional heat conduction equation by the fractional complex transform. Thermal Science, 16(2), 335-338.

Jumarie, G.(2007a). Fractional partial differential equations and modified Riemann-Liouville derivative new methods for solution. Journal of Applied Mathematics and Computing, 24(1-2), 31-48.

Jumarie, G.(2007b). The Minkowski's space–time is consistent with differential geometry of fractional order. Physics Letters A, 363(1), 5-11.

Jumarie, G. (2006). Modified Riemann-Liouville derivative and fractional Taylor series of nondifferentiable functions further results. Computers & Mathematics with Applications, 51(9), 1367-1376.

Wang, Q. L., He, J. H., & Li, Z. B. (2012). Fractional model for heat conduction in polar bear hairs. Thermal Science, 16(2), 339-342.

Wu, G. C. (2012a). Laplace transform overcoming principle drawbacks in application of the variational iteration method to fractional heat equations. Thermal Science, 16(4), 1257-1261.

Wu, G. C. (2012b). Variational iteration method for q-difference equations of second order. Journal of Applied Mathematics, 2012.

Song, Y. X., Tiangong Kaiwu (Exploitation of the Works of Nature, An Encyclopedia of Technology with illustrations), in ancient Chinese with modern Chinese translation, 1637, 2008 edition, Volumes Publishing Company, Shengyang.

Yang, X. J., & Baleanu, D. (2013). Fractal heat conduction problem solved by local fractional variation iteration method. Thermal Science, 17(2), 625-628.

Yang, X. J., Srivastava, H. M., He, J. H., & Baleanu, D. (2013). Cantor-type cylindrical-coordinate method for differential equations with local fractional derivatives. Physics Letters A, 377(28), 1696-1700.

Yang, X. J. (2012). Advanced local fractional calculus and its applications. World Science, New York, NY, USA.

Yang, X. J. (2011). Local Fractional Functional Analysis and Its Applications, Asian Academic publisher Limited, Hong Kong, 2011.

Mohamad Rafi Segi Rahmat, Dumitru Baleanu*, and Xiao-Jun Yang

Cantor-type spherical-coordinate Method for Differential Equations within Local Fractional Derivatives

Abstract: In this article, we utilize the Cantor-type spherical coordinate method to investigate a family of local fractional differential operators on Cantor sets. Some examples are discussed to show the capability of this method for the damped wave , Helmholtz and heat conduction equations defined on Cantor sets. We show that it is a powerful tool to convert differential equations on Cantor sets from Cantorian-coordinate systems to Cantor-type spherical-coordinate systems.

Keywords: Cantor sets; Local fractional derivatives; Local fractional dynamic equation; Cantor-type Spherical-coordinate

1 Introduction

In Euclidean space we observe several interesting physical phenomena by using differential equations in the styles of planar, cylindrical, and spherical geometries. There are many anysotropic models in cylindrical and spherical coordinates (Teixeira and Chew, 1997; Scheuer et al., 1990; Engheta, 1990; Engheta, 1999; Schetselaar, 1998; Petropoulos, 2000).

Fractional calculus has a long history and there exist many applications (Kilbas et al., 2006; Podlubny, 1999; Oldham and Spanier, 1970; Srivastava and Owa, 1989). However, fractional derivatives do not reflect the local geometric behaviors for specific functions. Therefore, attempts have been made recently to define a local version of the fractional derivative and there have been successively applied to model the non-differentiability of phenomena in fractal physical media (Adda and Cresson, 2001; Kolwankar and Gangal, 1996; Kolwankar and Gangal, 1997; Kolwankar and Gangal, 1998; Babakhani and Daftardar-Gejji, 2002; Chen et al., 2010; Yang, 2011a, b; Yang, 2012; Yang and Baleanu, 2013; Yang et al., 2013; Yan et al., 2014; Zhao et al., 2013; Yang et al., 2014; Srivastava et al., 2014).

***Corresponding Author: Dumitru Baleanu:** Department of Mathematics, Çankaya University, Öğretmenler Cad. 14 06530, Balgat, Ankara,Turkey
and Institute of Space Sciences, Magurele, Bucharest, Romania, E-mail: dumitru@cankaya.edu.tr
Mohamad Rafi Segi Rahmat: School of Applied Mathematics, The University of Nottingham Malaysia Campus, Jalan Broga, 43500 Semenyih, Selangor D.E, Malaysia
Xiao-Jun Yang: Department of Mathematics and Mechanics, China University of Mining and Technology, Xuzhou, Jiangsu, 221008, China

We recall that the Cantorian-coordinate system, firstly described in (Yang, 2011a, b; Yang, 2012), was set up on fractals. Based on this system, 3-dimensional fractal dynamical equations such as the heat conduction equation without heat generation in fractal media (Yang, 2012), the damped wave equation in fractal strings (Yang, 2012), Maxwell's equations on Cantor sets (Zhao et al., 2013), Helmholtz and diffusion equations on Cantor sets (Zhao et al., 2013) are presented. In the Refs. (Yang et al., 2013; Hao et al., 2013), the authors proposed to use Cantor-type cylindrical-coordinate method to investigate a family of local fractional differential equation. It is a powerful tool to convert the differential equations on Cantor sets from a Cantorian-coordinate system to Cantor-type cylindrical-coordinate system. We recall that many physical situations have spherical symmetry, e.g. the gravitational field of a single body and the electric field of a point charge exhibit spherical symmetry.

The aim of this chapter is to structure and investigate the Cantor-type spherical-coordinate method within framework of the local fractional vector operator. The layout of this chapter is given below. In Section 2, we propose and describe the Cantor-type spherical-coordinate method. In Section 3, we investigate equivalent forms of certain differential equations on a Cantor set with the proposed method.

2 Mathematical Tools

In this section, we briefly present some of the concepts of the local derivative of a scalar function and a vector function on Cantor set.

Let $f(x)$ be a local fractional continuous function on the interval (a, b) or

$$f(x) \in C_\alpha(a, b) \tag{1}$$

if for $x \in (a, b)$, $|x - x_0| < \delta$, $\epsilon, \delta > 0$, the following condition

$$|f(x) - f(x_0)| < \epsilon^\alpha \tag{2}$$

is valid (see (Yang, 2011 a,b)).

The local fractional derivative of $f(x) \in C_\alpha(a, b)$ of order α $(0 < \alpha \leq 1)$ in the interval $[a, b]$ is defined as (see (Yang, 2011a,b; Yang, 2012))

$$f^{(\alpha)}(x_0) = \frac{d^\alpha f(x_0)}{dx^\alpha} = \frac{\Delta^\alpha(f(x) - f(x_0))}{(x - x_0)^\alpha}, \tag{3}$$

where

$$\Delta^\alpha(f(x) - f(x_0)) \cong \Gamma(1 + \alpha)\Delta(f(x) - f(x_0))$$

with gamma function $\Gamma(1 + \alpha)$, and the condition defined by Eq. (1) holds.

The local fractional partial differential operator of order α $(0 < \alpha \leq 1)$ is given by (see (Yang, 2011a, b))

$$\frac{\partial^\alpha}{\partial t^\alpha} u(x_0, t) = \frac{\Delta^\alpha(u(x_0, t) - u(x_0, t_0))}{(t - t_0)^\alpha} \tag{4}$$

where

$$\Delta^{\alpha}(u(x_0,t)-u(x_0,t_0)) \cong \Gamma(1+\alpha)\Delta(u(x_0,t)-u(x_0,t_0)).$$

In a similar manner, for a given vector function

$$\mathbf{F}(t)=F_1(t)\mathbf{e}_1^{\alpha}+F_2(t)\mathbf{e}_2^{\alpha}+F_3(t)\mathbf{e}_3^{\alpha},$$

the local derivative is defined by (see (Yang,2012))

$$\mathbf{F}^{(\alpha)}(t_0)=\frac{d^{\alpha}\mathbf{F}(t_0)}{dt^{\alpha}}=\frac{\Delta^{\alpha}(\mathbf{F}(t)-\mathbf{F}(t_0))}{(t-t_0)^{\alpha}}, \tag{5}$$

where $\mathbf{e}_1^{\alpha}$, $\mathbf{e}_2^{\alpha}$, $\mathbf{e}_3^{\alpha}$ are the directions of the local fractional vector function.

The following product rule is valid (see (Yang, 2012))

$$D_x^{(\alpha)}[f(x)g(x)]=(D_x^{(\alpha)}f(x))g(x)+f(x)(D_x^{(\alpha)}g(x)). \tag{6}$$

3 Cantor-type Spherical-coordinate Method

We recall that the Cantor-type spherical-coordinates are defined by (Yang,2011b; Yang,2012):

$$\begin{cases} x^{\alpha}=\rho^{\alpha}\sin_{\alpha}\phi^{\alpha}\cos_{\alpha}\theta^{\alpha}, \\ y^{\alpha}=\rho^{\alpha}\sin_{\alpha}\phi^{\alpha}\sin_{\alpha}\theta^{\alpha}, \\ z^{\alpha}=\rho^{\alpha}\cos_{\alpha}\phi^{\alpha}, \end{cases} \tag{7}$$

with $\rho>0$, $0<\phi<\pi$, $0\le\theta\le 2\pi$ and $x^{2\alpha}+y^{2\alpha}+z^{2\alpha}=\rho^{2\alpha}$.

For these Cantor type spherical coordinates, we have the local fractional vector given by

$$\mathbf{r}=\rho^{\alpha}\sin_{\alpha}\phi^{\alpha}\cos_{\alpha}\theta^{\alpha}\mathbf{e}_1^{\alpha}+\rho^{\alpha}\sin_{\alpha}\phi^{\alpha}\sin_{\alpha}\theta^{\alpha}\mathbf{e}_2^{\alpha}+\rho^{\alpha}\cos_{\alpha}\phi^{\alpha}\mathbf{e}_3^{\alpha}, \tag{8}$$

so that

$$\begin{cases} \mathbf{C}_{\rho}^{\alpha}=\frac{1}{\Gamma(1+\alpha)}\frac{\partial^{\alpha}\mathbf{r}}{\partial\rho^{\alpha}} \\ =\sin_{\alpha}\phi^{\alpha}\cos_{\alpha}\theta^{\alpha}\mathbf{e}_1^{\alpha}+\sin_{\alpha}\phi^{\alpha}\sin_{\alpha}\theta^{\alpha}\mathbf{e}_2^{\alpha}+\cos_{\alpha}\phi^{\alpha}\mathbf{e}_3^{\alpha}, \\ \mathbf{C}_{\phi}^{\alpha}=\frac{1}{\Gamma(1+\alpha)}\frac{\partial^{\alpha}\mathbf{r}}{\partial\phi^{\alpha}} \\ =\frac{\rho^{\alpha}}{\Gamma(1+\alpha)}\cos_{\alpha}\phi^{\alpha}\cos_{\alpha}\theta^{\alpha}\mathbf{e}_1^{\alpha}+\frac{\rho^{\alpha}}{\Gamma(1+\alpha)}\cos_{\alpha}\phi^{\alpha}\sin_{\alpha}\theta^{\alpha}\mathbf{e}_2^{\alpha}-\frac{\rho^{\alpha}}{\Gamma(1+\alpha)}\sin_{\alpha}\phi^{\alpha}\mathbf{e}_3^{\alpha}, \\ \mathbf{C}_{\theta}^{\alpha}=\frac{1}{\Gamma(1+\alpha)}\frac{\partial^{\alpha}\mathbf{r}}{\partial\theta^{\alpha}} \\ =-\frac{\rho^{\alpha}}{\Gamma(1+\alpha)}\sin_{\alpha}\phi^{\alpha}\sin_{\alpha}\theta^{\alpha}\mathbf{e}_1^{\alpha}+\frac{\rho^{\alpha}}{\Gamma(1+\alpha)}\sin_{\alpha}\phi^{\alpha}\cos_{\alpha}\theta^{\alpha}\mathbf{e}_2^{\alpha}. \end{cases} \tag{9}$$

Therefore, we obtain

$$\begin{cases} \mathbf{e}_{\rho}^{\alpha}=\sin_{\alpha}\phi^{\alpha}\cos_{\alpha}\theta^{\alpha}\mathbf{e}_1^{\alpha}+\sin_{\alpha}\phi^{\alpha}\sin_{\alpha}\theta^{\alpha}\mathbf{e}_2^{\alpha}+\cos_{\alpha}\phi^{\alpha}\mathbf{e}_3^{\alpha}, \\ \mathbf{e}_{\phi}^{\alpha}=\cos_{\alpha}\phi^{\alpha}\cos_{\alpha}\theta^{\alpha}\mathbf{e}_1^{\alpha}+\cos_{\alpha}\phi^{\alpha}\sin_{\alpha}\theta^{\alpha}\mathbf{e}_2^{\alpha}-\sin_{\alpha}\phi^{\alpha}\mathbf{e}_3^{\alpha}, \\ \mathbf{e}_{\theta}^{\alpha}=-\sin_{\alpha}\theta^{\alpha}\mathbf{e}_1^{\alpha}+\cos_{\alpha}\theta^{\alpha}\mathbf{e}_2^{\alpha}, \end{cases} \tag{10}$$

where $\mathbf{C}_\rho^\alpha = \mathbf{e}_\rho^\alpha$, $\mathbf{C}_\phi^\alpha = \frac{\rho^\alpha}{\Gamma(1+\alpha)}\mathbf{e}_\phi^\alpha$, $\mathbf{C}_\theta^\alpha = \frac{\rho^\alpha}{\Gamma(1+\alpha)}\sin_\alpha \phi^\alpha \mathbf{e}_\theta^\alpha$.

Now, by making use of Eq. (10), we can write this result in matrix form as follows:

$$\begin{pmatrix} \mathbf{e}_\rho^\alpha \\ \mathbf{e}_\phi^\alpha \\ \mathbf{e}_\theta^\alpha \end{pmatrix} = \begin{pmatrix} \sin_\alpha \phi^\alpha \cos_\alpha \theta^\alpha & \sin_\alpha \phi^\alpha \sin_\alpha \theta^\alpha & \cos_\alpha \phi^\alpha \\ \cos_\alpha \phi^\alpha \cos_\alpha \theta^\alpha & \cos_\alpha \phi^\alpha \sin_\alpha \theta^\alpha & -\sin_\alpha \phi^\alpha \\ -\sin_\alpha \theta^\alpha & \cos_\alpha \theta^\alpha & 0 \end{pmatrix} \begin{pmatrix} \mathbf{e}_1^\alpha \\ \mathbf{e}_2^\alpha \\ \mathbf{e}_3^\alpha \end{pmatrix} \tag{11}$$

In view of Eq. (9) and (10), upon differentiating the Cantorian position vector with respect to the Cantor-type spherical coordinates it follows that

$$\begin{cases} \mathbf{e}_\rho^\alpha = \frac{1}{\Gamma(1+\alpha)}\frac{\partial^\alpha \mathbf{r}}{\partial \rho^\alpha} = \sin_\alpha \phi^\alpha \cos_\alpha \theta^\alpha \mathbf{e}_1^\alpha + \sin_\alpha \phi^\alpha \sin_\alpha \theta^\alpha \mathbf{e}_2^\alpha + \cos_\alpha \phi^\alpha \mathbf{e}_3^\alpha, \\ \mathbf{e}_\phi^\alpha = \frac{1}{\rho^\alpha}\frac{\partial^\alpha \mathbf{r}}{\partial \phi^\alpha} = \cos_\alpha \phi^\alpha \cos_\alpha \theta^\alpha \mathbf{e}_1^\alpha + \cos_\alpha \phi^\alpha \sin_\alpha \theta^\alpha \mathbf{e}_2^\alpha - \sin_\alpha \phi^\alpha \mathbf{e}_3^\alpha, \\ \mathbf{e}_\theta^\alpha = \frac{1}{\Gamma(1+\alpha)\sin_\alpha \phi^\alpha}\frac{\partial^\alpha \mathbf{r}}{\partial \theta^\alpha} = -\sin_\alpha \theta^\alpha \mathbf{e}_1^\alpha + \cos_\alpha \theta^\alpha \mathbf{e}_2^\alpha. \end{cases} \tag{12}$$

Eq. (12) is locally orthogonal and normalized everywhere (see (Yang, 2011a, b)). Hence, it is possible to define a local fractal basis with an orientation, which is derived from one fractal space to another fractal space. Based on this, a local fractional vector field can be defined as follows:

$$\mathbf{r}(\rho, \phi, \theta) = \mathbf{r} \cdot (\mathbf{e}_\rho^\alpha, \mathbf{e}_\phi^\alpha, \mathbf{e}_\theta^\alpha) \tag{13}$$

where the fractal vector coordinates are given by

$$r_\rho = \mathbf{r}(\rho, \phi, \theta) \cdot \mathbf{e}_\rho^\alpha, \quad r_\phi = \mathbf{r}(\rho, \phi, \theta) \cdot \mathbf{e}_\phi^\alpha, \quad r_\theta = \mathbf{r}(\rho, \phi, \theta) \cdot \mathbf{e}_\theta^\alpha \tag{14}$$

are the projections of $\mathbf{r}$ on the local fractal basis vectors.

The variations of orthogonal vectors with respect to the Cantor-type spherical coordinates are given as follows:

$$\begin{cases} \frac{\partial^\alpha \mathbf{e}_\rho^\alpha}{\partial \rho^\alpha} = 0; & \frac{\partial^\alpha \mathbf{e}_\rho^\alpha}{\partial \phi^\alpha} = \mathbf{e}_\phi^\alpha; & \frac{\partial^\alpha \mathbf{e}_\rho^\alpha}{\partial \theta^\alpha} = \sin_\alpha \phi^\alpha \mathbf{e}_\theta^\alpha ; \\ \frac{\partial^\alpha \mathbf{e}_\phi^\alpha}{\partial \rho^\alpha} = 0 ; & \frac{\partial^\alpha \mathbf{e}_\phi^\alpha}{\partial \phi^\alpha} = -\mathbf{e}_\rho^\alpha ; & \frac{\partial^\alpha \mathbf{e}_\phi^\alpha}{\partial \theta^\alpha} = \cos_\alpha \phi^\alpha \mathbf{e}_\theta^\alpha \\ \frac{\partial^\alpha \mathbf{e}_\theta^\alpha}{\partial \rho^\alpha} = 0 ; & \frac{\partial^\alpha \mathbf{e}_\theta^\alpha}{\partial \phi^\alpha} = 0 ; & \frac{\partial^\alpha \mathbf{e}_\theta^\alpha}{\partial \theta^\alpha} = -(\sin_\alpha \phi^\alpha \mathbf{e}_\rho^\alpha + \cos_\alpha \phi^\alpha \mathbf{e}_\phi^\alpha) \end{cases} \tag{15}$$

The local fractional derivatives with respect to the Cantor-type spherical coordinates are given by the local fractional differentiation through the Cantorian-coordinates as follows:

$$\begin{aligned} \frac{\partial^\alpha}{\partial \rho^\alpha} &= \left(\frac{\partial x}{\partial \rho}\right)^\alpha \frac{\partial^\alpha}{\partial x^\alpha} + \left(\frac{\partial y}{\partial \rho}\right)^\alpha \frac{\partial^\alpha}{\partial y^\alpha} + \left(\frac{\partial z}{\partial \rho}\right)^\alpha \frac{\partial^\alpha}{\partial z^\alpha} \\ &= \Gamma(1+\alpha)\left(\sin_\alpha \phi^\alpha \cos_\alpha \theta^\alpha \frac{\partial^\alpha}{\partial x^\alpha} + \sin_\alpha \phi^\alpha \sin_\alpha \theta^\alpha \frac{\partial^\alpha}{\partial y^\alpha} + \cos_\alpha \phi^\alpha \frac{\partial^\alpha}{\partial z^\alpha}\right) \\ &= \mathbf{e}_\rho^\alpha \cdot \Delta^\alpha = \Delta_\rho^\alpha, \end{aligned} \tag{16}$$

$$\begin{aligned}\frac{\partial^\alpha}{\partial\phi^\alpha} &= \left(\frac{\partial x}{\partial\phi}\right)^\alpha \frac{\partial^\alpha}{\partial x^\alpha} + \left(\frac{\partial y}{\partial\phi}\right)^\alpha \frac{\partial^\alpha}{\partial y^\alpha} + \left(\frac{\partial z}{\partial\phi}\right)^\alpha \frac{\partial^\alpha}{\partial z^\alpha}\\ &= \rho^\alpha\left(\cos_\alpha\phi^\alpha\cos_\alpha\theta^\alpha\frac{\partial^\alpha}{\partial x^\alpha} + \cos_\alpha\phi^\alpha\sin_\alpha\theta^\alpha\frac{\partial^\alpha}{\partial y^\alpha} - \sin_\alpha\phi^\alpha\frac{\partial^\alpha}{\partial z^\alpha}\right)\\ &= \rho^\alpha\mathbf{e}_\phi^\alpha\cdot\Delta^\alpha = \rho^\alpha\Delta_\phi^\alpha,\end{aligned} \tag{17}$$

$$\begin{aligned}\frac{\partial^\alpha}{\partial\theta^\alpha} &= \left(\frac{\partial x}{\partial\theta}\right)^\alpha \frac{\partial^\alpha}{\partial x^\alpha} + \left(\frac{\partial y}{\partial\theta}\right)^\alpha \frac{\partial^\alpha}{\partial y^\alpha} + \left(\frac{\partial z}{\partial\theta}\right)^\alpha \frac{\partial^\alpha}{\partial z^\alpha}\\ &= \Gamma(1+\alpha)\rho^\alpha\sin_\alpha\phi^\alpha\left(-\sin_\alpha\theta^\alpha\frac{\partial^\alpha}{\partial x^\alpha} + \cos_\alpha\theta^\alpha\frac{\partial^\alpha}{\partial y^\alpha}\right)\\ &= \rho^\alpha\sin_\alpha\phi^\alpha\mathbf{e}_\theta^\alpha\cdot\Delta^\alpha = \rho^\alpha\sin_\alpha\phi^\alpha\Delta_\theta^\alpha,\end{aligned} \tag{18}$$

where

$$\Delta_\rho^\alpha = \frac{\partial^\alpha}{\partial\rho^\alpha},\quad \Delta_\phi^\alpha = \frac{1}{\rho^\alpha}\frac{\partial^\alpha}{\partial\phi^\alpha},\quad \Delta_\theta^\alpha = \frac{1}{\rho^\alpha\sin_\alpha\phi^\alpha}\frac{\partial^\alpha}{\partial\theta^\alpha}. \tag{19}$$

In light of Eq. (19), the local fractional gradient operator is described as follows:

$$\begin{aligned}\Delta^\alpha &= \mathbf{e}_\rho^\alpha\Delta_\rho^\alpha + \mathbf{e}_\phi^\alpha\Delta_\phi^\alpha + \mathbf{e}_\theta^\alpha\Delta_\theta^\alpha\\ &= \mathbf{e}_\rho^\alpha\frac{\partial^\alpha}{\partial\rho^\alpha} + \mathbf{e}_\phi^\alpha\frac{1}{\rho^\alpha}\frac{\partial^\alpha}{\partial\phi^\alpha} + \mathbf{e}_\theta^\alpha\frac{1}{\rho^\alpha\sin_\alpha\phi^\alpha}\frac{\partial^\alpha}{\partial\theta^\alpha}\end{aligned} \tag{20}$$

A local fractional gradient operator in Cantor-type spherical coordinate system is expressed as follows:

$$\Delta^\alpha\Psi = \mathbf{e}_\rho^\alpha\frac{\partial^\alpha}{\partial\rho^\alpha}\Psi(\rho,\phi,\theta) + \mathbf{e}_\phi^\alpha\frac{1}{\rho^\alpha}\frac{\partial^\alpha}{\partial\phi^\alpha}\Psi(\rho,\phi,\theta) + \mathbf{e}_\theta^\alpha\frac{1}{\rho^\alpha\sin_\alpha\phi^\alpha}\frac{\partial^\alpha}{\partial\theta^\alpha}\Psi(\rho,\phi,\theta), \tag{21}$$

where

$$\Delta^\alpha\Psi = \mathbf{e}_\rho^\alpha\Delta_\rho^\alpha\Psi(\rho,\phi,\theta) + \mathbf{e}_\phi^\alpha\Delta_\phi^\alpha\Psi(\rho,\phi,\theta) + \mathbf{e}_\theta^\alpha\Delta_\theta^\alpha\Psi(\rho,\phi,\theta). \tag{22}$$

The local fractional divergence operator of

$$\mathbf{r} = r_\rho\mathbf{e}_\rho^\alpha + r_\phi\mathbf{e}_\phi^\alpha + r_\theta\mathbf{e}_\theta^\alpha$$

in Cantor-type spherical-coordinate systems is given by

$$\begin{aligned}\Delta^\alpha\cdot\mathbf{r} &= \left(\mathbf{e}_\rho^\alpha\frac{\partial^\alpha}{\partial\rho^\alpha} + \mathbf{e}_\phi^\alpha\frac{1}{\rho^\alpha}\frac{\partial^\alpha}{\partial\phi^\alpha} + \mathbf{e}_\theta^\alpha\frac{1}{\rho^\alpha\sin_\alpha\phi^\alpha}\frac{\partial^\alpha}{\partial\theta^\alpha}\right)\cdot(r_\rho\mathbf{e}_\rho^\alpha + r_\phi\mathbf{e}_\phi^\alpha + r_\theta\mathbf{e}_\theta^\alpha)\\ &= \mathbf{e}_\rho^\alpha\cdot\frac{\partial^\alpha\mathbf{r}}{\partial\rho^\alpha} + \mathbf{e}_\phi^\alpha\cdot\frac{1}{\rho^\alpha}\frac{\partial^\alpha\mathbf{r}}{\partial\phi^\alpha} + \mathbf{e}_\theta^\alpha\cdot\frac{1}{\rho^\alpha\sin_\alpha\phi^\alpha}\frac{\partial^\alpha\mathbf{r}}{\partial\theta^\alpha}\end{aligned} \tag{23}$$

With the help of the partial derivatives obtained in Eq. (15) and the product rule Eq. (6), the first term of Eq. (23) becomes

$$\begin{aligned}\mathbf{e}_\rho^\alpha \cdot \frac{\partial^\alpha \mathbf{r}}{\partial \rho^\alpha} &= \mathbf{e}_\rho^\alpha \cdot \left(\frac{\partial^\alpha r_\rho}{\partial \rho^\alpha}\mathbf{e}_\rho^\alpha + \frac{\partial^\alpha r_\phi}{\partial \rho^\alpha}\mathbf{e}_\phi^\alpha + \frac{\partial^\alpha r_\theta}{\partial \theta^\alpha}\mathbf{e}_\theta^\alpha + 0 + 0 + 0 \right) \\ &= \frac{\partial^\alpha r_\rho}{\partial \rho^\alpha}. \end{aligned} \tag{24}$$

The second term of Eq. (23) is given by

$$\begin{aligned}\mathbf{e}_\phi^\alpha \cdot \frac{1}{\rho^\alpha}\frac{\partial^\alpha \mathbf{r}}{\partial \phi^\alpha} &= \mathbf{e}_\phi^\alpha \cdot \frac{1}{\rho^\alpha} \left(\frac{\partial^\alpha r_\rho}{\partial \phi^\alpha}\mathbf{e}_\rho^\alpha + \frac{\partial^\alpha r_\phi}{\partial \phi^\alpha}\mathbf{e}_\phi^\alpha + \frac{\partial^\alpha r_\theta}{\partial \phi^\alpha}\mathbf{e}_\theta^\alpha + r_\rho \mathbf{e}_\phi^\alpha - r_\phi \mathbf{e}_\rho^\alpha \right) \\ &= \frac{1}{\rho^\alpha}\frac{\partial^\alpha r_\phi}{\partial \phi^\alpha} + \frac{r_\rho}{\rho^\alpha}. \end{aligned} \tag{25}$$

We notice that

$$\begin{aligned}\frac{\partial^\alpha \mathbf{r}}{\partial \theta^\alpha} =& \frac{\partial^\alpha r_\rho}{\partial \theta^\alpha}\mathbf{e}_\rho^\alpha + \frac{\partial^\alpha r_\phi}{\partial \theta^\alpha}\mathbf{e}_\phi^\alpha + \frac{\partial^\alpha r_\theta}{\partial \theta^\alpha}\mathbf{e}_\theta^\alpha \\ & - r_\theta \sin_\alpha \phi^\alpha \mathbf{e}_\rho^\alpha - r_\theta \cos_\alpha \phi^\alpha \mathbf{e}_\phi^\alpha + (r_\rho \sin_\alpha \phi^\alpha + r_\phi \cos_\alpha \phi^\alpha)\mathbf{e}_\theta^\alpha. \end{aligned}$$

Hence, the third term of Eq. (23) has the following form:

$$\begin{aligned}\mathbf{e}_\theta^\alpha \cdot \frac{1}{\rho^\alpha \sin_\alpha \phi^\alpha}\frac{\partial^\alpha \mathbf{r}}{\partial \theta^\alpha} &= \frac{1}{\rho^\alpha \sin_\alpha \phi^\alpha} \left(\frac{\partial^\alpha r_\theta}{\partial \theta^\alpha} + r_\rho \sin_\alpha \phi^\alpha + r_\phi \cos_\alpha \phi^\alpha \right) \\ &= \frac{1}{\rho^\alpha \sin_\alpha \phi^\alpha}\frac{\partial^\alpha r_\theta}{\partial \theta^\alpha} + \frac{r_\rho}{\rho^\alpha} + \frac{r_\phi \cos_\alpha \phi^\alpha}{\rho^\alpha \sin_\alpha \phi^\alpha} \end{aligned} \tag{26}$$

By combining Eqs. (24) to (26), Eq. (23) can be written as follows:

$$\begin{aligned}\Delta^\alpha \cdot \mathbf{r} &= \frac{\partial^\alpha r_\rho}{\partial \rho^\alpha} + \frac{1}{\rho^\alpha}\frac{\partial^\alpha r_\phi}{\partial \phi^\alpha} + \frac{r_\rho}{\rho^\alpha} + \frac{1}{\rho^\alpha \sin_\alpha \phi^\alpha}\frac{\partial^\alpha r_\theta}{\partial \theta^\alpha} + \frac{r_\rho}{\rho^\alpha} + \frac{r_\phi \cos_\alpha \phi^\alpha}{\rho^\alpha \sin_\alpha \phi^\alpha} \\ &= \frac{\partial^\alpha r_\rho}{\partial \rho^\alpha} + \left(\frac{1}{\rho^\alpha}\frac{\partial^\alpha r_\phi}{\partial \phi^\alpha} + \frac{r_\phi \cos_\alpha \phi^\alpha}{\rho^\alpha \sin_\alpha \phi^\alpha} \right) + \frac{1}{\rho^\alpha \sin_\alpha \phi^\alpha}\frac{\partial^\alpha r_\theta}{\partial \theta^\alpha} + 2\frac{r_\rho}{\rho^\alpha}. \end{aligned} \tag{27}$$

It follows that

$$\Delta^\alpha \cdot \mathbf{r} = \frac{\partial^\alpha r_\rho}{\partial \rho^\alpha} + \frac{1}{\rho^\alpha \sin_\alpha \phi^\alpha}\frac{\partial^\alpha}{\partial \phi^\alpha}(r_\phi \sin_\alpha \phi^\alpha) + \frac{1}{\rho^\alpha \sin_\alpha \phi^\alpha}\frac{\partial^\alpha r_\theta}{\partial \theta^\alpha} + 2\frac{r_\rho}{\rho^\alpha}. \tag{28}$$

Now, we derive the local fractional curl operator of

$$\mathbf{r} = r_\rho \mathbf{e}_\rho^\alpha + r_\phi \mathbf{e}_\phi^\alpha + r_\theta \mathbf{e}_\theta^\alpha$$

in Cantor-type spherical-coordinate systems.

$$\begin{aligned}\Delta^{\alpha}\times\mathbf{r}&=\left(\mathbf{e}_{\rho}^{\alpha}\frac{\partial^{\alpha}}{\partial\rho^{\alpha}}+\mathbf{e}_{\phi}^{\alpha}\frac{1}{\rho^{\alpha}}\frac{\partial^{\alpha}}{\partial\phi^{\alpha}}+\mathbf{e}_{\theta}^{\alpha}\frac{1}{\rho^{\alpha}\sin_{\alpha}\phi^{\alpha}}\frac{\partial^{\alpha}}{\partial\theta^{\alpha}}\right)\times\left(r_{\rho}\mathbf{e}_{\rho}^{\alpha}+r_{\phi}\mathbf{e}_{\phi}^{\alpha}+r_{\theta}\mathbf{e}_{\theta}^{\alpha}\right)\\&=\mathbf{e}_{\rho}^{\alpha}\times\frac{\partial^{\alpha}\mathbf{r}}{\partial\rho^{\alpha}}+\mathbf{e}_{\phi}^{\alpha}\times\frac{1}{\rho^{\alpha}}\frac{\partial^{\alpha}\mathbf{r}}{\partial\phi^{\alpha}}+\mathbf{e}_{\theta}^{\alpha}\times\frac{1}{\rho^{\alpha}\sin_{\alpha}\phi^{\alpha}}\frac{\partial^{\alpha}\mathbf{r}}{\partial\theta^{\alpha}}\end{aligned}\tag{29}$$

With the help of the partial derivatives obtained in Eq. (15) and the product rule Eq. (6), the first term of Eq. (29) becomes

$$\begin{aligned}\mathbf{e}_{\rho}^{\alpha}\times\frac{\partial^{\alpha}\mathbf{r}}{\partial\rho^{\alpha}}&=\mathbf{e}_{\rho}^{\alpha}\times\left(\frac{\partial^{\alpha}r_{\rho}}{\partial\rho^{\alpha}}\mathbf{e}_{\rho}^{\alpha}+\frac{\partial^{\alpha}r_{\phi}}{\partial\rho^{\alpha}}\mathbf{e}_{\phi}^{\alpha}+\frac{\partial^{\alpha}r_{\theta}}{\partial\theta^{\alpha}}\mathbf{e}_{\theta}^{\alpha}+0+0+0\right)\\&=\frac{\partial^{\alpha}r_{\phi}}{\partial\rho^{\alpha}}\mathbf{e}_{\theta}^{\alpha}-\frac{\partial^{\alpha}r_{\theta}}{\partial\theta^{\alpha}}\mathbf{e}_{\phi}^{\alpha}.\end{aligned}\tag{30}$$

The second term of Eq. (29) is given by

$$\begin{aligned}\mathbf{e}_{\phi}^{\alpha}\times\frac{1}{\rho^{\alpha}}\frac{\partial^{\alpha}\mathbf{r}}{\partial\phi^{\alpha}}&=\mathbf{e}_{\phi}^{\alpha}\times\frac{1}{\rho^{\alpha}}\left(\frac{\partial^{\alpha}r_{\rho}}{\partial\phi^{\alpha}}\mathbf{e}_{\rho}^{\alpha}+\frac{\partial^{\alpha}r_{\phi}}{\partial\phi^{\alpha}}\mathbf{e}_{\phi}^{\alpha}+\frac{\partial^{\alpha}r_{\theta}}{\partial\phi^{\alpha}}\mathbf{e}_{\theta}^{\alpha}+r_{\rho}\mathbf{e}_{\phi}^{\alpha}-r_{\phi}\mathbf{e}_{\rho}^{\alpha}\right)\\&=-\frac{1}{\rho^{\alpha}}\frac{\partial^{\alpha}r_{\rho}}{\partial\phi^{\alpha}}\mathbf{e}_{\theta}^{\alpha}+\frac{1}{\rho^{\alpha}}\frac{\partial^{\alpha}r_{\theta}}{\partial\phi^{\alpha}}\mathbf{e}_{\rho}^{\alpha}-\frac{r_{\phi}}{\rho^{\alpha}}\mathbf{e}_{\theta}^{\alpha}.\end{aligned}\tag{31}$$

We notice that

$$\begin{aligned}\frac{\partial^{\alpha}\mathbf{r}}{\partial\theta^{\alpha}}=&\frac{\partial^{\alpha}r_{\rho}}{\partial\theta^{\alpha}}\mathbf{e}_{\rho}^{\alpha}+\frac{\partial^{\alpha}r_{\phi}}{\partial\theta^{\alpha}}\mathbf{e}_{\phi}^{\alpha}+\frac{\partial^{\alpha}r_{\theta}}{\partial\theta^{\alpha}}\mathbf{e}_{\theta}^{\alpha}\\&-r_{\theta}\sin_{\alpha}\phi^{\alpha}\mathbf{e}_{\rho}^{\alpha}-r_{\theta}\cos_{\alpha}\phi^{\alpha}\mathbf{e}_{\phi}^{\alpha}+\left(r_{\rho}\sin_{\alpha}\phi^{\alpha}+r_{\phi}\cos_{\alpha}\phi^{\alpha}\right)\mathbf{e}_{\theta}^{\alpha}.\end{aligned}$$

Hence, the third term of Eq. (29) has the following form:

$$\begin{aligned}\mathbf{e}_{\theta}^{\alpha}\times\frac{1}{\rho^{\alpha}\sin_{\alpha}\phi^{\alpha}}\frac{\partial^{\alpha}\mathbf{r}}{\partial\theta^{\alpha}}&=\frac{1}{\rho^{\alpha}\sin_{\alpha}\phi^{\alpha}}\left(\frac{\partial^{\alpha}r_{\rho}}{\partial\theta^{\alpha}}\mathbf{e}_{\phi}^{\alpha}-\frac{\partial^{\alpha}r_{\phi}}{\partial\theta^{\alpha}}\mathbf{e}_{\rho}^{\alpha}-r_{\theta}\sin_{\alpha}\phi^{\alpha}\mathbf{e}_{\phi}^{\alpha}+r_{\theta}\cos_{\alpha}\phi^{\alpha}\mathbf{e}_{\rho}^{\alpha}\right)\\=\frac{1}{\rho^{\alpha}\sin_{\alpha}\phi^{\alpha}}\frac{\partial^{\alpha}r_{\rho}}{\partial\theta^{\alpha}}\mathbf{e}_{\phi}^{\alpha}&-\frac{1}{\rho^{\alpha}\sin_{\alpha}\phi^{\alpha}}\frac{\partial^{\alpha}r_{\phi}}{\partial\theta^{\alpha}}\mathbf{e}_{\rho}^{\alpha}-\frac{r_{\theta}}{\rho^{\alpha}}\mathbf{e}_{\phi}^{\alpha}+\frac{r_{\theta}\cos_{\alpha}\phi^{\alpha}}{\rho^{\alpha}\sin_{\alpha}\phi^{\alpha}}\mathbf{e}_{\rho}^{\alpha}.\end{aligned}\tag{32}$$

Substituting Eqs. (30) to (32) into Eq. (29), we obtain

$$\begin{aligned}\Delta^{\alpha}\times\mathbf{r}=&\left(\frac{\partial^{\alpha}r_{\phi}}{\partial\rho^{\alpha}}\mathbf{e}_{\theta}^{\alpha}-\frac{\partial^{\alpha}r_{\theta}}{\partial\theta^{\alpha}}\mathbf{e}_{\phi}^{\alpha}\right)+\left(-\frac{1}{\rho^{\alpha}}\frac{\partial^{\alpha}r_{\rho}}{\partial\phi^{\alpha}}\mathbf{e}_{\theta}^{\alpha}+\frac{1}{\rho^{\alpha}}\frac{\partial^{\alpha}r_{\theta}}{\partial\phi^{\alpha}}\mathbf{e}_{\rho}^{\alpha}-\frac{r_{\phi}}{\rho^{\alpha}}\mathbf{e}_{\theta}^{\alpha}\right)\\&+\left(\frac{1}{\rho^{\alpha}\sin_{\alpha}\phi^{\alpha}}\frac{\partial^{\alpha}r_{\rho}}{\partial\theta^{\alpha}}\mathbf{e}_{\phi}^{\alpha}-\frac{1}{\rho^{\alpha}\sin_{\alpha}\phi^{\alpha}}\frac{\partial^{\alpha}r_{\phi}}{\partial\theta^{\alpha}}\mathbf{e}_{\rho}^{\alpha}-\frac{r_{\theta}}{\rho^{\alpha}}\mathbf{e}_{\phi}^{\alpha}+\frac{r_{\theta}\cos_{\alpha}\phi^{\alpha}}{\rho^{\alpha}\sin_{\alpha}\phi^{\alpha}}\mathbf{e}_{\rho}^{\alpha}\right)\\=&\left(\frac{1}{\rho^{\alpha}}\frac{\partial^{\alpha}r_{\theta}}{\partial\phi^{\alpha}}-\frac{1}{\rho^{\alpha}\sin_{\alpha}\phi^{\alpha}}\frac{\partial^{\alpha}r_{\phi}}{\partial\theta^{\alpha}}+\frac{r_{\theta}\cos_{\alpha}\phi^{\alpha}}{\rho^{\alpha}\sin_{\alpha}\phi^{\alpha}}\right)\mathbf{e}_{\rho}^{\alpha}\\&+\left(\frac{1}{\rho^{\alpha}\sin_{\alpha}\phi^{\alpha}}\frac{\partial^{\alpha}r_{\rho}}{\partial\theta^{\alpha}}-\frac{\partial^{\alpha}r_{\theta}}{\partial\theta^{\alpha}}-\frac{r_{\theta}}{\rho^{\alpha}}\right)\mathbf{e}_{\phi}^{\alpha}+\left(\frac{\partial^{\alpha}r_{\phi}}{\partial\rho^{\alpha}}-\frac{1}{\rho^{\alpha}}\frac{\partial^{\alpha}r_{\rho}}{\partial\phi^{\alpha}}-\frac{r_{\phi}}{\rho^{\alpha}}\right)\mathbf{e}_{\theta}^{\alpha}\\=&\frac{1}{\rho^{\alpha}\sin_{\alpha}\phi^{\alpha}}\left(\frac{\partial^{\alpha}}{\partial\phi^{\alpha}}\left(r_{\theta}\sin_{\alpha}\phi^{\alpha}\right)-\frac{\partial^{\alpha}r_{\theta}}{\partial\phi^{\alpha}}\right)\mathbf{e}_{\rho}^{\alpha}\end{aligned}$$

$$+\left(\frac{1}{\rho^\alpha \sin_\alpha \phi^\alpha}\frac{\partial^\alpha r_\rho}{\partial\theta^\alpha}-\frac{\partial^\alpha r_\theta}{\partial\theta^\alpha}-\frac{r_\theta}{\rho^\alpha}\right)\mathbf{e}_\phi^\alpha+\left(\frac{\partial^\alpha r_\phi}{\partial\rho^\alpha}-\frac{1}{\rho^\alpha}\frac{\partial^\alpha r_\rho}{\partial\phi^\alpha}-\frac{r_\phi}{\rho^\alpha}\right)\mathbf{e}_\theta^\alpha. \tag{33}$$

Consequently, the local fractional Laplace operator in Cantor-type spherical-coordinates is given by

$$\begin{aligned}
\nabla^{2\alpha}\Psi &= \left(\mathbf{e}_\rho^\alpha\frac{\partial^\alpha}{\partial\rho^\alpha}+\mathbf{e}_\phi^\alpha\frac{1}{\rho^\alpha}\frac{\partial^\alpha}{\partial\phi^\alpha}+\mathbf{e}_\theta^\alpha\frac{1}{\rho^\alpha\sin_\alpha\phi^\alpha}\frac{\partial^\alpha}{\partial\theta^\alpha}\right)\\
&\quad\cdot\left(\mathbf{e}_\rho^\alpha\frac{\partial^\alpha\Psi}{\partial\rho^\alpha}+\mathbf{e}_\phi^\alpha\frac{1}{\rho^\alpha}\frac{\partial^\alpha\Psi}{\partial\phi^\alpha}+\mathbf{e}_\theta^\alpha\frac{1}{\rho^\alpha\sin_\alpha\phi^\alpha}\frac{\partial^\alpha\Psi}{\partial\theta^\alpha}\right)\\
&=\mathbf{e}_\rho^\alpha\cdot\frac{\partial^\alpha}{\partial\rho^\alpha}\left(\mathbf{e}_\rho^\alpha\frac{\partial^\alpha\Psi}{\partial\rho^\alpha}+\mathbf{e}_\phi^\alpha\frac{1}{\rho^\alpha}\frac{\partial^\alpha\Psi}{\partial\phi^\alpha}+\mathbf{e}_\theta^\alpha\frac{1}{\rho^\alpha\sin_\alpha\phi^\alpha}\frac{\partial^\alpha\Psi}{\partial\theta^\alpha}\right)\\
&\quad+\mathbf{e}_\phi^\alpha\cdot\frac{1}{\rho^\alpha}\frac{\partial^\alpha}{\partial\phi^\alpha}\left(\mathbf{e}_\rho^\alpha\frac{\partial^\alpha\Psi}{\partial\rho^\alpha}+\mathbf{e}_\phi^\alpha\frac{1}{\rho^\alpha}\frac{\partial^\alpha\Psi}{\partial\phi^\alpha}+\mathbf{e}_\theta^\alpha\frac{1}{\rho^\alpha\sin_\alpha\phi^\alpha}\frac{\partial^\alpha\Psi}{\partial\theta^\alpha}\right)\\
&\quad+\mathbf{e}_\theta^\alpha\cdot\frac{1}{\rho^\alpha\sin_\alpha\phi^\alpha}\frac{\partial^\alpha}{\partial\theta^\alpha}\left(\mathbf{e}_\rho^\alpha\frac{\partial^\alpha\Psi}{\partial\rho^\alpha}+\mathbf{e}_\phi^\alpha\frac{1}{\rho^\alpha}\frac{\partial^\alpha\Psi}{\partial\phi^\alpha}+\mathbf{e}_\theta^\alpha\frac{1}{\rho^\alpha\sin_\alpha\phi^\alpha}\frac{\partial^\alpha\Psi}{\partial\theta^\alpha}\right).
\end{aligned}\tag{34}$$

The first term of Eq. (34) is written as follows:

$$\begin{aligned}
&\mathbf{e}_\rho^\alpha\cdot\frac{\partial^\alpha}{\partial\rho^\alpha}\left(\mathbf{e}_\rho^\alpha\frac{\partial^\alpha\Psi}{\partial\rho^\alpha}+\mathbf{e}_\phi^\alpha\frac{1}{\rho^\alpha}\frac{\partial^\alpha\Psi}{\partial\phi^\alpha}+\mathbf{e}_\theta^\alpha\frac{1}{\rho^\alpha\sin_\alpha\phi^\alpha}\frac{\partial^\alpha\Psi}{\partial\theta^\alpha}\right)\\
&=\mathbf{e}_\rho^\alpha\cdot\left(\mathbf{e}_\rho^\alpha\frac{\partial^{2\alpha}\Psi}{\partial\rho^{2\alpha}}+\mathbf{e}_\phi^\alpha\frac{\partial^\alpha}{\partial\rho^\alpha}\left(\frac{1}{\rho^\alpha}\frac{\partial^\alpha\Psi}{\partial\phi^\alpha}\right)+\mathbf{e}_\theta^\alpha\frac{\partial^\alpha}{\partial\rho^\alpha}\left(\frac{1}{\rho^\alpha\sin_\alpha\phi^\alpha}\frac{\partial^\alpha\Psi}{\partial\theta^\alpha}\right)\right)\\
&=\frac{\partial^{2\alpha}\Psi}{\partial\rho^{2\alpha}}
\end{aligned}\tag{35}$$

The second term of Eq. (34) becomes

$$\begin{aligned}
&\mathbf{e}_\phi^\alpha\cdot\frac{1}{\rho^\alpha}\frac{\partial^\alpha}{\partial\phi^\alpha}\left(\mathbf{e}_\rho^\alpha\frac{\partial^\alpha\Psi}{\partial\rho^\alpha}+\mathbf{e}_\phi^\alpha\frac{1}{\rho^\alpha}\frac{\partial^\alpha\Psi}{\partial\phi^\alpha}+\mathbf{e}_\theta^\alpha\frac{1}{\rho^\alpha\sin_\alpha\phi^\alpha}\frac{\partial^\alpha\Psi}{\partial\theta^\alpha}\right)\\
&=\mathbf{e}_\phi^\alpha\cdot\frac{1}{\rho^\alpha}\left(\mathbf{e}_\rho^\alpha\frac{\partial^{2\alpha}\Psi}{\partial\phi^\alpha\partial\rho^\alpha}+\mathbf{e}_\phi^\alpha\frac{\partial^\alpha\Psi}{\partial\rho^\alpha}+\mathbf{e}_\phi^\alpha\frac{\partial^\alpha}{\partial\phi^\alpha}\left(\frac{1}{\rho^\alpha}\frac{\partial^\alpha\Psi}{\partial\phi^\alpha}\right)\right.\\
&\qquad\left.-\mathbf{e}_\rho^\alpha\frac{1}{\rho^\alpha}\frac{\partial^\alpha\Psi}{\partial\phi^\alpha}+\mathbf{e}_\theta^\alpha\frac{\partial^\alpha}{\partial\phi^\alpha}\left(\frac{1}{\rho^\alpha\sin_\alpha\phi^\alpha}\frac{\partial^\alpha\Psi}{\partial\theta^\alpha}\right)\right)\\
&=\frac{1}{\rho^\alpha}\frac{\partial^\alpha\Psi}{\partial\rho^\alpha}+\frac{1}{\rho^\alpha}\frac{\partial^\alpha}{\partial\phi^\alpha}\left(\frac{1}{\rho^\alpha}\frac{\partial^\alpha\Psi}{\partial\phi^\alpha}\right)\\
&=\frac{1}{\rho^{2\alpha}}\frac{\partial^{2\alpha}\Psi}{\partial\phi^{2\alpha}}+\frac{1}{\rho^\alpha}\frac{\partial^\alpha\Psi}{\partial\rho^\alpha}.
\end{aligned}\tag{36}$$

The third term of Eq. (34) is represented by

$$\mathbf{e}_\theta^\alpha\cdot\frac{1}{\rho^\alpha\sin_\alpha\phi^\alpha}\frac{\partial^\alpha}{\partial\theta^\alpha}\left(\mathbf{e}_\rho^\alpha\frac{\partial^\alpha\Psi}{\partial\rho^\alpha}+\mathbf{e}_\phi^\alpha\frac{1}{\rho^\alpha}\frac{\partial^\alpha\Psi}{\partial\phi^\alpha}+\mathbf{e}_\theta^\alpha\frac{1}{\rho^\alpha\sin_\alpha\phi^\alpha}\frac{\partial^\alpha\Psi}{\partial\theta^\alpha}\right)$$

$$= \mathbf{e}_\theta^\alpha \cdot \frac{1}{\rho^\alpha \sin_\alpha \phi^\alpha}\left(\mathbf{e}_\rho^\alpha \frac{\partial^{2\alpha}\Psi}{\partial\theta^\alpha\partial\rho^\alpha} + \frac{\partial^\alpha \mathbf{e}_\rho^\alpha}{\partial\theta^\alpha}\frac{\partial^\alpha\Psi}{\partial\rho^\alpha} + \frac{1}{\rho^\alpha}\frac{\partial^\alpha \mathbf{e}_\phi^\alpha}{\partial\theta^\alpha}\frac{\partial^\alpha\Psi}{\partial\phi^\alpha} + \mathbf{e}_\phi^\alpha \frac{\partial^\alpha}{\partial\theta^\alpha}\left(\frac{1}{\rho^\alpha}\frac{\partial^\alpha\Psi}{\partial\theta^\alpha}\right)\right.$$

$$\left. + \frac{1}{\rho^\alpha \sin_\alpha \phi^\alpha}\frac{\partial^\alpha \mathbf{e}_\theta^\alpha}{\partial\theta^\alpha}\frac{\partial^\alpha\Psi}{\partial\theta^\alpha} + \mathbf{e}_\theta^\alpha \frac{\partial^\alpha}{\partial\theta^\alpha}\left(\frac{1}{\rho^\alpha \sin_\alpha \phi^\alpha}\frac{\partial^\alpha\Psi}{\partial\theta^\alpha}\right)\right)$$

$$= \frac{1}{\rho^\alpha \sin_\alpha \phi^\alpha}\left(\sin_\alpha \phi^\alpha \frac{\partial^\alpha\Psi}{\partial\rho^\alpha} + \frac{\cos_\alpha \phi^\alpha}{\rho^\alpha}\frac{\partial^\alpha\Psi}{\partial\phi^\alpha} + \frac{\partial^\alpha}{\partial\theta^\alpha}\left(\frac{1}{\rho^\alpha \sin_\alpha \phi^\alpha}\frac{\partial^\alpha\Psi}{\partial\theta^\alpha}\right)\right)$$

$$= \frac{1}{\rho^\alpha}\frac{\partial^\alpha\Psi}{\partial\rho^\alpha} + \frac{\cos_\alpha \phi^\alpha}{\rho^{2\alpha}\sin_\alpha \phi^\alpha}\frac{\partial^\alpha\Psi}{\partial\phi^\alpha} + \frac{1}{\rho^\alpha \sin_\alpha \phi^\alpha}\frac{\partial^\alpha}{\partial\theta^\alpha}\left(\frac{1}{\rho^\alpha \sin_\alpha \phi^\alpha}\frac{\partial^\alpha\Psi}{\partial\theta^\alpha}\right)$$

$$= \frac{1}{\rho^\alpha}\frac{\partial^\alpha\Psi}{\partial\rho^\alpha} + \frac{\cos_\alpha \phi^\alpha}{\rho^{2\alpha}\sin_\alpha \phi^\alpha}\frac{\partial^\alpha\Psi}{\partial\phi^\alpha} + \frac{1}{\rho^{2\alpha}\sin_\alpha^2 \phi^\alpha}\frac{\partial^{2\alpha}\Psi}{\partial\theta^{2\alpha}}. \tag{37}$$

As a result, the expansion for the local fractional Laplace operator assumes the following form:

$$\begin{aligned}\nabla^{2\alpha}\Psi &= \frac{\partial^{2\alpha}\Psi}{\partial\rho^{2\alpha}} + \frac{1}{\rho^{2\alpha}}\frac{\partial^{2\alpha}\Psi}{\partial\phi^{2\alpha}} + \frac{2}{\rho^\alpha}\frac{\partial^\alpha\Psi}{\partial\rho^\alpha} + \frac{\cos_\alpha \phi^\alpha}{\rho^{2\alpha}\sin_\alpha \phi^\alpha}\frac{\partial^\alpha\Psi}{\partial\phi^\alpha} + \frac{1}{\rho^{2\alpha}\sin_\alpha^2 \phi^\alpha}\frac{\partial^{2\alpha}\Psi}{\partial\theta^{2\alpha}}\\ &= \frac{\partial^{2\alpha}\Psi}{\partial\rho^{2\alpha}} + \frac{1}{\rho^{2\alpha}\sin_\alpha \phi^\alpha}\frac{\partial^\alpha}{\partial\phi^\alpha}\left(\sin_\alpha \phi^\alpha \frac{\partial^\alpha\Psi}{\partial\phi^\alpha}\right) + \frac{2}{\rho^\alpha}\frac{\partial^\alpha\Psi}{\partial\rho^\alpha} + \frac{1}{\rho^{2\alpha}\sin_\alpha^2 \phi^\alpha}\frac{\partial^{2\alpha}\Psi}{\partial\theta^{2\alpha}}.\end{aligned} \tag{38}$$

4 Examples

Below we consider several differential equations on Cantor sets.

Example 30. *Let us consider the heat-conduction equation on Cantor sets without heat generation in fractal media (see (Yang et al., 2013)), namely*

$$K^{2\alpha}\nabla^{2\alpha}T(x,y,z,t) - \rho_\alpha c_\alpha \frac{\partial^\alpha T(x,y,z,t)}{\partial t^\alpha} = 0, \tag{39}$$

where

$$\nabla^{2\alpha} = \frac{\partial^{2\alpha}}{\partial x^{2\alpha}} + \frac{\partial^{2\alpha}}{\partial y^{2\alpha}} + \frac{\partial^{2\alpha}}{\partial z^{2\alpha}}$$

is the local Laplace operator (Yang,2011b; Yang,2012).

By using Eq. (38), Eq. (39) is transformed into

$$\begin{aligned}K^{2\alpha}&\left[\frac{\partial^{2\alpha}T(\rho,\phi,\theta,t)}{\partial\rho^{2\alpha}} + \frac{1}{\rho^{2\alpha}\sin_\alpha \phi^\alpha}\frac{\partial^\alpha}{\partial\phi^\alpha}\left(\sin_\alpha \phi^\alpha \frac{\partial^\alpha T(\rho,\phi,\theta,t)}{\partial\phi^\alpha}\right)\right.\\ &\left.+\frac{2}{\rho^\alpha}\frac{\partial^\alpha T(\rho,\phi,\theta,t)}{\partial\rho^\alpha} + \frac{1}{\rho^{2\alpha}\sin_\alpha^2 \phi^\alpha}\frac{\partial^{2\alpha}T(\rho,\phi,\theta,t)}{\partial\theta^{2\alpha}}\right] - \rho_\alpha c_\alpha \frac{\partial^\alpha T}{\partial t^\alpha} = 0\end{aligned} \tag{40}$$

which is the form of the heat-conduction equation on Cantor sets in Cantor-type spherical coordinate system.

Example 31. *Consider the damped wave equation in fractal strings as given below:*

$$\frac{\partial^{2\alpha} u(x,y,z,t)}{\partial t^{2\alpha}} - \frac{\partial^{\alpha} u(x,y,z,t)}{\partial t^{\alpha}} - \nabla^{2\alpha} u(x,y,z,t) = 0. \tag{41}$$

Applying Eq. (38) into Eq. (39), we get

$$\begin{aligned}&\frac{\partial^{2\alpha} u(\rho,\phi,\theta,t)}{\partial t^{2\alpha}} - \frac{\partial^{\alpha} u(\rho,\phi,\theta,t)}{\partial t^{\alpha}} - \left[\frac{\partial^{2\alpha} u(\rho,\phi,\theta,t)}{\partial \rho^{2\alpha}}\right.\\&+\frac{1}{\rho^{2\alpha}\sin_\alpha \phi^\alpha}\frac{\partial^\alpha}{\partial \phi^\alpha}\left(\sin_\alpha \phi^\alpha \frac{\partial^{\alpha} u(\rho,\phi,\theta,t)}{\partial \phi^{\alpha}}\right) + \frac{2}{\rho^\alpha}\frac{\partial^{\alpha} u(\rho,\phi,\theta,t)}{\partial \rho^{\alpha}}\\&\left.+\frac{1}{\rho^{2\alpha}\sin_\alpha^2 \phi^\alpha}\frac{\partial^{2\alpha} u(\rho,\phi,\theta,t)}{\partial \theta^{2\alpha}}\right] = 0\end{aligned} \tag{42}$$

which is the form of the damped wave equation in fractal strings in cantor-type spherical-coordinate system.

Example 32. *Let us consider the inhomogeneous Helmholtz equation on Cantor sets (Hao et al., 2013), namely*

$$\frac{\partial^{2\alpha} M(x,y,z)}{\partial x^{2\alpha}} + \frac{\partial^{2\alpha} M(x,y,z)}{\partial y^{2\alpha}} + \frac{\partial^{2\alpha} M(x,y,z)}{\partial z^{2\alpha}} + \omega^{2\alpha} M(x,y,z) = f(x,y,z), \tag{43}$$

where $f(x,y,z)$ is a local fractional continuous function. By using Eq. (34) into Eq. (35), we find that

$$\begin{aligned}&\frac{\partial^{2\alpha} M(\rho,\phi,\theta)}{\partial \rho^{2\alpha}} + \frac{1}{\rho^{2\alpha}\sin_\alpha \phi^\alpha}\frac{\partial^\alpha}{\partial \phi^\alpha}\left(\sin_\alpha \phi^\alpha \frac{\partial^{\alpha} M(\rho,\phi,\theta)}{\partial \phi^{\alpha}}\right) + \frac{2}{\rho^\alpha}\frac{\partial^{\alpha} M(\rho,\phi,\theta)}{\partial \rho^{\alpha}}\\&+\frac{1}{\rho^{2\alpha}\sin_\alpha^2 \phi^\alpha}\frac{\partial^{2\alpha} M(\rho,\phi,\theta)}{\partial \theta^{2\alpha}} + \omega^{2\alpha} M(\rho,\phi,\theta) = f(\rho,\phi,\theta),\end{aligned} \tag{44}$$

which is the form of the Helmholtz equation on Cantor sets in Cantor-type spherical coordinates system.

5 Conclusions

In this chapter, we discussed a new Cantor-type spherical coordinate method and the equivalent forms of differential equations on Cantor sets were investigated. Several examples of differential equations on Cantor sets, e.g. heat-conduction equation without heat generation on fractal media, damped wave equation in fractal strings, Helmholtz equation on Cantor sets were tested by applying the Cantor-type spherical coordinate method.

Bibliography

Adda F.B., Cresson J., About non-differentiable functions. J. Math. Anal. Appl., 263(2), (2001), 721-737.

Babakhani A., Daftardar-Gejji V., On calculus of local fractional derivatives. J. Math. Anal. Appl. 270, (2002), 66-69.

Chen Y., Yan Y., Zhang K., On the local fractional derivative. J. Math. Anal. Appl. 362. (2010), 17-23.

Engheta N., On fractional paradigm and intermediate zones in electromagnetism. I. Planar observation. Microwave and Optical Technology Letters, 22(4), (1990), 236-241.

Engheta N., On fractional paradigmand intermediate zones in electromagnetism. II. Cylindrical and spherical observations. Microwave and Optical Technology Letters, 23(2), (1999), 100-103.

Hao Y.-J., Srivastava H.M., Jafari H., Yang X.-J., Helmholtz and diffusion equations associated with local fractional operators involving the Cantorian and Cantor-type coordinates. Advance in Mathematical Physics, (2013), Article ID 754248, 5 pages.

Kilbas A.A., Srivastava H.M., Trujillo J.J., Theory and Applications of Fractional Differential Equations. (2006), Elsevier, North- Holland.

Kolwankar K.M., Gangal A.D., Fractional differentiability of nowhere differentiable functions and dimension. Chaos 6, (1996), 505-513.

Kolwankar K.M., Gangal A.D., Hölder exponents of irregular signals and local fractional derivatives. Pramana J. Phys. 48, (1997), 49-58.

Kolwankar K.M., Gangal A.D., Local fractional Fokker-Planck equation. Physical Review Letters, 80(2), (1998), 214-217.

Oldham K., Spanier J., The fractional Calculus. (1970), Academic Press, London.

Petropoulos P.G., Reflectionless sponge layers as absorbing boundary conditions for the numerical solution of Maxwell equations in rectangular, cylindrical, and spherical coordinates. SIAM Journal on Applied Mathematics, 60(3), (2000), 1037-1058.

Podlubny I., Fractional differential equation equations. (1999), Academic Press, USA.

Scheuer J.T., Shamim M., Conrad J.R., Model of plasma source ion implantation in planar, cylindrical, and spherical geometries. Journal of Applied Physics, 67(3), (1990), 1241-1245.

Schetselaar E.M., Fusion by the IHS transform: should we use cylindrical or spherical coordinates. International Journal of Remote Sensing, 19(4), (1998), 759-765.

Srivastava H.M., Owa S., Univalent Functions, Fractional Calculus and their Applications. (1989), Wiley, New York, 1989.

Srivastava H.M., Golmankhaneh A.K., Baleanu D., Yang X.-J., Local fractional Sumudu transform with applications to IVPs on Cantor sets. Abstr. Appl. Anal., (2014), Article ID 620529, 1-7.

Teixeira F.L., Chew W.C., Systematic derivation of anistropic PML absorbing media in cylindrical and spherical coordinates, IEEE Microwave and Guided Wave Letters, 7(11), (1997), 371-373.

Yan S.-H., Chen X.H., Xie G.-N., Cattani C., Yang X.-J., Solving Fokker-Planck equations on Cantor sets using local fractional decomposition method. Abstract and Applied Analysis, (2014), Article ID 396469, 6 pages.

Yang A.-M., Lie J., Srivastava H.M., Xie G.-N., Yang X.-J., Local fractional Laplace variational iteration method for solving linear partial differential equations with local fractional derivative. Discrete Dyn. Nature Soc., (2014), Article ID 365981, 1-8.

Yang X.-J., Local fractional integral transforms. Progr. Nonlinear Sci., 4(1), (2011), 225.

Yang X.-J., Local fractional functional analysis and its applications. Asian Academic Publisher, (2011), Hong Kong.

Yang X.-J., Advanced local fractional calculus and its applications. World Science Publisher, (2012), New York.

Yang X.-J., Baleanu D., Fractal heat conduction problem solved by local fractional variation iteration method. Thermal Science, 17(2), (2013), 625-628.

Yang Y.-J., Baleanu D., Yang X.-J., A local fractional variational iteration method for Laplace equation within local fractional operators. Abstract and Applied Analysis, (2013), Article ID 202650, 6 pages.

Yang X.-J., Srivastava H.M., He J.-H., Baleanu D., Cantor-type cylindrical-coordinate method for differential equations with local fractional derivatives. Physics Letters A, 377, (2013), 1696-1700.

Zhao Y., Baleanu D., Cattani C., Cheng D.-F., Yang X.-J., Maxwell's equations on Cantor sets: A local fractional approach. Advanced in High Energy Physics, (2013), Article ID 686371, 6 pages.

H. M. Srivastava, J. A. Tenreiro Machado, and Xiao-Jun Yang

Approximate Methods for Local Fractional Differential Equations

Abstract: This chapter presents new analytical methods for a family of the local fractional differential equations. Our attention focuses upon the local fractional variational iteration, Adomian decomposition and series expansion methods. Several examples with solutions and charts are also included.

1 Introduction

The theory of local fractional operators has powerful tools to deal with the non-differentiable problems that arise in science and engineering, such as those in operator theory (Yang, 2011; Kilbas et al., 2006; Yang et al., 2015; Ben and Cresson, 2001; Babakhani and Daftardar-Gejji, 2002; Chen et al., 2010, Anastassiou and Duman, 2013), theoretical physics (Kolwankar and Gangal, 1998; Yang et al., 2013a,b,c, 2014; Zhao et al., 2013; Hao et al., 2013), non-differentiable signals (Machado et al., Liao et al., 2013; Yang et al., 2013d; Baleanu and Yang, 2013; Liu et al., 2014; Chen et al., 2014), solid and fluid mechanics (Carpinteri et al., 2001; Carpinteri et al., 2004; Carpinteri et al., 2009; Yang, 2012; Yang et al., 2013e). Local fractional differential equations were utilized to describe various physical phenomena, such as Navier-Stokes (Yang et al., 2013e), damped and dissipative waves of fractal strings (Su et al., 2013a), waves (Su et al., 2013b; Baleanu et al., 2014), Helmholtz (Hao et al., 2013; Wang et al, 2014), diffusion (Hao et al., 2013; Yang et al., 2013f; Cao et al., 2014), heat (Yang, 2012; Liu et al., 2013), Laplace (Yang, 2012; Li et al., 2014a; Yang et al., 2014; Yan et al., 2014a; Yang et al., 2013g), Schrödinger (Yang et al., 2013b; Zhao et al., 2013), transport (Li et al., 2014b), Klein-Gordon (Yang et al., 2013h), Fokker-Planck (Yan et al., 2014b), Tricomi (Niu et al., 2014) equations defined on Cantor sets. Many studies for finding the solutions of differential equations defined on Cantor sets, such as variational iteration method (Su et al., 2013a; Baleanu et al., 2014; Yang et al., 2013g; Li et al., 2014b; Yang et al., 2013i; Yang et al., 2013j), Adomian decomposition method (Yang et al., 2013 f; Baleanu et al., 2014; Yan et al., 2014a; Niu et al., 2014; Yan et al., 2014b), functional method (Cao et al., 2014; Yan et al., 2014a), Laplace variational iteration

H. M. Srivastava: Department of Mathematics and Statistics, University of Victoria, Victoria, British Columbia V8W 3R4, Canada; E-Mail: harimsri@math.uvic.ca
J. A. Tenreiro Machado: Department of Electrical Engineering, Institute of Engineering, Polytechnic of Porto, 431 Rua Dr. António Bernardino de Almeida, PT-4249-015 Porto, Portugal; E-Mail: jtm@isep.ipp.pt
Xiao-Jun Yang: Department of Mathematics and Mechanics, China University of Mining and Technology, Xuzhou, Jiangsu 221008, People's Republic of China; E-Mail: dyangxiaojun@163.com

method (Liu et al., 2013; Yang et al., 2014a), series expansion method (Zhao et al., 2013; Yang et al., 2014h; Yang et al., 2013i), Picard successive approximation method (Yang et al., 2014b), Sumudu transform (Srivastava et al., 2014), Fourier series (Hu et al., 2012; Yang et al., 2013k; Yang et al., 2014c), Laplace transform (Zhao et al., 2014; He, 2012) and Fourier transform (He, 2012) involving the local fractional operators, were proposed. This chapter gives new applications of local fractional variational iteration, Adomian decomposition and series expansion methods to solve the partial differential equations defined on Cantor sets that arise in mathematical physics.

2 The Theory of Local Fractional Calculus

In this section, we present the basic theory of local fractional derivatives and local fractional integrals.

The local fractional derivative of $f(x)$ of order α $(\alpha \in R)$ is given as (Yang, 2011, 2012; Yang et al., 2013a,b,c, 2014; Zhao et al., 2013; Hao et al., 2013)

$$\frac{d^{\alpha} f(x_0)}{dx^{\alpha}} = \frac{\Delta^{\alpha}(f(x) - f(x_0))}{(x - x_0)^{\alpha}}, \tag{1}$$

where

$$\Delta^{\alpha}(f(x) - f(x_0)) \cong \Gamma(1+\alpha)\left[f(x) - f(x_0)\right]. \tag{2}$$

The local fractional integral of $f(x)$ of order α in the interval $[a, b]$ is defined through (Yang et al., 2013a,b,c, 2014; Zhao et al., 2013; Hao et al., 2013)

$$ {}_aI_b^{(\alpha)} f(x) = \frac{1}{\Gamma(1+\alpha)} \int_a^b f(t)(dt)^{\alpha} = \frac{1}{\Gamma(1+\alpha)} \lim_{\Delta t \to 0} \sum_{j=0}^{j=N-1} f(t_j)(\Delta t_j)^{\alpha}, \tag{3}$$

where the partitions of the interval $[a, b]$ is (t_j, t_{j+1}), with $\Delta t_j = t_{j+1} - t_j$, $t_0 = a$, $t_N = b$ and $\Delta t = \max\{\Delta t_0, \Delta t_1, \Delta t_j, \ldots\}$, $j = 0, \ldots, N-1$.

The local fractional series of non-differentiable functions are suggested as follows (Yang 2011, 2012):

$$E_{\alpha}(x^{\alpha}) = \sum_{k=0}^{\infty} \frac{x^{\alpha k}}{\Gamma(1+k\alpha)}, \tag{4}$$

$$\sin_{\alpha}(x^{\alpha}) = \sum_{k=0}^{\infty} (-1)^k \frac{x^{(2k+1)\alpha}}{\Gamma[1+(2k+1)\alpha]}, \tag{5}$$

$$\cos_{\alpha}(x^{\alpha}) = \sum_{k=0}^{\infty} (-1)^k \frac{x^{2\alpha k}}{\Gamma(1+2\alpha k)}, \tag{6}$$

$$\sinh_{\alpha}(x^{\alpha}) = \sum_{k=0}^{\infty} \frac{x^{(2k+1)\alpha}}{\Gamma[1+(2k+1)\alpha]}, \tag{7}$$

$$\cosh_\alpha\left(x^\alpha\right) = \sum_{k=0}^{\infty} \frac{x^{2\alpha k}}{\Gamma\left(1+2\alpha k\right)}. \tag{8}$$

The properties of local fractional derivatives and local fractional integrals of non-differentiable functions are as follows (Yang 2011; Yang 2012):

$$\frac{d^\alpha}{dx^\alpha}\frac{x^{n\alpha}}{\Gamma\left(1+n\alpha\right)} = \frac{x^{(n-1)\alpha}}{\Gamma\left(1+(n-1)\alpha\right)}, \tag{9}$$

$$\frac{d^\alpha}{dx^\alpha}E_\alpha\left(x^\alpha\right) = E_\alpha\left(x^\alpha\right), \tag{10}$$

$$\frac{d^\alpha}{dx^\alpha}\sin_\alpha\left(x^\alpha\right) = \cos_\alpha\left(x^\alpha\right), \tag{11}$$

$$\frac{d^\alpha}{dx^\alpha}\cos_\alpha\left(x^\alpha\right) = -\sin_\alpha\left(x^\alpha\right), \tag{12}$$

$$\frac{d^\alpha}{dx^\alpha}\sinh_\alpha\left(x^\alpha\right) = \cosh_\alpha\left(x^\alpha\right), \tag{13}$$

$$\frac{d^\alpha}{dx^\alpha}\cosh_\alpha\left(x^\alpha\right) = -\sinh_\alpha\left(x^\alpha\right) \tag{14}$$

and

$${}_0I_x^{(\alpha)}\frac{x^{n\alpha}}{\Gamma\left(1+n\alpha\right)} = \frac{x^{(n+1)\alpha}}{\Gamma\left(1+(n+1)\alpha\right)}. \tag{15}$$

3 Analysis of the Methods

In this section, three methods, namely the variational iteration method (Su et al., 2013a; Baleanu et al., 2014; Yang et al., 2013g; Li et al., 2014b; Yang et al., 2013i; Yang et al., 2013j), decomposition method (Yang et al., 2013 f; Baleanu et al., 2014; Yan et al., 2014a; Niu et al., 2014; Yan et al., 2014b) and series expansion method (Zhao et al., 2013; Yang et al., 2014h; Yang et al., 2013i) with local fractional derivative, are analyzed.

3.1 The local fractional variational iteration method

Let us consider the following differential equation with local fractional operator:

$$L_\alpha u + R_\alpha u = g\left(t\right), \tag{16}$$

where $L_\alpha = \frac{d^{k\alpha}}{ds^{k\alpha}}$ and $R_\alpha = \frac{d^{m\alpha}}{ds^{m\alpha}}$ $(k > m)$are two linear local fractional operators respectively and $g\left(t\right)$is the source term within the non-differentiable function.

The local fractional variational iteration algorithm is defined by

$$u_{n+1}\left(t\right) = u_n\left(t\right) + {}_0I_t^{(\alpha)}\left\{\eta\left[L_\alpha u_n\left(s\right) + R_\alpha u_n\left(s\right) - g\left(s\right)\right]\right\}, \tag{17}$$

where η is a fractal Lagrange multiplier.

Making use of Eq. (17), the iteration formula can be reformulates as follows:

$$\delta^{\alpha} u_{n+1}(t) = \delta^{\alpha} u_n(t) + \delta^{\alpha}{}_0 I_t^{(\alpha)} \{\eta [L_\alpha u_n(s) + R_\alpha \tilde{u}_n(s) - g(s)]\} \tag{18}$$

which leads to

$$(\eta(t))^{(n\alpha)} = 0 \quad 1 + \eta^{(n-1)\alpha}(t)\,|_{t=s} = 0, \tag{19}$$

where $\delta^{\alpha}\tilde{u}_n$ denotes a restricted local fractional variation, that is, $\delta^{\alpha}\tilde{u}_n = 0$.

Therefore, following (19), we have the fractal Lagrange multiplier, which can be identified as follows:

$$\eta(t) = \frac{(-1)^{(n-1)}(t-s)^{(n-1)\alpha}}{\Gamma(1+(n-1)\alpha)}. \tag{20}$$

From the equations (17) and (20), a local fractional variational iteration algorithm can be rewritten as follows:

$$u_{n+1}(t) = u_n(t) + {}_0 I_t^{(\alpha)} \left\{ \frac{(-1)^{(n-1)}(t-s)^{(n-1)\alpha}}{\Gamma(1+(n-1)\alpha)} [L_\alpha u_n(s) + R_\alpha u_n(s) - g(s)] \right\}. \tag{21}$$

For $n \geq 0$ the successive approximations u_{n+1} of the solution u can be given by using any selective local fractional function u_0. Consequently, we can obtain the solution in the following form:

$$u = \lim_{n\to\infty} u_n. \tag{22}$$

Finally, (22) is the solution of (16) with non-differentiable terms.

3.2 The local fractional Adomian decomposition method

We now consider the local fractional operator equation in the following form

$$L_t^{(2\alpha)} u(x,t) = L_x^{(k\alpha)} u(x,t) + g(x,t), \tag{23}$$

where

$$L_t^{(\alpha)} = \frac{\partial^{\alpha}}{\partial t^{\alpha}}$$

is a 2α-th local fractional differential operator with respect to t and

$$L_x^{(k\alpha)} = \frac{\partial^{k\alpha}}{\partial x^{k\alpha}}$$

is a $k\alpha$-th local fractional differential operator with respect to x.

The initial condition reads as follows:

$$u(x,0) = f(x) \quad (0 \leq x \leq l). \tag{24}$$

We now define the 2α-fold local fractional integral operator in the form:

$$L_t^{(-2\alpha)} m(t) = {}_0I_t^{(\alpha)} \left({}_0I_t^{(\alpha)} m(s) \right). \tag{25}$$

In view of Eq. (23), we construct

$$L_t^{(-2\alpha)} L_t^{(\alpha)} u(x,t) = L_t^{(-2\alpha)} \left\{ L_x^{(k\alpha)} u(x,t) + g(x,t) \right\}. \tag{26}$$

Making use of Eq. (26), we have

$$u(x,t) = r(x) + L_t^{(-2\alpha)} L_x^{(k\alpha)} u(x,t) + L_t^{(-2\alpha)} g(x,t), \tag{27}$$

where the non-differentiable term $r(x)$ is obtained from the initial condition.

Making use of Eq. (27), we give the recurrence relationsh in the following form:

$$u_{n+1}(x,t) = L_t^{(-2\alpha)} L_x^{(k\alpha)} u_n(x,t), \tag{28}$$

subject to the initial condition given by

$$u_0(x,t) = r(x) + L_t^{(-2\alpha)} g(x,t). \tag{29}$$

The approximation form of the solution is given by

$$u(x,t) = \lim_{n\to\infty} \varphi_n(x,t). \tag{30}$$

Finally, (30) is the solution of (23) with non-differentiable terms.

3.3 The local fractional series expansion method

We now consider the local fractional differential operator in the form:

$$u_{2s}^{(2\alpha)} = -\mathrm{I}_\alpha u, \tag{31}$$

where

$$\mathrm{I}_\alpha = \frac{\partial^{2\alpha}}{\partial \tau^{2\alpha}}$$

and u is a local fractional continuous function.

Making use of (31), the non-differentiable separated functions with respect to (s,τ) can be formulated as follows:

$$u(s,\tau) = \sum_{i=0}^{\infty} \theta_i(s) \vartheta_i(\tau), \tag{32}$$

where $\theta_i(s)$ and $\vartheta_i(\tau)$ are the local fractional continuous functions.

In view of (32), we consider

$$\theta_i(s) = \frac{s^{i\alpha}}{\Gamma(1+i\alpha)}, \tag{33}$$

which reduces to the form given by (33)

$$u(s,\tau) = \sum_{i=0}^{\infty} \frac{s^{i\alpha}}{\Gamma(1+i\alpha)} \vartheta_i(\tau). \tag{34}$$

Applying the formula (34), we get

$$u_{2s}^{(2\alpha)} = \sum_{i=0}^{\infty} \frac{s^{i\alpha}}{\Gamma(1+i\alpha)} \vartheta_{i+2}(\tau) \tag{35}$$

and

$$\mathrm{I}_\alpha u = \mathrm{I}_\alpha \left[\sum_{i=0}^{\infty} \frac{s^{i\alpha}}{\Gamma(1+i\alpha)} \vartheta_i(\tau) \right] = \sum_{i=0}^{\infty} \frac{s^{i\alpha}}{\Gamma(1+i\alpha)} (\mathrm{I}_\alpha \vartheta_i)(\tau), \tag{36}$$

which, by using (32), leads us to

$$\sum_{i=0}^{\infty} \frac{s^{i\alpha}}{\Gamma(1+i\alpha)} \vartheta_{i+2}(\tau) = - \sum_{i=0}^{\infty} \frac{s^{i\alpha}}{\Gamma(1+i\alpha)} (\mathrm{I}_\alpha \vartheta_i)(\tau). \tag{37}$$

By means of (37), we obtain the following relation:

$$\vartheta_{i+2}(\tau) = -(\mathrm{I}_\alpha \vartheta_i)(\tau). \tag{38}$$

Therefore, by using (38), we get

$$u(s,\tau) = \sum_{i=0}^{\infty} u_i(s,\tau) = \sum_{i=0}^{\infty} \frac{s^{i\alpha}}{\Gamma(1+i\alpha)} \vartheta_i(\tau). \tag{39}$$

Finally, the expression in (39) is the series solution of (31) with non-differentiable terms.

4 Applications to Solve Partial Differential Equations Involving Local Fractional Derivatives

In this section, some illustrative examples are given.

4.1 Solving the linear Boussinesq equation occurring in fractal long water waves with local fractional variational iteration method

The linear Boussinesq equation occurring in fractal long water waves with local fractional derivative is formulated as follows:

$$\frac{\partial^{2\alpha}u(x,t)}{\partial t^{2\alpha}}-\frac{\partial^{2\alpha}u(x,t)}{\partial x^{2\alpha}}-\frac{\partial^{4\alpha}u(x,t)}{\partial x^{2\alpha}\partial t^{2\alpha}}=0, \tag{40}$$

subject to the initial-boundary condition given by

$$u(x,0)=\frac{x^{2\alpha}}{\Gamma(1+2\alpha)}, 0\le x\le 1, \tag{41}$$

$$\frac{\partial^{\alpha}u(x,0)}{\partial t^{\alpha}}=0, 0\le x\le 1, \tag{42}$$

$$u(l,t)=u(0,t)=0 \qquad (t>0) \tag{43}$$

and

$$\frac{\partial^{\alpha}u(l,t)}{\partial x^{\alpha}}=\frac{\partial^{\alpha}u(0,t)}{\partial x^{\alpha}}=0 \qquad (t>0). \tag{44}$$

Following (21), we arrive at the iteration relation given by

$$u_{n+1}(x,t)=u_n(x,t)+{}_0I_t^{(\alpha)}\left\{\frac{(s-t)^{\alpha}}{\Gamma(1+\alpha)}\left[\frac{\partial^{2\alpha}u_n(x,t)}{\partial t^{2\alpha}}-\frac{\partial^{2\alpha}u_n(x,t)}{\partial x^{2\alpha}}-\frac{\partial^{4\alpha}u_n(x,t)}{\partial x^{2\alpha}\partial t^{2\alpha}}\right]\right\}, \tag{45}$$

where

$$u_0(x,t)=\frac{x^{2\alpha}}{\Gamma(1+2\alpha)}. \tag{46}$$

Utilizing (45) and (46), we get the following approximations:

$$\begin{aligned}&u_1(x,t)\\&=u_0(x,t)+{}_0I_t^{(\alpha)}\left\{\frac{(s-t)^{\alpha}}{\Gamma(1+\alpha)}\left[\frac{\partial^{2\alpha}u_0(x,t)}{\partial t^{2\alpha}}-\frac{\partial^{2\alpha}u_0(x,t)}{\partial x^{2\alpha}}-\frac{\partial^{4\alpha}u_0(x,t)}{\partial x^{2\alpha}\partial t^{2\alpha}}\right]\right\}\\&=\frac{x^{2\alpha}}{\Gamma(1+2\alpha)}-{}_0I_t^{(\alpha)}\left\{\frac{(s-t)^{\alpha}}{\Gamma(1+\alpha)}\right\}\\&=\frac{x^{2\alpha}}{\Gamma(1+2\alpha)}+\frac{t^{2\alpha}}{\Gamma(1+2\alpha)},\end{aligned} \tag{47}$$

$$\begin{aligned}&u_2(x,t)\\&=u_1(x,t)+{}_0I_t^{(\alpha)}\left\{\frac{(s-t)^{\alpha}}{\Gamma(1+\alpha)}\left[\frac{\partial^{2\alpha}u_1(x,t)}{\partial t^{2\alpha}}-\frac{\partial^{2\alpha}u_1(x,t)}{\partial x^{2\alpha}}-\frac{\partial^{4\alpha}u_1(x,t)}{\partial x^{2\alpha}\partial t^{2\alpha}}\right]\right\}\\&=\frac{x^{2\alpha}}{\Gamma(1+2\alpha)}+\frac{t^{2\alpha}}{\Gamma(1+2\alpha)},\end{aligned} \tag{48}$$

$$\begin{aligned}&u_3(x,t)\\&=u_2(x,t)-{}_0I_t^{(\alpha)}\left\{\frac{(s-t)^{\alpha}}{\Gamma(1+\alpha)}\left[\frac{\partial^{2\alpha}u_2(x,t)}{\partial t^{2\alpha}}-\frac{\partial^{2\alpha}u_2(x,t)}{\partial x^{2\alpha}}-\frac{\partial^{4\alpha}u_2(x,t)}{\partial x^{2\alpha}\partial t^{2\alpha}}\right]\right\}\\&=\frac{x^{2\alpha}}{\Gamma(1+2\alpha)}+\frac{t^{2\alpha}}{\Gamma(1+2\alpha)}\end{aligned} \tag{49}$$

and

$$\cdots \quad \cdots \quad \cdots$$

$$u_n(x,t) = \frac{x^{2\alpha}}{\Gamma(1+2\alpha)} + \frac{t^{2\alpha}}{\Gamma(1+2\alpha)}. \tag{50}$$

Hence, we obtain the non-differentiable solution of (40) as follows:

$$u(x,t) = \lim_{n\to\infty} u_n(x,t) = \frac{x^{2\alpha}}{\Gamma(1+2\alpha)} + \frac{t^{2\alpha}}{\Gamma(1+2\alpha)}. \tag{51}$$

The graph of the expression in (51) is given in Figure 1 when $\alpha = \frac{\ln 2}{\ln 3}$. The classical version of (40) with $\alpha = 1$ was discussed earlier (see Debnath, 2005).

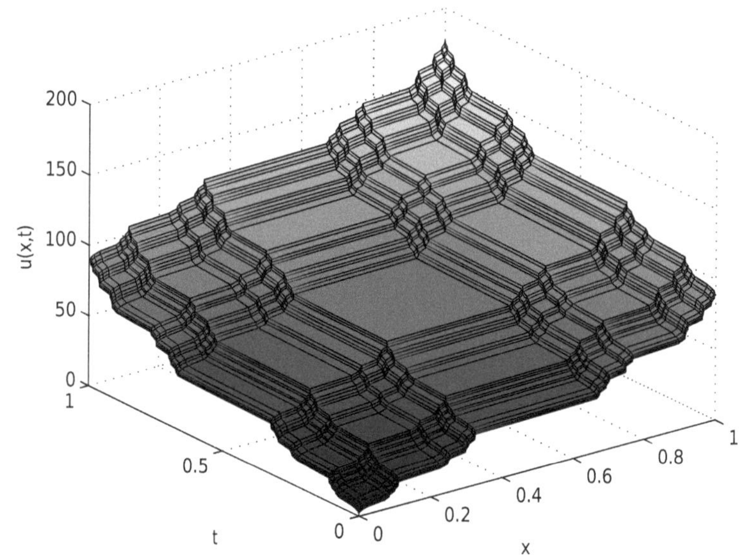

Fig. 1. The graph of (51) when $\alpha = \frac{\ln 2}{\ln 3}$.

4.2 Solving the equation of the fractal motion of a long string by the local fractional Adomian decomposition method

The partial differential equation of the fractal motion of a long string is written as follows:

$$\frac{\partial^{2\alpha}u(x,t)}{\partial t^{2\alpha}} = \frac{\partial^{2\alpha}u(x,t)}{\partial x^{2\alpha}} - g, \tag{52}$$

subject to the initial-boundary condition given by

$$u(x,0) = E_\alpha\left(x^\alpha\right), 0 \le x \le 1, \tag{53}$$

$$\frac{\partial^\alpha u(x,0)}{\partial t^\alpha} = 0, 0 \leq x \leq 1, \tag{54}$$

$$u(l,t) = u(0,t) = 0 \qquad (t > 0) \tag{55}$$

and

$$\frac{\partial^\alpha u(l,t)}{\partial x^\alpha} = \frac{\partial^\alpha u(0,t)}{\partial x^\alpha} = 0 \qquad (t > 0), \tag{56}$$

where g is the gravitational acceleration and $u(x, t)$ is the displacement of the fractal motion.

Using (28) and (29), we consider an iterative formula in the form:

$$u_{n+1}(x,t) = L_t^{(-2\alpha)}\left(\frac{\partial^{2\alpha} u_n(x,t)}{\partial x^{2\alpha}}\right), \tag{57}$$

subject to the initial condition given by

$$u_0(x,t) = E_\alpha(x^\alpha) + L_t^{(-2\alpha)} g = E_\alpha(x^\alpha) + \frac{g t^{2\alpha}}{\Gamma(1+2\alpha)}. \tag{58}$$

In view of (57) and (58), we have the following approximations:

$$u_1(x,t) = L_t^{(-2\alpha)}\left(\frac{\partial^{2\alpha} u_0(x,t)}{\partial x^{2\alpha}}\right) = \frac{t^{2\alpha}}{\Gamma(1+2\alpha)} E_\alpha(x^\alpha), \tag{59}$$

$$u_2(x,t) = L_t^{(-2\alpha)}\left(\frac{\partial^{2\alpha} u_1(x,t)}{\partial x^{2\alpha}}\right) = \frac{t^{4\alpha}}{\Gamma(1+4\alpha)} E_\alpha(x^\alpha), \tag{60}$$

$$u_3(x,t) = L_t^{(-2\alpha)}\left(\frac{\partial^{2\alpha} u_2(x,t)}{\partial x^{2\alpha}}\right) = \frac{t^{6\alpha}}{\Gamma(1+6\alpha)} E_\alpha(x^\alpha), \tag{61}$$

$$u_4(x,t) = L_t^{(-2\alpha)}\left(\frac{\partial^{2\alpha} u_3(x,t)}{\partial x^{2\alpha}}\right) = \frac{t^{8\alpha}}{\Gamma(1+8\alpha)} E_\alpha(x^\alpha) \tag{62}$$

and

$$u_n(x,t) = L_t^{(-2\alpha)}\left(\frac{\partial^{2\alpha} u_{n-1}(x,t)}{\partial x^{2\alpha}}\right) = \frac{t^{2n\alpha}}{\Gamma(1+2n\alpha)} E_\alpha(x^\alpha). \tag{63}$$

Finally, the solution of (52) with non-differentiable terms is given by

$$u(x,t) = \lim_{n\to\infty} \sum_{n=0}^{n} \varphi_n(x,t) = E_\alpha(x^\alpha) \sum_{n=0}^{n} \frac{t^{2n\alpha}}{\Gamma(1+2n\alpha)} + \frac{g t^{2\alpha}}{\Gamma(1+2\alpha)}. \tag{64}$$

The graph of the expression in (64) is illustrated in Figure 2 when $\alpha = \frac{\ln 2}{\ln 3}$ and $g = 10$.

The classical version of (52) with $\alpha = 1$ was presented earlier (see Myint-U and Debnath, 2007).

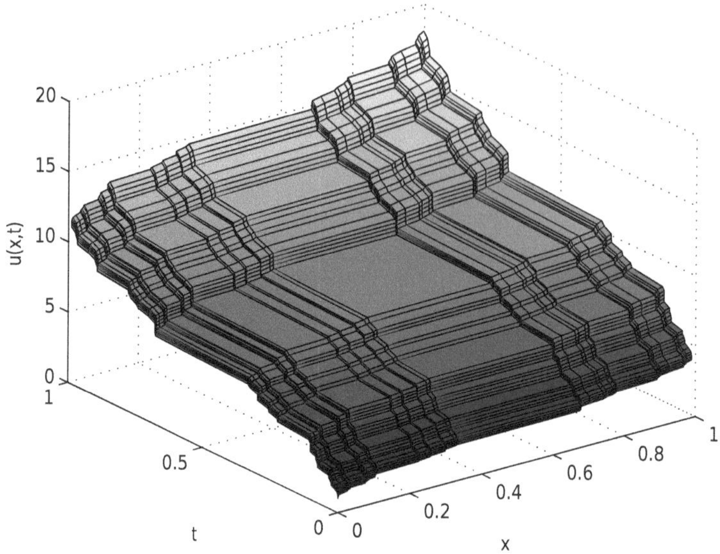

Fig. 2. The graph of solution of (64) when $\alpha = \frac{\ln 2}{\ln 3}$, $g = 10$ and $n = 50$.

4.3 Solving partial differential equations arising from the fractal transverse vibration of a beam with local fractional series expansion method

The partial differential equation arising from the fractal transverse vibration of a beam is recalled here as follows:

$$\frac{\partial^{2\alpha} u(x,t)}{\partial t^{2\alpha}} + \frac{\partial^{4\alpha} u(x,t)}{\partial x^{4\alpha}} = 0, \tag{65}$$

subject to the initial-boundary conditions given by

$$u(x,0) = \sin_\alpha\left(x^\alpha\right), 0 \le x \le 1, \tag{66}$$

$$\frac{\partial^{\alpha} u(x,0)}{\partial t^{\alpha}} = \cos_\alpha\left(x^\alpha\right), 0 \le x \le 1, \tag{67}$$

$$u(l,t) = u(0,t) = 0 \qquad (t > 0) \tag{68}$$

and

$$\frac{\partial^{2\alpha} u(l,t)}{\partial x^{2\alpha}} = \frac{\partial^{2\alpha} u(0,t)}{\partial x^{2\alpha}} = 0 \qquad (t > 0), \tag{69}$$

where $u(x,t)$ is the displacement of the fractal motion.

From (66), (67) and (38), we obtain the following iterative formula:

$$\vartheta_{i+2}(x) = -\frac{\partial^{4\alpha}\vartheta_i(x)}{\partial x^{4\alpha}}, \tag{70}$$

where

$$\vartheta_0(x) = \sin_\alpha(x^\alpha) \tag{71}$$

and

$$\vartheta_1(x) = \cos_\alpha(x^\alpha). \tag{72}$$

Following the equations (70) to (72), we get the approximations given by

$$\vartheta_2(x) = -\frac{\partial^{4\alpha}\vartheta_0(x)}{\partial x^{4\alpha}} = -\frac{\partial^{4\alpha}}{\partial x^{4\alpha}}\sin_\alpha(x^\alpha) = -\sin_\alpha(x^\alpha), \tag{73}$$

$$\vartheta_3(x) = -\frac{\partial^{4\alpha}\vartheta_1(x)}{\partial x^{4\alpha}} = -\frac{\partial^{4\alpha}}{\partial x^{4\alpha}}(-\sin x) = -\cos_\alpha(x^\alpha), \tag{74}$$

$$\vartheta_4(x) = -\frac{\partial^{4\alpha}\vartheta_2(x)}{\partial x^{4\alpha}} = -\frac{\partial^{4\alpha}(-\sin_\alpha(x^\alpha))}{\partial x^{4\alpha}} = \sin_\alpha(x^\alpha), \tag{75}$$

$$\vartheta_5(x) = -\frac{\partial^{4\alpha}\vartheta_3(x)}{\partial x^{4\alpha}} = -\frac{\partial^{4\alpha}(-\cos_\alpha(x^\alpha))}{\partial x^{4\alpha}} = \cos_\alpha(x^\alpha), \tag{76}$$

$$\vartheta_6(x) = -\frac{\partial^{4\alpha}\vartheta_4(x)}{\partial x^{4\alpha}} = -\frac{\partial^{4\alpha}(\sin_\alpha(x^\alpha))}{\partial x^{4\alpha}} = -\sin_\alpha(x^\alpha), \tag{77}$$

$$\vartheta_7(x) = -\frac{\partial^{4\alpha}\vartheta_5(x)}{\partial x^{4\alpha}} = -\frac{\partial^{4\alpha}(\cos_\alpha(x^\alpha))}{\partial x^{4\alpha}} = -\cos_\alpha(x^\alpha), \tag{78}$$

$$\vartheta_8(x) = -\frac{\partial^{4\alpha}\vartheta_7(x)}{\partial x^{4\alpha}} = -\frac{\partial^{4\alpha}(-\sin_\alpha(x^\alpha))}{\partial x^{4\alpha}} = \sin_\alpha(x^\alpha), \tag{79}$$

$$\vartheta_9(x) = -\frac{\partial^{4\alpha}\vartheta_7(x)}{\partial x^{4\alpha}} = -\frac{\partial^{4\alpha}(-\cos_\alpha(x^\alpha))}{\partial x^{4\alpha}} = \cos_\alpha(x^\alpha), \tag{80}$$

$$\ldots \quad \ldots \quad \ldots,$$

and so on.

Hence, the solution of (65) can be written as follows:

$$\begin{aligned} u(x,t) &= \sum_{i=0}^{\infty} u_i(x,t) = \sum_{i=0}^{\infty} \frac{t^{i\alpha}}{\Gamma(1+i\alpha)}\vartheta_i(x) \\ &= \sin_\alpha(x^\alpha)\sum_{i=0}^{\infty}\frac{(-1)^i t^{2i\alpha}}{\Gamma(1+2i\alpha)} + \cos_\alpha(x^\alpha)\sum_{i=0}^{\infty}\frac{(-1)^i t^{(2i+1)\alpha}}{\Gamma(1+(2i+1)\alpha)} \\ &= \sin_\alpha(x^\alpha)\cos_\alpha(t^\alpha) + \cos_\alpha(x^\alpha)\sin_\alpha(t^\alpha). \end{aligned} \tag{81}$$

The graph of (80) is shown in Figure 3 when $\alpha = \frac{\ln 2}{\ln 3}$. The classical version of (52) with $\alpha = 1$ was presented earlier (see Myint-U and Debnath, 2007).

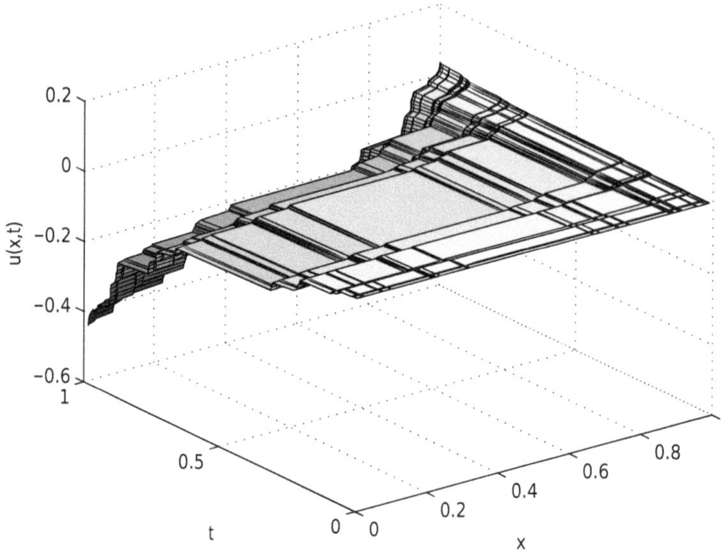

Fig. 3. The graph of the solution of (65) when $\alpha = \frac{\ln 2}{\ln 3}$.

5 Conclusions

In this chapter, we have successfully provided new applications of the local fractional variational iteration method, the local fractional decomposition method and the local fractional series expansion method for solving several partial differential equations occurring in mathematical physics and engineering. The graphs of some of the solutions with non-differentiable terms defined on Cantor sets are also presented.

Bibliography

Yang, X.-J. (2011). Local Fractional Functional Analysis and Its Applications, Asian Academic Publishers, Hong Kong, People's Republic of China, 2011.

Kilbas, A. A., Srivastava, H. M. and Trujillo, J. J. (2006). Theory and Applications of Fractional Differential Equations, North-Holland Mathematical Studies, Vol. 204, Elsevier (North-Holland) Science Publishers, Amsterdam, London and New York, 2006.

Yang, X.-J., Baleanu, D. and Srivastava, H. M. (2015). Local Fractional Integral Transforms and Their Applications, Elsevier (North-Holland) Science Publishers, Amsterdam, London and New York, 2015.

Ben Adda, F. and Cresson, J. (2001). About non-differentiable functions, Journal of Mathematical Analysis and Applications, 263 (2001), 721–737.

Babakhani, A. and Daftardar-Gejji, V. (2002). On calculus of local fractional derivatives, Journal of Mathematical Analysis and Applications, 270 (2002), 66–79.
Chen, Y., Yan, Y. and Zhang, K. (2010). On the local fractional derivative, Journal of Mathematical Analysis and Applications, 362 (2010), 17–33.
Anastassiou, G. A. and Duman, O. *Advances in Applied Mathematics and Approximation Theory*, Springer, New York, 2013.
Kolwankar, K. M. and Gangal, A. D. (1998). Local fractional Fokker-Planck equation, Physical Review Letters, 80 (1998), 214.
Yang, X.-J., Srivastava, H. M., He, J.-H. and Baleanu, D. (2013a). Cantor-type cylindrical-coordinate method for differential equations with local fractional derivatives, Physics Letters A, 377 (2013), 1696–1700.
Zhao, Y., Baleanu, D., Cattani, C., Cheng, D.-F. and Yang, X.-J. (2013). Maxwell's equations on Cantor sets: A local fractional approach, Advances in High Energy Physics, 2013 (2013), Article ID 686371, 1–6.
Hao, Y.-J., Srivastava, H. M., Jafari, H. and Yang, X.-J. (2013). Helmholtz and diffusion equations associated with local fractional derivative operators involving the Cantorian and Cantor-type cylindrical coordinates, Advances in Mathematical Physics, 2013 (2013), Article ID 754248, 1–5.
Yang, X.-J., Baleanu, D. and Machado, J. A. T. (2013b). Mathematical aspects of the Heisenberg uncertainty principle within local fractional Fourier analysis, Boundary Value Problems, 2013 (2013), 1–16.
Yang, X.-J., Baleanu, D. and J.-H. He, (2013c). Transport equations in fractal porous media within fractional complex transform method, *Proceedings of the Romanian Academy (Series A)*, 14 (2013), 287–292.
Yang, X.-J., Hristov, J., Srivastava, H. M. and Ahmad, B. (2014a). Modelling fractal waves on shallow water surfaces via local fractional Korteweg-de Vries equation, *Abstract and Applied Analysis*, 2014 (2014), Article ID 278672, 1–10.
Machado, J. A. T., Baleanu, D. and Luo, A. C.-J. Luo (Editors) (2014). Discontinuity and Complexity in Nonlinear Physical Systems, Springer, New York, 2014.
Liao, M., Yang, X.-J. and Yan, Q. (2013). A new viewpoint to Fourier analysis in fractal space, In: Advances in Applied Mathematics and Approximation Theory, pp. 397–409, Springer, New York, 2013.
Yang, X.-J., Baleanu, D., Srivastava, H. M. and Machado, J. A. T. (2013d). On local fractional continuous wavelet transform, Abstract and Applied Analysis, 2013 (2013), Article ID 725416, 1–5.
Baleanu, D. and Yang, X.-J. (2013). Local fractional Fourier series with applications to representations of fractal signals, In: Proceedings of the ASME 2013 International Design Engineering Technical Conferences and Computers and Information in Engineering Conference, Article ID DETC2013-12306, pp. 1–6, American Society of Mechanical Engineers, New York, 2013.
Liu, K., Hu, R.-J., Cattani, C., Xie, G.-N., Yang, X.-J. and Zhao, Y. (2014). Local fractional *Z*-transforms with applications to signals on Cantor sets, Abstract and Applied Analysis, 2014 (2014), Article ID 638648, 1–6.
Chen, Z.-Y., Cattani, C. and Zhong, W.-P. (2014). Signal processing for non-differentiable data defined on Cantor sets: A local fractional Fourier series approach, Advances in Mathematical Physics, 2014 (2014), Article ID 561434, 1–7.
Carpinteri, A., Chiaia, B. and Cornetti, P. (2001). Static-kinematic duality and the principle of virtual work in the mechanics of fractal media, Computer Methods in Applied Mechanics and Engineering, 191 (2001), 3–19.
Carpinteri, A., Chiaia, B. and Cornetti, P. (2004). A fractal theory for the mechanics of elastic materials, Materials Science and Engineering: A, 365 (2004), 235–240.

Carpinteri, A., Cornetti, P. and Sapora, A. (2009). Static-kinematic fractional operators for fractal and non-local solids, ZAMM (Zeitschrift für Angewandte Mathematik und Mechanik), 89 (2009), 207–217.

Yang, X.-J. (2012). Advanced Local Fractional Calculus and Its Applications, World Science Publisher, New York, 2012.

Yang, X.-J., Baleanu, D. and Machado, J. A. T. (2013e). Systems of Navier-Stokes equations on Cantor sets, Mathematical Problems in Engineering, 2013 (2013), Article ID 769724, 1–8.

Su, W.-H., Baleanu, D., Yang, X.-J. and Jafari, H. (2013a). Damped wave equation and dissipative wave equation in fractal strings within the local fractional variational iteration method, Fixed Point Theory and Applications, 2013 (2013), 1–11.

Su, W.-H., Yang, X.-J., Jafari, H. and Baleanu, D. (2013b). Fractional complex transform method for wave equations on Cantor sets within local fractional differential operator, Advances in Difference Equations, 2013 (2013), 1–8.

Baleanu, D., Machado, J. A. T., Cattani, C., Baleanu, M. C. and Yang, X.-J. (2014). Local fractional variational iteration and decomposition methods for wave equation on Cantor sets within local fractional operators, Abstract and Applied Analysis, 2014 (2014), Article ID 535048, 1–6.

Wang, X.-J., Zhao, Y., Cattani, C. and Yang, X.-J. (2014). Local fractional variational iteration method for inhomogeneous Helmholtz equation within local fractional derivative operator, Mathematical Problems in Engineering, 2014 (2014), Article ID 913202, 1–7.

Yang, X.-J., Baleanu, D. and Zhong, W.-P. (2013f). Approximate solutions for diffusion equations on Cantor space-time, Proceedings of the Romanian Academy (Series A), 14 (2013), 127–133.

Cao, Y., Ma, W.-G. and Ma, L.-C. (2014). Local fractional functional method for solving diffusion equations on Cantor sets, Abstract and Applied Analysis, 2014 (2014), Article ID 803693, 1–6.

Liu, C.-F., Kong, S.-S. and Yuan, S.-J. (2013). Reconstructive schemes for variational iteration method within Yang-Laplace transform with application to fractal heat conduction problem, Thermal Science, 17 (2013), 715–721.

Li, Y.-Y., Zhao, Y., Xie, G.-N., Baleanu, D., Yang, X.-J. and Zhao, K. (2014a). Local fractional Poisson and Laplace equations with applications to electrostatics in fractal domain, Advances in Mathematical Physics, 2014 (2014), Article ID 590574, 1–5.

Yang, Y.-J., Baleanu, D. and Yang, X.-J. (2013g). A local fractional variational iteration method for Laplace equation within local fractional operators, Abstract and Applied Analysis, 2013 (2013), Article ID 202650, 1–6.

Yan, S.-P., Jafari, H. and Jassim, H. K. (2014a). Local fractional Adomian decomposition and function decomposition methods for Laplace equation within local fractional operators, Advances in Mathematical Physics, 2014 (2014), Article ID 161580, 1–7.

Zhao, Y., Cheng, D.-F. and Yang, X.-J. (2013). Approximation solutions for local fractional Schrödinger equation in the one-dimensional Cantorian system, Advances in Mathematical Physics, 2013 (2013), Article ID 291386, 1–5.

Li, M., Hui, X.-F., Cattani, C., Yang, X.-J. and Zhao, Y. (2014b). Approximate solutions for local fractional linear transport equations arising in fractal porous media, Advances in Mathematical Physics, 2014 (2014), Article ID 487840, 1–8.

Yang, A.-M., Zhang, Y.-Z., Cattani, C., Xie, G.-N., Rashidi, M. M., Zhou, Y.-J. and Yang, X.-J. (2014h). Application of local fractional series expansion method to solve Klein-Gordon equations on Cantor sets, Abstract and Applied Analysis, 2014 (2014), Article ID 372741, 1–6.

Yan, S.-H., Chen, X.-H., Xie, G.-N., Cattani, C. and Yang, X.-J. (2014b). Solving Fokker-Planck equations on Cantor sets using local fractional decomposition method, Abstract and Applied Analysis, 2014 (2014), Article ID 396469, 1–6.

Niu, X.-F., Zhang, C.-L., Li, Z.-B. and Zhao, Y. (2014). Local fractional derivative boundary value problems for Tricomi equation arising in fractal transonic flow, Abstract and Applied Analysis, 2014 (2014), Article ID 872318, 1–5.

Yang, A.-M., Chen, Z.-S., Srivastava, H. M. and Yang, X. J. (2013i). Application of the local fractional series expansion method and the variational iteration method to the Helmholtz equation involving local fractional derivative operators, Abstract and Applied Analysis, 2013 (2013), Article ID 259125, 1–6.

Srivastava, H. M., A. K. Golmankhaneh, A. K., Baleanu, D. and Yang, X.-J. (2014). Local fractional Sumudu transform with applications to IVPs on Cantor sets, Abstract and Applied Analysis, 2014 (2014), Article ID 620529, 1–7.

Yang, A.-M., Li, J., Srivastava, H. M., Xie, G.-N. and Yang, X.-J. (2014a). The local fractional Laplace variational iteration method for solving linear partial differential equations with local fractional derivative, Discrete Dynamics in Nature and Society, 2014 (2014), Article ID 365981, 1–8.

Yang, A.-M., Zhang, C., Jafari, H., Cattani, C. and Jiao, Y. (2014b). Picard successive approximation method for solving differential equations arising in fractal heat transfer with local fractional derivative, Abstract and Applied Analysis, 2014 (2014), Article ID 395710, 1–5.

Yang, X.-J. and Baleanu, D. (2013j). Fractal heat conduction problem solved by local fractional variation iteration method, Thermal Science, 17 (2013), 625–628.

Hu, M.-S., Agarwal, R. P. and Yang, X.-J. (2012). Local fractional Fourier series with application to wave equation in fractal vibrating string, Abstract and Applied Analysis, 2012 (2012), Article ID 567401, 1–15.

Yang, Y.-J., Baleanu, D. and Yang, X.-J. (2013k). Analysis of fractal wave equations by local fractional Fourier series method, Advances in Mathematical Physics, 2013 (2013), Article ID 632309, 1–6.

Yang, A.-M., Cattani, C., Zhang, C., Xie, G.-N. and Yang, X.-J. (2014c). Local fractional Fourier series solutions for nonhomogeneous heat equations arising in fractal heat flow with local fractional derivative, Advances in Mechanical Engineering, 2014 (2014), Article ID 514639, 1–5.

Zhao, C.-G., Yang, A.-M., Jafari, H. and Haghbin, A. (2014). The Yang-Laplace transform for solving the IVPs with local fractional derivative, Abstract and Applied Analysis, 2014 (2014), Article ID 386459, 1–5.

He, J.-H. (2012). Asymptotic methods for solitary solutions and compactons, Abstract and Applied Analysis, 2012 (2012), Article ID 916793, 1–130.

Debnath, L. (2005). Nonlinear Partial Differential Equations for Scientists and Engineers, Springer, New York, 2005.

Myint-U, T. and Debnath, L. (2007). Linear Partial Differential Equations for Scientists and Engineers, Springer, New York, 2007.

Xiao-Jun Yang, Dumitru Baleanu, and J. A. Tenreiro Machado

Numerical Solutions for ODEs with Local Fractional Derivative

Abstract: In this chapter an efficient numerical algorithm for solving ODEs using the extended differential transform method via the generalized local fractional Taylor theorem is presented. Four examples are studied in order to illustrate the proposed technique.

1 Introduction

The differential transform method (DTM) was successfully utilized by Zhou (Zhou, 1986) to obtain the numerical solutions for differential equations arising in electrical circuits. This approach is extended to solve the partial differential equations of free vibration (Chen, 2009), Fokker–Planck (Hesam et al., 2012), nonlinear Klein–Gordon (Kanth and Aruna, 2009), Burgers and coupled Burgers (Abazari and Borhanifar, 2010), the KdV and mKdV (Kangalgil and Ayaz, 2009), and the Telegraph (Biazar and Eslami, 2010) equations.

The fractional DTM was structured (Arikoglu and Ozkol, 2007) in order to solve fractional differential equations. Elsaid considered the fractional DTM coupled within the Adomian polynomials (Elsaid, 2012). Nazari and Shahmorad used the fractional DTM to solve fractional-order integro-differential equations (Nazari and Shahmorad, 2010). There are other analytical methods for the fractional PDEs, such as the VIM (Momani et al., 2008), spectral Legendre–Gauss–Lobatto collocation method (Bhrawy and Baleanu, 2013), HDM (Atangana and Belhaouari, 2013), HBIM (Hristov, 2012) and other references therein. However, other classes of local fractional differential equations (Yang, 2011, 2012) arising in mathematical physics, such as the Fokker-Planck (Kolwankar and Gangal, 1998), diffusion (Carpinteri and Sapora, 2010), wave and heat (Yang et al., 2013) equations with the local fractional operator, were proposed. Since these equations are defined on Cantor sets, the traditional analytical and numerical approximation methods are not sufficient to deal with them. There are some analyti-

Xiao-Jun Yang: Department of Mathematics and Mechanics, China University of Mining and Technology, Xuzhou, Jiangsu 221008, China E-mail: dyangxiaojun@163.com
Dumitru Baleanu: Department of Mathematics and Computer Sciences, Faculty of Arts and Sciences, Cankaya University, TR-06530 Ankara, Turkey
and Institute of Space Sciences, Magurele, RO-077125 Bucharest, Romania E-mail: dumitru@cankaya.edu.tr
J. A. Tenreiro Machado: Department of Electrical Engineering, Institute of Engineering, Polytechnic of Porto Rua Dr. António Bernardino de Almeida, 431, 4249-015 Porto, Portugal E-mail: jtm@isep.ipp.pt

cal methods, such as the local fractional FTs (Zhong et al., 2012), local fractional VIM (Yang et al., 2014; Baleanu et al., 2014), local fractional LVITM (Li et al., 2014), local fractional DM (Yan et al., 2014), the local fractional FDM (Wang et al., 2014), the local fractional LTs (Zhang et al., 2014; He, 2014) and the local fractional VITM (Yang and Hua, 2014) for local fractional PDEs.

The local fractional DTM (Yang and Hua, 2014) is not sufficient because it should have been deduced from the generalized local fractional Taylor theorems (Yang, 2011, 2012). Also, an illustrative example of applying the local fractional DTM to solve local fractional ordinary differential equation (local fractional ODE) was not given. The objective of this work is to study the extended differential transform method (the local fractional DTM) (Yang and Hua, 2014) and to demonstrate the proofs of its theorems by means of the generalized local fractional Taylor theorem and illustrative examples. The chapter is organized as follows. In Section 2, the generalized local fractional Taylor theorems are presented. In Section 3, the local fractional DTM is studied. In Section 4, some illustrative examples are given. Finally, in Section 5 the conclusions are outlined.

2 The Generalized Local Fractional Taylor Theorems

In this section, the generalized local fractional Taylor theorems (Yang, 2011, 2012) are formulated to deduce the extended differential transform method.

Let $ffj_{\alpha}(a, b)$ denote the set of local fractional continuous functions, see (Yang et al., 2013; Zhong et al., 2012; Yang et al., 2014; Baleanu et al., 2014; Li et al., 2014; Yan et al, 2014; Wang et al., 2014; Zhang et al., 2014; He, 2014; Yang and Hua, 2014). Let $g(\tau) \in ffj_{\alpha}(a, b)$. The local fractional derivative of the non-differentiable function $g(\tau)$ of α ($\alpha \in R$) order is given as (Yang et al., 2013; Zhong et al., 2012; Yang et al., 2014; Baleanu et al., 2014; Li et al., 2014; Yan et al, 2014; Wang et al., 2014; Zhang et al., 2014; He, 2014; Yang and Hua, 2014):

$$D^{\alpha} g(\tau_0) = \frac{d^{\alpha} g(\tau_0)}{d\tau^{\alpha}} = \frac{\Delta^{\alpha}(g(\tau) - g(\tau_0))}{(\tau - \tau_0)^{\alpha}}, \tag{1}$$

where

$$\Delta^{\alpha}(g(\tau) - g(\tau_0)) \cong \Gamma(1+\alpha)[g(\tau) - g(\tau_0)]. \tag{2}$$

The local fractional derivative of $g(\tau)$ of $k\alpha$ order is (Yang, 2011, 2012)

$$g^{(k\alpha)}(\tau) = \overbrace{D_{\tau}^{(\alpha)} \ldots D_{\tau}^{(\alpha)}}^{k\text{-}times} g(\tau), \tag{3}$$

and the local fractional partial derivative of $g(s, \tau)$ of $k\alpha$ order is (Yang, 2011, 2012)

$$\frac{\partial^{k\alpha}}{\partial s^{k\alpha}} g(s, \tau) = \overbrace{\frac{\partial^{\alpha}}{\partial s^{\alpha}} \cdots \frac{\partial^{\alpha}}{\partial s^{\alpha}}}^{k\text{-}times} g(s, \tau). \tag{4}$$

Lemma 1 (Generalized local fractional Taylor theorems)

If $g^{((k+1)\alpha)}(\tau) \in \text{ffj}_{\alpha}(a, b)$, for $k = 0, 1, \cdots, n$ and $0 < \alpha \leq 1$, then

$$g(\tau) = \sum_{k=0}^{n} g^{(k\alpha)}(\tau_0) \frac{(\tau - \tau_0)^{k\alpha}}{\Gamma(1 + k\alpha)} + g^{((k+1)\alpha)}(\xi) \frac{(\tau - \tau_0)^{(k+1)\alpha}}{\Gamma(1 + (k+1)\alpha)}, \tag{5}$$

where $a < \tau_0 < \xi < \tau < b$, $\forall \tau \in (a, b)$.

Proof. See (Yang, 2011, 2012).

Lemma 2

If $g^{((k+1)\alpha)}(\tau) \in \text{ffj}_{\alpha}(a, b)$, for $k = 0, 1, \cdots, n$ and $0 < \alpha \leq 1$, then we obtain

$$g(\tau) = \sum_{k=0}^{n} g^{(k\alpha)}(\tau_0) \frac{(\tau - \tau_0)^{k\alpha}}{\Gamma(1 + k\alpha)} + R_{n\alpha}(\tau - \tau_0), \tag{6}$$

with $a < \tau_0 < \xi < \tau < b$, $\forall x \in (a, b)$, where the non-differentiable remainder term is $R_{n\alpha}(\tau - \tau_0) = O\left((\tau - \tau_0)^{n\alpha}\right)$.

Proof. See (Yang, 2011, 2012).

Lemma 3

If $g^{((k+1)\alpha)}(\tau) \in \text{ffj}_{\alpha}(a, b)$, for $k = 0, 1, \cdots, n$ and $0 < \alpha \leq 1$, then we obtain

$$g(\tau) = \sum_{k=0}^{n} g^{(k\alpha)}(0) \frac{\tau^{k\alpha}}{\Gamma(1 + k\alpha)} + g^{((k+1)\alpha)}(\theta\tau) \frac{\tau^{(k+1)\alpha}}{\Gamma(1 + (k+1)\alpha)}, \tag{7}$$

with $0 < \theta < 1$, $\forall \tau \in (a, b)$.

Proof. See (Yang, 2011, 2012).

Lemma 4

If $g^{((k+1)\alpha)}(\tau) \in \text{ffj}_{\alpha}(a, b)$, for $k = 0, 1, \cdots, n$ and $0 < \alpha \leq 1$, then we obtain

$$g(\tau) = \sum_{k=0}^{\infty} g^{(k\alpha)}(\tau_0) \frac{(\tau - \tau_0)^{k\alpha}}{\Gamma(1 + k\alpha)}, \tag{8}$$

where $a < \tau_0 < \tau < b$, $\forall \tau \in (a, b)$.

Proof. See (Yang, 2011, 2012).

Lemma 5

If $g^{((k+1)\alpha)}(\tau) \in \text{ffj}_{\alpha}(a, b)$, for $k = 0, 1, \cdots, n$ and $0 < \alpha \leq 1$, then we have

$$g(\tau) = \sum_{k=0}^{\infty} g^{(k\alpha)}(0) \frac{\tau^{k\alpha}}{\Gamma(1 + k\alpha)}, \tag{9}$$

with $0 < \theta < 1$, $\forall \tau \in (a, b)$.

Proof. See (Yang, 2011, 2012).

3 Extended DTM

In this section, the definition of the extended DTM (so-called local fractional DTM) and its theorems are given via the generalized local fractional Taylor theorems.

Definition 1

If $g(\tau)$ is a local fractional analytic function (Yang, 2011) in the domain Ω, then the local fractional differential transform method (local fractional DTM) is given as

$$G(\kappa) = \frac{1}{\Gamma(1+\kappa\alpha)} \frac{d^{\kappa\alpha} g(\tau)}{d\tau^{\kappa\alpha}} |_{\tau=\tau_0}, \quad \tau, \tau_0 \in \Omega, \tag{10}$$

where $\kappa = 0, 1, L, n$, $0 < \alpha \leq 1$, and the local fractional differential inverse transform of $G(\kappa)$ in the domain Ω is given by:

$$g(\tau) = \sum_{\kappa=0}^{\infty} G(\kappa)(\tau - \tau_0)^{\kappa\alpha}, \quad \kappa \in K. \tag{11}$$

The function $G(\kappa)$ is the so-called local fractional spectrum of $g(\tau)$.

When $\tau_0 = 0$, from (10) and (11) we get (Yang and Hua, 2014)

$$G(\kappa) = \frac{1}{\Gamma(1+\kappa\alpha)} \frac{d^{\kappa\alpha} g(\tau)}{d\tau^{\kappa\alpha}} |_{\tau=0}, \quad \tau \in \Omega, \tag{12}$$

and

$$g(\tau) = \sum_{\kappa=0}^{\infty} G(\kappa) x^{\kappa\alpha}, \quad \kappa \in K. \tag{13}$$

Using the expressions (10), (11), (12) and (13), we obtain

$$g(\tau) = \sum_{\kappa=0}^{\infty} g^{(\kappa\alpha)}(\tau_0) \frac{(\tau - \tau_0)^{\kappa\alpha}}{\Gamma(1+\kappa\alpha)}, \tag{14}$$

and

$$g(\tau) = \sum_{\kappa=0}^{\infty} g^{(\kappa\alpha)}(0) \frac{\tau^{\kappa\alpha}}{\Gamma(1+\kappa\alpha)}. \tag{15}$$

We notice that the above expressions are the formulas of the local fractional Taylor theorems. Meanwhile, when $\alpha = 1$, the expressions (14) and (15) are the classical DTM (Zhou, 1986; Chen, 2009; Hesam et al., 2012; Kanth and Aruna, 2009; Abazari and Borhanifar, 2010; Kangalgil and Ayaz, 2009; Biazar and Eslami, 2010). This method is different from the fractional DTM (Biazar and Eslami, 2010; Arikoglu and Ozkol, 2007; Elsaid, 2012; Nazari and Shahmorad, 2010) because of differential operator that was employed.

Applying (10) and (11), the theorems of the local fractional DTM are deduced as follows.

Theorem 1

Suppose that $s(\tau) = g(\tau) + h(\tau)$, then we have

$$S(\kappa) = G(\kappa) + H(\kappa). \tag{16}$$

Proof.

$$s(\tau) = \sum_{\kappa=0}^{\infty} S(\kappa)(\tau - \tau_0)^{\kappa\alpha} = \sum_{\kappa=0}^{\infty} G(\kappa)(\tau - \tau_0)^{\kappa\alpha} + \sum_{\kappa=0}^{\infty} H(\kappa)(\tau - \tau_0)^{\kappa\alpha},$$

$$s(\tau) = \sum_{\kappa=0}^{\infty} (G(\kappa) + H(\kappa))(\tau - \tau_0)^{\kappa\alpha}.$$

Using the local fractional DTM in (11), we obtain

$$S(\kappa) = G(\kappa) + H(\kappa).$$

□

Theorem 2

Suppose that $s(\tau) = ag(\tau)$, where a $(a \in R)$ is a constant. Then, there is

$$S(\kappa) = aG(\kappa). \tag{17}$$

Proof.

$$s(\tau) = \sum_{\kappa=0}^{\infty} S(\kappa)(\tau - \tau_0)^{k\alpha} = \sum_{\kappa=0}^{\infty} aG(\kappa)(\tau - \tau_0)^{\kappa\alpha},$$

$$s(\tau) = a\sum_{\kappa=0}^{\infty} G(\kappa)(\tau - \tau_0)^{\kappa\alpha}.$$

Using the local fractional DTM in (11), we obtain

$$S(k) = aG(k).$$

□

Theorem 3

Suppose that $s(\tau) = g(\tau)h(\tau)$. Then, there is

$$S(\kappa) = \sum_{l=0}^{\kappa} G(l)H(\kappa - l). \tag{18}$$

Proof.

$$\begin{aligned} s(\tau) &= \left(\sum_{\kappa=0}^{\infty} G(\kappa)(\tau - \tau_0)^{\kappa\alpha}\right) \otimes \left(\sum_{\kappa=0}^{\infty} H(\kappa)(\tau - \tau_0)^{\kappa\alpha}\right) \\ &= \left(G(0) + G(1)(\tau - \tau_0)^{\alpha} + G(2)(\tau - \tau_0)^{2\alpha} + \cdots\right) \\ &\otimes \left(H(0) + H(1)(\tau - \tau_0)^{\alpha} + H(2)(\tau - \tau_0)^{2\alpha} + \cdots\right) \\ &= G(0)H(0) + (G(1)H(0) + G(0)H(1))(\tau - \tau_0)^{\alpha} \\ &+ (G(2)H(0) + G(1)H(1) + G(0)H(2))(\tau - \tau_0)^{2\alpha} \\ &+ \cdots + (G(0)H(\kappa) + G(1)H(\kappa - 1) + \cdots + G(\kappa - 1)H(1) + G(\kappa)H(0))(\tau - \tau_0)^{\kappa\alpha}. \end{aligned}$$

Hence, we obtain

$$S(\kappa) = \sum_{l=0}^{\kappa} G(l) H(\kappa - l).$$

□

Theorem 4

Suppose that $s(\tau) = \frac{(\tau - \tau_0)^{n\alpha}}{\Gamma(1 + n\alpha)}$, where $n \in N$. Then, we have

$$S(\kappa) = \sum_{\kappa=0}^{\infty} \delta_\alpha(\kappa - p) \frac{(\tau - \tau_0)^{\kappa\alpha}}{\Gamma(1 + \kappa\alpha)}, \tag{19}$$

where the local fractional Dirac-delta function is (Yang, 2011)

$$\delta_\alpha(\kappa - p) = \begin{cases} 1, \kappa = p, \\ 0, \kappa \neq p. \end{cases}$$

Proof.

$$\begin{aligned} S(\kappa) &= \frac{1}{\Gamma(1 + \kappa\alpha)} \frac{d^{\kappa\alpha} s(\tau)}{d\tau^{\kappa\alpha}} |_{\tau=\tau_0} \\ &= \frac{1}{\Gamma(1 + \kappa\alpha)} \frac{d^{\kappa\alpha}\left(\frac{(\tau - \tau_0)^{n\alpha}}{\Gamma(1 + n\alpha)}\right)}{d\tau^{\kappa\alpha}} |_{\tau=\tau_0} \\ &= \frac{1}{\Gamma(1 + \kappa\alpha)} \delta_\alpha(\kappa - p). \end{aligned}$$

Hence, we obtain

$$S(\kappa) = \sum_{\kappa=0}^{\infty} \frac{\delta_\alpha(\kappa - p)}{\Gamma(1 + \kappa\alpha)}.$$

□

Theorem 5

Suppose that $s(\tau) = D^{n\alpha} g(\tau)$, where $n \in N$. Then, there is

$$S(\kappa) = \frac{\Gamma(1 + (\kappa + 1)\alpha)}{\Gamma(1 + \kappa\alpha)} G(\kappa + 1). \tag{20}$$

Proof.

$$\begin{aligned} S(\kappa) &= \frac{1}{\Gamma(1 + \kappa\alpha)} \frac{d^{\kappa\alpha}\left(D^{n\alpha} g(\tau)\right)}{d\tau^{\kappa\alpha}} |_{\tau=\tau_0} \\ &= \frac{1}{\Gamma(1 + \kappa\alpha)} \frac{d^{(\kappa+n)\alpha} g(\tau)}{d\tau^{(\kappa+n)\alpha}} |_{\tau=\tau_0}. \end{aligned}$$

Hence, we obtain

$$\begin{aligned} S(\kappa) &= \frac{1}{\Gamma(1 + \kappa\alpha)} \frac{d^{(\kappa+n)\alpha} g(\tau)}{d\tau^{(k+n)\alpha}} |_{\tau=\tau_0} \\ &= \sum_{\kappa=0}^{\infty} \frac{\Gamma(1 + (\kappa + n)\alpha)}{\Gamma(1 + \kappa\alpha)} G(\kappa + n), \end{aligned}$$

where

$$G(\kappa+n)=\frac{1}{\Gamma(1+(\kappa+n)\alpha)}\frac{d^{(\kappa+n)\alpha}g(\tau)}{d\tau^{(\kappa+n)\alpha}}.$$

□

Table 1. The basic operations of the local fractional DTM.

Original function	Transformed function
$s(\tau)=g(\tau)+h(\tau)$	$S(\kappa)=G(\kappa)+H(\kappa)$
$s(\tau)=ag(\tau)$	$S(\kappa)=aG(\kappa)$, a is a constant.
$s(\tau)=g(\tau)h(\tau)$	$S(\kappa)=\sum_{l=0}^{\kappa}G(l)H(\kappa-l)$
$s(\tau)=\frac{(\tau-\tau_0)^{n\alpha}}{\Gamma(1+n\alpha)}$	$S(\kappa)=\sum_{\kappa=0}^{\infty}\frac{\delta_\alpha(\kappa-p)}{\Gamma(1+\kappa\alpha)}$, $\delta_\alpha(\kappa-p)=\begin{cases}1, \kappa=p,\\ 0, \kappa\neq p.\end{cases}$
$s(\tau)=D^{n\alpha}g(\tau)$	$S(\kappa)=\frac{\Gamma(1+(\kappa+n)\alpha)}{\Gamma(1+n\alpha)}G(\kappa+n)$

4 Four Illustrative Examples

In this section, four examples of solutions of the local fractional ODEs are presented.

Example 1

The following local fractional ODE is presented in this first example:

$$m^{(\alpha)}(\tau)-m(\tau)=0 \tag{21}$$

with an initial value

$$m(0)=1. \tag{22}$$

Taking the local fractional DTM of (21), by using the basic operation in Table 1, yields

$$\frac{\Gamma(1+(\kappa+1)\alpha)}{\Gamma(1+\kappa\alpha)}M(\kappa+1)-M(\kappa)=0, \tag{23}$$

which reduce to the following formula

$$M(\kappa+1)=\frac{\Gamma(1+\kappa\alpha)}{\Gamma(1+(\kappa+1)\alpha)}M(\kappa), \tag{24}$$

where $M(0)=1$.

Following (24), we have the following relations:

$$M(0)=1, \tag{25}$$

$$M(1) = \frac{1}{\Gamma(1+\alpha)}, \tag{26}$$

$$M(2) = \frac{1}{\Gamma(1+2\alpha)}, \tag{27}$$

$$\vdots$$

$$M(n) = \frac{1}{\Gamma(1+n\alpha)}. \tag{28}$$

Hence, $m(\tau)$ is evaluated as follows:

$$\begin{aligned} m(\tau) &= \sum_{\kappa=0}^{\infty} F(\kappa)\tau^{\kappa\alpha} \\ &= 1 + \frac{\tau^{\alpha}}{\Gamma(1+\alpha)} + \frac{\tau^{2\alpha}}{\Gamma(1+2\alpha)} + L + \frac{\tau^{n\alpha}}{\Gamma(1+n\alpha)} + L \\ &= \sum_{\kappa=0}^{\infty} \frac{\tau^{\kappa\alpha}}{\Gamma(1+\kappa\alpha)} \\ &= E_{\alpha}\left(\tau^{\alpha}\right), \end{aligned} \tag{29}$$

and its graph is illustrated in Fig. 1.

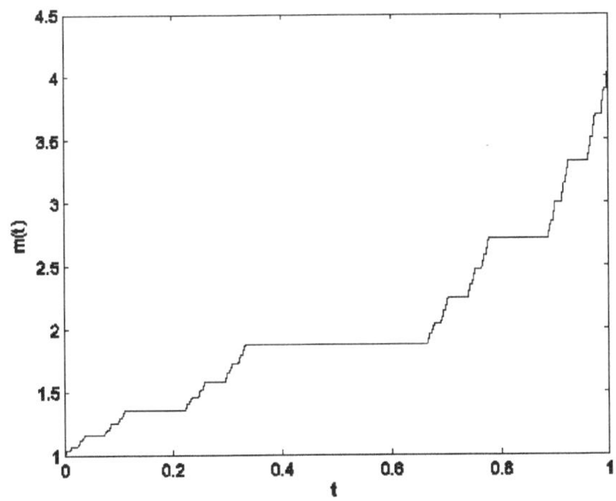

Fig. 1. Plot of non-differentiable solution to local fractional ODE (21) when $\kappa = 100$.

Example 2

The following local fractional ODE is our second example:

$$m^{(2\alpha)}(\tau) + m(\tau) = 0 \tag{30}$$

with initial values

$$m^{(\alpha)}(0) = 0, \quad m(0) = 1. \tag{31}$$

By using the basic operation in Table 1, expressions (30) and (31) become

$$\frac{\Gamma(1+(\kappa+2)\alpha)}{\Gamma(1+\kappa\alpha)}M(\kappa+2)+M(\kappa)=0, \tag{32}$$

$$M(\kappa+1)=0, \tag{33}$$

$$M(0)=1. \tag{34}$$

Following (32) and (34), we get

$$M(\kappa+2)=-\frac{\Gamma(1+\kappa\alpha)}{\Gamma(1+(\kappa+2)\alpha)}M(\kappa), \tag{35}$$

which reduces to

$$M(0)=1, \tag{36}$$

$$M(2)=-\frac{1}{\Gamma(1+2\alpha)}, \tag{37}$$

$$M(4)=\frac{1}{\Gamma(1+4\alpha)}, \tag{38}$$

$$M(6)=-\frac{1}{\Gamma(1+6\alpha)}, \tag{39}$$

Taking the local fractional differential inverse transform of (35), we have

$$\begin{aligned} m(\tau) &= \sum_{\kappa=0}^{\infty} M(\kappa)\tau^{\kappa\alpha} \\ &= 1-\frac{\tau^{2\alpha}}{\Gamma(1+2\alpha)}+\frac{\tau^{4\alpha}}{\Gamma(1+4\alpha)}-\frac{\tau^{6\alpha}}{\Gamma(1+6\alpha)}+L \\ &= \sum_{\kappa=0}^{\infty}(-1)^{\kappa}\frac{\tau^{2\alpha}}{\Gamma(1+2\kappa\alpha)} \\ &= \cos_{\alpha}\left(\tau^{\alpha}\right), \end{aligned} \tag{40}$$

and its graph is given in Fig. 2.

Example 3

The following local fractional ODE comprises our third example:

$$m^{(3\alpha)}(\tau)+m(\tau)=0 \tag{41}$$

with initial values

$$m^{(2\alpha)}(0)=0,\ m^{(\alpha)}(0)=0,\ m(0)=1. \tag{42}$$

By using the basic operation in Table 1, we transform (34) and (35) into

$$\frac{\Gamma(1+(\kappa+3)\alpha)}{\Gamma(1+3k\alpha)}M(\kappa+3)+M(\kappa)=0, \tag{43}$$

$$M(\kappa+2)=0, \tag{44}$$

$$M(\kappa+1)=0, \tag{45}$$

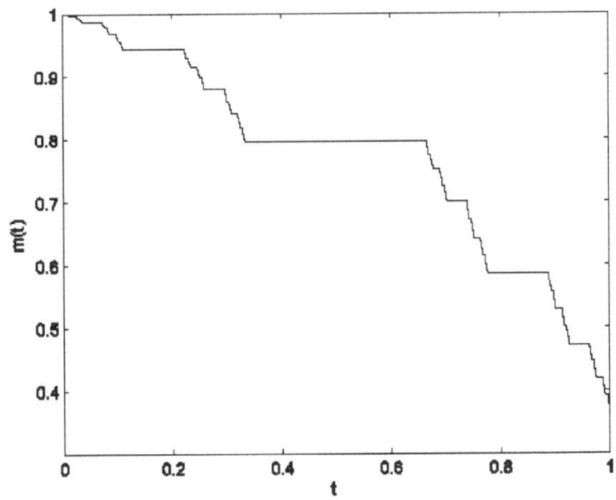

Fig. 2. Plot of non-differentiable solution to local fractional ODE (40) when $\kappa = 100$.

$$M(\kappa) = 1. \tag{46}$$

Making use of (43) and (46), we have

$$M(0) = 1, \tag{47}$$

$$M(3) = -\frac{1}{\Gamma(1+3\alpha)}, \tag{48}$$

$$M(6) = \frac{1}{\Gamma(1+6\alpha)}, \tag{49}$$

$$M(9) = -\frac{1}{\Gamma(1+9\alpha)}, \tag{50}$$

$$\vdots$$

Taking the local fractional differential inverse transform, $m(\tau)$ is evaluated as

$$\begin{aligned} m(\tau) &= \sum_{\kappa=0}^{\infty} M(\kappa)\tau^{\kappa\alpha} \\ &= 1 - \frac{\tau^{3\alpha}}{\Gamma(1+3\alpha)} + \frac{\tau^{6\alpha}}{\Gamma(1+6\alpha)} - \frac{\tau^{9\alpha}}{\Gamma(1+9\alpha)} + L \\ &= \sum_{\kappa=0}^{\infty} (-1)^{\kappa} \frac{\tau^{3\alpha}}{\Gamma(1+3\kappa\alpha)} \end{aligned} \tag{51}$$

and it graph is shown in Fig. 3.

Example 4

The next local fractional ODE is analyzed in this last example:

$$m^{(4\alpha)}(\tau) + m(\tau) = 0 \tag{52}$$

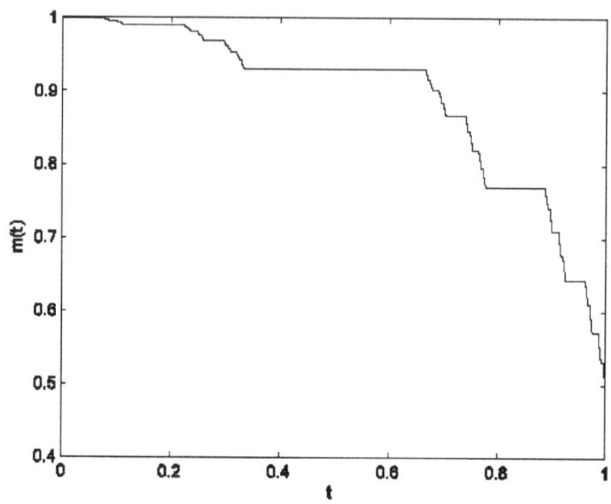

Fig. 3. Plot of non-differentiable solution to local fractional ODE (51) when $\kappa = 100$.

with an initial value

$$m^{(3\alpha)}(0) = 0,\ m^{(2\alpha)}(0) = 0,\ m^{(\alpha)}(0) = 0,\ m(0) = 1. \tag{53}$$

By using the basic operation in Table 1, we change (52) and (53) into

$$\frac{\Gamma(1+(\kappa+4)\alpha)}{\Gamma(1+4k\alpha)} M(\kappa+4) + M(\kappa) = 0, \tag{54}$$

$$M(\kappa+3) = 0, \tag{55}$$

$$M(\kappa+2) = 0, \tag{56}$$

$$M(\kappa+1) = 0, \tag{57}$$

$$M(\kappa) = 1. \tag{58}$$

Making use of (43) and (46), we have

$$M(0) = 1, \tag{59}$$

$$M(4) = -\frac{1}{\Gamma(1+4\alpha)}, \tag{60}$$

$$M(8) = \frac{1}{\Gamma(1+8\alpha)}, \tag{61}$$

$$M(12) = -\frac{1}{\Gamma(1+12\alpha)}, \tag{62}$$

The local fractional differential inverse transform leads to

$$\begin{aligned} m(\tau) &= \sum_{\kappa=0}^{\infty} M(\kappa)\tau^{\kappa\alpha} \\ &= 1 - \frac{\tau^{4\alpha}}{\Gamma(1+4\alpha)} + \frac{\tau^{8\alpha}}{\Gamma(1+8\alpha)} - \frac{\tau^{12\alpha}}{\Gamma(1+12\alpha)} + \cdots \\ &= \sum_{\kappa=0}^{\infty} (-1)^{\kappa} \frac{\tau^{4\alpha}}{\Gamma(1+4\kappa\alpha)} \end{aligned} \tag{63}$$

and it graph is shown in Fig. 4.

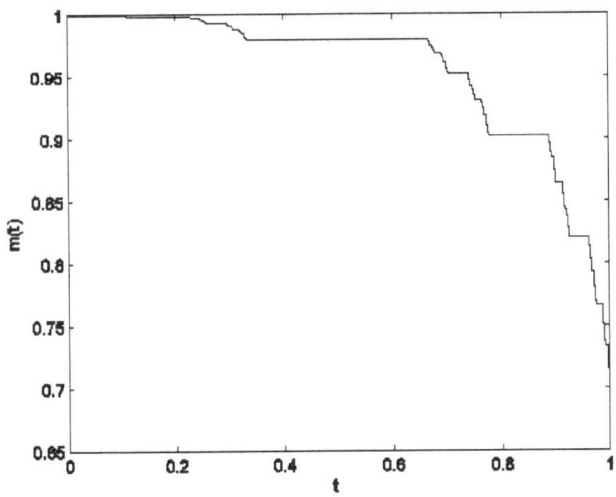

Fig. 4. Plot of non-differentiable solution to local fractional ODE (63) when $\kappa = 100$.

5 Conclusions

In this work the local fractional DTM was studied through the framework of the local fractional Taylor theorems. We presented and proved the basic theorems of the local fractional DTM. The non-differentiable solutions for local fractional ODEs were obtained and four examples were given, in order to illustrate the high efficiency and accuracy of the proposed technique to solve the local fractional ODEs.

Bibliography

Zhou, J. K. (1986). *Differential Transformation and Its Applications for Electrical Circuits*, Huazhong University Press, Wuhan, China.

Chen, S. S. (2009). Application of the differential transformation method to the free vibrations of strongly non-linear oscillators. *Nonlinear analysis: Real world applications*, 10(2), 881-888.

Hesam, S., Nazemi, A. R., Haghbin, A. (2012). Analytical solution for the Fokker–Planck equation by differential transform method. *Scientia Iranica*, 19(4), 1140-1145.

Kanth, A. R., Aruna, K. (2009). Differential transform method for solving the linear and nonlinear Klein–Gordon equation. *Computer Physics Communications*, 180(5), 708-711.

Abazari, R., Borhanifar, A. (2010). Numerical study of the solution of the Burgers and coupled Burgers equations by a differential transformation method. *Computers & Mathematics with Applications*, 59(8), 2711-2722.

Kangalgil, F., Ayaz, F. (2009). Solitary wave solutions for the KdV and mKdV equations by differential transform method. *Chaos, Solitons & Fractals*, 41(1), 464-472.

Biazar, J., Eslami, M. (2010). Analytic solution for Telegraph equation by differential transform method. *Physics Letters A*, 374(29), 2904-2906.

Arikoglu, A., Ozkol, I. (2007). Solution of fractional differential equations by using differential transform method. *Chaos, Solitons & Fractals*, 34(5), 1473-1481.

Elsaid, A. (2012). Fractional differential transform method combined with the Adomian polynomials. *Applied Mathematics and Computation*, 218(12), 6899-6911.

Nazari, D., Shahmorad, S. (2010). Application of the fractional differential transform method to fractional-order integro-differential equations with nonlocal boundary conditions. *Journal of Computational and Applied Mathematics*, 234(3), 883-891.

Momani, S., Odibat, Z., Alawneh, A. (2008). Variational iteration method for solving the space- and time- fractional KdV equation. *Numerical Methods for Partial Differential Equations*, 24(1), 262-271.

Bhrawy, A. H., Baleanu, D. (2013). A spectral Legendre–Gauss–Lobatto collocation method for a space-fractional advection diffusion equations with variable coefficients. *Reports on Mathematical Physics*, 72(2), 219-233.

Atangana, A., Belhaouari, S. B. (2013). Solving partial differential equation with space-and time-fractional derivatives via homotopy decomposition method. *Mathematical Problems in Engineering*, 2013, 318590, 9 pages.

Hristov, J. (2012). The heat-balance integral: 2. Parabolic profile with a variable exponent: The concept, analysis and numerical experiments. *Comptes Rendus Mécanique*, 340(7), 493-500.

Yang, X. -J. (2012). *Advanced Local Fractional Calculus and Its Applications*, World Science, New York, NY, USA.

Yang, X. -J. (2011). *Local Fractional Functional Analysis and Its Applications*, Asian Academic publisher Limited, HK.

Kolwankar, K. M., Gangal, A. D. (1998). Local fractional Fokker-Planck equation. *Physical Review Letters*, 80(2), 214–217.

Carpinteri, A., Sapora, A. (2010). Diffusion problems in fractal media defined on Cantor sets. *ZAMM Journal of Applied Mathematics and Mechanics*, 90(3), 203–210.

Yang, X. J., Srivastava, H. M., He, J. H., Baleanu, D. (2013). Cantor-type cylindrical-coordinate method for differential equations with local fractional derivatives. *Physics Letters A*, 377(28), 1696-1700.

Zhong, W. P., Gao, F., Shen, X. M. (2012). Applications of Yang-Fourier transform to local fractional equations with local fractional derivative and local fractional integral. *Advanced Materials Research*, 461, 306-310.

Yang, X. J., Baleanu, D., Khan, Y., Mohyud-Din, S. T. (2014). Local fractional variational iteration method for diffusion and wave equations on Cantor sets. *Romanian Journal of Physics*, 2014, 59(1-2), 36-48.

Baleanu, D., Tenreiro Machado, J. A., Cattani, C., Baleanu, M. C., Yang, X. J. (2014). Local fractional variational iteration and decomposition methods for wave equation on Cantor sets within local fractional operators. *Abstract and Applied Analysis*, (2014), 2014, Article ID 535048, 6 pages.

Li, Y., Wang, L. F., Zeng, S. D., Zhao, Y. (2014). Local fractional Laplace variational iteration method for fractal vehicular traffic flow. *Advances in Mathematical Physics*, 2014, 649318, 7 pages.

Yan, S. P., Jafari, H., Jassim, H. K. (2014). Local fractional Adomain decomposition and function decomposition methods for Laplace equation within local fractional operators. *Advances in Mathematical Physics*, 2014, 161580, 7 pages.

Wang, S. Q., Yang, Y. J., Jassim, H. K. (2014). Local fractional function decomposition method for solving inhomogeneous wave equations with local fractional derivative. *Abstract and Applied Analysis*, 2014, 176395, 7 pages.

Zhang, Y. Z., Yang, A. M., Long, Y. (2014). Initial boundary value problem for fractal heat equation in the semi-infinite region by Yang-Laplace transform. *Thermal Science*, 18(2), 677-681.

He, J. H. (2014). A tutorial review on fractal spacetime and fractional calculus. *International Journal of Theoretical Physics*, 53(11), 3698-3718.

Y Yang, Y. J., Hua, L. Q. (2014). Variational iteration transform method for fractional differential equations with local fractional derivative. *Abstract and Applied Analysis*, 2014, 760957, 9 pages.

Xiao-Jun Yang, Carlo Cattani, and Gongnan Xie

Local Fractional Calculus Application to Differential Equations Arising in Fractal Heat Transfer

Abstract: This chapter presents an application of local fractional calculus to differential equations arising in fractal heat transfer. The non-differentiable problems comprising the homogeneous and non-homogeneous heat, Poisson and Laplace equations of fractal heat transfer are investigated. The 2D partial differential equations of fractal heat transfer in Cantor-type circle coordinate systems are also discussed.

1 Introduction

Fractals are important in nature because often the behavior on the surfaces and boundary of materials are structurally similar. Fractal geometry is an efficient tool to describe real world problems of dynamical systems in nature (see Mandelbrot, 1983; Falconer, 1986; Family, 1991; Nottale, 1993; Peters, 1994; West, 1994; Massopust, 1994; Edgar, 2007, Lapidus and Frankenhuysen, 2000).

Heat transfer is one such behavior that can be studying fractal geometry. For example, the heat transfer characteristics of the Koch island fractal heat exchanger was studied in (Meyer and Vyver, 2005). The fractal heat transfer analysis of interfacial temperature distribution in the slow sliding regime was reported in (Wang and Komvopoulos, 1994). The fractal heat flux of the pool boiling was considered in (Xiao and Yu, 2007). A fractal model approach for heat and mass transfer of the randomly packed hollow fiber membrane module was presented in (Zhang, 2011). A fractal approach for fluid–solid coupling heat transfer in porous medium was reported in (Cai and Huai, 2010). The fractal heat transfer in microchannel heat sink was discussed in (Liu et al.).

Recently, local fractional vector calculus (Yang, 2012; Yang, 2013; Zhao et al., 2013; Li et al. 2014; Yang et al., 2014) has found many applications in science and engineering. Local fractional derivatives were applied to describe the heat equations arising in fractal heat flow (Yang et al., 2014; Yang et al., 2013; Hu et al, 2013) and silk cocoon hierarchy (He and Liu, 2013). There are many analytical methods to find

Xiao-Jun Yang: Department of Mathematics and Mechanics, China University of Mining and Technology, Xuzhou, Jiangsu 221008, China, E-mail: dyangxiaojun@163.com
Carlo Cattani: School of Engineering (DEIM), Largo dell'Università –Riello, 01100 Viterbo, Italy, E-Mail: cattani@unitus.it
Gongnan Xie: Department of Mechanical and Power Engineering, School of Marine Science and Technology, Northwestern Polytechnical University, Xi'an, 710072, China

non-differentiable solutions for the partial differential equations of fractal heat flow, such as local fractional variation iteration method (Yang and Baleanu, 2013), coupling method of local fractional variation iteration method and Laplace transform (Liu et al., 2013), local fractional Fourier transform (Yang et al., 2013), local fractional Fourier series method (Yang et al., 2013), and the local fractional Adomian decomposition method (Yang et al., 2014; Yang et al., 2013).

The main goal of this chapter is to apply local fractional calculus to differential equations arising in fractal heat transfer, including the homogeneous and non-homogeneous heat, Poisson and Laplace equations.

2 Theory of Local Fractional Vector Calculus

In this section, the basic theory of the local fractional vector calculus will be introduced and the basic theorems of the local fractional vector integrals are presented.

Definition 1 (Yang, 2012; Yang, 2013; Zhao et al., 2013; Li et al., 2014; Yang et al., 2014)

For $1 > \alpha > 0$, the local fractional line integral of the function $\mathbf{u}(x_P, y_P, z_P)$ along a fractal line l^{α} is defined as

$$\int_{l^{(\alpha)}} \mathbf{u}(x_P, y_P, z_P) \cdot d\mathbf{l}^{(\alpha)} = \lim_{N\to\infty} \sum_{P=1}^{N} \mathbf{u}(x_P, y_P, z_P) \cdot \Delta \mathbf{l}_P^{(\alpha)} \tag{1}$$

where each part of the line $\Delta \mathbf{l}_P^{(\alpha)}$ satisfies $\left|\Delta l_P^{\alpha}\right| \to 0$ as $N \to \infty$ and $\beta = 2\alpha$.

Definition 2 (Yang, 2012; Yang, 2013; Zhao et al., 2013; Li et al., 2014; Yang et al., 2014)

For $\gamma = \frac{3}{2}\beta = 3\alpha$, $1>\alpha>0$, the local fractional surface integral of $u(r_P)$ is defined as

$$\iint u(r_P)\, d\mathbf{S}^{(\beta)} = \lim_{N\to\infty} \sum_{P=1}^{N} u(r_P)\, \mathbf{n}_P \Delta S_P^{(\beta)} \tag{2}$$

where $d\mathbf{S}^{(\beta)}$ represents N elements of area with a unit normal local fractional vector n_P, $\Delta S_P^{(\beta)} \to 0$ as $N \to \infty$.

Definition 3 (Yang, 2012; Yang, 2013; Zhao et al., 2013; Li et al., 2014; Yang et al., 2014)

For $\gamma = \frac{3}{2}\beta = 3\alpha$, $1>\alpha>0$, the local fractional volume integral of the function $\mathbf{u}(r_P)$ is defined as

$$\iiint \mathbf{u}(r_P)\, dV^{(\gamma)} = \lim_{N\to\infty} \sum_{P=1}^{N} \mathbf{u}(r_P)\, \Delta V_P^{(\gamma)}, \tag{3}$$

where $\Delta V_P^{(\gamma)}$ are the volume elements $\Delta V_P^{(\gamma)} \to 0$ as $N \to \infty$.

Basic operators of local fractional vector integrals are listed in Table 1. More details of the local fractional vector integrals, see Yang, 2012.

Table 1. Basic operators of local fractional vector integrals

Basic operators of local fractional vector integrals	Conditions
$\int_{l^{(\alpha)}} (\mathbf{u}_1 + \mathbf{u}_2) \cdot d\mathbf{l}^{(\alpha)} = \int_{l^{(\alpha)}} \mathbf{u}_1 \cdot d\mathbf{l}^{(\alpha)} + \int_{l^{(\alpha)}} \mathbf{u}_2 \cdot d\mathbf{l}^{(\alpha)}$	
$\int_{l^{(\alpha)}} \mathbf{u} \cdot d\mathbf{l}^{(\alpha)} = \int_{l_1^{(\alpha)}} \mathbf{u} \cdot d\mathbf{l}^{(\alpha)} + \int_{l_2^{(\alpha)}} \mathbf{u} \cdot d\mathbf{l}^{(\alpha)}$	$\mathbf{l}^{(\alpha)} = \mathbf{l}_1^{(\alpha)} + \mathbf{l}_2^{(\alpha)}$
$\iint_{S^{(\beta)}} (\mathbf{u}_1 + \mathbf{u}_2) \cdot d\mathbf{S}^{(\beta)} = \iint_{S^{(\beta)}} \mathbf{u}_1 \cdot d\mathbf{S}^{(\beta)} + \iint_{S^{(\beta)}} \mathbf{u}_2 \cdot d\mathbf{S}^{(\beta)}$	
$\iint_{S^{(\beta)}} \mathbf{u} \cdot d\mathbf{S}^{(\beta)} = \iint_{S_1^{(\beta)}} \mathbf{u} \cdot d\mathbf{S}^{(\beta)} + \iint_{S_2^{(\beta)}} \mathbf{u} \cdot d\mathbf{S}^{(\beta)}$	$\mathbf{S}^{(\beta)} = \mathbf{S}_1^{(\beta)} + \mathbf{S}_2^{(\beta)}$
$\iiint_{V^{(\gamma)}} (\mathbf{u}_1 + \mathbf{u}_2) \cdot dV^{(\gamma)} = \iiint_{V^{(\gamma)}} \mathbf{u}_1 \cdot dV^{(\gamma)} + \iiint_{V^{(\gamma)}} \mathbf{u}_2 \cdot dV^{(\gamma)}$	
$\iiint_{V^{(\gamma)}} (\mathbf{u}_1 + \mathbf{u}_2) \cdot dV^{(\gamma)} = \iiint_{V_1^{(\gamma)}} \mathbf{u} \cdot dV^{(\gamma)} + \iiint_{V_2^{(\gamma)}} \mathbf{u} \cdot dV^{(\gamma)}$	$V^{(\gamma)} = V_1^{(\gamma)} + V_2^{(\gamma)}$

Definition 4 (Yang, 2012)

For $\gamma = \frac{3}{2}\beta = 3\alpha$, $1>\alpha>0$, the local fractional gradient of the scalar function ϕ is given as

$$\nabla^\alpha \phi = \lim_{dV^{(\gamma)}\to 0} \left(\frac{1}{dV^{(\gamma)}} \oiint_{S^{(\beta)}} \phi d\mathbf{S}^{(\beta)} \right) = \frac{\partial^\alpha \phi}{\partial x_1^\alpha} \mathbf{e}_1^\alpha + \frac{\partial^\alpha \phi}{\partial x_2^\alpha} \mathbf{e}_2^\alpha + \frac{\partial^\alpha \phi}{\partial x_3^\alpha} \mathbf{e}_3^\alpha, \tag{4}$$

where $V^{(\gamma)}$ is a small fractal volume enclosing P, $S^{(\beta)}$ is its bounding fractal surface, and ∇^α is a local fractional Hamilton operator. Basic operators of local fractional vector integrals are listed in Table 2. For more details of the local fractional vector integrals, see (Yang, 2012).

Definition 5 (Yang, 2012)

For $\gamma = \frac{3}{2}\beta = 3\alpha$, $1>\alpha>0$, the local fractional divergence of the vector function $\mathbf{u}$ is given as

$$\begin{aligned} \nabla^\alpha \bullet \mathbf{u} &= \lim_{dV^{(\gamma)}\to 0} \left(\frac{1}{dV^{(\gamma)}} \oiint_{S^{(2\alpha)}} \mathbf{u} \bullet d\mathbf{S}^{(\beta)} \right) \\ &= \frac{\partial^\alpha u_1}{\partial x_1^\alpha} + \frac{\partial^\alpha u_2}{\partial x_2^\alpha} + \frac{\partial^\alpha u_3}{\partial x_3^\alpha} \end{aligned} \tag{5}$$

where $\mathbf{u} = u_1\mathbf{e}_1^\alpha + u_2\mathbf{e}_2^\alpha + u_3\mathbf{e}_3^\alpha$.

Definition 6 (Yang, 2012)

For $\gamma = \frac{3}{2}\beta = 3\alpha$, $1>\alpha>0$, the local fractional curl of the vector function $\mathbf{u}$ is defined as follows:

$$\begin{aligned} \nabla^\alpha \times \mathbf{u} &= \lim_{dS^{(\beta)}\to 0} \left(\frac{1}{dS^{(\beta)}} \oint_{l^{(\alpha)}} \mathbf{u} \cdot d\mathbf{l}^{(\alpha)} \right) \mathbf{n}_P \\ &= \left(\frac{\partial^\alpha u_3}{\partial x_2^\alpha} - \frac{\partial^\alpha u_2}{\partial x_3^\alpha} \right) \mathbf{e}_1^\alpha + \left(\frac{\partial^\alpha u_1}{\partial x_3^\alpha} - \frac{\partial^\alpha u_3}{\partial x_1^\alpha} \right) \mathbf{e}_2^\alpha + \left(\frac{\partial^\alpha u_2}{\partial x_1^\alpha} - \frac{\partial^\alpha u_1}{\partial x_2^\alpha} \right) \mathbf{e}_3^\alpha \end{aligned}, \tag{6}$$

where $\mathbf{u} = u_1\mathbf{e}_1^\alpha + u_2\mathbf{e}_2^\alpha + u_3\mathbf{e}_3^\alpha$.

Table 2. Basic operators of local fractional Hamilton operator.

$\nabla^\alpha(\varphi_1\varphi_2) = (\nabla^\alpha\varphi_1)\varphi_2 + \varphi_1\nabla^\alpha\varphi_2$	$\nabla^\alpha C = 0$
$\nabla^\alpha \times (\mathrm{u}_1 + \mathrm{u}_2) = \nabla^\alpha \times \mathrm{u}_1 + \nabla^\alpha \times \mathrm{u}_2$	$\nabla^\alpha (C\varphi) = C\nabla^\alpha\varphi$
$\nabla^\alpha \bullet (\mathrm{u}_1 + \mathrm{u}_2) = \nabla^\alpha \bullet \mathrm{u}_1 + \nabla^\alpha \bullet \mathrm{u}_2$	$\nabla^\alpha(\varphi_1 \pm \varphi_2) = \nabla^\alpha\varphi_1 \pm \nabla^\alpha\varphi_2$
$\nabla^\alpha \bullet (\mathrm{u}_1 \times \mathrm{u}_2) = \mathrm{u}_2 \bullet (\nabla^\alpha \times \mathrm{u}_1) - \mathrm{u}_1 \bullet (\nabla^\alpha \times \mathrm{u}_2)$	$\nabla^\alpha \bullet (\varphi\mathrm{u}) = \varphi\nabla^\alpha \bullet \mathrm{u} + \nabla^\alpha\varphi \bullet \mathrm{u}$
$\nabla^\alpha \times (\varphi\mathrm{u}) = \varphi\nabla^\alpha \times \mathrm{u} + \nabla^\alpha\varphi \times \mathrm{u}$	$\nabla^\alpha \times (C\mathrm{u}) = C\nabla^\alpha \times \mathrm{u}$
$\nabla^\alpha \times (\mathrm{u}_1 \times \mathrm{u}_2) = (\mathrm{u}_2 \bullet \nabla^\alpha)\mathrm{u}_1 - (\mathrm{u}_1 \bullet \nabla^\alpha)\mathrm{u}_2 - \mathrm{u}_2(\nabla^\alpha \bullet \mathrm{u}_1) + \mathrm{u}_1(\nabla^\alpha \bullet \mathrm{u}_2)$	$\nabla^\alpha \bullet (\nabla^\alpha \mathrm{u}) = 0$
$\nabla^\alpha(\mathrm{u}_1 \bullet \mathrm{u}_2) = \mathrm{u}_1 \times (\nabla^\alpha \times \mathrm{u}_2) + (\mathrm{u}_1 \bullet \nabla^\alpha)\mathrm{u}_2 + \mathrm{u}_2 \times (\nabla^\alpha \times \mathrm{u}_1) + (\mathrm{u}_2 \bullet \nabla^\alpha)\mathrm{u}_1$	$\nabla^\alpha \times (\nabla^\alpha \mathrm{u}) = 0$
$\nabla^{2\alpha}(\varphi_1 + \varphi_2) = \nabla^{2\alpha}\varphi_1 + \nabla^{2\alpha}\varphi_2$	$(\mathrm{u} \bullet \nabla^\alpha)\varphi = \mathrm{u} \bullet \nabla^\alpha\varphi$
$\nabla^\alpha \times (\nabla^\alpha \times \mathrm{u}) = \nabla^\alpha(\nabla^\alpha \bullet \mathrm{u}) - \nabla^{2\alpha}\mathrm{u}$	$\nabla^\alpha \bullet (\nabla^\alpha \times \mathrm{u}) = 0$

Theorem 1 (Local fractional Gauss theorem) (Yang, 2012; Zhao et al., 2013)

For $\gamma = \frac{3}{2}\beta = 3\alpha,\ 1>\alpha>0$, the local fractional Gauss theorem of the fractal vector field states that

$$\iiint\limits_{V^{(\gamma)}} \nabla^\alpha \cdot \mathbf{u} dV^{(\gamma)} = \oiint\limits_{S^{(\beta)}} \mathbf{u} \cdot d\mathbf{S}^{(\beta)}. \tag{7}$$

Theorem 2 (Local fractional Stokes' theorem) (Yang, 2012; Yang, 2013; Zhao et al., 2013; Li et al., 2014; Yang et al., 2014)

For $\beta = 2\alpha,\ 1>\alpha>0$, the local fractional Stokes' theorem for a fractal field states that

$$\oint\limits_{l^{(\alpha)}} \mathbf{u} \cdot d\mathbf{l}^\alpha = \iint\limits_{S^{(\beta)}} (\nabla^\alpha \times \mathbf{u}) \cdot d\mathbf{S}^{(\beta)}. \tag{8}$$

3 The Local Fractional Heat Equations Arising in Fractal Heat Transfer

In this section, we will discuss the partial differential equations arising in fractal heat transfer.

Let us consider the temperature $T(x, y, z, t)$ at a point $(x, y, z) \in V^{(\gamma)}$, and time $t \in T$.

The first law of thermodynamics in a fractal medium states that (Yang, 2012)

$$\mathrm{W}_1(x, y, z, t) + W_2(x, y, z, t) = W_3(x, y, z, t), \tag{9}$$

where $\mathrm{W}_1(x, y, z, t)$ is the heat entering in unit time through the fractal surface $\mathrm{S}^{(\beta)}$, $\mathrm{W}_2(x, y, z, t)$ is the energy generation in unit time in the fractal volume $\mathrm{V}^{(\gamma)}$, and $\mathrm{W}_3(x, y, z, t)$ is the change in unit time of storage energy in the fractal volume $\mathrm{V}^{(\gamma)}$.

The first term of 11 is equal to

$$W_1(x, y, z, t) = \iint_{S^{(\beta)}} \mathbf{u}(x, y, z, t) \cdot d\mathbf{S}^{(\beta)}, \tag{10}$$

where the local fractional Fourier law can be written as

$$\mathbf{u}(x, y, z, t) = K^{2\alpha} \nabla^{\alpha} T(x, y, z, t). \tag{11}$$

The second term of 9 is

$$W_2(x, y, z, t) = \iiint_{V^{(\gamma)}} g(x, y, z, t)\, dV^{(\gamma)}. \tag{12}$$

The third term of 9 can be written as

$$W_2(x, y, z, t) = \iiint_{V^{(\gamma)}} \rho_\alpha c_\alpha \frac{\partial^\alpha T(x, y, z, t)}{\partial t^\alpha} dV^{(\gamma)}. \tag{13}$$

whereρ_α and c_α are the density and the specific heat of the material, respectively.

Substituting the above formulas into the first law of thermodynamics in fractal media yields

$$\iint_{S^{(\beta)}} \mathbf{u}(x, y, z, t) \cdot d\mathbf{S}^{(\beta)} + \iiint_{V^{(\gamma)}} g(x, y, z, t)\, dV^{(\gamma)} = \iiint_{V^{(\gamma)}} \rho_\alpha c_\alpha \frac{\partial^\alpha T(x, y, z, t)}{\partial t^\alpha} dV^{(\gamma)}. \tag{14}$$

Using the local fractional Gauss theorem, we obtain

$$\begin{aligned} &\iiint_{V^{(\gamma)}} \nabla^\alpha \cdot \left(K^{2\alpha} \nabla^\alpha T(x, y, z, t)\right) dV^{(\gamma)} + \iiint_{V^{(\gamma)}} g(x, y, z, t)\, dV^{(\gamma)} \\ &= \iiint_{V^{(\gamma)}} \rho_\alpha c_\alpha \frac{\partial^\alpha T(x, y, z, t)}{\partial t^\alpha} dV^{(\gamma)}, \end{aligned} \tag{15}$$

which leads to

$$\iiint_{V^{(\gamma)}} \left(\nabla^\alpha \cdot \left(K^{2\alpha} \nabla^\alpha T\right) + g(x, y, z, t) - \rho_\alpha c_\alpha \frac{\partial^\alpha T(x, y, z, t)}{\partial t^\alpha}\right) dV^{(\gamma)} = 0. \tag{16}$$

3.1 The Non-homogeneous Heat Problems Arising in Fractal Heat Flow

The energy generation unit time in the fractal volume $V^{(\gamma)}$ is not equal to zero, namely

$$W_2(x, y, z, t) = \iiint_{V^{(\gamma)}} g(x, y, z, t)\, dV^{(\gamma)} \neq 0, \tag{17}$$

which leads to

$$g(x, y, z, t) \neq 0. \tag{18}$$

From 16, the local fractional non-homogeneous heat equation arising in fractal heat flow reads as follows:

$$\nabla^\alpha \cdot \left(K^{2\alpha} \nabla^\alpha T(x, y, z, t)\right) + g(x, y, z, t) - \rho_\alpha c_\alpha \frac{\partial^\alpha T(x, y, z, t)}{\partial t^\alpha} = 0. \tag{19}$$

Using 19, the local fractional non-homogeneous linear heat equation arising from fractal heat flow is suggested as follows (Yang, 2012):

$$K^{2\alpha} \nabla^{2\alpha} T(x, y, z, t) + g(x, y, z, t) - \rho_\alpha c_\alpha \frac{\partial^\alpha T(x, y, z, t)}{\partial t^\alpha} = 0, \tag{20}$$

where $\nabla^{2\alpha} = \nabla^\alpha \cdot \nabla^\alpha$.

The local fractional dimensionless non-homogeneous heat equation in fractal 3D time-space can be written as

$$\nabla^{2\alpha} T(x, y, z, t) + g(x, y, z, t) - \frac{\partial^\alpha T(x, y, z, t)}{\partial t^\alpha} = 0, \tag{21}$$

subject the initial-boundary conditions

$$\frac{\partial^\alpha}{\partial x^\alpha} T(x, y, z, t) = \vartheta_x(y, z, t), \tag{22}$$

$$\frac{\partial^\alpha}{\partial y^\alpha} T(x, y, z, t) = \vartheta_y(x, z, t), \tag{23}$$

$$\frac{\partial^\alpha}{\partial z^\alpha} T(x, y, z, t) = \vartheta_z(x, y, t), \tag{24}$$

$$T(x, y, z, 0) = \theta(x, y, z). \tag{25}$$

The local fractional dimensionless non-homogeneous heat equation in fractal 2D time-space can be written as

$$\frac{\partial^{2\alpha}}{\partial x^{2\alpha}} T(x, y, t) + \frac{\partial^{2\alpha}}{\partial y^{2\alpha}} T(x, y, t) + g(x, y, t) - \frac{\partial^\alpha T(x, y, t)}{\partial t^\alpha} = 0, \tag{26}$$

subject the initial-boundary conditions

$$\frac{\partial^\alpha}{\partial x^\alpha} T(x, y, t) = \vartheta_x(y, t), \tag{27}$$

$$\frac{\partial^\alpha}{\partial y^\alpha} T(x, y, t) = \vartheta_y(x, t), \tag{28}$$

$$T(x, y, 0) = \theta(x, y). \tag{29}$$

The local fractional dimensionless non-homogeneous heat equation in fractal 1D time-space can be written as (Yang, 2012)

$$\frac{\partial^{2\alpha}}{\partial x^{2\alpha}} T(x, t) + g(x, t) - \frac{\partial^\alpha T(x, t)}{\partial t^\alpha} = 0, \tag{30}$$

subject the initial-boundary conditions

$$\frac{\partial^{\alpha}}{\partial x^{\alpha}} T(x, t) = \vartheta_x(t), \tag{31}$$

$$T(x, 0) = \theta(x). \tag{32}$$

When the fractal dimension α is equal to $\ln 2/\ln 3$, the graph of the heat source term $g(x, t) = x^{\alpha} t^{\alpha} / \Gamma^2(1 + \alpha)$ is shown in Fig. 1.

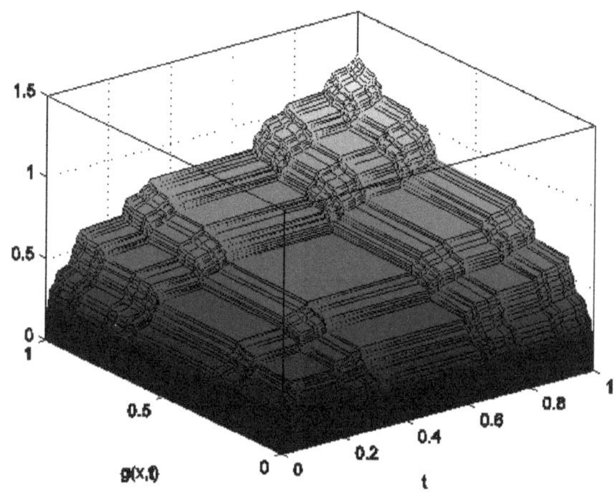

Fig. 1. The plot of the heat source term $g(x, t) = x^{\alpha} t^{\alpha} / x^{\alpha} t^{\alpha} \Gamma^2(1 + \alpha)$ with parameter$\alpha = \ln 2/\ln 3$.

3.2 The Homogeneous Heat Problems Arising in Fractal Heat Flow

The energy generation unit time in the fractal volume $V^{(\gamma)}$ is equal to zero, namely

$$W_2(x, y, z, t) = \iiint_{V^{(\gamma)}} g(x, y, z, t)\, dV^{(\gamma)} = 0, \tag{33}$$

which yields

$$g(x, y, z, t) = 0. \tag{34}$$

Making use of 16, the local fractional homogeneous heat equation arising from fractal heat flow is given as follows:

$$\nabla^{\alpha} \cdot \left(K^{2\alpha} \nabla^{\alpha} T(x, y, z, t)\right) - \rho_{\alpha} c_{\alpha} \frac{\partial^{\alpha} T(x, y, z, t)}{\partial t^{\alpha}} = 0. \tag{35}$$

Applying 19, the local fractional homogeneous linear heat equation arising from fractal heat flow is suggested as follows (Yang, 2012):

$$K^{2\alpha}\nabla^{2\alpha}T(x,y,z,t)-\rho_\alpha c_\alpha\frac{\partial^\alpha T(x,y,z,t)}{\partial t^\alpha}=0, \tag{36}$$

where $\nabla^{2\alpha}=\nabla^\alpha\cdot\nabla^\alpha$.

The local fractional dimensionless homogeneous heat equation in fractal 3D time-space can be written as (Yang et al., 2014)

$$\nabla^{2\alpha}T(x,y,z,t)-\frac{\partial^\alpha T(x,y,z,t)}{\partial t^\alpha}=0, \tag{37}$$

and the initial-boundary conditions are as follows:

$$\frac{\partial^\alpha}{\partial x^\alpha}T(x,y,z,t)=\vartheta_x(y,z,t), \tag{38}$$

$$\frac{\partial^\alpha}{\partial y^\alpha}T(x,y,z,t)=\vartheta_y(x,z,t), \tag{39}$$

$$\frac{\partial^\alpha}{\partial z^\alpha}T(x,y,z,t)=\vartheta_z(x,y,t), \tag{40}$$

$$T(x,y,z,0)=\theta(x,y,z). \tag{41}$$

The local fractional dimensionless non-homogeneous heat equation in fractal 2D time-space is written as

$$\frac{\partial^{2\alpha}}{\partial x^{2\alpha}}T(x,y,t)+\frac{\partial^{2\alpha}}{\partial y^{2\alpha}}T(x,y,t)+g(x,y,t)-\frac{\partial^\alpha T(x,y,t)}{\partial t^\alpha}=0, \tag{42}$$

and the initial-boundary conditions are as follows:

$$\frac{\partial^\alpha}{\partial x^\alpha}T(x,y,t)=\vartheta_x(y,t), \tag{43}$$

$$\frac{\partial^\alpha}{\partial y^\alpha}T(x,y,t)=\vartheta_y(x,t), \tag{44}$$

$$T(x,y,0)=\theta(x,y). \tag{45}$$

The local fractional dimensionless non-homogeneous heat equations in fractal 1D time-space can be written as

$$\frac{\partial^{2\alpha}}{\partial x^{2\alpha}}T(x,t)+g(x,t)-\frac{\partial^\alpha T(x,t)}{\partial t^\alpha}=0, \tag{46}$$

and the initial-boundary conditions are as follows:

$$\frac{\partial^\alpha}{\partial x^\alpha}T(x,t)=\vartheta_x(t), \tag{47}$$

$$T(x,0)=\theta(x). \tag{48}$$

When the fractal dimension α is equal to $\ln 2/\ln 3$, the graph of the initial value condition $\vartheta_x(t)=5t^\alpha/\Gamma(1+\alpha)$ is shown in Fig. 2.

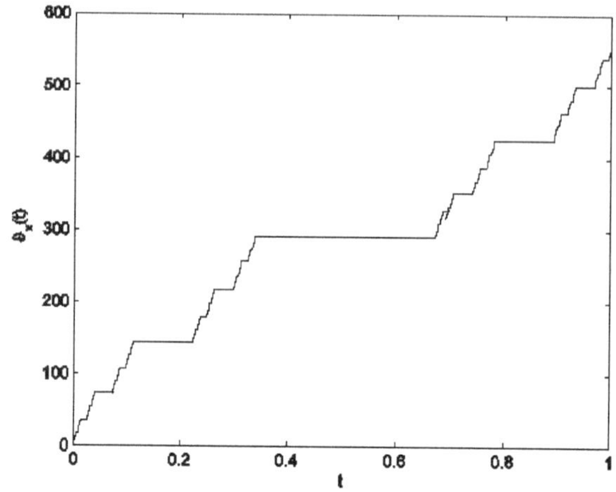

Fig. 2. The graph of initial value condition $\vartheta_x(t) = 5t^{\alpha}/\Gamma(1+\alpha)$ with parameter $\alpha = \ln 2/\ln 3$.

4 Local Fractional Poisson Problems Arising in Fractal Heat Flow

The energy generation in unit time in the fractal volume $V^{(\gamma)}$ can be written as follows:

$$W_2(x,y,z,t) = \iiint\limits_{V^{(\gamma)}} \rho_\alpha c_\alpha \frac{\partial^\alpha T(x,y,z,t)}{\partial t^\alpha} dV^{(\gamma)} = 0. \tag{49}$$

The first law of thermodynamics in fractal media is defined as follows:

$$\iiint\limits_{V^{(\gamma)}} \nabla^\alpha \cdot \left(K^{2\alpha}\nabla^\alpha T(x,y,z,t)\right) dV^{(\gamma)} + \iiint\limits_{V^{(\gamma)}} g(x,y,z,t)\, dV^{(\gamma)} = 0, \tag{50}$$

or

$$\iiint\limits_{V^{(\gamma)}} \left[\nabla^\alpha \cdot \left(K^{2\alpha}\nabla^\alpha T(x,y,z,t)\right) + g(x,y,z,t)\right] dV^{(\gamma)} = 0, \tag{51}$$

which leads to

$$\nabla^\alpha \cdot \left(K^{2\alpha}\nabla^\alpha T(x,y,z,t)\right) + g(x,y,z,t) = 0. \tag{52}$$

From 52, the local fractional Poisson equation arising from fractal heat flow is written as

$$K^{2\alpha}\nabla^{2\alpha} T(x,y,z,t) + g(x,y,z,t) = 0, \tag{53}$$

which yields the local fractional dimensionless Poisson equation of heat flow in fractal 3D time-space (Li et al., 2014)

$$\nabla^{2\alpha} T(x,y,z,t) + g(x,y,z,t) = 0 \tag{54}$$

subject to the initial-boundary conditions given as follows:

$$\frac{\partial^{\alpha}}{\partial x^{\alpha}} T(x, y, z, t) = \vartheta_x(y, z, t), \tag{55}$$

$$\frac{\partial^{\alpha}}{\partial y^{\alpha}} T(x, y, z, t) = \vartheta_y(x, z, t), \tag{56}$$

$$\frac{\partial^{\alpha}}{\partial z^{\alpha}} T(x, y, z, t) = \vartheta_z(x, y, t). \tag{57}$$

The local fractional dimensionless Poisson equation of heat flow in fractal 2D time-space is given as follows (Li et al., 2014):

$$\frac{\partial^{2\alpha}}{\partial x^{2\alpha}} T(x, y, t) + \frac{\partial^{2\alpha}}{\partial y^{2\alpha}} T(x, y, t) + g(x, y, t) = 0, \tag{58}$$

subject to the initial-boundary conditions given as follows:

$$\frac{\partial^{\alpha}}{\partial x^{\alpha}} T(x, y, t) = \vartheta_x(y, t), \tag{59}$$

$$\frac{\partial^{\alpha}}{\partial y^{\alpha}} T(x, y, t) = \vartheta_y(x, t). \tag{60}$$

5 Local Fractional Laplace Problems Arising From Fractal Heat Flow

We now consider

$$W_2(x, y, z, t) = \iiint_{V^{(\gamma)}} \rho_\alpha c_\alpha \frac{\partial^{\alpha} T(x, y, z, t)}{\partial t^{\alpha}} dV^{(\gamma)} = 0, \tag{61}$$

$$\mathrm{W}_2(x, y, z, t) = \iiint_{V^{(\gamma)}} g(x, y, z, t)\, dV^{(\gamma)} = 0. \tag{62}$$

The first law of thermodynamics in fractal media is presented as follows:

$$\iiint_{V^{(\gamma)}} \nabla^{\alpha} \cdot \left(K^{2\alpha} \nabla^{\alpha} T(x, y, z, t)\right) dV^{(\gamma)} = 0, \tag{63}$$

which reduces

$$\nabla^{\alpha} \cdot \left(K^{2\alpha} \nabla^{\alpha} T(x, y, z, t)\right) = 0. \tag{64}$$

Using 64, we get the local fractional Laplace problems arising in fractal heat flow

$$K^{2\alpha} \nabla^{2\alpha} T(x, y, z, t) = 0. \tag{65}$$

We rewrite 65 as the local fractional dimensionless Laplace equation of heat flow in fractal 3D time-space given as (Li et al., 2014)

$$\nabla^{2\alpha} T(x, y, z, t) = 0 \tag{66}$$

and the initial-boundary conditions is reported as

$$\frac{\partial^\alpha}{\partial x^\alpha} T(x, y, z, t) = \vartheta_x (y, z, t), \tag{67}$$

$$\frac{\partial^\alpha}{\partial y^\alpha} T(x, y, z, t) = \vartheta_y (x, z, t), \tag{68}$$

$$\frac{\partial^\alpha}{\partial z^\alpha} T(x, y, z, t) = \vartheta_z (x, y, t). \tag{69}$$

The local fractional dimensionless Laplace equation of heat flow in fractal 2D time-space is given as follows (Li et al., 2014):

$$\frac{\partial^{2\alpha}}{\partial x^{2\alpha}} T(x, y, t) + \frac{\partial^{2\alpha}}{\partial y^{2\alpha}} T(x, y, t) = 0, \tag{70}$$

and the initial-boundary conditions are defined as

$$\frac{\partial^\alpha}{\partial x^\alpha} T(x, y, t) = \vartheta_x (y, t), \tag{71}$$

$$\frac{\partial^\alpha}{\partial y^\alpha} T(x, y, t) = \vartheta_y (x, t). \tag{72}$$

6 The 2D Partial Differential Equations of Fractal Heat Transfer in Cantor-type Circle Coordinate Systems

For $R \in (0, +\infty)$, $\theta \in (0, 2\pi]$ and $x^{2\alpha} + y^{2\alpha} = R^{2\alpha}$, the Cantor-type circle coordinates are given as (Li et al., 2014)

$$\begin{cases} x^\alpha = R^\alpha \cos_\alpha \theta^\alpha, \\ y^\alpha = R^\alpha \sin_\alpha \theta^\alpha. \end{cases} \tag{73}$$

In view of 73, we get the relations

$$\nabla^\alpha \phi (R, \theta) = e_{\mathrm{R}}^\alpha \frac{\partial^\alpha}{\partial R^\alpha} \phi + e_\theta^\alpha \frac{1}{R^\alpha} \frac{\partial^\alpha}{\partial \theta^\alpha} \phi, \tag{74}$$

$$\nabla^{2\alpha} \phi (R, \theta) = \frac{\partial^{2\alpha}}{\partial R^{2\alpha}} \phi + \frac{1}{R^{2\alpha}} \frac{\partial^{2\alpha}}{\partial \theta^{2\alpha}} \phi + \frac{1}{R^\alpha} \frac{\partial^\alpha}{\partial R^\alpha} \phi, \tag{75}$$

with

$$\begin{cases} e_R^\alpha = \cos_\alpha \theta^\alpha e_1^\alpha + \sin_\alpha \theta^\alpha e_2^\alpha, \\ e_\theta^\alpha = -\sin_\alpha \theta^\alpha e_1^\alpha + \cos_\alpha \theta^\alpha e_2^\alpha, \end{cases} \tag{76}$$

$$r = R^\alpha \cos_\alpha \theta^\alpha e_1^\alpha + R^\alpha \sin_\alpha \theta^\alpha e_2^\alpha = r_R e_R^\alpha + r_\theta e_\theta^\alpha. \tag{77}$$

Using 75, the local fractional dimensionless non-homogeneous heat equation in fractal 2D time-space is transferred into (Yang et al., 2013)

$$\frac{\partial^{2\alpha}}{\partial R^{2\alpha}}\psi(R,\theta,t) + \frac{1}{R^{2\alpha}}\frac{\partial^{2\alpha}}{\partial \theta^{2\alpha}}\psi(R,\theta,t) + \frac{1}{R^\alpha}\frac{\partial^\alpha}{\partial R^\alpha}\psi(R,\theta,t) + g(R,\theta,t) - \frac{\partial^\alpha T(R,\theta,t)}{\partial t^\alpha} = 0 \tag{78}$$

Using 75, the local fractional dimensionless homogeneous heat equation in fractal 2D time-space is changed into

$$\frac{\partial^{2\alpha}}{\partial R^{2\alpha}}\psi(R,\theta,t) + \frac{1}{R^{2\alpha}}\frac{\partial^{2\alpha}}{\partial \theta^{2\alpha}}\psi(R,\theta,t) + \frac{1}{R^\alpha}\frac{\partial^\alpha}{\partial R^\alpha}\psi(R,\theta,t) - \frac{\partial^\alpha T(R,\theta,t)}{\partial t^\alpha} = 0. \tag{79}$$

Using 75, the local fractional dimensionless Poisson equation of heat flow in fractal 2D time-space become into (Li et al., 2014)

$$\frac{\partial^{2\alpha}}{\partial R^{2\alpha}}\psi(R,\theta,t) + \frac{1}{R^{2\alpha}}\frac{\partial^{2\alpha}}{\partial \theta^{2\alpha}}\psi(R,\theta,t) + \frac{1}{R^\alpha}\frac{\partial^\alpha}{\partial R^\alpha}\psi(R,\theta,t) + g(R,\theta,t) = 0. \tag{80}$$

Using 75, the local fractional dimensionless Laplace equation of heat flow in fractal 2D time-space become (Li et al., 2014)

$$\frac{\partial^{2\alpha}}{\partial R^{2\alpha}}\psi(R,\theta,t) + \frac{1}{R^{2\alpha}}\frac{\partial^{2\alpha}}{\partial \theta^{2\alpha}}\psi(R,\theta,t) + \frac{1}{R^\alpha}\frac{\partial^\alpha}{\partial R^\alpha}\psi(R,\theta,t) = 0. \tag{81}$$

7 Conclusions

Local fractional vector calculus was applied to model the partial differential equations arising from fractal heat transfer. The homogeneous and non-homogeneous heat, Poisson and Laplace equations of fractal heat transfer were discussed and the 2D partial differential equations of fractal heat transfer in Cantor-type circle coordinate systems were also considered.

Bibliography

Mandelbrot, B. B. (1983). The fractal geometry of nature. New York, NY: Macmillan.

Falconer, K. J. (1986). The geometry of fractal sets. Cambridge University Press.

Family, F., & Vicsek, T. (Eds.). (1991). Dynamics of fractal surfaces. New York, NY: World Scientific. Singapore.

Nottale, L. (1993). Fractal space-time and microphysics. New York, NY: World Scientific, Singapore.

Peters, E. E. (1994). Fractal market analysis: applying chaos theory to investment and economics. New York, NY: John Wiley & Sons.

West, B. J. (1994). Fractal physiology. New York, NY: Oxford University Press.

Massopust, P. R. (1994). Fractal functions, fractal surfaces, and wavelets. New York, SD: Academic Press.

Edgar, G. (2007). Measure, topology, and fractal geometry. New York, NY: Springer.

Lapidus, M. L., Van Frankenhuysen, M. (2000). Fractal Geometry and Number Theory: Complex dimensions of fractal strings and zeros of zeta functions. Boston: NY: Birkhäuser.

Meyer, J. P., van der Vyver, H. (2005). Heat transfer characteristics of a quadratic Koch island fractal heat exchanger. *Heat transfer engineering*, 26(9), 22-29.

Wang, S., Komvopoulos, K. (1994). A fractal theory of the interfacial temperature distribution in the slow sliding regime: Part I—elastic contact and heat transfer analysis. *Journal of Tribology*, 116(4), 812-822.

Xiao, B., Yu, B. (2007). A fractal model for critical heat flux in pool boiling. *International Journal of Thermal Sciences*, 46(5), 426-433.

Zhang, L. Z. (2011). Heat and mass transfer in a randomly packed hollow fiber membrane module: a fractal model approach. *International Journal of Heat and Mass Transfer*, 54(13), 2921-2931.

Cai, J., Huai, X. (2010). Study on fluid–solid coupling heat transfer in fractal porous medium by lattice Boltzmann method. *Applied Thermal Engineering*, 30(6), 715-723.

Liu, S., Zhang, Y., Liu, P. (2007). Heat transfer and pressure drop in fractal microchannel heat sink for cooling of electronic chips. *Heat and Mass Transfer*, 44(2), 221-227.

Yang, X. J. (2012). Advanced local fractional calculus and its applications. World Science, New York, NY, USA.

Yang, X. J., Baleanu, D., Tenreiro Machado, J. A. (2013). Systems of Navier-Stokes equations on Cantor sets. *Mathematical Problems in Engineering*, 2013, Article ID 769724, 8 pages.

Zhao, Y., Baleanu, D., Cattani, C., et al. (2013). Maxwell's equations on Cantor sets: a local fractional approach. *Advances in High Energy Physics*, 2013, Article ID 686371, 6 pages.

Li, Y. Y., Zhao, Y., Xie, G. N., et al. (2014). Local fractional Poisson and Laplace equations with applications to electrostatics in fractal domain. *Advances in Mathematical Physics*, 2014, 5.

Yang, A. M., Cattani, C., Zhang, et al. (2014). Local fractional Fourier series solutions for nonhomogeneous heat equations arising in fractal heat flow with local fractional derivative. *Advances in Mechanical Engineering*, 2014, Article ID 514639, 5 pages.

Yang, X. J., Srivastava, H. M., He, J. H., et al. (2013). Cantor-type cylindrical-coordinate method for differential equations with local fractional derivatives. *Physics Letters A*, 377(28), 1696-1700.

Hu, M. S., Baleanu, D., Yang, X. J. (2013). One-phase problems for discontinuous heat transfer in fractal media. *Mathematical Problems in Engineering*, 2013, Article ID 358473, 3 pages.

He, J. H. J. F. Liu, (2013). Local fractional variational iteration method for fractal heat transfer in silk cocoon hierarchy. *Nonlinear Science Letters A*, 4(1), 15-20.

Yang, X. J., Baleanu, D. (2013). Fractal heat conduction problem solved by local fractional variation iteration method. *Thermal Science*, 17(2), 625-628.

Liu, C. F., Kong, S. S., Yuan, S. J. (2013). Reconstructive schemes for variational iteration method within Yang-Laplace transform with application to fractal heat conduction problem. *Thermal Science*, 17(3), 715-721.

Yang, A. M., Zhang, Y. Z., Long, Y. (2013). The Yang-Fourier transforms to heat-conduction in a semi-infinite fractal bar. *Thermal Science*, 17(3), 707-713.

Yang, X. J., Zhang, Y., & Yang, A. (2013). 1-D heat conduction in a fractal medium: a solution by the local fractional Fourier series method. *Thermal Science*, 17(3), 953-956.

Yang, X. J., Baleanub, D., Lazarević, M. P. et al. (2014). Fractal boundary value problems for integral and differential equations with local fractional operators. *Thermal Science*, 2014.

Yang, A. M., Cattani, C., Jafari, H., et al. (2013). Analytical solutions of the one-dimensional heat equations arising in fractal transient conduction with local fractional derivative. *Abstract and Applied Analysis*, 2013, Article ID 462535, 5 pages.

Hossein Jafari, Hassan Kamil Jassim, and Syed Tauseef Mohyud-Din

Local Fractional Laplace Decomposition Method for Solving Linear Partial Differential Equations with Local Fractional Derivative

Abstract: In this chapter, the local fractional Laplace decomposition method (LFLDM), coupled with the local fractional Adomian decomposition method and Laplace transform, is first proposed in order to solve linear partial differential equations with local fractional derivatives. Some examples are given to illustrate that this method provides us with a convenient way to find non-differentiable solutions to local fractional differential equations.

1 Introduction

The Adomian decomposition method was applied to solve the differential, integral and integro-differential equations arising in mathematical physics (Adomian, 1994; Adomian, 1990; Adomian, 1991; Cherruault et al. 1992; Babolian et al. 2004; Hashim, 2006; Abbasbandy, 2006). Based on it, the Laplace decomposition method (Syam and Hamdan, 2006) has found successful applications in physics and applied mathematics. For example, the numerical solution for the Duffing equation was considered (Yusufoğlu, 2006). An effective modification for solving nonlinear equations was presented in (Khan, 2009). The solutions for nonlinear Volterra integro–differential equation (Wazwaz, 2010) and diffusion–wave (Jafari et al., 2011) and advection equations (Khan and Austin, 2010) were reported.

The local fractional calculus theory (Yang, 2011; Yang, 2012) has attracted a lot of interest for scientists and engineers because it is applied to model problems in fractal mathematics and engineering. It plays a key role in many applications in several fields, such as physics (Yang, 2012; Kolwankar and Gangal, 1998; Yang et al., 2013; Jafari et al. (2014), Zhao et al., 2013; Yang et al., 2013; Li et al., 2014), heat conduction theory (Yang, 2012; Yang et al., 2013; Yang et al., 2013; Xu et al., 2014), fracture and elasticity mechanics (Yang, 2012; Carpinteri et al., 2001; Carpinteri et al., 2004), fluid mechan-

Hossein Jafari: Department of Mathematics, Faculty of Mathematical Sciences, University of Mazandaran, Babolsar, Iran, E-mail: Jafari@umz.ac.ir
Hassan Kamil Jassim: Department of Mathematics, Faculty of Mathematical Sciences, University of Mazandaran, Babolsar, Iran,
and Department of Mathematics, Faculty of Education for Pure Sciences, University of Thi-Qar, Nasiriyah, Iraq, E-mail: hassan.kamil28@yahoo.com
Syed Tauseef Mohyud-Din: Faculty of Sciences, HITEC University, Taxila Cantt, Pakistan, E-mail: syedtauseefs@hotmail.com

ics (Yang, 2012; Yang et al., 2013) and so on. The local fractional partial differential equations arising in mathematical physics describe the non-differentiable behaviors of physical laws. Finding non-differentiable solutions is the hot topics. Useful techniques have been successfully applied to local fractional differential equations. The main techniques include the decomposition method (Yang et al., 2013; Jafari et al., (2014), Baleanu et al., 2014; Yan et al., 2014) and Laplace transform method (He, 2012; Zhao et al., 2013; Liu et al., 2013; Wang et al., 2014; Jafari et al., 2014; Li et al., 2014; Cao et al., 2014; He, 2014; Yan et al., 2014) with local fractional operator. In this chapter, our aim is to use the local fractional Laplace decomposition method to solve linear local fractional partial differential equations. The structure of the chapter is suggested as follows. In Section 2, the basic theory of local fractional calculus and local fractional Laplace transform are introduced. In Section 3, we give analysis of the methods used. In Section 4, some examples for the local fractional partial differential equations are given. Finally, conclusions are offered in Section 5.

2 Mathematical Fundamentals

In this section, we present the basic theory of local fractional calculus and concept of the local fractional Laplace transform (Yang, 2011; Yang, 2012; Baleanu et al., 2014; He, 2012; Zhao et al., 2013; Liu et al., 2013; Wang et al., 2014; Yang et al., 2014; Li et al., 2014; Cao et al., 2014; He, 2014; Yan et al., 2014).

Definition 1 (Yang, 2012; Yang et al., 2013; Baleanu et al., 2014; Yan et al., 2014)

Suppose that one has the relation

$$\left|f(x) - f(x_0)\right| < \varepsilon^{\alpha} \ , \ 0 < \alpha \leq 1 \tag{1}$$

with $|x - x_0| < \delta$, for $\varepsilon, \delta > 0$ and $\varepsilon, \delta \in R$, then the function $f(x)$ is called local fractional continuous at $x = x_0$ and it is denoted by $\lim_{x \to x_0} f(x) = f(x_0)$.

Definition 2 (Yang, 2012; Yang et al., 2013; Baleanu et al., 2014; Yan et al., 2014)

Suppose that the function $f(x)$ satisfies condition 1, for $x \in (a, b)$; it is then called local fractional continuous on the interval (a, b), denoted by $f(x) \in C_\alpha(a, b)$.

Definition 3 (Yang, 2012; Yang et al., 2013; Baleanu et al., 2014; Yan et al., 2014)

In fractal space, let $f(x) \in C_\alpha(a, b)$, then the local fractional derivative of $f(x)$ of order α at $x = x_0$ is given by

$$D_x^{\alpha} f(x_0) = \frac{d^{\alpha}}{dx^{\alpha}} f(x)\,|_{x=x_0} = f^{(\alpha)}(x_0) = \lim_{x \to x_0} \frac{\Delta^{\alpha}(f(x) - f(x_0))}{(x - x_0)^{\alpha}}, \tag{2}$$

where $\Delta^{\alpha}(f(x) - f(x_0)) \cong \Gamma(\alpha + 1)(f(x) - f(x_0))$.

Definition 4 (Yang, 2012; Yang et al., 2013; Baleanu et al., 2014; Yan et al., 2014)

A partition of the interval $[a, b]$ is denoted as (t_j, t_{j+1}), $j = 0, \ldots, N-1$, $t_0 = a$ and $t_N = b$ with $\Delta t_j = t_{j+1} - t_j$ and $\Delta t = \max\{\Delta t_0, \Delta t_1, \ldots..\}$. Local fractional integral of $f(x)$ in the interval $[a, b]$ is given by

$$ {}_aI_b^{(\alpha)} f(x) = \frac{1}{\Gamma(1+\alpha)} \int_a^b f(t)\,(dt)^\alpha = \frac{1}{\Gamma(1+\alpha)} \lim_{\Delta t \to 0} \sum_{j=0}^{N-1} f(t_j)\,(\Delta t_j)^\alpha . \tag{3} $$

Definition 5 (He, 2012; Zhao et al., 2013; Liu et al., 2013; Wang et al., 2014; Yang et al., 2014; Li et al., 2014; Cao et al., 2014; He, 2014; Yan et al., 2014)

Let $\frac{1}{\Gamma(1+\alpha)} \int_0^\infty |f(x)|\,(dx)^\alpha < k < \infty$. The local fractional Laplace transform of $f(x)$ is given by

$$ \tilde{L}_\alpha\{f(x)\} = f_s^{L,\alpha}(s) = \frac{1}{\Gamma(1+\alpha)} \int_0^\infty E_\alpha(-s^\alpha x^\alpha) f(x)\,(dx)^\alpha , \quad 0 < \alpha \le 1, \tag{4} $$

where the latter integral converges.

Definition 6 (He, 2012; Zhao et al., 2013; Liu et al., 2013; Wang et al., 2014; Yang et al., 2014; Li et al., 2014; Cao et al., 2014; He, 2014; Yan et al., 2014)

The inverse formula of the local fractional Laplace transforms of $f(x)$is given as follows:

$$ \tilde{L}_\alpha^{-1}\{f_s^{L,\alpha}(s)\} = f(t) = \frac{1}{(2\pi)^\alpha} \int_{\beta - i\omega}^{\beta + i\omega} E_\alpha(s^\alpha x^\alpha) f_s^{L,\alpha}(s)\,(ds)^\alpha , \quad 0 < \alpha \le 1 , \tag{5} $$

where $s^\alpha = \beta^\alpha + i^\alpha \omega^\alpha$, fractal imaginary unit i^α and $Re(s) = \beta > 0$.

The properties for local fractional Laplace transform used in the paper are given as (Yang, 2011):

$$ \tilde{L}_\alpha \left\{ af(x) + bg(x) \right\} = af_s^{L,\alpha}(s) + bg_s^{L,\alpha}(s), \tag{6} $$

$$ \tilde{L}_\alpha \left\{ E_\alpha(c^\alpha x^\alpha) f(x) \right\} = f_s^{L,\alpha}(s - c), \tag{7} $$

$$ \tilde{L}_\alpha \left\{ f^{(k\alpha)}(x) \right\} = s^{k\alpha} f_s^{L,\alpha}(s) - s^{(k-1)\alpha} f(0) - s^{(k-2)\alpha} f^{(\alpha)}(0) - \cdots - f^{((k-1)\alpha)}(0), \tag{8} $$

$$ \tilde{L}_\alpha \left\{ E_\alpha(a^\alpha x^\alpha) \right\} = \frac{1}{s^\alpha - a^\alpha}, \tag{9} $$

$$ \tilde{L}_\alpha \left\{ \sin_\alpha(a^\alpha x^\alpha) \right\} = \frac{a^\alpha}{s^{2\alpha} + a^{2\alpha}}, \tag{10} $$

$$ \tilde{L}_\alpha \left\{ x^{k\alpha} \right\} = \frac{\Gamma(1 + k\alpha)}{s^{(k+1)\alpha}}. \tag{11} $$

3 Local Fractional Laplace Decomposition Method

The local fractional decomposition method has been developed and applied to solve a class of local fractional partial differential equations (Yang et al., 2013; Baleanu et al., 2014; Yan et al., 2014). Based on it, we suggest a new analytical method.

Let us consider the following linear operator with local fractional derivative:

$$L_\alpha u(x,t) + R_\alpha u(x,t) = h(x,t), \tag{12}$$

where $L_\alpha = \frac{\partial^{k\alpha}}{\partial x^{k\alpha}}$ denotes the linear local fractional differential operator, R_α is the remaining linear operator, and $h(x,t)$ is a source term.

Taking local fractional Laplace transform on Eq. 12, we obtain

$$\tilde{L}_\alpha \left\{L_\alpha u(x,t)\right\} + \tilde{L}_\alpha \left\{R_\alpha u(x,t)\right\} = \tilde{L}_\alpha \left\{h(x,t)\right\} \tag{13}$$

By applying the local fractional Laplace transform differentiation property, we have that:

$$\begin{aligned} &s^{k\alpha}\tilde{L}_\alpha \left\{u(x,t)\right\} - s^{(k-1)\alpha}u(0,t) - s^{(k-2)}u^{(\alpha)}(0,t) - \cdots - u^{((k-1)\alpha)}(0,t) + \tilde{L}_\alpha \left\{R_\alpha u(x,t)\right\} \\ &= \tilde{L}_\alpha \left\{h(x,t)\right\}, \end{aligned} \tag{14}$$

or

$$\begin{aligned} &\tilde{L}_\alpha \left\{u(x,t)\right\} \\ &= \frac{1}{s^\alpha}u(0,t) + \frac{1}{s^{2\alpha}}u^{(\alpha)}(0,t) + \cdots + \frac{1}{s^{k\alpha}}u^{((k-1)\alpha)}(0,t) + \frac{1}{s^{k\alpha}}\tilde{L}_\alpha \left\{h(x,t)\right\} - \frac{1}{s^{k\alpha}}\tilde{L}_\alpha \left\{R_\alpha(x,t)\right\} \end{aligned} \tag{15}$$

Taking the inverse of local fractional Laplace transform of 15, we obtain

$$\begin{aligned} &u(x,t) \\ &= u(0,t) + \frac{x^\alpha}{\Gamma(1+\alpha)}u^{(\alpha)}(0,t) + \cdots + \frac{x^{(k-1)}}{\Gamma\left[1+(k-1)\alpha\right]}u^{((k-1)\alpha)}(0,t) + \tilde{L}_\alpha^{-1}\left(\frac{1}{s^{k\alpha}}\tilde{L}_\alpha \left\{h(x,t)\right\}\right) \\ &-\tilde{L}_\alpha^{-1}\left(\frac{1}{s^{k\alpha}}\tilde{L}_\alpha \left\{R_\alpha u(x,t)\right\}\right). \end{aligned} \tag{16}$$

We are going to represent the solution in an infinite series given by:

$$u(x,t) = \sum_{n=0}^{\infty} u_n(x,t). \tag{17}$$

Substituting 17 into 16) gives us the result that

$$\begin{aligned} &\sum_{n=0}^{\infty} u_n(x,t) \\ &= u(0,t) + \frac{x^\alpha}{\Gamma(1+\alpha)}u^{(\alpha)}(0,t) + \cdots + \frac{x^{(k-1)\alpha}}{\Gamma\left[1+(k-1)\alpha\right]}u^{((k-1)\alpha)}(0,t) + \tilde{L}_\alpha^{-1}\left(\frac{1}{s^{k\alpha}}\tilde{L}_\alpha \left\{h(x,t)\right\}\right) \\ &-\tilde{L}_\alpha^{-1}\left(\frac{1}{s^{k\alpha}}\tilde{L}_\alpha \left\{R_\alpha \sum_{n=0}^{\infty} u_n(x,t)\right\}\right). \end{aligned} \tag{18}$$

When we compare the left and right hand sides of 18 we obtain

$$\begin{aligned} u_0(x,t) =& u(0,t) + \frac{x^{\alpha}}{\Gamma(1+\alpha)} u^{(\alpha)}(0,t) + \cdots \\ &+ \frac{x^{(k-1)\alpha}}{\Gamma\left[1+(k-1)\alpha\right]} u^{((k-1)\alpha)}(0,t) + \tilde{L}_{\alpha}^{-1}\left(\frac{1}{s^{k\alpha}}\tilde{L}_{\alpha}\left\{h(x,t)\right\}\right) \end{aligned} \tag{19}$$

$$u_1(x,t) = -\tilde{L}_{\alpha}^{-1}\left(\frac{1}{s^{k\alpha}}\tilde{L}_{\alpha}\left\{R_{\alpha}u_0(x,t)\right\}\right), \tag{20}$$

$$u_2(x,t) = -\tilde{L}_{\alpha}^{-1}\left(\frac{1}{s^{k\alpha}}\tilde{L}_{\alpha}\left\{R_{\alpha}u_1(x,t)\right\}\right), \tag{21}$$

$$\cdots\cdots$$

and so on.

The local fractional recursive relation in its general form is

$$\begin{aligned} u_0(x,t) =& u(0,t) + \frac{x^{\alpha}}{\Gamma(1+\alpha)} u^{(\alpha)}(0,t) + \cdots \\ &+ \frac{x^{(k-1)\alpha}}{\Gamma\left[1+(k-1)\alpha\right]} u^{((k-1)\alpha)}(0,t) + \tilde{L}_{\alpha}^{-1}\left(\frac{1}{s^{k\alpha}}\tilde{L}_{\alpha}\left\{h(x,t)\right\}\right), \end{aligned} \tag{22}$$

$$u_{n+1}(x,t) = -\tilde{L}_{\alpha}^{-1}\left(\frac{1}{s^{k\alpha}}\tilde{L}_{\alpha}\left\{R_{\alpha}u_n(x,t)\right\}\right). \tag{23}$$

We notice that the above formulas are only valid for non-differentiable functions.

4 Illustrative Examples

In this section, some examples for linear local fractional differential equations are presented in order to demonstrate the simplicity and the efficiency of the above method.

Example 1. The local fractional differential equation is given by:

$$\frac{\partial^{3\alpha}u(x,t)}{\partial t^{3\alpha}} = \frac{\partial^{2\alpha}u(x,t)}{\partial x^{2\alpha}}, \tag{24}$$

subject to the initial values given as

$$u(0,t) = E_{\alpha}(-t^{\alpha}), \quad \frac{\partial^{\alpha}u(0,t)}{\partial x^{\alpha}} = 0. \tag{25}$$

In view of 24 and 25 the local fractional iteration algorithms can be written as follows:

$$u_0(x,t) = E_{\alpha}(-t^{\alpha}), \tag{26}$$

$$u_{n+1}(x,t) = \tilde{L}_{\alpha}^{-1}\left(\frac{1}{s^{2\alpha}}\tilde{L}_{\alpha}\left\{\frac{\partial^{3\alpha}u_n(x,t)}{\partial t^{3\alpha}}\right\}\right), \quad n \geq 0. \tag{27}$$

Therefore, from 26 and 27 we give the components as follows:

$$u_0(x,t) = E_\alpha(-t^\alpha), \tag{28}$$

$$u_1(x,t) = \tilde{L}_\alpha^{-1}\left(\frac{1}{s^{2\alpha}}\tilde{L}_\alpha\left\{\frac{\partial^{3\alpha}u_0(x,t)}{\partial t^{3\alpha}}\right\}\right) = \tilde{L}_\alpha^{-1}\left(-\frac{1}{s^{3\alpha}}E_\alpha(-t^\alpha)\right) = -\frac{x^{2\alpha}}{\Gamma(1+2\alpha)}E_\alpha(-t^\alpha), \tag{29}$$

$$u_2(x,t) = \tilde{L}_\alpha^{-1}\left(\frac{1}{s^{2\alpha}}\tilde{L}_\alpha\left\{\frac{\partial^{3\alpha}u_1(x,t)}{\partial t^{3\alpha}}\right\}\right) = \tilde{L}_\alpha^{-1}\left(\frac{1}{s^{5\alpha}}E_\alpha(-t^\alpha)\right) = \frac{x^{4\alpha}}{\Gamma(1+4\alpha)}E_\alpha(-t^\alpha), \tag{30}$$

$$u_3(x,t) = \tilde{L}_\alpha^{-1}\left(\frac{1}{s^{2\alpha}}\tilde{L}_\alpha\left\{\frac{\partial^{3\alpha}u_2(x,t)}{\partial t^{3\alpha}}\right\}\right) = \tilde{L}_\alpha^{-1}\left(-\frac{1}{s^{7\alpha}}E_\alpha(-t^\alpha)\right) = -\frac{x^{6\alpha}}{\Gamma(1+6\alpha)}E_\alpha(-t^\alpha),$$
$$\ldots\ldots \tag{31}$$

and so on.

Consequently, we obtain

$$\begin{aligned} u(x,t) &= E_\alpha(-t^\alpha)\left(1 - \frac{x^{2\alpha}}{\Gamma(1+2\alpha)} + \frac{x^{4\alpha}}{\Gamma(1+4\alpha)} - \frac{x^{6\alpha}}{\Gamma(1+6\alpha)} + \cdots\right) \\ &= E_\alpha(-t^\alpha)\sum_{k=0}^{\infty}(-1)^k\frac{x^{2k\alpha}}{\Gamma(1+2k\alpha)} \\ &= E_\alpha(-t^\alpha)\cos_\alpha(x^\alpha). \end{aligned} \tag{32}$$

The non-differentiable solution of 24 is shown in Fig. 1 where $\alpha = \ln 2/\ln 3$

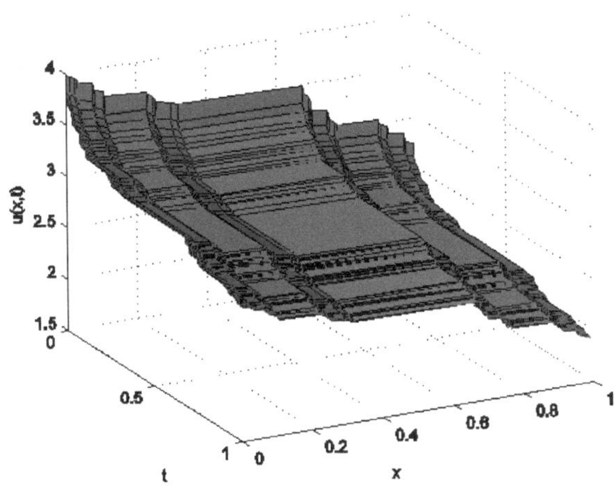

Fig. 1. The plot of solution of 24 where $\alpha = \ln 2/\ln 3$.

Example 2. We report the following local fractional partial differential equation:

$$\frac{\partial^{2\alpha}u(x,t)}{\partial t^{2\alpha}} = \frac{\partial^{3\alpha}u(x,t)}{\partial x^{3\alpha}}, \tag{33}$$

subject to the initial values

$$u(0,t)=0\ ,\quad \frac{\partial^{\alpha}u(0,t)}{\partial x^{\alpha}}=E_{\alpha}(t^{\alpha})\ ,\quad \frac{\partial^{2\alpha}u(0,t)}{\partial x^{2\alpha}}=0 \tag{34}$$

Making use of 33 and 34, the local fractional iteration algorithm reads as follows:

$$u_0(x,t)=\frac{x^{\alpha}}{\Gamma(1+\alpha)}E_{\alpha}(t^{\alpha})\,, \tag{35}$$

$$u_{n+1}(x,t)=\tilde{L}_{\alpha}^{-1}\left(\frac{1}{s^{3\alpha}}\tilde{L}_{\alpha}\left\{\frac{\partial^{2\alpha}u_n(x,t)}{\partial t^{2\alpha}}\right\}\right)\ ,\quad n\geq 0\,. \tag{36}$$

Therefore, from 36 we obtain the components as follows:

$$u_0(x,t)=\frac{x^{\alpha}}{\Gamma(1+\alpha)}E_{\alpha}(t^{\alpha})\,, \tag{37}$$

$$u_1(x,t)=\tilde{L}_{\alpha}^{-1}\left(\frac{1}{s^{3\alpha}}\tilde{L}_{\alpha}\left\{\frac{\partial^{2\alpha}u_0(x,t)}{\partial t^{2\alpha}}\right\}\right)=\tilde{L}_{\alpha}^{-1}\left(\frac{1}{s^{5\alpha}}E_{\alpha}(t^{\alpha})\right)=\frac{x^{4\alpha}}{\Gamma(1+4\alpha)}E_{\alpha}(t^{\alpha})\,, \tag{38}$$

$$u_2(x,t)=\tilde{L}_{\alpha}^{-1}\left(\frac{1}{s^{3\alpha}}\tilde{L}_{\alpha}\left\{\frac{\partial^{2\alpha}u_1(x,t)}{\partial t^{2\alpha}}\right\}\right)=\tilde{L}_{\alpha}^{-1}\left(\frac{1}{s^{8\alpha}}E_{\alpha}(t^{\alpha})\right)=\frac{x^{7\alpha}}{\Gamma(1+7\alpha)}E_{\alpha}(t^{\alpha})\,, \tag{39}$$

$$u_3(x,t)=\tilde{L}_{\alpha}^{-1}\left(\frac{1}{s^{3\alpha}}\tilde{L}_{\alpha}\left\{\frac{\partial^{2\alpha}u_2(x,t)}{\partial t^{2\alpha}}\right\}\right)=\tilde{L}_{\alpha}^{-1}\left(\frac{1}{s^{11\alpha}}E_{\alpha}(t^{\alpha})\right)=\frac{x^{10\alpha}}{\Gamma(1+10\alpha)}E_{\alpha}(t^{\alpha})\,,$$

$$\ldots\ldots \tag{40}$$

and so on. Their plots are given in Fig. 2, 3, 4 and 5.

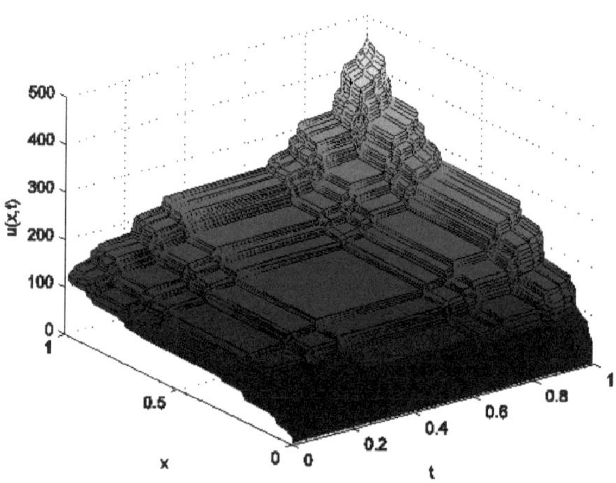

Fig. 2. The plot of solution of $u_0(x,t)$ where $\alpha=\ln 2/\ln 3$.

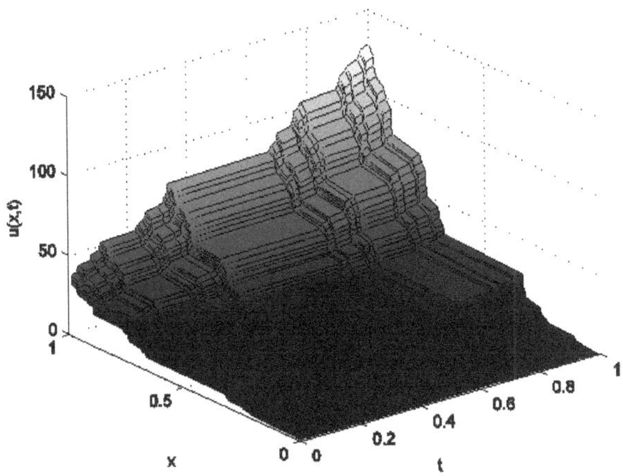

Fig. 3. The plot of solution of $u_1(x, t)$ where $\alpha = \ln 2/\ln 3$.

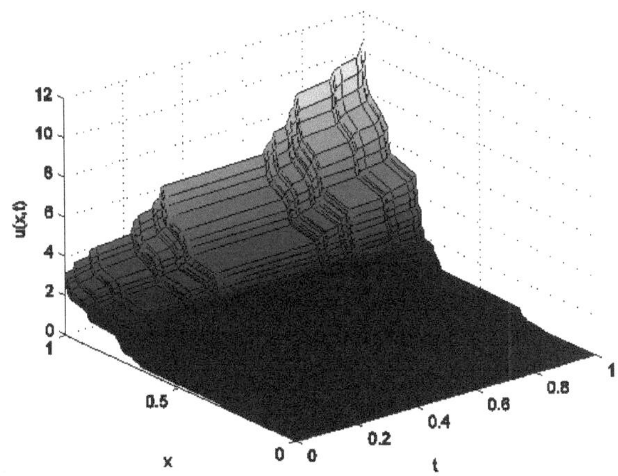

Fig. 4. The plot of solution of $u_2(x, t)$ where $\alpha = \ln 2/\ln 3$.

Hence, the approximate solution of 33 is given by

$$\begin{aligned} u(x,t) & \\ &= E_\alpha(t^\alpha)\left(\frac{x^\alpha}{\Gamma(1+\alpha)}+\frac{x^{4\alpha}}{\Gamma(1+4\alpha)}+\frac{x^{7\alpha}}{\Gamma(1+7\alpha)}+\frac{x^{11\alpha}}{\Gamma(1+11\alpha}+\cdots\right) \\ &= E_\alpha(t^\alpha)\sum_{k=0}^{\infty}\frac{x^{(3k+1)\alpha}}{\Gamma(1+(3k+1)\alpha)}. \end{aligned} \tag{41}$$

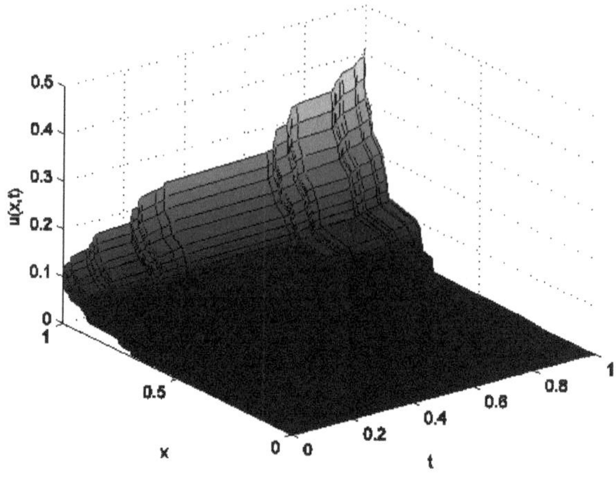

Fig. 5. The plot of $u_3(x, t)$ where $\alpha = \ln 2/\ln 3$.

Example 3. The following local fractional partial differential equation

$$\frac{\partial^{2\alpha}u(x,t)}{\partial t^{2\alpha}} - \frac{\partial^{4\alpha}u(x,t)}{\partial x^{4\alpha}} = 0 \tag{42}$$

is presented and its initial values are defined as follows:

$$u(0,t) = 0\,, \quad \frac{\partial^{\alpha}u(0,t)}{\partial x^{\alpha}} = 0\,, \quad \frac{\partial^{2\alpha}u(0,t)}{\partial x^{2\alpha}} = 0\,, \quad \frac{\partial^{3\alpha}u(0,t)}{\partial x^{3\alpha}} = E_\alpha(t^\alpha)\,. \tag{43}$$

In view of 42 and 43, the local fractional iteration algorithms are as follows:

$$u_0(x,t) = \frac{x^{3\alpha}}{\Gamma(1+3\alpha)}E_\alpha(t^\alpha)\,, \tag{44}$$

$$u_{n+1}(x,t) = \tilde{L}_\alpha^{-1}\left(\frac{1}{s^{4\alpha}}\tilde{L}_\alpha\left\{\frac{\partial^{2\alpha}u_n(x,t)}{\partial t^{2\alpha}}\right\}\right)\,, \quad n \geq 0\,. \tag{45}$$

Therefore, using 45 we give the components as follows:

$$u_0(x,t) = \frac{x^{3\alpha}}{\Gamma(1+3\alpha)}E_\alpha(t^\alpha)\,, \tag{46}$$

$$u_1(x,t) = \tilde{L}_\alpha^{-1}\left(\frac{1}{s^{4\alpha}}\tilde{L}_\alpha\left\{\frac{\partial^{2\alpha}u_0(x,t)}{\partial t^{2\alpha}}\right\}\right) = \tilde{L}_\alpha^{-1}\left(\frac{1}{s^{8\alpha}}E_\alpha(t^\alpha)\right) = \frac{x^{7\alpha}}{\Gamma(1+7\alpha)}E_\alpha(t^\alpha)\,, \tag{47}$$

$$u_2(x,t) = \tilde{L}_\alpha^{-1}\left(\frac{1}{s^{4\alpha}}\tilde{L}_\alpha\left\{\frac{\partial^{2\alpha}u_1(x,t)}{\partial t^{2\alpha}}\right\}\right) = \tilde{L}_\alpha^{-1}\left(\frac{1}{s^{12\alpha}}E_\alpha(t^\alpha)\right) = \frac{x^{11\alpha}}{\Gamma(1+11\alpha)}E_\alpha(t^\alpha)\,, \tag{48}$$

$$u_3(x,t) = \tilde{L}_\alpha^{-1}\left(\frac{1}{s^{4\alpha}}\tilde{L}_\alpha\left\{\frac{\partial^{2\alpha}u_2(x,t)}{\partial t^{2\alpha}}\right\}\right) = \tilde{L}_\alpha^{-1}\left(\frac{1}{s^{16\alpha}}E_\alpha(t^\alpha)\right) = \frac{x^{15\alpha}}{\Gamma(1+15\alpha)}E_\alpha(t^\alpha),$$
$$\ldots\ldots \tag{49}$$

and so on, Their plots are illustrated in Fig. 6, 7, 8 and 9.

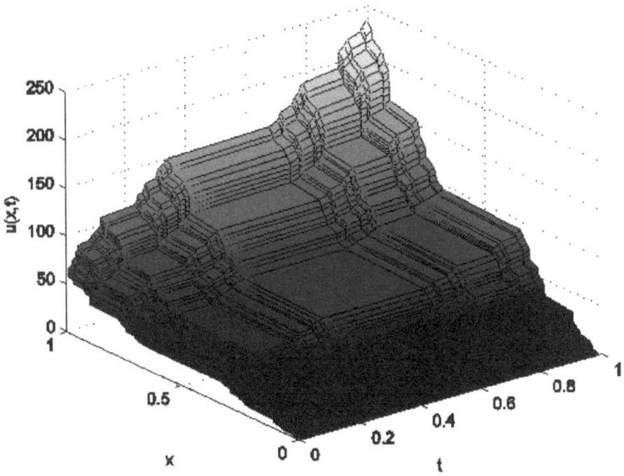

Fig. 6. The plot of solution of $u_0(x,t)$ where $\alpha = \ln 2/\ln 3$.

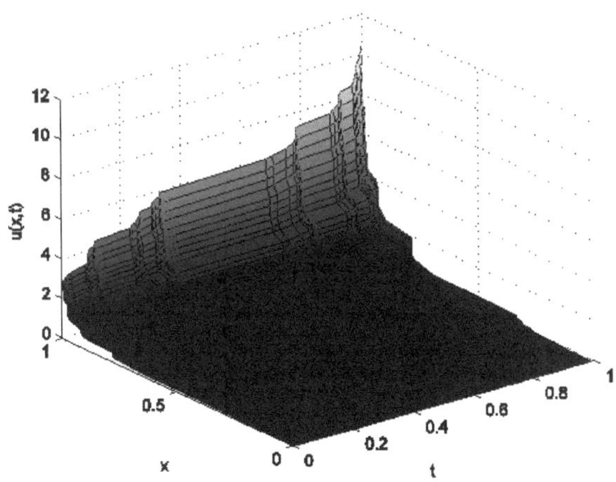

Fig. 7. The plot of solution of $u_1(x,t)$ where $\alpha = \ln 2/\ln 3$.

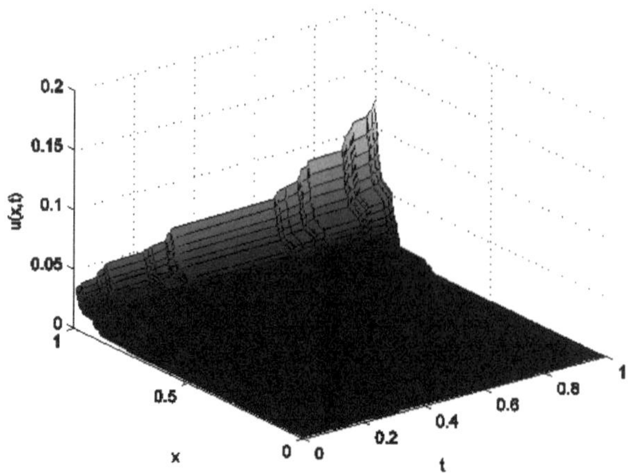

Fig. 8. The plot of solution of $u_2(x, t)$ where $\alpha = \ln 2/\ln 3$.

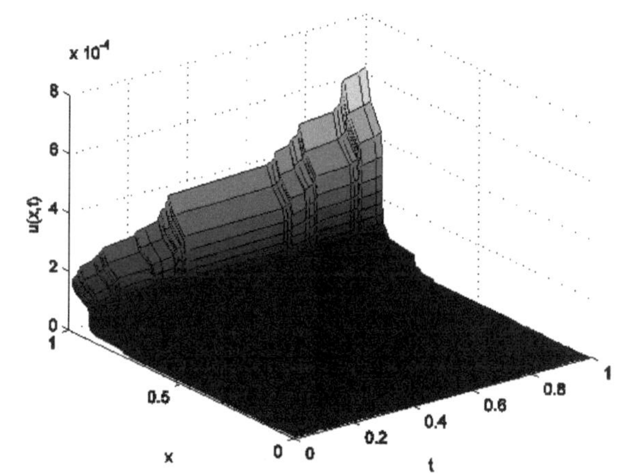

Fig. 9. The plot of $u_3(x, t)$ where $\alpha = ln2/\ln 3$.

Thereby, the non-differentiable solution of 42 is as follows:

$$\begin{aligned} &u(x,t) \\ &= E_\alpha(t^\alpha)\left(\frac{x^{3\alpha}}{\Gamma(1+3\alpha)}+\frac{x^{7\alpha}}{\Gamma(1+7\alpha)}+\frac{x^{11\alpha}}{\Gamma(1+11\alpha)}+\frac{x^{15\alpha}}{\Gamma(1+15\alpha}+\cdots\right) \\ &= E_\alpha(t^\alpha)\sum_{k=0}^{\infty}\frac{x^{(4k+3)\alpha}}{\Gamma(1+(4k+3)\alpha)}. \end{aligned} \tag{50}$$

Example 4. Let us consider the following local fractional partial differential equation

$$\frac{\partial^{\alpha} u(x,t)}{\partial t^{\alpha}} - \frac{\partial^{4\alpha} u(x,t)}{\partial x^{4\alpha}} = 0, \tag{51}$$

with initial values as follows:

$$u(0,t) = E_{\alpha}(t^{\alpha})\,, \quad \frac{\partial^{\alpha} u(0,t)}{\partial x^{\alpha}} = 0\,, \quad \frac{\partial^{2\alpha} u(0,t)}{\partial x^{2\alpha}} = 0\,, \quad \frac{\partial^{3\alpha} u(0,t)}{\partial x^{3\alpha}} = 0\,. \tag{52}$$

Appling 51 and 52, we write the local fractional iteration algorithms as

$$u_0(x,t) = E_{\alpha}(t^{\alpha})\,, \tag{53}$$

$$u_{n+1}(x,t) = \tilde{L}_{\alpha}^{-1}\left(\frac{1}{s^{4\alpha}}\tilde{L}_{\alpha}\left\{\frac{\partial^{\alpha} u_n(x,t)}{\partial t^{\alpha}}\right\}\right)\,, \quad n \geq 0\,. \tag{54}$$

Hence, from 54 we give the components as follows:

$$u_0(x,t) = E_{\alpha}(t^{\alpha})\,, \tag{55}$$

$$u_1(x,t) = \tilde{L}_{\alpha}^{-1}\left(\frac{1}{s^{4\alpha}}\tilde{L}_{\alpha}\left\{\frac{\partial^{\alpha} u_0(x,t)}{\partial t^{\alpha}}\right\}\right) = \tilde{L}_{\alpha}^{-1}\left(\frac{1}{s^{5\alpha}}E_{\alpha}(t^{\alpha})\right) = \frac{x^{4\alpha}}{\Gamma(1+4\alpha)}E_{\alpha}(t^{\alpha}), \tag{56}$$

$$u_2(x,t) = \tilde{L}_{\alpha}^{-1}\left(\frac{1}{s^{4\alpha}}\tilde{L}_{\alpha}\left\{\frac{\partial^{\alpha} u_1(x,t)}{\partial t^{\alpha}}\right\}\right) = \tilde{L}_{\alpha}^{-1}\left(\frac{1}{s^{9\alpha}}E_{\alpha}(t^{\alpha})\right) = \frac{x^{8\alpha}}{\Gamma(1+8\alpha)}E_{\alpha}(t^{\alpha}), \tag{57}$$

$$\begin{aligned} u_3(x,t) &= \tilde{L}_{\alpha}^{-1}\left(\frac{1}{s^{4\alpha}}\tilde{L}_{\alpha}\left\{\frac{\partial^{\alpha} u_2(x,t)}{\partial t^{\alpha}}\right\}\right) = \tilde{L}_{\alpha}^{-1}\left(\frac{1}{s^{13\alpha}}E_{\alpha}(t^{\alpha})\right) = \frac{x^{12\alpha}}{\Gamma(1+12\alpha)}E_{\alpha}(t^{\alpha}), \\ &\ldots\ldots \end{aligned} \tag{58}$$

and so on. Their plots are given in Fig. 10, 11, 12 and 13.

Therefore, the non-differentiable solution of 51 is:

$$\begin{aligned} u(x,t) &= E_{\alpha}(t^{\alpha})\left(1 + \frac{x^{4\alpha}}{\Gamma(1+4\alpha)} + \frac{x^{8\alpha}}{\Gamma(1+8\alpha)} + \frac{x^{12\alpha}}{\Gamma(1+12\alpha)} + \cdots\right) \\ &= E_{\alpha}(t^{\alpha})\sum_{k=0}^{\infty}\frac{x^{4k\alpha}}{\Gamma(1+4k\alpha)}\,. \end{aligned} \tag{59}$$

Example 5. The following local fractional partial differential equation is reported as

$$\frac{\partial^{\alpha} u(x,t)}{\partial t^{\alpha}} = \frac{\partial^{3\alpha} u(x,t)}{\partial x^{3\alpha}}\,, \tag{60}$$

subject to the initial values

$$u(0,t) = 0\,, \quad \frac{\partial^{\alpha} u(0,t)}{\partial x^{\alpha}} = 0\,, \quad \frac{\partial^{2\alpha} u(0,t)}{\partial x^{2\alpha}} = E_{\alpha}(t^{\alpha})\,. \tag{61}$$

Making use of 60 and 61, the local fractional iteration algorithms are suggested as

$$u_0(x,t) = \frac{x^{2\alpha}}{\Gamma(1+2\alpha)}E_{\alpha}(t^{\alpha})\,, \tag{62}$$

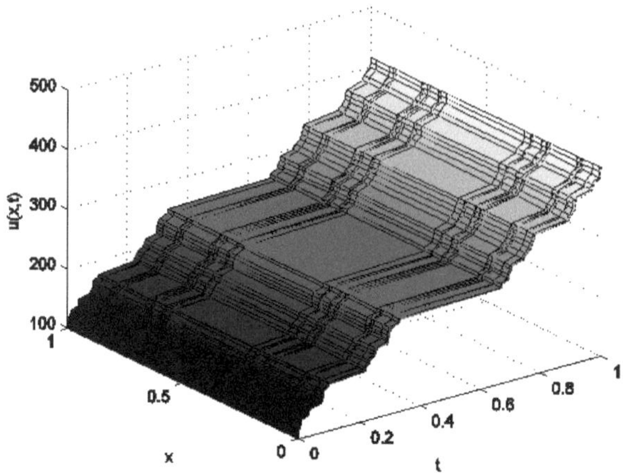

Fig. 10. The plot of solution of $u_0(x, t)$ where $\alpha = \ln 2/ \ln 3$.

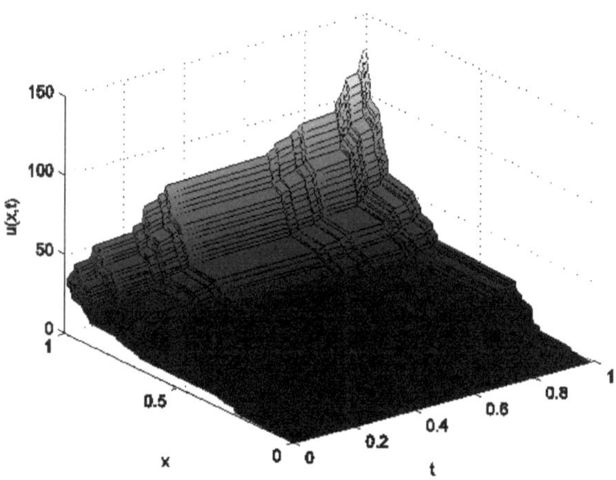

Fig. 11. The plot of solution of $u_1(x, t)$ where $\alpha = \ln 2/ \ln 3$.

$$u_{n+1}(x,t) = \tilde{L}_\alpha^{-1}\left(\frac{1}{s^{3\alpha}}\tilde{L}_\alpha\left\{\frac{\partial^\alpha u_n(x,t)}{\partial t^\alpha}\right\}\right), \quad n \geq 0. \tag{63}$$

Therefore, from 63 the components with non-differentiable terms are as follows:

$$u_0(x,t) = \frac{x^{2\alpha}}{\Gamma(1+2\alpha)} E_\alpha(t^\alpha), \tag{64}$$

$$u_1(x,t) = \tilde{L}_\alpha^{-1}\left(\frac{1}{s^{3\alpha}}\tilde{L}_\alpha\left\{\frac{\partial^\alpha u_0(x,t)}{\partial t^\alpha}\right\}\right) = \tilde{L}_\alpha^{-1}\left(\frac{1}{s^{6\alpha}}E_\alpha(t^\alpha)\right) = \frac{x^{5\alpha}}{\Gamma(1+5\alpha)}E_\alpha(t^\alpha), \tag{65}$$

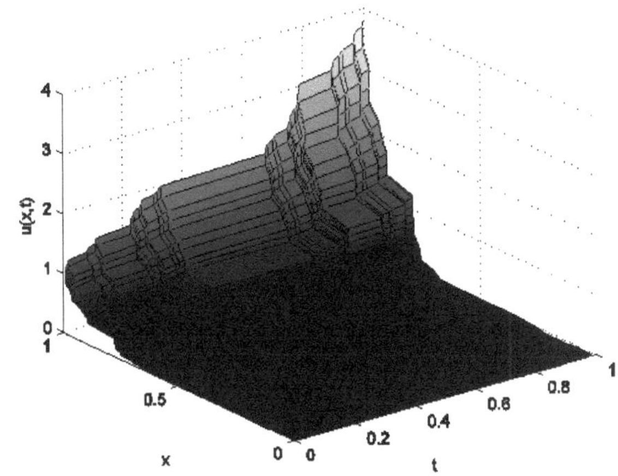

Fig. 12. The plot of solution of $u_2(x, t)$ where $\alpha = \ln 2/\ln 3$.

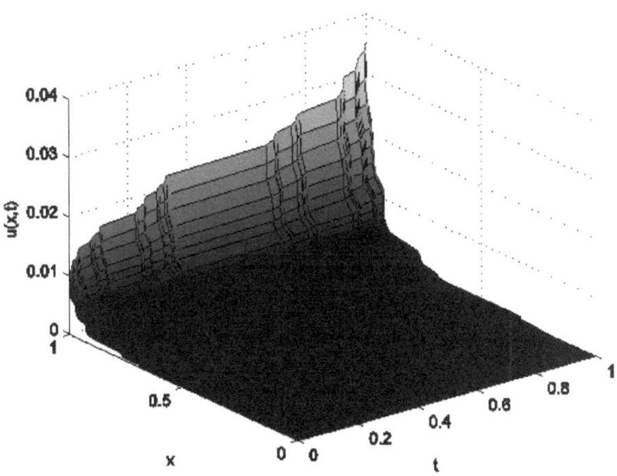

Fig. 13. The plot of $u_3(x, t)$ where $\alpha = \ln 2/\ln 3$.

$$u_2(x,t) = \tilde{L}_\alpha^{-1}\left(\frac{1}{s^{3\alpha}}\tilde{L}_\alpha\left\{\frac{\partial^\alpha u_1(x,t)}{\partial t^\alpha}\right\}\right) = \tilde{L}_\alpha^{-1}\left(\frac{1}{s^{9\alpha}}E_\alpha(t^\alpha)\right) = \frac{x^{8\alpha}}{\Gamma(1+8\alpha)}E_\alpha(t^\alpha), \quad (66)$$

$$u_3(x,t) = \tilde{L}_\alpha^{-1}\left(\frac{1}{s^{3\alpha}}\tilde{L}_\alpha\left\{\frac{\partial^\alpha u_2(x,t)}{\partial t^\alpha}\right\}\right) = \tilde{L}_\alpha^{-1}\left(\frac{1}{s^{12\alpha}}E_\alpha(t^\alpha)\right) = \frac{x^{11\alpha}}{\Gamma(1+11\alpha)}E_\alpha(t^\alpha),$$

$$\ldots\ldots \quad (67)$$

and so on. Their plots are given in Fig. 14, 15, 16 and 17.

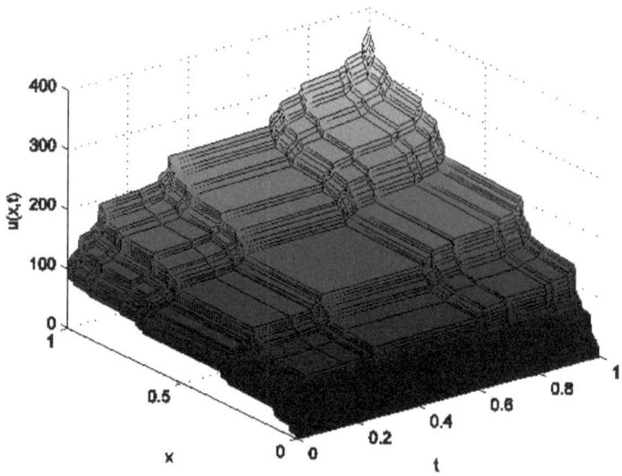

Fig. 14. The plot of solution of $u_0(x, t)$ where $\alpha = \ln 2/\ln 3$.

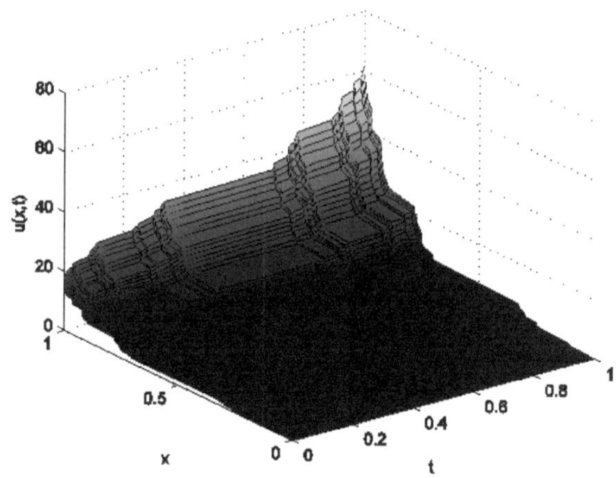

Fig. 15. The plot of solution of $u_1(x, t)$ where $\alpha = \ln 2/\ln 3$.

Hence, the approximate solution of 60 is given by:

$$\begin{aligned} u(x,t) \\ &= E_\alpha(t^\alpha)\left(\frac{x^{2\alpha}}{\Gamma(1+2\alpha)}+\frac{x^{5\alpha}}{\Gamma(1+5\alpha)}+\frac{x^{8\alpha}}{\Gamma(1+8\alpha)}+\frac{x^{11\alpha}}{\Gamma(1+11\alpha}+\cdots\right) \\ &= E_\alpha(t^\alpha)\sum_{k=0}^{\infty}\frac{x^{(3k+2)\alpha}}{\Gamma(1+(3k+2)\alpha)}. \end{aligned} \tag{68}$$

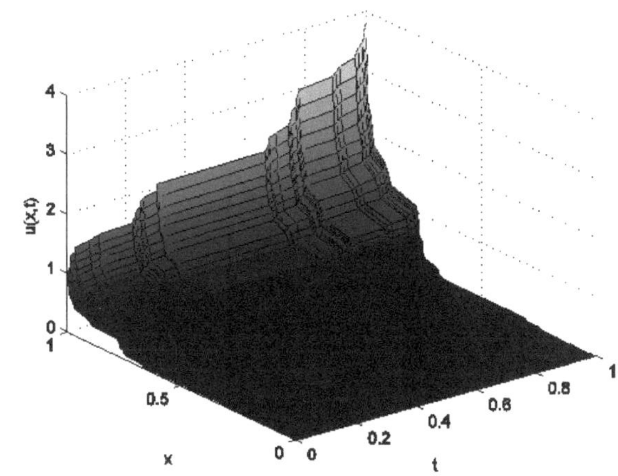

Fig. 16. The plot of solution of $u_2(x, t)$ where $\alpha = \ln 2/\ln 3$.

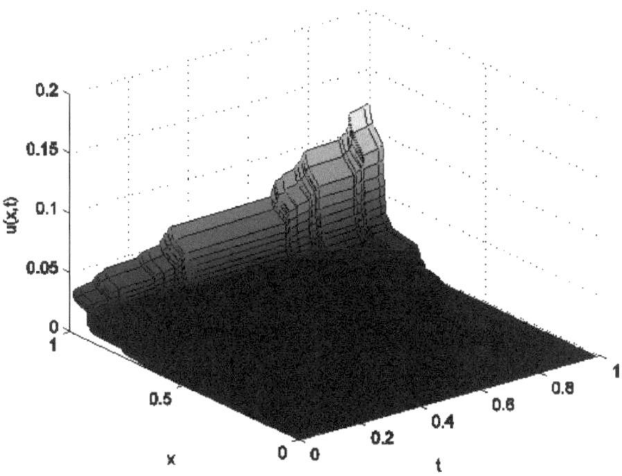

Fig. 17. The plot of $u_3(x, t)$ where $\alpha = \ln 2/\ln 3$.

Example 6. Consider the following local fractional partial differential equation

$$\frac{\partial^{2\alpha} u(x,t)}{\partial t^{2\alpha}} - \frac{\partial^{5\alpha} u(x,t)}{\partial x^{5\alpha}} = 0, \tag{69}$$

with initial values as follows:

$$u(0,t) = 0\,, \quad \frac{\partial^{\alpha} u(0,t)}{\partial x^{\alpha}} = 0\,, \quad \frac{\partial^{2\alpha} u(0,t)}{\partial x^{2\alpha}} = 0\,, \quad \frac{\partial^{3\alpha} u(0,t)}{\partial x^{3\alpha}} = 0\,, \quad \frac{\partial^{4\alpha} u(0,t)}{\partial x^{4\alpha}} = E_\alpha(t^\alpha)\,. \tag{70}$$

Using 20 and 46, we report the local fractional iteration algorithms in the form

$$u_0(x,t) = \frac{x^{4\alpha}}{\Gamma(1+4\alpha)} E_\alpha(t^\alpha), \tag{71}$$

$$u_{n+1}(x,t) = \tilde{L}_\alpha^{-1}\left(\frac{1}{s^{5\alpha}}\tilde{L}_\alpha\left\{\frac{\partial^{2\alpha}u_n(x,t)}{\partial t^{2\alpha}}\right\}\right), \quad n \geq 0. \tag{72}$$

From 72 we can show the components given as follows:

$$u_0(x,t) = \frac{x^{4\alpha}}{\Gamma(1+4\alpha)} E_\alpha(t^\alpha), \tag{73}$$

$$u_1(x,t) = \tilde{L}_\alpha^{-1}\left(\frac{1}{s^{5\alpha}}\tilde{L}_\alpha\left\{\frac{\partial^{2\alpha}u_0(x,t)}{\partial t^{2\alpha}}\right\}\right) = \tilde{L}_\alpha^{-1}\left(\frac{1}{s^{10\alpha}}E_\alpha(t^\alpha)\right) = \frac{x^{9\alpha}}{\Gamma(1+9\alpha)}E_\alpha(t^\alpha), \tag{74}$$

$$u_2(x,t) = \tilde{L}_\alpha^{-1}\left(\frac{1}{s^{5\alpha}}\tilde{L}_\alpha\left\{\frac{\partial^{2\alpha}u_1(x,t)}{\partial t^{2\alpha}}\right\}\right) = \tilde{L}_\alpha^{-1}\left(\frac{1}{s^{15\alpha}}E_\alpha(t^\alpha)\right) = \frac{x^{14\alpha}}{\Gamma(1+14\alpha)}E_\alpha(t^\alpha), \tag{75}$$

$$u_3(x,t) = \tilde{L}_\alpha^{-1}\left(\frac{1}{s^{5\alpha}}\tilde{L}_\alpha\left\{\frac{\partial^{2\alpha}u_2(x,t)}{\partial t^{2\alpha}}\right\}\right) = \tilde{L}_\alpha^{-1}\left(\frac{1}{s^{20\alpha}}E_\alpha(t^\alpha)\right) = \frac{x^{19\alpha}}{\Gamma(1+19\alpha)}E_\alpha(t^\alpha)$$

$$\cdots\cdots \tag{76}$$

and so on. Their plots are given in Fig. 18, 19, 20 and 21.

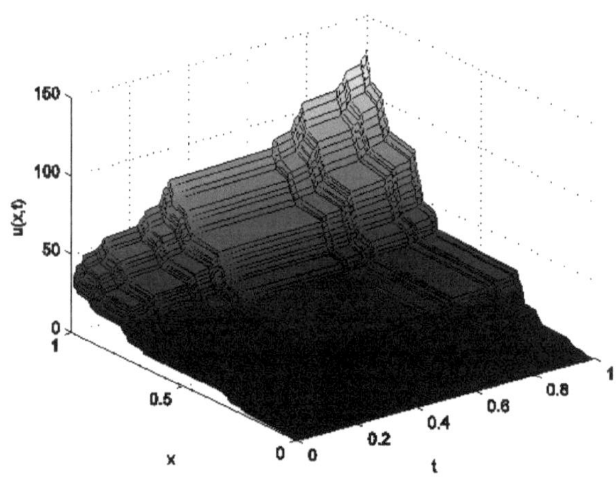

Fig. 18. The plot of solution of $u_0(x,t)$ where $\alpha = \ln 2/\ln 3$.

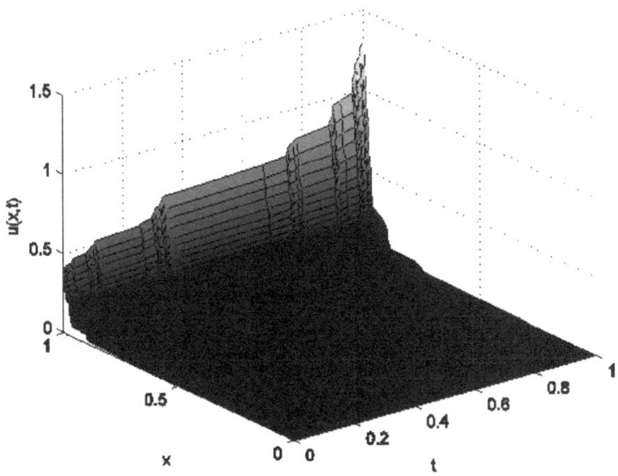

Fig. 19. The plot of solution of $u_1(x, t)$ where $\alpha = \ln 2/\ln 3$.

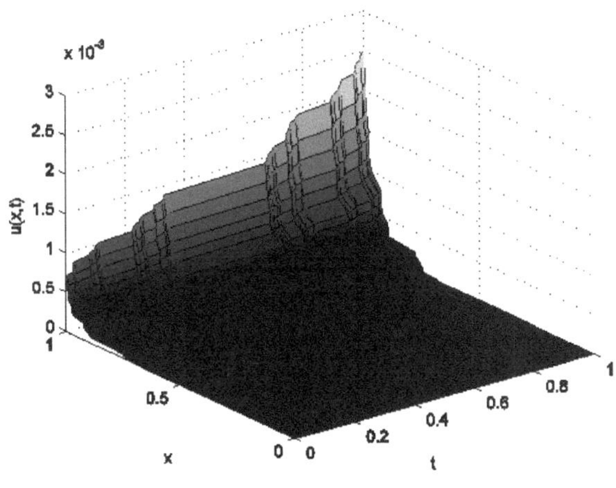

Fig. 20. The plot of solution of $u_2(x, t)$ where $\alpha = \ln 2/\ln 3$.

Therefore, the non-differentiable solution of 69 is:

$$\begin{aligned} &u(x,t) \\ &= E_\alpha(t^\alpha)\left(\frac{x^{4\alpha}}{\Gamma(1+4\alpha)} + \frac{x^{9\alpha}}{\Gamma(1+9\alpha)} + \frac{x^{14\alpha}}{\Gamma(1+14\alpha)} + \frac{x^{19\alpha}}{\Gamma(1+19\alpha} + \cdots\right) \\ &= E_\alpha(t^\alpha)\sum_{k=0}^{\infty}\frac{x^{(5k+4)\alpha}}{\Gamma(1+(5k+4)\alpha)}\,. \end{aligned} \tag{77}$$

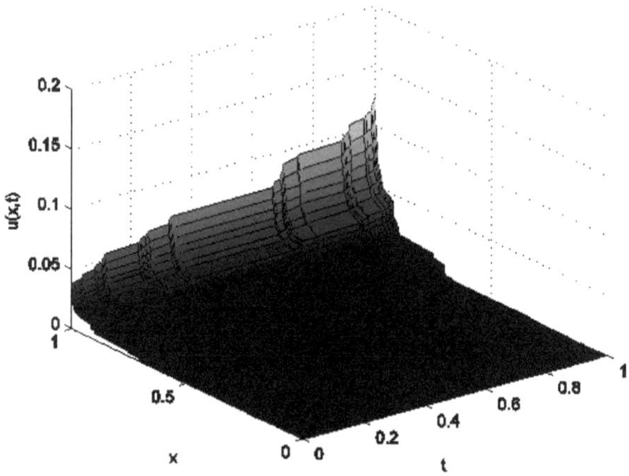

Fig. 21. The plot of $u_3(x, t)$ where $\alpha = \ln 2 / \ln 3$.

5 Conclusions

In this work we considered the coupling method of the local fractional decomposition method and Laplace transform to solve linear local fractional partial differential equations and their nondifferentiable solutions were obtained. The results include an efficient implement of the local fractional Laplace decomposition method to solve the partial differential equations with local fractional derivative.

Bibliography

Adomian, G. (1994). Solving frontier problems of physics the decomposition method. Springer.

Adomian, G. (1990). A review of the decomposition method and some recent results for nonlinear equations. *Mathematical and Computer Modelling*, 13(7), 17-43.

Adomian, G. (1991). A review of the decomposition method and some recent results for nonlinear equations. *Computers & Mathematics with Applications*, 21(5), 101-127.

Cherruault, Y., Saccomandi, G., & Some, B. (1992). New results for convergence of Adomian's method applied to integral equations. *Mathematical and Computer Modelling*, 16(2), 85-93.

Babolian, E. S. M. A. E. I. L., Biazar, J., & Vahidi, A. R. (2004). The decomposition method applied to systems of Fredholm integral equations of the second kind. *Applied Mathematics and Computation*, 148(2), 443-452.

Abbasbandy, S. (2006). Numerical solutions of the integral equations: Homotopy perturbation method and Adomian's decomposition method. *Applied Mathematics and Computation*, 173(1), 493-500.

Hashim, I. (2006). Adomian decomposition method for solving BVPs for fourth-order integro-differential equations. *Journal of Computational and Applied Mathematics*, 193(2), 658-664.

Syam, M. I., & Hamdan, A. (2006). An efficient method for solving Bratu equations. *Applied Mathematics and Computation*, 176(2), 704-713.

Yusufoğlu, E. (2006). Numerical solution of Duffing equation by the Laplace decomposition algorithm. *Applied Mathematics and Computation*, 177(2), 572-580.

Khan, Y. (2009). An effective modification of the Laplace decomposition method for nonlinear equations. *International Journal of Nonlinear Sciences and Numerical Simulation*, 10(11-12), 1373-1376.

Wazwaz, A. M. (2010). The combined Laplace transform–Adomian decomposition method for handling nonlinear Volterra integro–differential equations. *Applied Mathematics and Computation*, 216(4), 1304-1309

Jafari, H., Khalique, C. M., Nazari, M. (2011). Application of the Laplace decomposition method for solving linear and nonlinear fractional diffusion–wave equations. *Applied Mathematics Letters*, 24(11), 1799-1805.

Khan, Y., & Austin, F. (2010). Application of the Laplace decomposition method to nonlinear homogeneous and non-homogenous advection equations. *Zeitschrift für Naturforschung-A*, 65(10), 849.

Yang, X. J. (2011). Local Fractional Functional Analysis and Its Applications. Asian Academic Publisher Limited, Hong Kong.

Yang, X. J. (2012). Advanced Local Fractional Calculus and Its Applications. World Science Publisher, New York.

Kolwankar, K. M., & Gangal, A. D. (1998). Local fractional Fokker-Planck equation. *Physical Review Letters*, 80(2), 214.

Yang, X. J., Baleanu, D., & Machado, J. A. T. (2013). Mathematical aspects of the Heisenberg uncertainty principle within local fractional Fourier analysis. *Boundary Value Problems*, 2013(1), 1-16.

Zhao, Y., Baleanu, D., Cattani, C., et al. (2013). Maxwell's equations on Cantor sets: a local fractional approach. *Advances in High Energy Physics*, 2013, Article ID 686371, 6 pages.

Li, Y. Y., Zhao, Y., Xie, G. N., et al. (2014). Local fractional Poisson and Laplace equations with applications to electrostatics in fractal domain. *Advances in Mathematical Physics*, 2014, Article ID 590574, 5 pages.

Yang, X. J., Srivastava, H. M., He, J. H., et al. (2013). Cantor-type cylindrical-coordinate method for differential equations with local fractional derivatives. *Physics Letters A*, 377(28), 1696-1700.

Xu, S., Ling, X., Cattani, C., Xie, G. N., Yang, X. J., Zhao, Y. (2014). Local fractional Laplace variational iteration method for nonhomogeneous heat equations arising in fractal heat flow. *Mathematical Problems in Engineering*, 2014, Article ID 914725, 7 pages.

Carpinteri, A., Chiaia, B., & Cornetti, P. (2001). Static–kinematic duality and the principle of virtual work in the mechanics of fractal media. *Computer Methods in Applied Mechanics and Engineering*, 191(1), 3-19.

Carpinteri, A., Chiaia, B., & Cornetti, P. (2004). A fractal theory for the mechanics of elastic materials. *Materials Science and Engineering: A*, 365(1), 235-240.

Yang, X. J., Baleanu, D., & Tenreiro Machado, J. A. (2013). Systems of Navier-Stokes equations on Cantor sets. *Mathematical Problems in Engineering*, 2013, Article ID 769724, 8 pages.

Yang, X. J., Baleanu, D., & Zhong, W. P. (2013). Approximate solutions for diffusion equations on cantor space-time. *Proceedings of the Romanian Academy, Series A*, 14(2), 127-133.

Baleanu, D., Tenreiro Machado, J. A., Cattani, C., Baleanu, M. C., & Yang, X. J. (2014). Local fractional variational iteration and decomposition methods for wave equation on Cantor sets within local fractional operators. *Abstract and Applied Analysis*, 2014, Article ID 535048, 6 pages.

Jafari, H., & Jassim, H. K. (2014). Local fractional Adomian decomposition method for solving two dimensional heat conduction equations within local fractional operators, *Journal of Advance in Mathematics*, 9(4) 2574-2582.

He, J. H. (2012). Asymptotic methods for solitary solutions and compactons. In *Abstract and Applied Analysis*, 2012, 130.

Zhao, Y., Baleanu, D., Baleanu, M. C., Cheng, D. F., & Yang, X. J. (2013). Mappings for special functions on Cantor sets and special integral transforms via local fractional operators. In *Abstract and Applied Analysis*, 2013, Article ID 316978, 6 pages.

Liu, C. F., Kong, S. S., & Yuan, S. J. (2013). Reconstructive schemes for variational iteration method within Yang-Laplace transform with application to fractal heat conduction problem. *Thermal Science*, 17(3), 715-721.

Wang, S. Q., Yang, Y. J., & Jassim, H. K. (2014). Local fractional function decomposition method for solving inhomogeneous wave equations with local fractional derivative. *Abstract and Applied Analysis*, 2014, Article ID 176395, 7 pages.

Yang, A. M., Li, J., Srivastava, H. M., Xie, G. N., & Yang, X. J. (2014). The local fractional Laplace variational iteration method for solving linear partial differential equations with local fractional derivatives. *Discrete Dynamics in Nature and Society*, 2014, Article ID 365981, 8 pages.

Jafari, H., & Jassim, H. K. (2014). Local fractional Laplace variational iteration method for solving nonlinear partial differential equations on Cantor sets within local fractional operators, *Journal of Zankoy Sulaimani-Part A*, 16(4), 49-57.

Jafari, H., & Jassim, H. K. (2014). Local fractional series expansion method for solving Laplace and Schrodinger equations on Cantor sets within local fractional operators, *International Journal of Mathematics and Computer Research*, 2(11) 736-744.

He, J. H. (2014). A Tutorial Review on Fractal Spacetime and Fractional Calculus. *International Journal of Theoretical Physics*, 53(11), 3698-3718.

Yan, S. P., Jafari, H., & Jassim, H. K. (2014). Local fractional adomian decomposition and function decomposition methods for Laplace equation within local fractional operators. *Advances in Mathematical Physics*, 2014, Article ID 161580, 7 pages.

Alireza K. Golmankhaneh and D. Baleanu

Calculus on Fractals

Abstract: In this chapter we present a framework and a calculus on fractals. The suggested equation has been solved and applied in physics and dynamics.

1 Introduction

It is well known that fractals can models many phenomena in science and engineering (Mandelbrot, 1977; Bunde and Havlin, 1985; Falconer, 1985a; Falconer, 1990b; Falconer, 1997c; Edgar, 1998;). Geometry has important role to play in physics (Jurgen, 2009; Theodore, 2011; Mikio, 2003). Since mathematical physics was first developed by introducing calculus on sets. Analysis on vector space is the second stage in which it was introduced. Calculus on manifolds as third stage has been studied and it leads to as basic tool in general relativity. Recently, Fractal geometry has been suggested as the geometry of the real word (Mandelbrot, 1977; Bunde and Havlin, 1985; Falconer, 1985a; Falconer, 1990b; Falconer, 1997c; Edgar, 1998). Scientists and mathematicians have tried to establish a calculus or analysis on fractals (Freiberg and M. Zähle, 2000; Dalrymple, Strichartz and Vinson, 1999; Kigami, 2000; Freiberg, 2003; Strichartz, 2000a; Strichartz, 2000b). Measure theory has been applied to generalize calculus to fractals but it is not algorithmic (Freiberg and M. Zähle, 2000; Dalrymple, Strichartz and Vinson, 1999; Kigami, 2000; Freiberg, 2003; Strichartz, 2000a; Strichartz, 2000b). The Riemann like method has applied by A. Parvate and A. D. Gangal to construct a calculus for functions with support on Cantor sets, such as fractal curves. The suggested methods are useful since they are algorithmic (Parvate and Gangal, 2009a; Parvate and Gangal, 2011b, Parvate, Satin and Gangal, 2011). Local fractional derivatives on Cantor set have been suggested and applied in science (Yang, 2012a; Yang, 2011b; Kolwankar and Gangal 1998). Alireza K. Golmankaneh and D.Baleanu has suggested a fractional derivative on fractal sets embedded in R^3 (Golmankhaneh and Baleanu, 2013). Using F^{α}-calculus the classical mechanics has been generalized on Cantor set (Golmankhaneh, Fazlollahi and Baleanu, 2013; Golmankhaneh and Baleanu, 2013). The Schrödinger equation on a fractal curve is derived using the Feynman path method (Golmankhaneh and Baleanu, 2014). Fokker Planck Equation and Langevin

Alireza K. Golmankhaneh: Department of Physics, Urmia Branch, Islamic Azad University, Urmia, Iran, E-mail: a.khalili@iaurmia.ac.ir
D. Baleanu: Department of Mathematics and Computer Science, ankaya University, 06530 Ankara, Turkey
and Institute of Space Sciences, P.O.BOX, MG-23, R 76900, Magurele-Bucharest, Romania E-mail: dumitru@cankaya.edu.tr

Equation on Fractal Curves is obtained as a model for anomalous diffusion (Satin, Parvate and Gangal, 2013; Satin, Gangal, 2014).

2 Calculus on Fractal Subset of Real-Line

In this section, we summarize the F^{α}-calculus (Parvate and Gangal, 2009a; Parvate and Gangal, 2011b, Parvate, Satin and Gangal, 2011). This calculus is algorithmic and it looks like Riemann method. Fractional and locality of derivative is one the important properties of it. Since in most cases phenomena in nature have local properties then F^{α}-calculus may be the correct proper mathematical framework since it does not violate causality. We begin by summarizing the F^{α}-calculus (Parvate and Gangal, 2009a; Parvate and Gangal, 2011b, Parvate, Satin and Gangal, 2011).

2.1 Staircase Functions

Definition 1. Let $P_{[a,b]}$ be a subdivision of a interval $I = [a, b]$ which is a set of points $\{a = x_0, x_1, ..., x_n = b\}$, such that $x_i < x_{i+1}$ (Parvate and Gangal, 2009a; Parvate and Gangal, 2011b, Parvate, Satin and Gangal, 2011).
Definition 2. Suppose $F \subset R$ is a fractal set and $P_{[a,b]}$ a subdivision. The mass function is defined by:

$$\gamma^{\alpha}(F, a, b) = \lim_{\delta\to 0} \inf_{\{P_{[a,b]}:|P|\leqslant\delta\}} \sum_{i=0}^{n-1} \frac{(x_{i+1} - x_i)^{\alpha}}{\Gamma(\alpha + 1)} \theta(F, [x_{i+1} - x_i]), \tag{1}$$

here $\theta(F, [x_{i+1} - x_i]) = 1$ if $F \cap [x_{i+1} - x_i]$ in non-empty, and zero otherwise

$$|P| = \max_{0\leq i\leq n}(x_{i+1} - x_i).$$

Definition 2. is deduced from Hausdorff measure and local fractional derivatives.
Some of the important properties of mass function are as follows:
1. (Interval-wise additivity). If $a < b < c$ and $\gamma^{\alpha}(F, a, b) < \infty$. Then $\gamma^{\alpha}(F, a, c) = \gamma^{\alpha}(F, a, b) + \gamma^{\alpha}(F, b, c)$.
2. (Translation). For $F \subset R$ and $\lambda \in R$, let $F + \lambda$ denote the set $F + \lambda = \{x + \lambda : x \in F\}$. Then, $\gamma^{\alpha}(F + \lambda, a + \lambda, b + \lambda) = \gamma^{\alpha}(F, a, b)$.
3. (Scaling). For $F \subset R$ and $\lambda \geq 0$, let λF denote the set $\lambda F = \{\lambda x : x \in F\}$. Then, $\gamma^{\alpha}(\lambda F, a\lambda, b\lambda) = \lambda^{\alpha}\gamma^{\alpha}(F, a, b)$.
Definition 3. Let a_0 be a arbitrary real number. The staircase function of order α is defined as

$$S_F^{\alpha}(x) = \begin{cases} \gamma^{\alpha}(F, a, b), & \text{if } x \geq a_0, \\ -\gamma^{\alpha}(F, a, b), & \text{othewise.} \end{cases}$$

Some of the properties of the staircase function are

1. S_F^α is increasing in x.
2. If $F \cap (x, y) = \emptyset$, then S_F^α is a constant in $[x, y]$.
3. $S_F^\alpha(y) - S_F^\alpha(x) = \gamma^\alpha(F, x, y)$.
4. S_F^α is continuous on (a, b).

Definition 4. The γ-dimension is defined using the mass function as follows

$$\begin{aligned} dim_\gamma(F \cap [a, b]) &= \inf\{\alpha : \gamma^\alpha(F, a, b) = 0\}, \\ &= \sup\{\alpha : \gamma^\alpha(F, a, b) = \infty\}. \end{aligned} \tag{2}$$

Remark: It is shown that $dim_H(F \cap [a, b]) \leq dim_\gamma(F \cap [a, b]) \leq dim_B(F \cap [a, b])$ where dim_H and dim_B denote the Hausdorff dimension and the box dimension respectively.

2.2 *F*-Limit and *F*-Continuity

Let $F \subset R$, $f : R \to R$ and $x \in F$. A number l is said to be the limit of f through the points of F or simply F-limit of f as $y \to x$ if given any $\epsilon > 0$ there exists $\delta > 0$ such that

$$y \in F \text{ and } |y - x| < \delta \Rightarrow |f(y) - l| < \epsilon. \tag{3}$$

If such a number exists then it is denoted by

$$l = F - \lim_{y \to x} f(y). \tag{4}$$

A function $f : R \to R$ is said to be F-continuous at $x \in F$ if

$$f(x) = F - \lim_{y \to x} f(x). \tag{5}$$

2.3 F^α-Integration

In this section, we review F^α-integration for function whose support is fractal set. If f be a bounded function on F and I be a closed interval (Parvate and Gangal, 2009a; Parvate and Gangal, 2011b, Parvate, Satin and Gangal, 2011) . Then

$$\begin{aligned} M[f, F, I] &= \sup_{x \in F \cap I} f(x) \quad \text{if } F \cap I \neq 0, \\ &= 0 \quad \text{otherwise}, \end{aligned} \tag{6}$$

and similarly

$$m[f, F, I] = \inf_{x \in F \cap I} f(x) \quad \text{if } F \cap I \neq 0,$$

$$= \quad 0 \quad \text{otherwise.} \tag{7}$$

Let $S_F^\alpha(x)$ be finite for $x \in [a, b]$ and P be a subdivision of $[a, b]$ with points $x_0, \ldots, x_n$. The upper F^α-sum and the lower F^α-sum for a function f over the subdivision P are defined respectively (Parvate and Gangal, 2009a; Parvate and Gangal, 2011b, Parvate, Satin and Gangal, 2011)

$$U^\alpha[f, F, P] = \sum_{i=1}^{n} M[f, F, [x_i, x_{i-1}]](S_F^\alpha(x_i) - S_F^\alpha(x_{i-1})), \tag{8}$$

and

$$L^\alpha[f, F, P] = \sum_{i=1}^{n} m[f, F, [x_i, x_{i-1}]](S_F^\alpha(x_i) - S_F^\alpha(x_{i-1})). \tag{9}$$

If f be a bounded function $f \in B(F)$. f is F^α-integrable on $[a, b]$ if

$$\underline{\int_a^b} f(x) d_F^\alpha x = \sup_{P_{[a,b]}} L^\alpha[f, F, P] = \overline{\int_a^b} f(x) d_F^\alpha x = \inf_{P_{[a,b]}} L^\alpha[f, F, P]. \tag{10}$$

In that case the F^α-integral of f on $[a, b]$ denoted by $\int_a^b f(x) d_F^\alpha x$ is given by the common value. Here some of the important properties of F^α-calculus are presented:
If $c \in [a, b]$ then

$$\int_a^b f(x) d_F^\alpha x = \int_a^c f(x) d_F^\alpha x + \int_c^b f(x) d_F^\alpha x. \tag{11}$$

F^α-integration is a linear operation

$$\int_a^b \lambda f(x) d_F^\alpha x = \lambda \int_a^b f(x) d_F^\alpha x \quad \text{where } \lambda \text{ is constant,} \tag{12}$$

$$\int_a^b [m f(x) + n g(x)] d_F^\alpha x = m \int_a^b f(x) d_F^\alpha x + n \int_a^b g(x) d_F^\alpha x, \tag{13}$$

here m, n are constant.

$$\int_a^b f(x)\, d_F^\alpha x \geq \int_a^b g(x) d_F^\alpha x \quad \text{if} \quad f(x) \geq g(x), \tag{14}$$

and

$$\int_b^a f(x)\, d_F^\alpha x = -\int_a^b g(x)\, d_F^\alpha x \quad \text{if } f(x) \geq g(x). \tag{15}$$

If $f(x) = \chi_F(x)$ is the characteristic function of F, then

$$\int_b^a \chi_F(x)\, d_F^\alpha x = S_F^\alpha(b) - S_F^\alpha(a). \tag{16}$$

Remark: The ordinary Riemann integral of fractal support functions is zero or undefined.

2.4 F^α-Differentiation

We must mention that the F^α-differentiation is suitable for functions whose change set is $Schf \subset F$ (set of all points such that in their open neighborhood the function f is not constant). If F is an α-perfect set (closed and every point is its limit) then the F^α-derivative of f at x is

$$D_F^\alpha f(x) = \begin{cases} \text{F}-\lim_{y\to x} \frac{f(y)-f(x)}{S_F^\alpha(y)-S_F^\alpha(x)} & \text{if } x \in F, \\ 0 & otherwise, \end{cases} \tag{17}$$

if the limit exists.
Remark: The ordinary derivative of such functions is zero almost everywhere or undefined.
Some of the properties of F^α-derivative are as follows:

1. The F^α-derivative is a linear operator.
2. The F^α-derivative of a constant function is zero.
3. The derivative of the staircase function is the characteristic function χ_F of F

$$D_F^\alpha(S_F^\alpha(x)) = \chi_F(x).$$

2.5 First Fundamental Theorem of F^α-calculus

Suppose $F \subset R$ be an α-perfect set. If $f \in B(F)$ is an F-continuous function on $F \cap [a, b]$ and

$$g(x) = \int_a^x f(y) d_F^\alpha y,$$

for all $x \in [a, b]$, then

$$D_F^\alpha(g(x)) = f(x)\chi_F(x),$$

2.6 Second Fundamental Theorem of F^α-calculus

Let $f : R \to R$ be a continuous, F^α-differentiable function such that $Sch(f)$ is an contained in an α-perfect set F and $h : R \to R$ be F-continuous such that

$$h(x)\chi_F(x) = D_F^\alpha(f(x)).$$

Therefore

$$\int_a^b h(x)d_F^\alpha x = f(b) - f(a).$$

For proof one can see ref. (Parvate and Gangal, 2009a; Parvate and Gangal, 2011b, Parvate, Satin and Gangal, 2011).

2.7 Taylor Series on Fractal Sets

The fractal Taylor series under some conditions for functions f is defined as

$$f(w) = \sum_{n=0}^{\infty} \frac{(S_F^\alpha(w) - S_F^\alpha(x))^n}{n!}(D_F^\alpha f(x))^n.$$

2.8 Integration by Parts in F^α-calculus

Let the functions $u : R \to R$ and $v : R \to R$ be such that
1. $u(x)$ is continuous on $[a, b]$ and $\mathbf{S}ch(u) \subset F$,
2. $D_F^\alpha u(x)$ exists and is F-continuous on $[a, b]$,
3. $v(x)$ is F-continuous on $[a, b]$.

Then

$$\int_a^b u(x)v(x)d_F^\alpha x = [u(x)\int_a^x v(x')d_F^\alpha x']_a^b - \int_a^b D_F^\alpha u(x)\int_a^x v(x')d_F^\alpha x' d_F^\alpha x. \tag{18}$$

There some analogous rules and theorems in F^α-calculus, which include Rolle, mean value theorem, and the Leibniz rule.

3 Fractal F^α-differential Equation

The F^α-differential equations contain at least one F^α-derivative. In the following we present some examples (Parvate and Gangal, 2009a; Parvate and Gangal, 2011b, Parvate, Satin and Gangal, 2011).

Example 1. The equation

$$D_{F,t}^\alpha y(t) = \chi_F(t), \tag{19}$$

satisfying the initial condition $y(t) = 0$ at $t = 0$ has solution $y(t) = S_F^\alpha(t)$.
Example 2. The linear F^α-differential equation

$$D_{F,t}^\alpha x = \chi_F(t)Ax, \tag{20}$$

where A is an $n \times n$ constant matrix and $x \in R^n$ has as its solution:

$$x(t) = e^{S_F^\alpha(t)Ax_0}. \tag{21}$$

Example 3. Consider the following differential equation which is model of a particle undergoing friction in a fractal medium

$$D_{F,x}^\alpha v(x) = -k(x).$$

Then we have

$$v(x) = v_0 - \int_{x_0}^{x} k(x')d_F^\alpha x'.$$

If $k(x) = k\chi_F(x)$ here k is constant

$$v(x) = v_0 - k(S_F^\alpha(x) - S_F^\alpha(x_0)),$$

Example 4. Relaxation phenomena in glassy materials is modelled as

$$D_{F,x}^\alpha \mu(t) = -k\mu(t)\chi_F(t),$$

where $\mu(t)$ is polarization or magnetic moment that changes with time on the fractal subset F. Then the solution is

$$\mu(t) = Ae^{-kS_F^\alpha(t)}.$$

Relaxation function for this stochastic process is defined as

$$w(t) = < \frac{\mu(t)\mu(0)}{\mu^2(0)} >, \quad w(t) = e^{-(t/\tau)^\alpha}, \quad 0 < \alpha < 1,$$

here τ is called relaxation time (Shlesinger, 1988).
Example 5. Consider the following diffusion equation with fractal time as follows

$$D_{F,t}^\alpha W(x,t) = \frac{\chi_F(t)}{2}\frac{\partial^2}{\partial x^2}W(x,t),$$

where W is the density function. The solution is

$$W(x,t) = \frac{1}{(2\pi S_F^\alpha(t))^2} e^{-\frac{x^2}{2S_F^\alpha(t)}}, \quad W(x,0) = \delta(x).$$

Finally, some of the important analogies between F^α-calculus and ordinary calculus are given in Table 1 .

Table 1. Comparison between ordinary calculus and F^α-Calculus

Ordinary Calculus	**F^α-Calculus**
R	**An α-perfect set F**
Limit	**F-Limit**
Continuity	**F- Continuity**
$\int_0^y x^n dx = \frac{1}{n+1} y^{n+1}$	$\int_0^y (S_F^\alpha(x))^n d_F^\alpha x = \frac{1}{n+1}(S_F^\alpha(y))^{n+1}$
$\frac{d}{dx} x^n = nx^{n-1}$	$D_F^\alpha (S_F^\alpha(x))^n = n(S_F^\alpha(x))^{n-1} \chi_F(x)$
Leibniz rule	**F^α-Leibniz rule**
Fundamental theorem	**Fundamental theorem of F^α-Calculus**
Integration by part ordinary calculus	**Integration by part in F^α-calculus**
Taylor expansion	**Fractal Taylor expansion**

4 Calculus on Fractal Curves

In this section we review calculus on fractal curves (Parvate and Gangal, 2009a; Parvate and Gangal, 2011b, Parvate, Satin and Gangal, 2011).

4.1 Staircase Function on Fractal Curves

Consider a fractal curve $F \subset R^3$ which is continuously parameterizable i.e there exists a function $\mathbf{w} : [a_0, b_0] \to F \subset R^3$ which is continuous. We also assume $\mathbf{w}$ to be invertible. A subdivision $P_{[a,b]}$ of interval $[a, b]$, $a < b$, is a finite set of points $\{a = v_0 < v_1, \ldots < v_n = b\}$. For $a_0 \le a < b < b_0$ and appropriate α:

$$\gamma^\alpha(F, a, b) = \lim_{\delta \to 0} \inf_{\{P_{[a,b]}:|P| \le \delta\}} \sum_{i=0}^{n-1} \frac{|\mathbf{w}(v_{i+1}) - \mathbf{w}(v_i)|^\alpha}{\Gamma(\alpha + 1)}, \tag{22}$$

where $|.|$ denotes the Euclidean norm on R^3 and $|P| = \max\{v_{i+1} - v_i; i = 0, \ldots, n-1\}$. The γ-dimension of F, is defined as (Parvate and Gangal, 2009a; Parvate and Gangal, 2011b, Parvate, Satin and Gangal, 2011)

$$\dim_\gamma(F) = \inf\{\alpha : \gamma^\alpha(F, a, b) = 0\} = \sup\{\alpha : \gamma^\alpha(F, a, b) = \infty\}. \tag{23}$$

After this definition α is equal to $dim_\gamma(F)$. The staircase function $S_F^\alpha : [a_0, b_0] \to R$ of order α for a set F, is defined as

$$S_F^\alpha(v) = \begin{cases} \gamma^\alpha(F, p_0, v) & v \ge p_0, \\ -\gamma^\alpha(F, v, p_0) & v < p_0, \end{cases} \tag{24}$$

where $a_0 \le p_0 \le b_0$ are arbitrary but fixed, and $v \in [a_0, b_0]$. It is monotonic function. The $\theta = \mathbf{w}(v)$, denote a point on fractal curve F

$$J(\theta) = S_F^\alpha(\mathbf{w}^{-1}(\theta)), \quad \theta \in F. \tag{25}$$

We suppose that fractal curves whose S_F^α is finite are invertible on $[a, b]$.

4.2 F-Limit and F-Continuity on Fractal Curves

Here we study F-limit and F-continuity along a fractal curve (Parvate and Gangal, 2009a; Parvate and Gangal, 2011b, Parvate, Satin and Gangal, 2011).
Let $F \subset R^n$ be a fractal curve, $f : F \to R$ and $\theta \in F$. A number l is said to be the limit of f through points of F

$$\forall \epsilon, \exists \delta, \theta' \in F, \text{ and } |\theta' - \theta| < \delta \Rightarrow |f(\theta') - f(\theta)| < \epsilon.$$

A function $f : F \to R$ is said to be F-continuous at $\theta \in F$ if

$$f(\theta) = F - \lim_{\theta' \to \theta} f(\theta').$$

4.3 F^α-integration on Fractal Curves

For a fractal curve, the segment $C(t_1, t_2)$ where $t_1, t_2 \in [a_0, b_0], t_1 \leq t_2$ is defined (Parvate and Gangal, 2009a; Parvate and Gangal, 2011b, Parvate, Satin and Gangal, 2011)

$$C(t_1, t_2) = \{w(t') : t' \in [t_1, t_2]\},$$

and

$$M[f, C(t_1, t_2)] = \sup_{\theta \in C(t_1, t_2)} f(\theta), \quad m[f, C(t_1, t_2)] = \inf_{\theta \in C(t_1, t_2)} f(\theta).$$

Then the upper sum $U^\alpha[f, F, P]$ and lower sum $L^\alpha[f, F, P]$ are defined as:

$$U^\alpha[f, F, P] = \sum_{i=0}^{n-1} M[f, C(t_i, t_{i+1})][S_F^\alpha(t_{i+1}) - S_F^\alpha(t_i)],$$

and

$$L^\alpha[f, F, P] = \sum_{i=0}^{n-1} m[f, C(t_i, t_{i+1})][S_F^\alpha(t_{i+1}) - S_F^\alpha(t_i)].$$

Suppose F is defined so that S_F^α in finite on $[a, b]$. For $f \in B(f)$ using lower and upper sum F^α-integral of the function f is given on the section $C(a, b)$ as follows

$$\underline{\int_{C(a,b)}} f(\theta) d_F^\alpha \theta = \sup_{P_{[a,b]}} L^\alpha[f, F, P] = \inf_{P_{[a,b]}} U^\alpha[f, F, P] = \overline{\int_{C(a,b)}} f(\theta) d_F^\alpha \theta,$$

and is denoted by

$$\int_{C(a,b)} f(\theta) d_F^\alpha \theta.$$

Remark Let $\gamma^\alpha(F, a, b)$ be finite and $f(\theta) = 1, \theta \in F$ denote the constant function. Then

$$\int_{C(a,b)} f(\theta) d_F^\alpha \theta = \int_{C(a,b)} 1 d_F^\alpha \theta = S_F^\alpha(b) - S_F^\alpha(a) = J((w(b)) - J((w(a)). \tag{26}$$

4.4 F^{α}-Differentiation on Fractal Curves

For the function f on fractal curve the F^{α}-derivative at $\theta \in F$ is defined

$$(D_F^{\alpha} f)(\theta) = F - \lim_{\theta' \to \theta} \frac{f(\theta') - f(\theta)}{J(\theta') - J(\theta)},$$

if the limit exist.

4.5 First Fundamental Theorem on Fractal Curve

If f in F-continuous and $f \in B(f)$ on $C(a, b)$. Let $g : f \to R$ be defined as (Parvate and Gangal, 2009a; Parvate and Gangal, 2011b, Parvate, Satin and Gangal, 2011)

$$g(w(t)) = \int_{C(a,t)} f(\theta) d_F^{\alpha} \theta,$$

for all $t \in [a, b]$. Then

$$(D_F^{\alpha} g)(\theta) = f(\theta).$$

4.6 Second Fundamental Theorem on Fractal Curve

Suppose $f : F \to R$ is F^{α}-differentiable function and $h : F \to R$ is F-continuous, such that $h(\theta) = (D_F^{\alpha} f)(\theta))$ then (Parvate and Gangal, 2009a; Parvate and Gangal, 2011b, Parvate, Satin and Gangal, 2011)

$$\int_{C(a,t)} h(\theta) d_F^{\alpha} \theta = f(w(b)) - f(w(a)). \tag{27}$$

5 Gradient, Divergent, Curl and Laplacian on Fractal Curves

In this section we generalized the F^{α}-calculus by defining the gradient, divergent, curl and Laplacian on fractal curves imbedded in R^3 (Golmankhaneh and Baleanu 2014).

5.1 Gradient on Fractal Curves

Let us consider the $f \in B(F)$ as an F-continuous function on $C(a, b) \subset F$ and $\mathbf{w}(v, w_i(v)) : R \to R^3, i = 1, 2, 3$, so the gradient of the $f(\mathbf{w}) : F \to R$ is

$$\nabla^{\alpha}_{F} f(\mathbf{w}) = {}^{w_i}\mathfrak{D}^{\alpha}_{F} f(\mathbf{w})\hat{e}^i \quad i = 1, 2, 3, \dots \tag{28}$$

where the $\hat{e}^i$ is the basis of R^n.

5.2 Divergent on Fractal Curves

Let the $\mathbf{f}(\mathbf{w}(v)) = f_i(\mathbf{w}(v))\ \hat{e}^i \ \ i = 1, 2, 3 \dots$, be a vector field on a fractal curve. Then we define the divergence of the $\mathbf{f} : F \to R^n$ as follows

$$\nabla^{\alpha}_{F}.\mathbf{f}(\mathbf{w}(v)) = {}^{w_i}\mathfrak{D}^{\alpha}_{F} f_i(\mathbf{w}(v)), \tag{29}$$

where $f_i(\mathbf{w}(v))$ are components of vector field.

5.3 Laplacian on Fractal Curves

Consider the $\mathbf{w}(v, w_i(v)) : R \to R^3$ on the fractal curve, then the Laplacian is defined as

$$\triangle^{\alpha}_{F} f = (\nabla^{\alpha}_{F})^2 f = ({}^{w_i}\mathfrak{D}^{\alpha}_{F})^2 f(\mathbf{w}(v)), \tag{30}$$

where the $\triangle^{\alpha}_{F}$ is called the Laplacian on fractal curve (Golmankhaneh and Baleanu 2014).

Analogous Taylor series is defined for $h(\theta) \in B(F)$ as (Golmankhaneh and Baleanu 2014)

$$f(\mathbf{w}(v)) = \sum_{n=0}^{\infty} \frac{(S^{\alpha}_{F}(v) - S^{\alpha}_{F}(v'))^n}{n!} (\mathfrak{D}^{\alpha}_{F})^n f(\mathbf{w}(v')), \tag{31}$$

where $h(\theta)$ is F^{α}-differentiable any number of times on $C(a, b)$. That is $(\mathfrak{D}^{\alpha}_{F})^n h \in B(F)$, $\forall n > 0$.

6 Function Spaces in F^{α}-calculus

In section we review function spaces in the F^{α}-calculus (Parvate and Gangal, 2009a; Parvate and Gangal, 2011b, Parvate, Satin and Gangal, 2011) .

6.1 Spaces of F^{α}-differentiable Functions

The space of all function k-times continuously ordinary differentiable on $[c, d]$ is defined by $C^k[c, d]$.

$C^0(F)$ is set of functions which are F-continuous.
$C^k,\ k \in N$ is shown all the functions $f : F \to R$ such that

$$(D_F^\alpha)^n f \in C^0(F), \quad for\ all\ n \le k.$$

The norm on $C^k(F)$ is defined

$$||f|| = \sum_{0 \le k} \sup |[D_F^\alpha)^n f](\theta)|, \quad f \in C^k(F). \tag{32}$$

Spaces of $C^k(F)$ are complete and separable with respect to this norm (Parvate and Gangal, 2009a; Parvate and Gangal, 2011b, Parvate, Satin and Gangal, 2011).

6.2 Spaces of F^α-Integrable Functions

$L(F)$ is the space of the F^α-integrable functions vector space with usual addition, multiplication and the following norm

$$N_p(u) = ||u||_p = \left[\int_a^b |u(x)|^p d_F^\alpha x \right]^{1/p}, \quad 1 \le p < \infty. \tag{33}$$

This definition is used to generalize analogues of abstract Sobolev spaces (Parvate and Gangal, 2009a; Parvate and Gangal, 2011b, Parvate, Satin and Gangal, 2011).

7 Calculus on Fractal Subsets of R^3

In this section we introduce calculus on fractal subsets of R^3 (Golmankhaneh and Baleanu, 2013; Golmankhaneh and Baleanu, 2014;). We begin by defining the integral of the staircase function.

7.1 Integral Staircase for Fractal Subsets of R^3

For a set F and I= $[a, b] \times [c, d] \times [e, f]$, with subdivision $P_{[a,b]\times[c,d]\times[e,f]}$, $a < b, c < d, e < f$ we define

$$\Lambda^\xi[F, I] = \sum_{i=1}^{n} \frac{(x_i - x_{i-1})^\alpha}{\Gamma(\alpha+1)} \frac{(y_i - y_{i-1})^\beta}{\Gamma(\beta+1)} \frac{(z_i - z_{i-1})^\epsilon}{\Gamma(\epsilon+1)} \tag{34}$$

$$\times \Theta(F, [x_{i-1}, x_i] \times [y_{i-1}, y_i] \times [z_{i-1}, z_i]), \tag{35}$$

where $\xi = \alpha + \beta + \epsilon$. Let $0 < \alpha \le 1, 0 < \beta \le 1,\ and\ 0 < \epsilon \le 1$. Given $\delta > 0$ and $a \le b, c \le d, e \le f$ the coarse-grained mass $\gamma_\delta^\xi(F, a, b, c, d, e, f)$ of $F \cap [a, b] \times [c, d] \times [e, f]$

is given by

$$\gamma_\delta^\xi(F, a, b, c, d, e, f) = \inf_{P_{[a,b]\times[c,d]\times[e,f]}:|P|\le\delta} \Lambda^\xi[F, I], \tag{36}$$

where $|P| = \max_{1\le i\le n}(x_i - x_{i-1})(y_i - y_{i-1})(z_i - z_{i-1})$. Taking an infimum over all subdivisions P of $[a, b] \times [c, d] \times [e, f]$ satisfying $|P| \le \delta$.
The mass function $\gamma^\xi(F, a, b, c, d, e, f)$ is given by

$$\gamma^\xi(F, a, b, c, d, e, f) = \lim_{\delta\to 0} \gamma_\delta^\xi(F, a, b, c, d, e, f). \tag{37}$$

Let a_0, b_0, c_0 be fixed and real numbers. The integral staircase function $S_F^\xi(x, y, z)$ of order ξ for a set F is

$$S_F^\xi(x, y, z) = \begin{cases} \gamma^\xi(F, a_0, b_0, c_0, x, y, z) & \text{if} \quad x \ge a_0, y \ge b_0, z \ge c_0, \\ -\gamma^\xi(F, a_0, b_0, c_0, x, y, z) & \text{otherwise.} \end{cases} \tag{38}$$

The η-dimension of $F\cap[a, b]\times[c, d]\times[e, f]$ is denoted by $\dim_\eta(F\cap[a, b]\times[c, d]\times[e, f])$ and define

$$\begin{aligned} \dim_\eta(F \cap [a, b] \times [c, d] \times [e, f]) &= \inf\{\xi : \gamma^\xi(F, a, b, c, d, e, f) = 0\}, \\ &= \sup\{\xi : \gamma^\xi(F, a, b, c, d, e, f) = \infty\}. \end{aligned}$$

Let $F \subset R^3$ be such that $S_F^\xi(x, y, z)$ is finite for all $(x, y, z) \in R^3$ for $\xi = \dim_\eta F$. Then the $Sch(S_F^\xi)$ is said to be ξ-perfect (Closed and every point of $Sch(S_F^\xi)$ is its limit point).

7.2 F^α-integration on Fractal Subset of R^3

Let f be a bounded function on F and I be a closed ball (Parvate and Gangal, 2009; Golmankhaneh and Baleanu, 2013). Then

$$M[f, F, I] = \sup_{(x,y,z)\in F\cap I} f(x, y, z) \quad \text{if } F \cap I \neq 0, \tag{39}$$

$$= 0 \quad \text{otherwise,} \tag{40}$$

and similarly

$$m[f, F, I] = \inf_{(x,y,z)\in F\cap I} f(x, y, z) \quad \text{if } F \cap I \neq 0, \tag{41}$$

$$= 0 \quad \text{otherwise.} \tag{42}$$

Let $S_F^\xi(x, y, z)$ be finite for $(x, y, z) \in [a, b] \times [c, d] \times [e, f]$. Let P be a subdivision of $[a, b]\times[c, d]\times[e, f]$ with points $x_0, y_0, z_0 \ldots, x_n, y_n, z_n$ The upper F^ξ-sum and the lower F^ξ-sum for the function f over the subdivision P are given respectively by (Parvate and Gangal, 2009; Golmankhaneh and Baleanu, 2013)

$$U^\xi[f, F, P] = \sum_{i=1}^{n} M[f, F, [(x_{i-1}, y_{i-1}, z_{i-1}), (x_i, y_i, z_i)]]$$

$$\times(S_F^\xi(x_i, y_i, z_i) - S_F^\xi(x_{i-1}, y_{i-1}, z_{i-1})) \tag{43}$$

and

$$L^\xi[f, F, P] = \sum_{i=1}^{n} m[f, F, [(x_{i-1}, y_{i-1}, z_{i-1}), (x_i, y_i, z_i)]]$$
$$\times(S_F^\xi(x_i, y_i, z_i) - S_F^\xi(x_{i-1}, y_{i-1}, z_{i-1})). \tag{44}$$

If f be a bounded function on F. We say that f is F^ξ-integrable on on $I = [a, b] \times [c, d] \times [e, f]$ if (Parvate and Gangal, 2009; Golmankhaneh and Baleanu, 2013)

$$\underline{\iiint_I} f(x, y, z) d_F^\xi x d_F^\xi y d_F^\xi z = \sup_{P_{[a,b]\times[c,d]\times[e,f]}} L^\xi[f, F, P] \tag{45}$$

$$= \overline{\iiint_I} f(x, y, z) d_F^\xi x d_F^\xi y d_F^\xi z = \inf_{P_{[a,b]\times[c,d]\times[e,f]}} U^\xi[f, F, P]. \tag{46}$$

In that case the F^ξ-integral of f on $[a, b] \times [c, d] \times [e, f]$ denoted by

$$\iiint_I f(x, y, z) d_F^\xi x d_F^\xi y d_F^\xi z$$

is given by the common value (Parvate and Gangal, 2009; Golmankhaneh and Baleanu, 2013).

7.3 F^α-differentiation on Fractal Subsets of R^3

If F is an ξ-perfect set then the F^ξ-partial derivative of f respect to x is

$${}^{x}D_F^\xi f(x, y, z) = \begin{cases} \text{F-}\lim_{(x',y,z)\to(x,y,z)} \frac{f(x',y,z)-f(x,y,z)}{S_F^\xi(x',y,z)-S_F^\xi(x,y,z)} & \text{if } (x, y, z) \in F, \\ 0 & otherwise, \end{cases} \tag{47}$$

if the limit exists. Likewise the ${}^{y}D_F^\xi f(x, y, z)$ and ${}^{z}D_F^\xi f(x, y, z)$ can be defined.

8 F^α-differential Form

In this section we generalize the F^ξ-fractional calculus on fractals subset of R^3 (Parvate and Gangal, 2009; Golmankhaneh and Baleanu, 2013).

8.1 F^α-Fractional 1-forms

A differential fractional 1-form on an set F subset of R^3 is a expression $H(x,y,z)d_F^\alpha x + G(x,y,z)d_F^\beta y + N(x,y,z)d_F^\epsilon z$ where H, G, N are functions on the open set. If $f(x,y,z)$ is C_ξ^1 function, then its F^ξ-fractional total differential (or exterior derivative) is

$$d_F^\xi f(x,y,z) = {}^xD_F^\alpha f(x,y,z)d_F^\alpha x + {}^yD_F^\beta f(x,y,z)d_F^\beta y + {}^zD_F^\epsilon f(x,y,z)d_F^\epsilon z,$$
$$\xi = \alpha + \beta + \epsilon. \tag{48}$$

In the same manner Eq. (48) can generalized to higher dimensions.

8.2 F^α- Fractional Exactness

Suppose that $Hd_F^\alpha x + Gd_F^\beta y + Nd_F^\epsilon z$ is a F^ξ-fractional differential on F with C_ξ^1 coefficients. We will say that it is exact if one can find a C_ξ^2 function $f(x,y,z)$ with $d_F^\xi f = Hd_F^\alpha x + Gd_F^\beta y + Nd_F^\epsilon z$. We will call a F^ξ-fractional differential closed if

$${}^xD_F^\xi f = H, \quad {}^yD_F^\xi f = G, \qquad {}^zD_F^\beta f = N. \tag{49}$$

Therefor $Hd_F^\alpha x + Gd_F^\beta y + Nd_F^\epsilon z$ is exact if we have

$${}^yD_F^\beta N = {}^zD_F^\epsilon G, \qquad {}^xD_F^\alpha G = {}^yD_F^\beta H, \qquad {}^zD_F^\epsilon H = {}^xD_F^\alpha N. \tag{50}$$

8.3 F^α-Fractional 2-forms

A F^ξ-Fractional 2-forms is like:

$$M(x,y,z)d_F^\alpha x \bigwedge d_F^\beta y + W(x,y,z)d_F^\beta y \bigwedge d_F^\epsilon z + L(x,y,z)d_F^\epsilon z \bigwedge d_F^\alpha x,$$

where M, W and L are functions. And $\bigwedge$ wedge product of two F^ξ-Fractional 1-forms with some properties (Parvate and Gangal, 2009; Golmankhaneh and Baleanu, 2013).

9 Gauge Integral and F^α-calculus

In this section we generalize F^α-calculus for unbounded functions with support on Cantor set (Golmankhaneh and Baleanu, 2013). Suppose $\dot{P}$ is a δ-fine partition. Similarly as in the previous case with tags we define $\sigma_\star^\alpha[F,I]$ as

$$\sigma_\star^\alpha[F,I] = \sum_{i=1}^{n} \frac{(x_i - x_{i-1})^\alpha}{\Gamma(\alpha+1)}\,\theta(F,[x_{i-1},x_i]). \tag{51}$$

The generalized coarse grained mass using gauge function is ${}^{\star}\gamma_{\delta}^{\alpha}(F,a,b)$ of $F \cap [a,b]$ is (Golmankhaneh and Baleanu, 2013)

$$ {}^{\star}\gamma_{\delta(t)}^{\alpha}(F,a,b) = \inf_{|\dot{P}|<\sup\{\delta(t_i);\ t_i\in[x_{i-1},x_i]\}} \sigma_{\star}^{\alpha}[F,I], \tag{52} $$

where $|\dot{P}| = \max_{1\le i\le n}(x_i - x_{i-1})$. We define the generalized mass function ${}^{\star}\gamma^{\alpha}(F,a,b)$ as follows (Golmankhaneh and Baleanu, 2013)

$$ {}^{\star}\gamma^{\alpha}(F,a,b) = \lim_{\sup\{\delta(t_i):\ t_i\in[x_{i-1},x_i]\}\to 0} {}^{\star}\gamma_{\delta(t)}^{\alpha}(F,a,b). $$

The generalized integral staircase function ${}^{\star}S_F^{\alpha}(x)$ of order α for a set F using gauge functions is given by (Golmankhaneh and Baleanu, 2013)

$$ {}^{\star}S_F^{\alpha}(x) = \begin{cases} {}^{\star}\gamma^{\alpha}(F,a_0,x) & \text{if } x \ge a_0, \\ -{}^{\star}\gamma^{\alpha}(F,a_0,x) & \text{otherwise.} \end{cases} \tag{53} $$

The ${}^{\star}\gamma$-dimension of $F\cap[a,b]$, denote by $\dim_{{}^{\star}\gamma}(F\cap[a,b])$ and define (Golmankhaneh and Baleanu, 2013)

$$ \begin{aligned} \dim_{{}^{\star}\gamma}(F\cap[a,b]) &= \inf\{\alpha : {}^{\star}\gamma^{\alpha}(F,a,b) = 0\}, \\ &= \sup\{\alpha : {}^{\star}\gamma^{\alpha}(F,a,b) = \infty\}. \end{aligned} $$

Example 1. Let us consider a discontinuous function as follows:

$$ f(x) = \begin{cases} 1 & \text{if } x \in \{[a,b] \cap Q \cap F(\text{Cantor set})\}, \\ 0 & \text{otherwise} \end{cases} \tag{54} $$

where Q is set of rational number and $f(x)$ is discontinues at every point of $[a,b]$. If $\{r_k = k \in N\}$ is an enumeration of the rational number in $[a,b]$ and $\epsilon > 0$ we define the gauge as follows

$$ \delta_{\epsilon}(t) = \begin{cases} \epsilon/2^{k+1} & \text{if } t = r_k, \\ 1 & \text{otherwise.} \end{cases} \tag{55} $$

Then by this gauge we can control contribution of ${}^{\star}S_F^{\alpha}(x_i) - {}^{\star}S_F^{\alpha}(x_{i-1})$ in the Riemann sum $S(f,\dot{P})$. Then this function is ${}^{\star}F^{\alpha}$-integrable but it is not F^{α}-integrable (Golmankhaneh and Baleanu, 2013).

10 Application of F^{α}-calculus

In this section we present the F^{α}-calculus in dynamics and physics.

10.1 Lagrangian and Hamiltonian Mechanics on Fractals

Our goal in this section is to derive the equations of motion for a particle whose position is $x(t) : F \to R$ that varies on a fractal subset of real F (Golmankhaneh and Baleanu, 2013). We aim to find a function L_F^α such that the paths of the particles between times $t_1 \in F$ and $t_2 \in F$ extremize the integral

$$A_F^\alpha = \int_{t_1}^{t_2} L_F^\alpha(t, x(t), {}^tD_F^\alpha x(t))d_F^\alpha t, \quad L_F^\alpha : F \times R \times R \to R,$$

where integral A_F^α is the action of the particle and the function L_F^α its Lagrangian. So we have

$$\delta A_F^\alpha = \int_{t_1}^{t_2} [{}^xD_F^\alpha\, L_F^\alpha\, \delta x + {}^{{}^tD_F^\alpha x}D_F^\alpha\, L_F^\alpha \delta({}^tD_F^\alpha x(t))]d_F^\alpha t = 0,$$

$$= \int_{t_1}^{t_2} [{}^xD_F^\alpha\, L_F^\alpha\, \delta x + {}^{{}^tD_F^\alpha x}D_F^\alpha\, L_F^\alpha\, {}^tD_F^\alpha \delta(x(t))]d_F^\alpha t = 0. \tag{56}$$

Using Eq. (18) and $\delta(x(t_2)) = \delta(x(t_1)) = 0$ and let ${}^tD_F^\alpha x = \varrho$ we have

$$= \int_{t_1}^{t_2} [{}^xD_F^\alpha\, L_F^\alpha\, - \, {}^tD_F^\alpha({}^\varrho D_F^\alpha\, L_F^\alpha)]\, \delta(x)d_F^\alpha t = 0. \tag{57}$$

We arrive at a Euler-Lagrange equation on fractal time sets as

$${}^xD_F^\alpha L_F^\alpha - {}^tD_F^\alpha({}^\varrho D_F^\alpha\, L_F^\alpha) = 0. \tag{58}$$

Now if we define the generalized momentum $p_F^\alpha = {}^\varrho D_F^\alpha\, L_F^\alpha$ so that we have a Hamiltonian defined as

$$H_F^\alpha = {}^tD_F^\alpha x\, p_F^\alpha - L_F^\alpha. \tag{59}$$

Applying d_F^α to both side of Eq. (59) we obtain

$$d_F^\alpha H_F^\alpha = {}^tD_F^\alpha x\, d_F^\alpha p_F^\alpha - {}^tD_F^\alpha\, L_F^\alpha\, d_F^\alpha t - {}^xD_F^\alpha\, L_F^\alpha\, d_F^\alpha x, \tag{60}$$

Eq. (60) shows that H_F^α is function of t, p_F^α, x so we have Hamilton equation on a fractal subset of real line F as follows (Golmankhaneh and Baleanu, 2013)

$${}^tD_F^\alpha x = {}^\zeta D_F^\alpha H_F^\alpha; \;\; \zeta = p_F^\alpha; \;\; {}^tD_F^\alpha L_F^\alpha = {}^tD_F^\alpha H_F^\alpha; \;\; {}^tD_F^\alpha\, p_F^\alpha = -\, {}^xD_F^\alpha H_F^\alpha. \tag{61}$$

Example 1. Suppose that the Lagrangian of a particle is

$$L_F^\alpha(t, x(t), {}^tD_F^\alpha x) = a({}^tD_F^\alpha x)^2 - b(x(t))^2, \quad c,\ e,\ are\ constant. \tag{62}$$

Therefore, using Eq. (58) the Lagrange equation will be defined as the following

$$-2b(x(t)) = 2a({}^{t}D_F^{\alpha})^2 x(t). \tag{63}$$

Example 2. Let the Hamiltonian of a particle be given by:

$$H_F^{\alpha} = c(p_F^{\alpha})^2 + e(x(t))^2, \quad c,\ e,\ are\ constant.$$

So the Hamilton's equation is

$${}^{t}D_F^{\alpha} p_F^{\alpha} = -2ex(t), \qquad {}^{t}D_F^{\alpha} x = 2cp_F^{\alpha}. \tag{64}$$

11 Quantum Mechanics on Fractal Curve

Classical mechanics is reformulated in terms of a minimum principle. The Euler-Lagrange equations of motion is derived from the least action. The Feynman paths for a particle in quantum mechanics are fractals with dimension 2 (Abbott and Wise, 1981). In this section, we obtain the Schrödinger equation on fractal curves.

11.1 Generalized Feynman Path Integral Method

The Feynman method for studying quantum mechanics using classical Lagrangian and action is presented in Ref. (Feynman and Hibbs, 1965). Now we want to generalize the Feynman (Schulman, 1981) method using Lagrangian on fractals curves. Consider the generalized action:

$$\mathfrak{A}_F^{\alpha} = \int_{t_1}^{t_2} L_F^{\alpha}(t, \mathbf{w}(v), {}^{t}D_F^{\alpha}\mathbf{w}(v)) d_F^{\alpha}v\ d_F^{\alpha}t, \quad L_F^{\alpha} : F \times F \times F \to R. \tag{65}$$

In view of the Feynman method, if a wave function on fractal in $t_1, \mathbf{w}_a(v_1)$ is $\psi_F^{\alpha}(t_1, \mathbf{w}_a(v_1))$. It gives the total probability amplitude at t_2, $\mathbf{w}_b(v_2)$ as

$$\psi_F^{\alpha}(t_2, \mathbf{w}_b(v_2)) = \int_{-\infty}^{\infty} K_F^{\alpha}(t_2, \mathbf{w}_b(v_2), t_1, \mathbf{w}_a(v_1))(\psi_F^{\alpha}(t_1, \mathbf{w}_a(v_1)) d_F^{\alpha}\mathbf{w}(v), \tag{66}$$

where K_F^{α} is the propagator which is defined as follows (Golmankhaneh and Baleanu, 2014)

$$K_F^{\alpha}(t_2, \mathbf{w}_b(v_2), t_1, \mathbf{w}_a(v_1)) = \int_{w_a}^{wb} \exp[\frac{i}{\hbar}\mathfrak{A}_F^{\alpha}]\mathfrak{D}_F^{\alpha}\mathbf{w}(v). \tag{67}$$

Where symbol $\mathfrak{D}_F^\alpha$ indicates the integration over all fractal paths from $\mathbf{w}_a(v_1)$ to $\mathbf{w}_b(v_2)$.

Now we derive the Schrödinger equation for a free particle on fractal curves, which describes the evolution of the wave function from $\mathbf{w}_a(v_1)$ to $\mathbf{w}_b(v_2)$, when t_2 differs from t_1 or an infinitesimal amount ϵ. Supposing that $S_F^\alpha(v_2) = S_F^\alpha(v_1) + \epsilon$, leads to a Lagrangian for free particles as

$$L_F^\alpha(t, \mathbf{w}(v), {}^tD_F^\alpha \mathbf{w}(v)) \simeq \frac{m(\mathbf{w}(v) - \mathbf{w}(v_0))^2}{2(S_F^\alpha(v_2) - S_F^\alpha(v_1))}. \tag{68}$$

The generalized action on a fractal $\mathfrak{A}_F^\alpha$ is approximately

$$\mathfrak{A}_F^\alpha \sim \epsilon L_F^\alpha = \frac{m(\mathbf{w}(v) - \mathbf{w}(v_0))^2}{2\epsilon}. \tag{69}$$

As a consequence, we obtain

$$\psi_F^\alpha(t+\epsilon, \mathbf{w}(v)) = \int_{-\infty}^{+\infty} \frac{1}{A} \exp[\frac{i}{\hbar} \frac{m(\mathbf{w}(v) - \mathbf{w}_0(v_0))^2}{2\epsilon}] \psi_F^\alpha(t, \mathbf{w}_0(v_0)) \mathfrak{D}_F^\alpha \mathbf{w}_0(v_0). \tag{70}$$

Here, because of properties of exponential functions only those fractal paths give significant contributions which are very close to $\mathbf{w}(v)$. Changing the variable in the integral $\delta = \mathbf{w}(v) - \mathbf{w}_0(v_0)$ we have $\psi_F^\alpha(t, \mathbf{w}_0(v_0)) = \psi_F^\alpha(t, \mathbf{w}(v) + \delta)$. Since both ϵ and δ are small quantities, so that $\psi_F^\alpha(t, \mathbf{w}(v) + \delta)$ and $\psi_F^\alpha(t+\epsilon, \mathbf{w}(v))$ can be expanded using Eq. (31). We only keep terms of up to second order of ϵ and δ. As a result we get

$$\begin{aligned}\psi_F^\alpha(t, \mathbf{w}(v)) + \epsilon({}^t\mathfrak{D}_F^\alpha)\psi_F^\alpha(t, \mathbf{w}(v)) \simeq \chi_F(t) \int_{-\infty}^{+\infty} \frac{1}{A} \exp[\frac{i}{\hbar}\frac{m\delta^2}{2\epsilon}](\psi_F^\alpha(t, \mathbf{w}(v)) \\ +\delta({}^{w_i}\mathfrak{D}_F^\alpha)\psi_F^\alpha(t, \mathbf{w}(v))) + \frac{\delta^2}{2}({}^{w_i}\mathfrak{D}_F^\alpha)^2\psi_F^\alpha(t, \mathbf{w}(v)))d_F^\alpha\delta,\end{aligned} \tag{71}$$

where $\chi_F(t)$ is the characteristic function for Cantor like sets. The second term in the right hand side vanishes up on integration. It follows by equating the leading terms on both sides

$$\psi_F^\alpha(t, \mathbf{w}(v)) = \int_{-\infty}^{+\infty} \frac{1}{A} \exp[\frac{i}{\hbar}\frac{m\delta^2}{2\epsilon}]\psi_F^\alpha(t, \mathbf{w}(v)) d_F^\alpha\delta. \tag{72}$$

Also, we arrive at

$$A = \int_{-\infty}^{+\infty} \frac{1}{A} \exp[\frac{i}{\hbar}\frac{m\delta^2}{2\epsilon}] d_F^\alpha\delta = \sqrt{\frac{2i\pi\hbar\epsilon}{m}}, \tag{73}$$

and

$$\int_{-\infty}^{+\infty} \frac{1}{A} \exp[\frac{i}{\hbar}\frac{m\delta^2}{2\epsilon}](\frac{\delta^2}{2}({}^{w_i}\mathfrak{D}_F^\alpha)^2\psi_F^\alpha(t, \mathbf{w}(v)))d_F^\alpha\delta = \epsilon\frac{i\hbar}{2m}({}^{w_i}\mathfrak{D}_F^\alpha)^2\psi_F^\alpha(t, \mathbf{w}(v))). \tag{74}$$

Finally, equating the remaining terms, we get Schrödinger equation on fractal curves for free particle as

$$(i\hbar\,{}^{t}\mathfrak{D}_F^\alpha)\psi_F^\alpha(t,\mathbf{w}(v)) = \chi_F(t)\frac{-\hbar^2}{2m}({}^{w_i}\mathfrak{D}_F^\alpha)^2\psi_F^\alpha(t,\mathbf{w}(v))). \tag{75}$$

The Eq. (75) leads to the definition of the Hamiltonian and momentum operator on fractal curves as

$$\hat{H}_F^\alpha = i\hbar\,{}^{t}\mathfrak{D}_F^\alpha, \quad \hat{P}_F^\alpha = -i\hbar\nabla_F^\alpha. \tag{76}$$

The solution of Eq. (75) can be found using the conjugate equation as

$$i\hbar\frac{\partial\theta(t,\xi)}{\partial t} = \frac{-\hbar^2}{2m}\frac{\partial^2}{\partial\xi^2}\theta(t,\xi) \quad \theta(\xi,t) = \phi[\psi_F^\alpha(t,\mathbf{w}(v)))]. \tag{77}$$

Since the solution Eq. (77) is

$$\theta(t,\xi) = (Ae^{ik\xi} + Be^{-ik\xi})e^{-i\beta t}, \tag{78}$$

where $k = \frac{\sqrt{2mE}}{\hbar}$ and $\omega = \frac{E}{\hbar}$ are constants. Now by applying ϕ^{-1} we have the solutions as

$$\psi_F^\alpha(t,\mathbf{w}(v))) = (Ae^{ikS_F^\alpha(v)} + Be^{-ikS_F^\alpha(v)})e^{-i\beta S_F^\alpha(t)}. \tag{79}$$

It is straight forward to extended to the case of a free particle to motion involving a potential. In this case the Lagrangian will be $L_F^\alpha = T_F^\alpha - V_F^\alpha(t,\mathbf{w}(v))$. By substituting the Lagrangian in the Eq. (71) one can derive the Schrödinger equation as

$$\begin{aligned}\psi_F^\alpha(t,\mathbf{w}(v)) + \epsilon({}^{t}\mathfrak{D}_F^\alpha)\psi_F^\alpha(t,\mathbf{w}(v)) &\simeq \chi_F(t)\int_{-\infty}^{+\infty}\frac{1}{A}\exp[\frac{i}{\hbar}\frac{m\delta^2}{2\epsilon}]\\ &\times[1-\frac{i\epsilon}{\hbar}V_F^\alpha(t,\mathbf{w}(v))](\psi_F^\alpha(t,\mathbf{w}(v)) + \delta({}^{w_i}\mathfrak{D}_F^\alpha)\psi_F^\alpha(t,\mathbf{w}(v)))\\ &+\frac{\delta^2}{2}({}^{w_i}\mathfrak{D}_F^\alpha)^2\psi_F^\alpha(t,\mathbf{w}(v)))d_F^\alpha\delta.\end{aligned} \tag{80}$$

The same manner we worked out above Eq. (80) becomes

$$\begin{aligned}(i\hbar\,{}^{t}\mathfrak{D}_F^\alpha)\psi_F^\alpha(t,\mathbf{w}(v)) &= \chi_F(t)\frac{-\hbar^2}{2m}({}^{w_i}\mathfrak{D}_F^\alpha)^2\psi_F^\alpha(t,\mathbf{w}(v)))\\ &+\chi_F(t)V_F^\alpha(t,\mathbf{w}(v))\psi_F^\alpha(t,\mathbf{w}(v)).\end{aligned} \tag{81}$$

12 Continuity Equation and Probability on Fractal

It is well known that the continuity equation is a important concept in quantum mechanics. Therefore, the probability density on the fractal for a particle is defined as

$$\rho_F^\alpha(t,\mathbf{w}(v)) = ({}^{*}\psi_F^\alpha(t,\mathbf{w}(v)))\,\psi_F^\alpha(t,\mathbf{w}(v)). \tag{82}$$

The complex conjugate wave function of Eq. (81) is

$$(-i\hbar\, {}^{t}\mathfrak{D}_F^{\alpha})\, {}^{*}\psi_F^{\alpha}(t,\mathbf{w}(v)) = \chi_F(t)\frac{-\hbar^2}{2m}({}^{w_i}\mathfrak{D}_F^{\alpha})^2\, {}^{*}\psi_F^{\alpha}(t,\mathbf{w}(v)))$$
$$+\chi_F(t)V(t,\mathbf{w}(v))\, {}^{*}\psi_F^{\alpha}(t,\mathbf{w}(v)), \tag{83}$$

where $V_F^{\alpha}(t,\mathbf{w}(v)) = {}^{*}V_F^{\alpha}(t,\mathbf{w}(v))$. Applying this identity is given below

$${}^{t}\mathfrak{D}_F^{\alpha}(\psi_F^{\alpha}(t,\mathbf{w}(v))\, {}^{*}\psi_F^{\alpha}(t,\mathbf{w}(v))) = {}^{t}\mathfrak{D}_F^{\alpha}(\psi_F^{\alpha}(t,\mathbf{w}(v)))^{*}\psi_F^{\alpha}(t,\mathbf{w}(v))$$
$$+\psi_F^{\alpha}(t,\mathbf{w}(v))\, {}^{t}\mathfrak{D}_F^{\alpha}({}^{*}\psi_F^{\alpha}(t,\mathbf{w}(v))), \tag{84}$$

and substituting Eq. (81) and Eq. (83), into Eq.(84) yield us

$$i\hbar\, {}^{t}\mathfrak{D}_F^{\alpha}\rho_F^{\alpha}(t,\mathbf{w}(v)) = \chi_F(t)\frac{\hbar^2}{2m}[\psi_F^{\alpha}(t,\mathbf{w}(v))({}^{w_i}\mathfrak{D}_F^{\alpha})^2\, {}^{*}\psi_F^{\alpha}(t,\mathbf{w}(v))$$
$$-\, {}^{*}\psi_F^{\alpha}(t,\mathbf{w}(v))({}^{w_i}\mathfrak{D}_F^{\alpha})^2\, \psi_F^{\alpha}(t,\mathbf{w}(v))]. \tag{85}$$

As a consequence the definition of a probability current density on fractal curve is

$$J_F^{\alpha}(t,\mathbf{w}(v)) = \chi_F(t)\frac{\hbar}{2mi}[\psi_F^{\alpha}(t,\mathbf{w}(v))({}^{w_i}\mathfrak{D}_F^{\alpha})^2\, {}^{*}\psi_F^{\alpha}(t,\mathbf{w}(v))$$
$$-\, {}^{*}\psi_F^{\alpha}(t,\mathbf{w}(v))({}^{w_i}\mathfrak{D}_F^{\alpha})^2\, \psi_F^{\alpha}(t,\mathbf{w}(v))]. \tag{86}$$

13 Newtonian Mechanics on Fractals

13.1 Kinematics of Motion

Generalized average velocity: Suppose $x_F^{\alpha}(t)$ is position function of a particle. Such that $Sch(x_F^{\alpha}(t))$is α-perfect set then we can define the generalized average velocity between $[t_1, t_2]$ as follows:

$$\overline{v_F^{\alpha}}(t) = \frac{x_F^{\alpha}(t_2) - x_F^{\alpha}(t_1)}{S_F^{\alpha}(t_2) - S_F^{\alpha}(t_1)}. \tag{87}$$

Generalized velocity: Let $x_F^{\alpha}(t)$ is position function of a particle. Let $Sch(x_F^{\alpha}(t))$ be an α-perfect set. The velocity of particle on a fractal set F is defined

$$v_F^{\alpha}(t) = D_F^{\alpha}x(t). \tag{88}$$

Generalized average acceleration: Let $x_F^{\alpha}(t)$ be the velocity function of a particle. We define average acceleration of a particle as follows

$$\overline{a_F^{\alpha}}(t) = \frac{v_F^{\alpha}(t_2) - v_F^{\alpha}(t_1)}{S_F^{\alpha}(t_2) - S_F^{\alpha}(t_1)}. \tag{89}$$

Here $Sch(v_F^{\alpha}(t))$ is an α-perfect set.
Generalized acceleration: Consider $x_F^{\alpha}(t)$ as position function of a particle. So we can define the generalized acceleration as

$$a_F^{\alpha}(t) = (D_F^{\alpha})^2 x_F^{\alpha}(t) = D_F^{\alpha}v_F^{\alpha}(t), \tag{90}$$

where $Sch(x_F^\alpha(t))$, $v_F^\alpha(t)$ are α-perfect sets.

Example 1. Suppose a particle is moving with constant acceleration $k\chi_F$. The position and velocity function of particle can be derived as follows

$$a_F^\alpha(t) = D_F^\alpha v_F^\alpha(t) = k\chi_F, \tag{91}$$

then by applying F^α-integration in both side we have

$$v_F^\alpha(t) = v_F^\alpha(t_1) + \chi_F(S_F^\alpha(t) - S_F^\alpha(t_1)), \tag{92}$$

likewise using F^α-integration in both side we obtain

$$x_F^\alpha(t) = \frac{\chi_F}{2}(S_F^\alpha(t) - S_F^\alpha(t_1))^2 + x_F^\alpha(t_1) + \chi_F\, v_F^\alpha(t_1)\,(S_F^\alpha(t) - S_F^\alpha(t_1)). \tag{93}$$

13.2 Dynamics of Motion

Generalized momentum: If $x_F^\alpha(t)$ is position function of particle such that $Sch(x_F^\alpha(t))$ is α-perfect set. Let $v_F^\alpha(t)$

$$p_F^\alpha = mv_F^\alpha. \tag{94}$$

Generalized Newton's second law: Consider a particle with mass m whose position function $x_F^\alpha(t)$ is such that $Sch(x_F^\alpha(t)$ and α-perfect set. Then the generalized Newton's 2nd law is

$$f_F^\alpha = m\, a_F^\alpha(t), \tag{95}$$

or

$$f_F^\alpha = D_F^\alpha\, p_F^\alpha. \tag{96}$$

Example 2. Suppose that the force applied to a particle is $f_F^\alpha = (S_F^\alpha(t))^2$ then the equation of the motion is as follows:

$$a_F^\alpha(t) = \frac{1}{m}(S_F^\alpha(t))^2, \tag{97}$$

then velocity will be

$$v_F^\alpha(t) = v_F^\alpha(t_1) + \frac{1}{3m}(S_F^\alpha(t) - S_F^\alpha(t_1))^3, \tag{98}$$

position function

$$x_F^\alpha(t) = x_F^\alpha(t_1) + v_F^\alpha(t_1)\,(S_F^\alpha(t) - S_F^\alpha(t_1)) + \frac{1}{12m}((S_F^\alpha(t) - S_F^\alpha(t_1))^4. \tag{99}$$

14 Work and Energy Theorem on Fractals

Generalized kinetic energy: If $x_F^\alpha(t)$ is position function of particle such that $Sch(x_F^\alpha(t))$ is α-perfect set. Let $v_F^\alpha(t)$ be velocity of particle

$$k_F^\alpha = \frac{1}{2} m \, (v_F^\alpha)^2. \tag{100}$$

Generalized potential energy: Let f_F^α be generalized conservative force (i.e. F^α-integral of f_F^α in the closed curve is zero). Then we can derive the generalized potential u_F^α as follows

$$u_F^\alpha(t) = u_F^\alpha(a) - \chi_F \int_a^t f_F^\alpha d_F^\alpha x, \tag{101}$$

or

$$f_F^\alpha = -D_F^\alpha \, u_F^\alpha(t). \tag{102}$$

Example 3. Let the conservative force $f_C^\alpha = x\chi_C(x)$ here C stands for a Cantor set. then the corresponding potential will be

$$u_C^\alpha(t) = u_C^\alpha(a) - \chi_C \int_a^t x d_C^\alpha x, \tag{103}$$

$$= u_C^\alpha(a) - \chi_C \sum_{n=1}^{\infty} I_n(t), \tag{104}$$

where

$$I_n(t) = \begin{cases} 0 & \text{if } \; y_n(t) = 0 \; \text{ or } \; y_i(t) = 1 \; \text{ for some } \; i < n \\ \frac{1}{\Gamma(\alpha+1)}\left[\frac{T_{n-1}(y)}{2^n} + \frac{1}{2.6^n}\right] & \text{otherwise.} \end{cases} \tag{105}$$

An approximation of $y \in [0, 1]$ by a finite number of digits is denoted by

$$T_0(t) = 0 \qquad \text{and} \qquad T_n(t) = \sum_{i=1}^{n} \frac{y_i(t)}{3^i}. \tag{106}$$

Generalized work and energy theorem on fractals: If $x_F^\alpha(t)$ is the position function of particle such that $Sch(x_F^\alpha(t))$ is α-perfect set. Let $v_F^\alpha(t)$ be velocity of particle

$$\frac{1}{2} m \, (v_F^\alpha(t_2))^2 - \frac{1}{2} m \, (v_F^\alpha(t_1))^2 = \int_{t_1}^{t_2} f_F^\alpha d_F^\alpha x. \tag{107}$$

Example 4. Consider a particle modelled by a simple harmonic oscillator. The conservation of energy implies

$$\frac{1}{2} m \, (v_F^\alpha(t_2))^2 - \frac{1}{2} m \, (v_F^\alpha(t_1))^2 = \chi_C \sum_{n=1}^{\infty} I_n(t). \tag{108}$$

15 Langevin F^{α}-Equation on Fractals

The theory of Brownian motion is the simplest way to treat the dynamics of a non-equilibrium system. The fundamental equation is called the Langevin equation, it contains both frictional forces and random forces. The fluctuation-dissipation theorem relates these forces to each other. While the motion of a dust particle under Brownian motion appears to be quite random, it must nevertheless be describable by the same equations of motion as any other dynamical system. The equations of motion for a Brownian particle are the Langevinian equations. The Langevinian equations on the fractals subset real line is introduced. Generalized the Langevinian F^{α}-equation of a Brownian particle of mass m, with the surrounding medium represented by a generalized force $-\epsilon\chi_F v(t)$ and random density fluctuations in the fluid is

$$\frac{m}{2}(D_F^{\alpha})^2 x(t) = -\epsilon\chi_F v(t) + \zeta\ \xi(t), \tag{109}$$

where $\xi(t)$ random force and ζ is constant.

16 Maxwell's Equation on Fractals

We obtain the fractional Maxwell's equation on fractals as follows (Golmankhaneh and Baleanu, 2013)

$$\begin{aligned}\omega_F^{\xi} = (E_1 d_F^{\alpha} x_1 + E_2 d_F^{\beta} x_2 + E_3 d_F^{\epsilon} x_3) \bigwedge d_F^{\kappa} t + B_1 d_F^{\beta} x_2 \bigwedge d_F^{\epsilon} x_3 \\ + B_2 d_F^{\epsilon} x_3 \bigwedge d_F^{\alpha} x_1 + B_3 d_F^{\beta} x_1 \bigwedge d_F^{\alpha} x_2,\end{aligned} \tag{110}$$

where $E_i,\ B_j,\ i,j = 1,2,3$ are components of electromagnetic field. Applying d_F^{ξ} that is

$$d_F^{\xi} = {}^{x_1}D_F^{\mu} d_F^{\mu} x_1 + {}^{x_2}D_F^{\mu} d_F^{\mu} x_2 + {}^{x_3}D_F^{\mu} d_F^{\mu} x_3 + {}^{t}D_F^{\mu} d_F^{\mu} t, \tag{111}$$

to Eq. (110) and supposing $\alpha = \beta = \epsilon = \kappa = \mu$. Since $d_F^{\xi}\omega_F^{\xi} = 0$, so Eq. (110) leads to:

$$\begin{aligned}&\{{}^{x_1}D_F^{\mu}E_2 - {}^{x_2}D_F^{\mu}E_1\} d_F^{\mu} x_1 \bigwedge d_F^{\mu} x_2 \bigwedge d_F^{\mu} t \\ &+\{{}^{x_1}D_F^{\mu}E_3 - {}^{x_3}D_F^{\mu}E_1\} d_F^{\mu} x_1 \bigwedge d_F^{\mu} x_3 \bigwedge d_F^{\mu} t \\ &+\{{}^{x_1}D_F^{\mu}E_2 - {}^{x_2}D_F^{\mu}E_3\} d_F^{\mu} x_3 \bigwedge d_F^{\mu} x_2 \bigwedge d_F^{\mu} t\end{aligned} \tag{112}$$

$$\begin{aligned}&= -{}^{t}D_F^{\mu}B_3 d_F^{\mu} x_1 \bigwedge d_F^{\mu} x_2 \bigwedge d_F^{\mu} t - {}^{t}D_F^{\mu}B_2 d_F^{\mu} x_1 \bigwedge d_F^{\mu} x_3 \bigwedge d_F^{\mu} t \\ &-{}^{t}D_F^{\mu}B_1 d_F^{\mu} x_3 \bigwedge d_F^{\mu} x_2 \bigwedge d_F^{\mu} t,\end{aligned} \tag{113}$$

and

$$({}^{x_1}D_F^{\mu}B_1 + {}^{x_2}D_F^{\mu}B_2 + {}^{x_3}D_F^{\mu}B_3)d_F^{\mu}x_1 d_F^{\mu}x_2 d_F^{\mu}x_3 = 0. \tag{114}$$

In vector notation it is given by:

$$curl_F^{\mu}E = -{}^{t}D_F^{\mu}B \tag{115}$$

$$div_F^{\mu}B = 0. \tag{116}$$

Now let define a fractional form as

$$\pi_F^{\mu} = (J_1 d_F^{\mu}x_2 \bigwedge d_F^{\mu}x_3 + J_2 d_F^{\mu}x_3 \bigwedge d_F^{\mu}x_1 + J_3 d_F^{\mu}x_1 \bigwedge d_F^{\mu}x_2) \bigwedge d_F^{\mu}t$$
$$-\rho d_F^{\mu}x_1 \bigwedge d_F^{\mu}x_2 \bigwedge d_F^{\mu}x_3, \tag{117}$$

where J_i, $i = 1, 2, 3$ is components of current and ρ density of charge. Likewise applying d_F^{μ} to left side of Eq. (117) and $d_F^{\mu}\pi_F^{\mu} = 0$ we have

$$({}^{x_1}D_F^{\mu}J_1 +{}^{x_2} D_F^{\mu}J_2 + {}^{x_3}D_F^{\mu}J_3 + {}^{t}D_F^{\mu}\rho)d_F^{\mu}x_1 \bigwedge d_F^{\mu}x_2 \bigwedge d_F^{\mu}x_3 \bigwedge d_F^{\mu}t = 0. \tag{118}$$

Furthermore conservation of charge on fractals is

$$div_F^{\mu}\mathbf{J} + {}^{t}D_F^{\mu}\rho = 0. \tag{119}$$

Let us consider the following fractional form

$$\zeta_F^{\mu} = A_1 d_F^{\mu}x_1 + A_2 d_F^{\mu}x_2 + A_3 d_F^{\mu}x_3 + \varphi d_F^{\mu}t, \tag{120}$$

where A_i, $i = 1, 2, 3$ is vector potential. Using $d_F^{\mu}\zeta_F^{\mu} = 0$ one can obtain

$$curl_F^{\mu}\mathbf{A} = \mathbf{B}, \qquad grad_F^{\mu}\varphi - {}^{t}D_F^{\mu}\mathbf{A} = \mathbf{E}. \tag{121}$$

Therefore we arrive at the wave equation as following

$$c^2 curl_F^{\mu}\mathbf{B} = {}^{t}D_F^{\mu}\mathbf{E}, \tag{122}$$

where c is speed of light.

Acknowledgment: Alireza K. Golmankhaneh would like to thank Prof. A. D. Gangal for useful discussion on the topics, Prof. Dumitru Baleanu for rigorous discussions and collaboration, and Prof. Xiao-Jun Yang and Prof. Carlo Cattani for their invitation to write a chapter for this book, as well as all the people who helped to make this book possible.

Bibliography

Abbott L.F., Wise M.B., Dimension of a quantum-mechanical path. American Journal of Physics, (1981), 49, 37.

Bunde A., Havlin S., (eds), Fractal in Science. (1995), Springer.

Dalrymple K., Strichartz R.S., Vinson J.P., Fractal differential equations on the Sierpinski gasket. J. Fourier Anal. Appl. (1999), 5, 205.

Edgar G.A., Integral, Probability and Fractal Measures. (1998), Springer-Verlag, New York.

Falconer K., The Geometry of fractal sets. (1985), Cambridge University Press.

Falconer K., Fractal Geometry: Mathematical foundations and applications. (1990), John Wiley and Sons.

Falconer K., Techniques in Fractal Geometry. (1997) John Wiley and Sons.

Feynman R.P., Hibbs A.R., Quantum Mechanics and Path Integrals. (1965), McGraw-Hill, New York, USA.

Freiberg U., Analytical properties of measure geometric Krein-Feller-operators on the real line. (2003), Math. Nachr. 260, 34.

Freiberg U., Zähle M., Harmonic calculus on fractals-a measure geometric approach I. Potential Analysis, (2002), 16(3), 265-277.

Golmankhaneh A.K., Baleanu D., On a new measure on fractals. Journal of Inequalities and Applications, (2013), 2013.1, 1-9.

Golmankhaneh A.K., Fazlollahi V., Baleanu D., Newtonian mechanics on fractals subset of real-line. Rom. Rep. Phys., (2013), 65, (84-93).

Golmankhaneh A.K., Golmankhaneh A.K., Baleanu D., About Maxwell's equations on fractal subsets of R^3. Cent. Eur. J. of Phys., (2013), 11(6), 863-867.

Golmankhaneh A.K., Golmankhaneh A.K., Baleanu D., Lagrangian and Hamiltonian mechanics on fractals subset of real-line. Int. J. Theo. Phys., (2013), 52(11), 4210-4217.

Golmankhaneh A.K., Golmankhaneh A.K., Baleanu D., Schrödinger Equation on Fractals Curves Imbedding in R^3. Int. J. Theo. Phys., (2015), 54(4), 1275-1282.

Jurgen J., Geometry and physics. (2009), Springer.

Kigami J., Analysis on Fractals. (2000), Cambrigde University Press.

Kolwankar K.M., Gangal A.D., Local fractional Fokker-Planck equation. Phys Rev. Lett., (1998), 80, 214.

Mandelbrot B.B., The Fractal Geometry of Nature. (1977), Freeman and company.

Mikio N., Geometry, topology and physics. (2003), CRC Press. Parvate A., Gangal A.D., Calculus on fractal subsets of real line-I: formulation. Fractals, (2009), 17, 53-81.

Parvate A., Gangal A.D., Calculus on fractal subsets of real line–II: Conjugacy with ordinary calculus. Fractals, (2011), 19(03), 271-290.

Parvate A., Satin S., Gangal A.D., Calculus on Fractal Curves in R^n Fractals, (2011), 19(1), 15-27.

Satin S., Parvate A., Gangal A.D., Fokker–Planck equation on fractal curves. Chaos, Solitons Fractals, (2013), 52, 30-35.

Satin S., Gangal A.D., Langevin Equation on Fractal Curves. (2014), arXiv preprint arXiv:1404.6896.

Schulman L.S., Techniques and Applications of Path Integrations. (1981), Wiley Inter science, New York.

Shlesinger M.F., Fractal time in condensed matter. Annual Review of Physical Chemistry, (1988), 39(1), 269-290.

Strichartz R.S., Taylor approximations on Sierpinski gasket type fractals, J. funct. Anal., (2000), 174, 76.

Strichartz R.S., Differential Equations of fractals. (2006), Princeton University Press New Jersey.

Theodore F., The geometry of physics: an introduction. (2011), Cambridge University Press.

Yang X.-J., Advanced Local Fractional Calculus and Its Applications. World Science, (2012), New York, USA.

Yang X.-J., Local Fractional Functional Analysis and Its Applications. Asian Academic Publisher Limited, (2011), Hong Kong.

Haci Mehmet Baskonus, Fethi Bin Muhammad Belgacem, and Hasan Bulut

Solutions of Nonlinear Fractional Differential Equations Systems through an Implementation of the Variational Iteration Method

Abstract: In this book chapter, the Variational Iteration Method is used to solve systems of nonlinear fractional differential equations. Implementations with two separate nonlinear fractional differential equations systems illustrate a high degree of effectiveness. In the absence of explicit analytical methods, the Variational Iterative Method turns out to be a powerful tool to understand the behavior of nonlinear fractional differential equations and systems. Numerical solutions plots are obtained for each system via the Mathematica Symbolic Algebra Software 9.

Keywords: Variational Iterational method; Fractional differential equation; Nonlinear fractional differential equation systems.

1 Introduction

Fractional differential equations and systems have come to earn a high standing as important mathematical tools and as a means for describing and modeling linear and non-linear phenomena. (See for instance, Goswami & Belgacem, 2012; Gupta, Sharma, & Belgacem, 2011; and Katatbeh & Belgacem, 2011, as well as references therein). Such importance can be measured through the growing attention of scientists to nonlinear studies, thereby affecting nearly every aspect of an intertwined world of nonlinear sciences, both at the physical and theoretical level (Belgacem et al. 2014; Bulut, Belgacem & Baskonus, 2013). The ramifications of this increased focus have deeply influenced engineering applications, the physical sciences and applied mathematics, (Bulut, Baskonus & Belgacem, 2013; Dubey, Goswami & Belgacem, 2014; Tuluce, Bulut &Belgacem, 2014). This is particularly true with Nonlinear Fractional Differential Equations and Systems (NFDEs & NFDES's).

Haci Mehmet Baskonus: Department of Computer Engineering, University of Tunceli, Tunceli, Turkey, E-mail: hmbaskonus@gmail.com
Fethi Bin Muhammad Belgacem: Department of Mathematics, Arap Open University, Al-Ardhyia, Kuwait, E-mail: fbmbelgacem@gmail.com
Hasan Bulut: Department of Mathematics, Firat University, Elazig, Turkey, E-mail: hbulut@firat.edu.tr

In physics and mathematics new methods are constantly needed to discover exact or approximate solutions for NFDEs and NFDESs. Nonlinear mathematical theory remains scarce and researchers need to specify the type of nonlinearity at hand in order to proceed. NFDEs and NFDES's may not always be solved exactly or explicitly due to the lack of analytical schemes either due to the difficulty of the problems or the specificity of the nonlinearities. Fortunately, such problems may at least be treated numerically and approximate solutions of equations and systems can most often be obtained.

Numerical methods, more than analytical ones, are therefore at the forefront of solving NFDE's, and NFDES's (Momani & Abuasad, 2006; He & Wu, 2007; Sweilam, 2007).

Most of the nonlinear fractional differential equation systems do not have known exact analytical solutions. There are however some analytical techniques for solving nonlinear fractional differential equation systems. Some of the available methods include iteration techniques (Mousa, Kaltayev & Ragab, 2009). Many applications in this area are based on He's Variational Iterative Method (VIM), (He, 1997; He, 1998). The application of VIM can be found in many works from the turn of the century. For instance, we refer the esteemed readers' attention to the treatments in (Abdou and Soliman, 2005; Odibat and Momani, 2006). In 2007, new and important developments and advanced applications of the VIM took place, in particular, in the works by Abbasbandy, 2007; He and Wu, 2007; Sweilam, 2007; Wazwaz, 2007. In this work, we implement the VIM and deliver numerical solutions and plots by means of the Mathematica Symbolic Algebra Package 9.

2 The Variational Iteration Method

In order to illustrate the basic concepts of VIM, the following general nonlinear partial differential equation can be considered as a prototype,

$$Lu(x,t) + Ru(x,t) + Nu(x,t) = f(x,t), \tag{1}$$

where L is a linear time derivative operator, R is a linear operator which has derivatives with respect to x and t, N is a nonlinear differential operator, and $f(x,t)$ is an inhomogeneous source term (also see Bulut, Baskonus, & Tuluce, 2013). To implement the VIM we construct the next iteration for Eq. 1,

$$u_{n+1}(x,t) = u_n(x,t) + \int_0^t \lambda(\tau)\left\{Lu_n(x,\tau) + R\widetilde{u_n}(x,\tau) + N\widetilde{u_n}(x,\tau) - f(x,\tau)\right\} d\tau, \tag{2}$$

where the parameter λ is a general Lagrange multiplier, which can be identified optimally via variational theory, the subscript n denotes the nth-order approximation, and $\widetilde{u_n}$ is considered as a restricted variation which implies that $\delta\widetilde{u_n} = 0$, (following for instance the treatments in Sweilam, 2007 and Abbasbandy, 2007).

Clearly, the main steps of He's VIM require first the determination of the Lagrange multiplier λ, which needs to be identified optimally. Once λ is determined, the successive approximations, u_{n+1}, $n \geq 0$ of the solution u are obtained by using a suitably selected function u_0, which a priority satisfies the boundary conditions.

3 VIM Application to NFDES

Here we implement the VIM for two separate NFDES, both of which were treated in Khan et al. 2011, by alternative means. The fractional derivations are in the sense of Riemann-Liouville fractional derivation throughout paper.

3.1 Example 1

First, we consider and treat the following NFDES,

$$\begin{aligned} D_t^{\alpha} u(t) &= k(-1-\varepsilon)\, u(t) + k(1-\varepsilon)\, v(t), \\ D_t^{\alpha} v(t) &= k(1-\varepsilon)\, u(t) + k(-1-\varepsilon)\, v(t), \end{aligned} \tag{3}$$

with the conditions

$$u(0) = 1,\ \ v(0) = 3,\ \ 0 < \alpha < 1. \tag{4}$$

We start by defining a structure of iteration for Eq. 3 in the following form,

$$\begin{aligned} u_{k+1} &= u_k + \int_0^t \lambda(\tau) \left[\frac{\partial^{\alpha} u_k(\tau)}{\partial \tau^{\alpha}} + k(1+\varepsilon)\, u_k(\tau) + k(\varepsilon-1)\, v_k(\tau) \right] d\tau, \quad k = 0, 1, 2, 3 \cdots, \\ v_{k+1} &= v_k + \int_0^t \lambda(\tau) \left[\frac{\partial^{\alpha} v_k(\tau)}{\partial \tau^{\alpha}} + k(-1+\varepsilon)\, u_k(\tau) + k(\varepsilon+1)\, v_k(\tau) \right] d\tau. \end{aligned} \tag{5}$$

Calculating variation with respect to u_n, we have the stationary conditions as follows;

$$\begin{cases} 1 + \lambda(\tau)\,|_{\tau=x} = 0, \\ \lambda'(\tau) = 0. \end{cases} \tag{6}$$

The Lagrange multiplier can be easily identified as

$$\lambda(\tau) = -1. \tag{7}$$

Substituting Eq. 7 in Eq. 5, we have the iteration formula as follows;

$$\begin{aligned} u_{k+1} &= u_k - \int_0^t \left[\frac{\partial^{\alpha} u_k(\tau)}{\partial \tau^{\alpha}} + k(1+\varepsilon)\, u_k(\tau) + k(\varepsilon-1)\, v_k(\tau) \right] d\tau, \quad k = 0, 1, 2, 3 \cdots \\ v_{k+1} &= v_k - \int_0^t \left[\frac{\partial^{\alpha} v_k(\tau)}{\partial \tau^{\alpha}} + k(-1+\varepsilon)\, u_k(\tau) + k(\varepsilon+1)\, v_k(\tau) \right] d\tau. \end{aligned} \tag{8}$$

We choose the initial approximate solution in the form $u(0) = 1,\ v(0) = 3$ and substitute in Eq. 8 and construct a simple algorithm in any computational software such as Mathematica 9. Then, we obtain the approximate solution as follows,

$$\begin{aligned} k = 0 \Rightarrow u_1 &= u_0 - \int_0^t \left[\frac{\partial^\alpha u_0}{\partial t^\alpha} + k(1+\varepsilon)u_0 + k(\varepsilon - 1)v_0 \right] dt, \\ &= 1 - \int_0^t \left[\frac{\partial^\alpha 1}{\partial t^\alpha} \right] dt - k(4\varepsilon - 2)t = 1 - \frac{t^{1-\alpha}}{(1-\alpha)\Gamma(1-\alpha)} - k(4\varepsilon - 2)t, \\ &= 1 - k(4\varepsilon - 2)t - \frac{t^{1-\alpha}}{(1-\alpha)\Gamma(1-\alpha)}, \end{aligned} \tag{9}$$

$$\begin{aligned} \Rightarrow v_1 &= v_0 - \int_0^t \left[\frac{\partial^\alpha v_0}{\partial t^\alpha} + k(-1+\varepsilon)u_0 + k(\varepsilon + 1)v_0 \right] dt, \\ &= 3 - \int_0^t \left[\frac{\partial^\alpha 3}{\partial t^\alpha} + k(-1+\varepsilon) + k(\varepsilon+1)3 \right] dt = 3 - \int_0^t \left[\frac{\partial^\alpha 3}{\partial t^\alpha} + k(4\varepsilon + 2) \right] dt, \\ &= 3 - \int_0^t \left[\frac{\partial^\alpha 3}{\partial t^\alpha} \right] dt - k(4\varepsilon+2)t = 3 - \frac{3t^{1-\alpha}}{(1-\alpha)\Gamma(1-\alpha)} - k(4\varepsilon + 2)t, \\ &= 3 - k(4\varepsilon+2)t - \frac{3t^{1-\alpha}}{(1-\alpha)\Gamma(1-\alpha)}. \end{aligned} \tag{10}$$

Therefore, we get the following approximate solution,

$$(u_1, v_1) = \left(1 - k(4\varepsilon - 2)t - \frac{t^{1-\alpha}}{(1-\alpha)\Gamma(1-\alpha)},\ 3 - k(4\varepsilon + 2)t - \frac{3t^{1-\alpha}}{(1-\alpha)\Gamma(1-\alpha)} \right). \tag{11}$$

$$\begin{aligned} &k = 1 \Rightarrow u_2 = \\ &u_1 - \int_0^t \left[\frac{\partial^\alpha u_1}{\partial t^\alpha} + k(1+\varepsilon)u_1 + k(\varepsilon - 1)v_1 \right] dt, \\ &= 1 - k(4\varepsilon - 2)t - \frac{t^{1-\alpha}}{(1-\alpha)\Gamma(1-\alpha)} - \int_0^t \left[\frac{\partial^\alpha u_1}{\partial t^\alpha} \right] dt - \int_0^t [k(1+\varepsilon)u_1]\, dt - \int_0^t [k(\varepsilon-1)v_1]\, dt, \\ &= 1 - k(4\varepsilon - 2)t - \frac{t^{1-\alpha}}{(1-\alpha)\Gamma(1-\alpha)} - \underbrace{\int_0^t \left[\frac{\partial^\alpha u_1}{\partial t^\alpha} \right] dt}_{I_1} - k(1+\varepsilon)\underbrace{\int_0^t [u_1]\, dt}_{I_2} - k(\varepsilon - 1)\underbrace{\int_0^t [v_1]\, dt}_{I_3}, \end{aligned}$$

We calculate the terms of I_1, I_2 and I_3 as follows;

$$
\begin{aligned}
I_1 &= \int_0^t \left[\frac{\partial^\alpha u_1}{\partial t^\alpha}\right] dt = \int_0^t \left[\frac{\partial^\alpha}{\partial t^\alpha}\left(1 - k(4\varepsilon - 2)\,t - \frac{t^{1-\alpha}}{(1-\alpha)\Gamma(1-\alpha)}\right)\right] dt,\\
&= \int_0^t \left[\frac{\partial^\alpha 1}{\partial t^\alpha}\right] dt + 2k(1-2\varepsilon)\int_0^t \left[\frac{\partial^\alpha t}{\partial t^\alpha}\right] dt - \frac{1}{(1-\alpha)\Gamma(1-\alpha)}\int_0^t \left[\frac{\partial^\alpha t^{1-\alpha}}{\partial t^\alpha}\right] dt,\\
&= \frac{t^{1-\alpha}}{(1-\alpha)\Gamma(1-\alpha)} + \frac{2k(1-2\varepsilon)\,t^{2-\alpha}}{(2-\alpha)\Gamma(2-\alpha)} - \frac{\Gamma(2-\alpha)\,t^{2-2\alpha}}{2(1-\alpha)^2\Gamma(1-\alpha)\Gamma(2-3\alpha)},\\
I_2 &= \int_0^t [u_1]\, dt = \int_0^t \left(1 - k(4\varepsilon - 2)\,t - \frac{t^{1-\alpha}}{(1-\alpha)\Gamma(1-\alpha)}\right) dt\\
&= t - k(2\varepsilon - 1)\,t^2 - \frac{t^{2-\alpha}}{(2-\alpha)(1-\alpha)\Gamma(1-\alpha)},\\
I_3 &= \int_0^t [v_1]\, dt = \int_0^t \left[3 - k(4\varepsilon + 2)\,t - \frac{3t^{1-\alpha}}{(1-\alpha)\Gamma(1-\alpha)}\right] dt\\
&= 3t - k(2\varepsilon + 1)\,t^2 - \frac{3t^{2-\alpha}}{(2-\alpha)(1-\alpha)\Gamma(1-\alpha)}.
\end{aligned}
$$

thus

$$
\begin{aligned}
u_2 &= 1 - k(4\varepsilon - 2)\,t - k(1+\varepsilon)t + k^2(1+\varepsilon)(2\varepsilon - 1)t^2 - \frac{2t^{1-\alpha}}{(1-\alpha)\Gamma(1-\alpha)}\\
&- \frac{2k(1-2\varepsilon)t^{2-\alpha}}{(2-\alpha)\Gamma(2-\alpha)} + \frac{\Gamma(2-\alpha)t^{2-2\alpha}}{2(1-\alpha)^2\Gamma(1-\alpha)\Gamma(2-3\alpha)} + \frac{k(1+\varepsilon)t^{2-\alpha}}{(2-\alpha)(1-\alpha)\Gamma(1-\alpha)}\\
&-3k(\varepsilon - 1)t + (\varepsilon - 1)(2\varepsilon + 1)k^2t^2 + \frac{3k(\varepsilon - 1)t^{2-\alpha}}{(2-\alpha)(1-\alpha)\Gamma(1-\alpha)},
\end{aligned}
\tag{12}
$$

$$
\begin{aligned}
\Rightarrow v_2 &= v_1 - \int_0^t \left[\frac{\partial^\alpha v_1}{\partial t^\alpha} + k(-1+\varepsilon)\,u_1 + k(\varepsilon+1)\,v_1\right] dt,\\
&= 3 - k(4\varepsilon + 2)\,t - \frac{3t^{1-\alpha}}{(1-\alpha)\Gamma(1-\alpha)} - \int_0^t \left[\frac{\partial^\alpha}{\partial t^\alpha}\left(3 - k(4\varepsilon+2)\,t - \frac{3t^{1-\alpha}}{(1-\alpha)\Gamma(1-\alpha)}\right)\right] dt\\
&- \int_0^t \left[k(-1+\varepsilon)\left(1 - k(4\varepsilon - 2)\,t - \frac{t^{1-\alpha}}{(1-\alpha)\Gamma(1-\alpha)}\right)\right] dt\\
&- \int_0^t \left[k(\varepsilon+1)\left(3 - k(4\varepsilon + 2)\,t - \frac{3t^{1-\alpha}}{(1-\alpha)\Gamma(1-\alpha)}\right)\right] dt,\\
&= 3 - k(4\varepsilon + 2)\,t - \frac{3t^{1-\alpha}}{(1-\alpha)\Gamma(1-\alpha)} - \int_0^t \left[\frac{\partial^\alpha}{\partial t^\alpha}\left(3 - k(4\varepsilon+2)\,t - \frac{3t^{1-\alpha}}{(1-\alpha)\Gamma(1-\alpha)}\right)\right] dt\\
&-k(-1+\varepsilon)t + (\varepsilon - 1)(2\varepsilon - 1)k^2t^2 + \frac{k(\varepsilon - 1)t^{2-\alpha}}{(2-\alpha)(1-\alpha)\Gamma(1-\alpha)}\\
&-3k(\varepsilon + 1)t + k(\varepsilon + 1)(2\varepsilon + 1)k^2t^2 + \frac{3k(\varepsilon + 1)t^{2-\alpha}}{(2-\alpha)(1-\alpha)\Gamma(1-\alpha)},
\end{aligned}
$$

$$
\begin{aligned}
&= 3 - k(4\varepsilon + 2)\,t - k(-1+\varepsilon)t + (\varepsilon - 1)(2\varepsilon - 1)k^2t^2 - 3k(\varepsilon + 1)t\\
&- \frac{6t^{1-\alpha}}{(1-\alpha)\Gamma(1-\alpha)} + \frac{k(4\varepsilon + 2)t^{2-\alpha}}{(2-\alpha)\Gamma(2-\alpha)} + \frac{3\Gamma(2-\alpha)t^{2-2\alpha}}{2(1-\alpha)^2\Gamma(1-\alpha)\Gamma(2-3\alpha)}\\
&+k(\varepsilon + 1)(2\varepsilon + 1)k^2t^2 + \frac{k(\varepsilon - 1)t^{2-\alpha}}{(2-\alpha)(1-\alpha)\Gamma(1-\alpha)} + \frac{3k(\varepsilon + 1)t^{2-\alpha}}{(2-\alpha)(1-\alpha)\Gamma(1-\alpha)},
\end{aligned}
\tag{13}
$$

$$\vdots$$

The remainder components of this iteration formula Eq. 8 are obtained in the same manner using Mathematica 9.

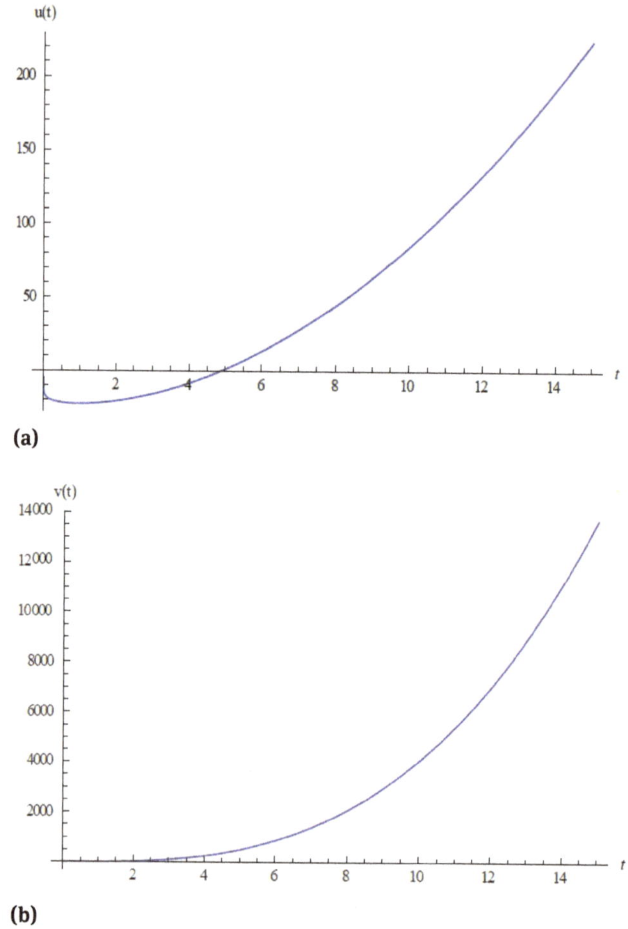

Fig. 1. The surfaces generated from the VIM result of $u(t)$ in Eq. 12 $\alpha = k = \varepsilon = 0.91$ over the interval $0 < t < 15$ shown in (a), shown in (b) over the same intervals for approximate solution compound $v(t)$ in Eq. 13.

3.2 Example 2

Our second example was also treated in Khan et al. 2011 by alternative methods. We apply the VIM to the NFDES,

$$\begin{aligned} D_t^{\alpha} u(t) &= -1002u(t) + 1000v^2(t), \\ D_t^{\alpha} v(t) &= u(t) - v(t) - v^2(t), \end{aligned} \tag{14}$$

with the initial conditions

$$u(0) = 1, \;\; v(0) = 1, \;\; 0 < \alpha < 1. \tag{15}$$

When we apply the VIM, we define the structure of Eq. 14 in the following form,

$$\begin{aligned} u_{k+1} &= u_k + \int_0^t \lambda(\lambda) \left[\frac{\partial^\alpha u_k(\tau)}{\partial \tau^\alpha} + 1002 u_k(\tau) - 1000 v_k^2(\tau) \right] d\tau, \quad k = 0, 1, 2, 3 \cdots, \\ v_{k+1} &= v_k + \int_0^t \lambda(\lambda) \left[\frac{\partial^\alpha v_k(\tau)}{\partial \tau^\alpha} - u_k(\tau) + v_k(\tau) + v_k^2(\tau) \right] d\tau . \end{aligned} \tag{16}$$

Calculating variation with respect to u_n, we have the stationary conditions as follows;

$$\begin{cases} 1 + \lambda(\tau)\,|_{\tau=x} = 0, \\ \lambda'(\tau) = 0. \end{cases} \tag{17}$$

in this case, the Lagrange multiplier can be easily identified as,

$$\lambda(\tau) = -1. \tag{18}$$

Substituting Eq. 18 in Eq. 16, we obtain the following iteration formula,

$$\begin{aligned} u_{k+1} &= u_k - \int_0^t \left[\frac{\partial^\alpha u_k(\tau)}{\partial \tau^\alpha} + 1002 u_k(\tau) - 1000 v_k^2(\tau) \right] d\tau, \quad k = 0, 1, 2, 3 \cdots, \\ v_{k+1} &= v_k - \int_0^t \left[\frac{\partial^\alpha v_k(\tau)}{\partial \tau^\alpha} - u_k(\tau) + v_k(\tau) + v_k^2(\tau) \right] d\tau . \end{aligned} \tag{19}$$

When we choose the initial approximate solution to be $u(0) = 1,\ v(0) = 1$ and substitute in Eq. 19, we get the next approximate solution,

$$\begin{aligned} k = 0 \Rightarrow u_1 = u_0 &- \int_0^t \left[\frac{\partial^\alpha u_0(t)}{\partial t^\alpha} + 1002 u_0 - 1000 v_0^2 \right] dt, \\ &= 1 - \int_0^t \left[\frac{\partial^\alpha 1}{\partial t^\alpha} + 2 \right] dt = 1 - 2t - \int_0^t \left[\frac{\partial^\alpha 1}{\partial t^\alpha} \right] dt, \\ &= 1 - 2t - \frac{t^{1-\alpha}}{(1-\alpha)\Gamma(1-\alpha)}, \end{aligned} \tag{20}$$

$$\begin{aligned} \Rightarrow v_1 = v_0 - \int_0^t \left[\frac{\partial^\alpha v_0(t)}{\partial t^\alpha} - u_0(t) + v_0(t) + v_0^2(t) \right] dt &= 1 - \int_0^t \left[\frac{\partial^\alpha 1}{\partial t^\alpha} + 1 \right] dt, \\ = 1 - t - \frac{t^{1-\alpha}}{(1-\alpha)\Gamma(1-\alpha)}. \end{aligned} \tag{21}$$

Therefore, we get an approximate solution as follows;

$$(u_1, v_1) = \left(1 - 2t - \frac{t^{1-\alpha}}{(1-\alpha)\Gamma(1-\alpha)},\ 1 - t - \frac{t^{1-\alpha}}{(1-\alpha)\Gamma(1-\alpha)} \right), \tag{22}$$

$$\begin{aligned} k = 1 \Rightarrow u_2 = u_1 &- \int_0^t \left[\frac{\partial^\alpha u_1}{\partial t^\alpha} + 1002 u_1 - 1000 v_1^2 \right] dt, \\ &= 1 - 2t + 1000t - 1000t^2 + 1000t^3 - \frac{2t^{1-\alpha}}{(1-\alpha)\Gamma(1-\alpha)} + \frac{2000 t^{3-\alpha}}{(3-\alpha)(1-\alpha)\Gamma(1-\alpha)} \\ &-1002 \left[t - t^2 - \frac{t^{2-\alpha}}{(2-\alpha)(1-\alpha)\Gamma(1-\alpha)} \right] + \frac{1000 t^{3-2\alpha}}{(3-2\alpha)[(1-\alpha)\Gamma(1-\alpha)]^2}, \\ &- \frac{2000 t^{2-\alpha}}{(2-\alpha)(1-\alpha)\Gamma(1-\alpha)} + \frac{t^{2-\alpha}}{(2-\alpha)\Gamma(1-\alpha)} + \frac{\Gamma(2-\alpha) t^{2-2\alpha}}{2(1-\alpha)^2\Gamma(1-\alpha)\Gamma(2-3\alpha)} \end{aligned} \tag{23}$$

$$\Rightarrow v_2 = v_1 - \int_0^t \left[\frac{\partial^\alpha v_1}{\partial t^\alpha} - u_1 + v_1 + v_1^2 \right] dt,$$
$$= 1 - 2t + \frac{t^2}{2} - \frac{t^3}{3} - \frac{2t^{1-\alpha}}{(1-\alpha)\Gamma(1-\alpha)} - \frac{t^{3-2\alpha}}{(3-2\alpha)[(1-\alpha)\Gamma(1-\alpha)]^2}$$
$$+ \frac{2t^{2-\alpha}}{(2-\alpha)(1-\alpha)\Gamma(1-\alpha)} - \frac{2t^{3-\alpha}}{(3-\alpha)(1-\alpha)\Gamma(1-\alpha)} + \frac{t^{2-\alpha}}{(2-\alpha)\Gamma(2-\alpha)} \qquad (24)$$
$$+ \frac{\Gamma(2-\alpha)t^{2-2\alpha}}{2(1-\alpha)^2\Gamma(1-\alpha)\Gamma(2-3\alpha)}.$$

The rest components of the iteration formula Eq. 19, can easily be obtained, in the same manner using software such as Mathematica 9.

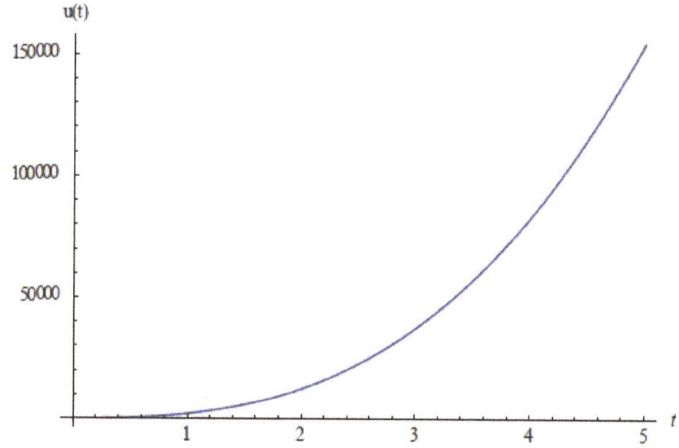

(a)

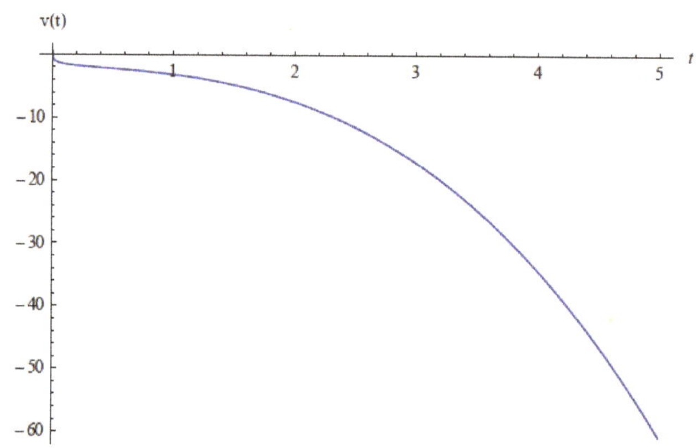

(b)

Fig. 2. The surfaces generated from the VIM result of $u(t)$ in Eq. 23 $\alpha = 0.91$ over the intervals $0 < t < 5$ shown in (a), shown in (b) over the same intervals for Eq. 24

4 Conclusions and Future Work

In this research, the Variational Iteration Method was successfully applied for solving two systems of nonlinear fractional differential equations systems. Approximate solutions were obtained and plotted by using the Mathematica symbolic algebra package software 9. The VIM set up, despite its simplicity, turns out to be a valuable tool to treat NFDES. Consequently, we recommend the use of VIM for such systems when direct analytical approaches are absent, which are mostly the case. For future work, we plan consider the Hyper Chaotic NFDES treated in (Mohamed & Ghany 2011), by different means, where we hope to enrich the understanding of such four by four chaos engendering systems, thanks to this now celebrated method, the VIM. We look forward to any communications, and remarks, and collaborations from esteemed fellow readers.

Acknowledgment: F. B. M. Belgacem wishes to acknowledge the continued support of the Public Authority for Applied Education and Training Research Department, (PAAET RD), Kuwait.

The Authors wish to acknowledge the helpful support of Editor Prof. Carlo Cattani, and DGO Managing Editor, Mathematics, Agnieszka Bednarczyk-Drag.

Bibliography

Abbasbandy, S. A. (2007), New application of He's variational iteration method for quadratic Riccati differential equation by using Adomian's polynomials, *J. Comput. Appl. Math.*, 207(1) 59-63.

Abdou M. A. and Soliman, A. A. (2005). Variational iteration method for solving Burger's and coupled Burger's equations, *J. Comput. Appl. Math.*, 181(2), 245–51.

Belgacem, F.B.M., Bulut H., Baskonus M.H., & Akurk T. (2014). Mathematical Analysis of the Generalized Benjamin Kdv Equations via the Extended Trial Equations Method, *Journal of the Associations of Arab Universities for Basic and App. Sciences*, Vol.16, 91-100.

Bulut H., Baskonus, H. M., & Tuluce S. (2013), The solutions of homogenous and nonhomogeneous linear fractional differential equations by variational iteration method, *Acta Universitatis Apulensis. Mathematics-Informatics*, 36, 235-243.

Bulut H., Baskonus M.H., & Belgacem, F.B.M. (2013); The Analytical Solution of Some Fractional Ordinary Differential Equations by the Sumudu transform Method, *Applied and Abstract Analysis*, Article ID 203875, 1-6.

Bulut H., Belgacem, F.B.M, & Baskonus, H.M. (2013); Fractional Partial Differential Equations Systems Solutions by Adomian Decomposition Method Implementation. ICMCA 2013: 4th International Conference on Mathematical and computational Applications, Manisa, Turkey, June, 11-13,

Dubey, R.S., Goswami, P., & Belgacem, F.B.M. (2014). Generalized Time-Fractional Telegraph Equation Analytical Solution by Sumudu and Fourier Transforms. *Journal of Fractional Calculus and Applications*, 5(2), 52-58.

Goswami, P., &, Belgacem, F.B.M. (2012). Fractional Differential Equations Solutions through a Sumudu Rational, *Nonlinear Studies Journal*, 19(4), 591-598.

Gupta, V.G.; Sharma, B.; & Belgacem, F. B. M. (2011). On the Solutions of Generalized Fractional Kinetic Equations. *Journal of Applied Math. Sciences*, Vol.5, No. 19. 899 – 910.
He J.H. (1997). Variational iteration method for delay differential equations, *Commun. Nonlinear Sci. Numer. Simul.* 2(4), 235–236.
He J.H. (1998), Approximate solution of nonlinear differential equations with convolution product nonlinearities, *Comput. Methods Appl. Mech. Eng.* 167 (1–2), 69–73.
He J.H. & Wu, X.-H. (2007). Variational iteration method: new development and applications, *Comput. Math. Appl.* 54 (7-8), 881–894.
Katatbeh, Q.K. & Belgacem, F.B.M. (2011). Applications of the Sumudu Transform to Fractional Diff. Equations, *Nonlinear Studies*, Vol. 18, No.1, 99-112.
Khan, N.A., Jamil M., Ara, A., & Khan, N.U. (2011), On Efficient Method for System of fractional Differential Equations. *Advances in Difference Equations*, (2011), 15 pages .
Mohamed S. M., & Ghany H.A. (2011), Analytic approximations for fractional-order hyperchaotic system, *Journal of Advanced Research in Dynamical and Control Systems*, 3(3), 1-12.
Momani S., & Abuasad S. (2006). Application of He's variational iteration method to Helmholtz equation. *Chaos Soliton Fractals*, 27, 1119–1123.
Mousa, M.M. Kaltayev A., & Ragab S.F., (2009). Investigation of a transition from steady convection to chaos in porous media using piecewise variational iteration method, world Academy of Science, *Engineering and Technology*, 58, 1088–1097.
Odibat Z. M. and Momani S., (2006) Application of variational iteration method to nonlinear differential equations of fractional order, *Int. J. Non-linear Sci. Numer. Simul.* 7(1), 27–34.
Sweilam, N. H. (2007), Fourth order integro-differential equations using variational iteration method, *Comput. Math. Appl.* 54, 7-81086–1091.
Tuluce D. S., Bulut H., & Belgacem F. B. M., (2014). Sumudu Transform Method for Analytical Solutions of Fractional Type Ordinary Differential Equations, *Mathematical Problems in Engineering*, in a Special Issue on: *Partial Fractional Equations and their Applications.*
Wazwaz, A.-M. (2007), The variational iteration method for exact solutions of Laplace equation, *Phys. Lett. A*, 363(4), 260–262.

Toufik Mekkaoui, Zakia Hammouch, Fethi B.M.Belgacem*, and Ahmed El Abbassi

Fractional-order Nonlinear Systems: Chaotic Dynamics, Numerical Simulation and Circuits Design

Abstract: Based on fractional transfer function approximation in frequency domain and a Predictor-Corrector numerical algorithm, this chapter aims to prove the existence of chaos in two fractional-order dynamical systems. Analog circuits are designed to confirm their chaotic dynamics.

Keywords: Chaotic system; Circuit design; Fractional-order calculus; Numerical simulation; Stability

1 Introduction

During the last decade, fractional differential calculus (FDC) has attracted the attention of engineers, physicists and mathematicians because many phenomena can be modeled by applying fractional derivatives and integrals Oldham et al. (1974), Podlubny (1999). Consequently, considerable attention has been devoted to the solutions of systems of fractional differential equations. The study of nonlinear dynamical systems and chaos has become the subject of great interest and it attracted enormous research interest after the numerical demonstration of chaos by Lorenz (1963). Chaotic systems are associated with complex dynamical behaviors that possess some special features including bounded trajectories with positive Lyapunov exponents and sensitive dependence on initial conditions. The study of chaotic dynamics has found application in various fields of scientific and engineering disciplines, including meteorology, physics, chemistry, engineering, economics, biology, secure communication, encryption and control processing e.g. Li et al. (2004), Petras (2011). On the other hand, the development of models based on fractional-order differential systems have recently received growing attention in the investigation of dynamical systems. The interest in the study of fractional-order nonlinear systems lies in the fact that fractional derivatives provide an excellent tool for the description of memory and hereditary properties, which are not taken into account in the classical integer-order models.

***Corresponding Author: Fethi B.M.Belgacem:** Department of Mathematics, Faculty of Basic Education, PAAET, Al-Aardhia, Kuwait, E-mail: fbmbelgacem@gmail.com
Toufik Mekkaoui, Zakia Hammouch: E3MI, Département de mathématique, FST Errachidia, Université Moulay Ismail, BP.509 Boutalamine 52000 Errachidia, Morocco
Ahmed El Abbassi: E3MI, Département de Physique, FST Errachidia, Université Moulay Ismail, BP.509 Boutalamine 52000 Errachidia, Morocco

Further fractional versions of many systems have been investigated in Baleanu et al. (2010), Sun et al. (2010b).
Recently, extensive work has been carried out in order to understand chaos in fractional order dynamical systems. In general, it is a difficult task to obtain the exact solution for nonlinear FODEs. Some analytical and numerical methods have been proposed to overcome this difficulty. The classical approaches include Laplace transform method, Mellin transform method and power series method. Recently, some analytical methods such as variational iteration method (VIM) (Goh et al. 2009), Adomian decomposition method (ADM) (Caponetto et al. 2013) and homotopy perturbation method (HPM) (Wang et al. 2012) have been applied to obtain numerical-analytical solutions of fractional-order chaotic systems. On the other hand, two approaches have been developed for numerically studing chaotic systems with fractional-order derivatives: the frequency-domain method (Ahmad et al. 2003) and the time-domain methods, (Liu et al. 2011). One of the best methods in the second category is the Adams-Bashforth-Moulton numerical algorithm (Diethelm et al. 2002).
In the present work, we consider two fractional-order dynamical systems, the first is a new Lorenz-like system and the second is the well-known Nosé-Hoover system. A numerical scheme of Adams-Bashforth-Moulton has been used for investigating the dynamics of those systems. Moreover, simulation circuit experiments results are obtained by means of the Electronic Workbench Multisim13.
The remainder of this chapter is organized as follows: Preliminaries regarding fractional calculus and stability criteria are described in Section 2. In Section 3, fractional-order models are described and their dynamical behaviours are studied. In Section 4 numerical simulations and circuit implementations are presented and discussed. Concluding remarks are drawn in the last section.

2 Preliminaries

2.1 Basic definitions for Fractional Calculus

Fractional calculus is a generalization of integration and differentiation. The concept has been known since the development of regular calculus; the first reference is associated with Leibniz and L'Hôpital in 1695 where the half-order derivative was mentioned. A historical introduction to FDEs can be found in Oldham et al. (1974), Podlubny (1999). Commonly used definitions for fractional derivatives are due to Riemann-Liouville, and Caputo (1967). In what follows, we consider Caputo derivatives, and take advantage of traditional initial and boundary conditions in the formulation of these problems.

Definition 33. *A real function $f(x)$, $x > 0$, is said to be in the space C_μ, $\mu \in \mathbb{R}$ if there exists a real number $\lambda > \mu$ such that $f(x) = x^\lambda g(x)$, where $g(x) \in C[0, \infty)$ and it is said to be in the space C_μ^m if and only if $f^{(m)} \in C_\mu$ for $m \in \mathbf{N}$.*

Definition 34. *The Riemann-Liouville fractional integral operator of order α of a real function $f(x) \in C_\mu$, $\mu \geq -1$, is defined as*

$$J^\alpha f(x) = \frac{1}{\Gamma(\alpha)} \int_0^x (x-t)^{\alpha-1} f(t)dt, \quad \alpha > 0, x > 0 \quad and \quad J^0 f(x) = f(x). \tag{1}$$

The operators J^α has some properties, for $\alpha, \beta \geq 0$ and $\xi \geq -1$:
- $J^\alpha J^\beta f(x) = J^{\alpha+\beta} f(x)$,
- $J^\alpha J^\beta f(x) = J^\beta J^\alpha f(x)$,
- $J^\alpha x^\xi = \frac{\Gamma(\xi+1)}{\Gamma(\alpha+\xi+1)} x^{\alpha+\xi}$.

Definition 35. *The Caputo fractional derivative D^α of a function $f(x)$ of any real number α such that $m-1 < \alpha \leq m$, $m \in \mathbf{N}$, for $x > 0$ and $f \in C_{-1}^m$ in the terms of J^α is*

$$D^\alpha f(x) = J^{m-\alpha} D^m f(x) = \frac{1}{\Gamma(m-\alpha)} \int_0^x (x-t)^{m-\alpha-1} f^{(m)}(t)dt \tag{2}$$

and has the following properties for $m-1 < \alpha \leq m$, $m \in \mathbf{N}$, $\mu \geq -1$ and $f \in C_\mu^m$
- $D^\alpha J^\alpha f(x) = f(x)$,
- $J^\alpha D^\alpha f(x) = f(x) - \sum_{k=0}^{m-1} f^{(k)}(0^+)\frac{x^k}{k!}$, *for $x > 0$*

2.2 Stability Criterion

In order to investigate the dynamics and to control the chaotic behaviour of a fractional-order dynamic system

$$D_t^\alpha X(t) = F(X(t)), \tag{3}$$

we need the following indispensable stability theorem

Theorem 36 (See Matignon (1996)). *For a given commensurate fractional order system (3), the equilibria can be obtained by calculating $F(X) = 0$. These equilibrium points are locally asymptotically stable if all the eigenvalues λ of the Jacobian matrix $J = \frac{\partial F}{\partial X}$ at the equilibrium points satisfy*

$$|arg(\lambda)| > \frac{\pi}{2}\alpha. \tag{4}$$

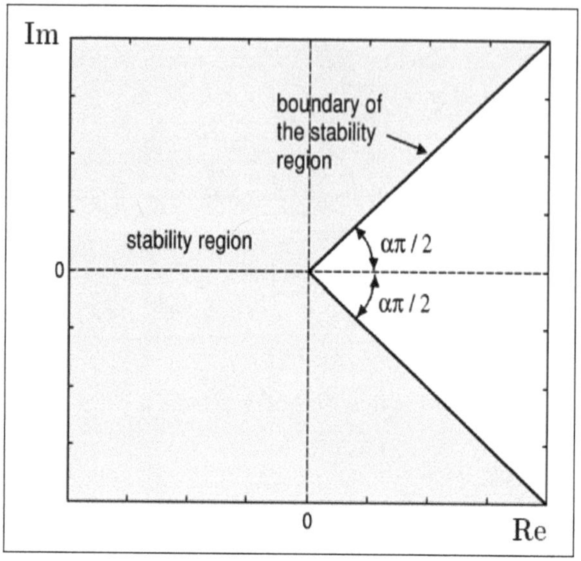

Fig. 1. Stability region of the fractional-order system (3).

3 Circuits Implementation and Numerical Simulations

3.1 Adams-Bashforth (PECE) Algorithm

Here we recall here an improved improved version of Adams-Bashforth-Moulton algorithm (Diethelm et al. 2005) for the fractional-order systems. Consider the fractional-order initial value problem

$$\begin{cases} D_t^{\alpha} x = f(x(t)) & 0 \le t \le T, \\ x^{(k)}(0) = x_0^{(k)}, & k = 0, 1, \dots, m-1. \end{cases} \tag{5}$$

It is equivalent to the Volterra integral equation

$$x(t) = \sum_{k=0}^{[\alpha]-1} x_0^{(k)} \frac{t^k}{k!} + \frac{1}{\Gamma(\alpha)} \int_0^t (t-s)^{\alpha-1} f(s, x(s)) ds. \tag{6}$$

Diethelm et al. have given a predictor-corrector scheme (see Diethelm et al. (2005)), based on the Adams-Bashforth-Moulton algorithm to integrate Equation (6). By applying this scheme to the fractional-order system (5), and setting

$$h = \frac{T}{N}, \quad t_n = nh, \quad n = 0, 1, \dots, N,$$

equation (6) can be discretized as follows:

$$x_h(t_{n+1}) = \sum_{k=0}^{[\alpha]-1} x_0^{(k)} \frac{t^k}{k!} + \frac{h^\alpha}{\Gamma(\alpha+2)} f(t_{n+1}, x_h^p(t_{n+1})) + \frac{h^\alpha}{\Gamma(\alpha+2)} \sum_{j=0}^{n} a_{j,n+1} f(t_j, x_h(t_j)), \quad (7)$$

where

$$a_{j,n+1} = \begin{cases} n^{\alpha+1} - (n-\alpha)(n+1)^\alpha, \; j = 0, \\ (n-j+2)^{\alpha+1} + (n-j)^{\alpha+1} - 2(n-j+1)^{\alpha+1}, \quad 1 \leq j \leq n \\ 1, \quad j = n+1, \end{cases} \quad (8)$$

and the predictor is given by

$$x_h^p(t_{n+1}) = \sum_{k=0}^{[\alpha]-1} x_0^{(k)} \frac{t^k}{k!} + \frac{1}{\Gamma(\alpha)} \sum_{j=0}^{n} b_{j,n+1} f(t_j, x_h(t_j)), \quad (9)$$

where $b_{j,n+1} = \frac{h^\alpha}{\alpha}((n+1)-j)^\alpha - (n-j)^\alpha$.
The error estimate of the above scheme is

$$\max_{j=0,1,..,N} \left\{ |x(t_j) - x_h(t_j)| \right\} = O(h^p),$$

in which $p = min(2, 1+\alpha)$.

3.2 The Fractional Frequency Domain Approximation

The standard definition of fractional derivative and integral does not allow the direct implementation of the fractional operators in time-domain simulations. In order to study such systems, it is necessary to develop approximations to the fractional operators by using the standard integer order operators. In Liu et al. (2011), an effective algorithm was developed to approximate the fractional-order transfer functions by utilizing frequency-domain techniques based on Bode diagrams. According the circuit theory in the Laplace domain, the approximation formulation of α, can be realized by the complex frequency-domain of the chain shape equivalent circuit between A and B in Figure.2, when $\alpha = 0.98$, as

$$\frac{1}{s^{0.98}} = \frac{1.2974(s+1125)}{(s+1423)(s+0.01125)}. \quad (10)$$

The transfer function $H(s)$ between A and B can be obtained as follows

$$H(s) = \frac{R_1}{sR_1C_1+1} + \frac{R_2}{sR_2C_2+1} = \frac{1}{C_0} \frac{\left(\frac{C_0}{C_1} + \frac{C_0}{C_2}\right)\left[s + \frac{\frac{1}{R_1}+\frac{1}{R_2}}{C_1+C_2}\right]}{\left(s+\frac{1}{R_1C_1}\right)\left(s+\frac{1}{R_2C_2}\right)}, \quad (11)$$

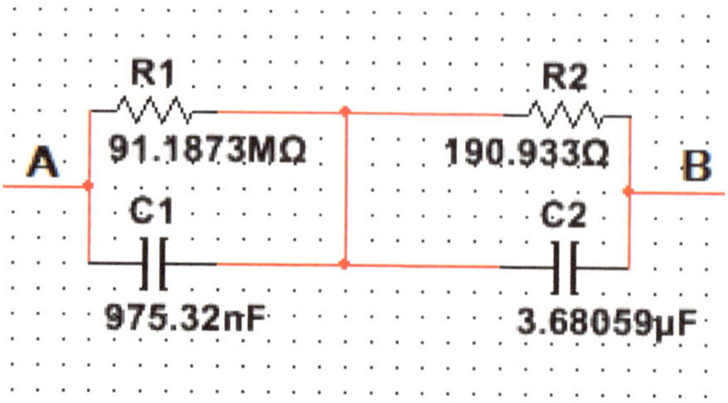

Fig. 2. The chain shape unit of $\frac{1}{s^{0.98}}$

where C_0 is a unit parameter. Taking $C_0 = 1\mu\text{F}$, since $H(s)C_0 = \frac{1}{s^{0.98}}$, we can reach

$$R_1 = 91.1873\text{M}\Omega,\ R_1 = 190.933\Omega,\ C_1 = 975.32n\text{F},\ C_2 = 3.6806\mu\text{F}. \tag{12}$$

3.3 A New Fractional-Order Chaotic System

Now, we introduce the following new system

$$\begin{cases} D^\alpha x = y - x, \\ D^\alpha y = |x| - xz, \\ D^\alpha z = xy - 1, \end{cases} \tag{13}$$

where the fractional-order $\alpha \in (0, 1]$. The corresponding eigenvalues are as given in Table.1.

Table 1. Equilibrium points and corresponding eigenvalues.

Equilibrium points	Eigenvalues	
$E_1(1, 1, 1)$	$\lambda_1 = -1.353209965,$	$\lambda_{2,3} = 0.1766049821 \pm 1.202820820I.$
$E_2(-1, -1, -1)$	$\lambda_1 = -1.353209965,$	$\lambda_{2,3} = 0.1766049821 \pm 1.202820820I.$

Hence $E_{1,2}$ are saddle points of index 2. According to Tavazoei et al. (2007) the necessary condition for system (13) to remain chaotic is that the two conjugate eigenvalues of the Jacobian matrix evaluated at each equilibrium point lie in the unstable region. Therefore, to ensure that system (13) keeps exhibiting chaotic behaviour, we have to maintain

$$\alpha > \frac{2}{\pi} \max_i |\lambda_i| = \frac{2}{\pi} \max_i \left(\arctan \left| \frac{Im(\lambda_i)}{Re(\lambda_i)} \right| \right) = 0.9071909975. \tag{14}$$

3.4 A Fractional-order Chaotic System without Equilibrium Points

Chaotic systems without equilibrium points have been introduced in many papers e.g Cafagna et al. (2014), Pham et al. (2014). Because they can have neither homoclinic nor heteroclinic orbits, the Shilnikov method for verifying chaos cannot be applied to such systems. Non-equilibrium chaotic systems are categorized as chaotic systems with hidden attraction since their basins of attraction do not intersect with small neighborhoods of any equilibrium points (Cafagna et al. 2014). Their dynamics are explored through Lyapunov exponents which measure the exponential rates of divergence and convergence of nearby trajectories in the phase space.

Though there exist a large volume of research on the analysis of chaotic dynamics in fractional-order systems such as Chua Hartley et al. (1995), Lorenz Grigorenko et al. (2003), Genesio Jun-Guo et al. (2005), etc., to our best knowledge, no study has been performed that analyzes the fractional-order Nosé-Hoover dynamical system (Sprott 2014). In this paper a new fractional-order chaotic system with no equilibrium points is presented. The proposed system can be considered elegant in the sense given by Sprott (2010).

The equations of the proposed fractional-order Nosé-Hoover system are readily derived from the equations of the integer-order counterpart as

$$\begin{cases} D_t^\alpha x = y, \\ D_t^\alpha y = yz - x, \\ D_t^\alpha z = 1 - y^2. \end{cases} \tag{15}$$

System (15) has no equilibria. A numerical calculation using the Gram-Schmidt orthonormalization procedure (Mekkaoui et al. 2015) for the initial conditions $(x_0, y_0, z_0) = (0.7, 1, 0.2)$, reveals that system (15) has the following Lyapunov exponents. System (15) turns out to be hyperchaoic since two of its Lyapunov exponents are positive.

3.5 Numerical Simulations

Applying the improved version of Adams-Bashforth-Moulton numerical algorithm described above with a step size $h = 0.01$, systems (13) and (15) can be discretized. It is found that chaos exists in the proposed fractional-order systems for some values

Table 2. Lyapunov exponents to system (15).

L_1 :	0.0111790106468204,
L_2 :	0.0049601416269987,
L_3 :	−0.006770042386138.

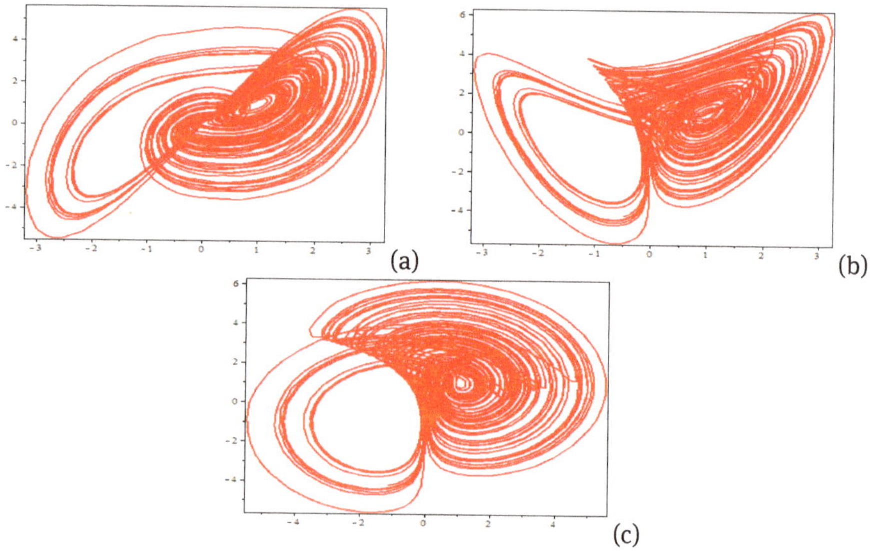

Fig. 3. The chaotic trajectories of fractional-order system (13): (a) $x - y$ plane, (b) $x - z$ plane and (c) $y - z$ plane.

of α. For example, we find that chaos exists in the fractional-order system (13) when $\alpha \geq 0.91$ with the initial condition $(x_0, y_0, z_0) = (0, 1, 0)$. Figures 3 and 4 demonstrate that the system has chaotic behavior for $\alpha = 0.98$.

3.6 Circuit Designs

Implementations of fractional-order systems by means of electronic circuits provides an efficient approach for investigating the dynamics of such systems and for illustrating the correctness of the theoretical models.

In this subsection, the analog circuits are designed to realize the chaotic behaviour of the fractional-order systems (13) and (15). In each circuit, three state variables x, y and z are implemented by three channels, respectively. The implementations use resistors, capacitors, analog multipliers and analog operational amplifiers, as shown in Figures

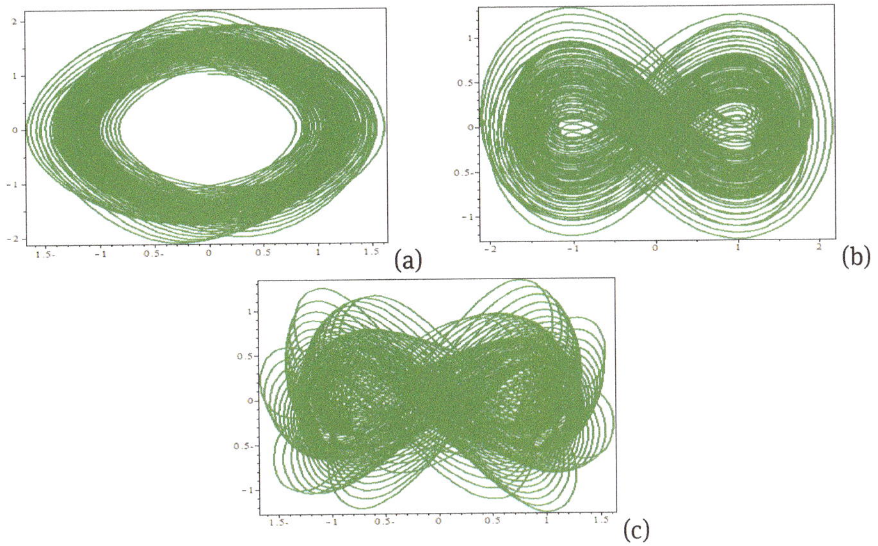

Fig. 4. The chaotic trajectories of fractional-order system (15): (a) $x - y$ plane, (b) $x - z$ plane and (c) $y - z$ plane.

5 and 7. By using Multisim software 13 to conduct simulation on the systems (13) and (15), we can illustrate the phase diagrams through the circuit simulation. This shows that these circuit simulation diagrams have the same chaotic attractor. It verifies the existence and the validity of the proposed systems.

4 Concluding Remarks

The chaotic dynamics of two interesting fractional order dynamical systems are investigated. Lyapunov exponents and analytical conditions given in the literature have been used to ensure the existence of chaos. Numerical calculations are performed for the interested readers consideration with Maple 13. Moreover, using the time frequency-domain approximation method analog circuits have been implemented for each system. That is, the chaotic dynamics of the systems are confirmed by physical implementation with the help pf Electronic Workbench Multisim. For future studies, we plan to investigate further the mannerisms of chaos synchronization in the considered systems and related ones. In the mean time, we look foreward to any remarks and feedback from fellow readers.

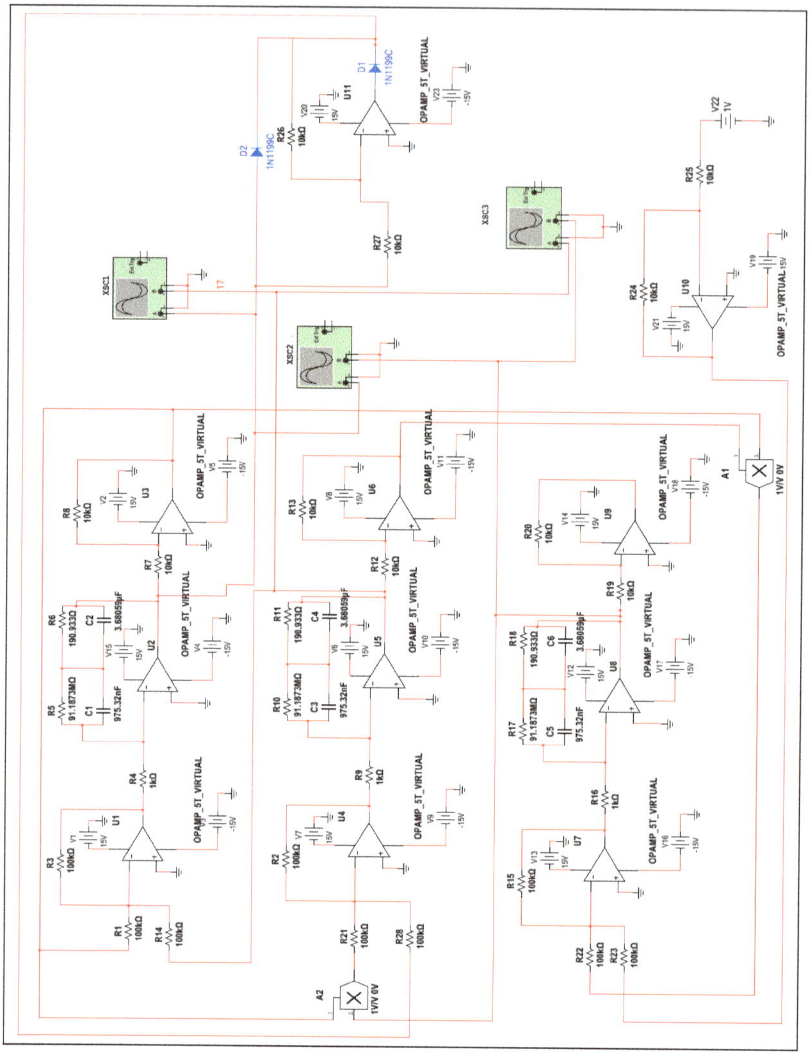

Fig. 5. Circuit diagram for the realization of the fractional-order chaotic system (13).

Acknowledgment: F. B. M. Belgacem wishes to acknowledge the continued support of the Public Authority for Applied Education and Training Research Department, (PAAET RD), Kuwait.

The Authors wish to acknowledge the helpful support of Editor Prof. Carlo Cattani, and DGO Managing Editor Mathematics, Agnieszka Bednarczyk-Drag.

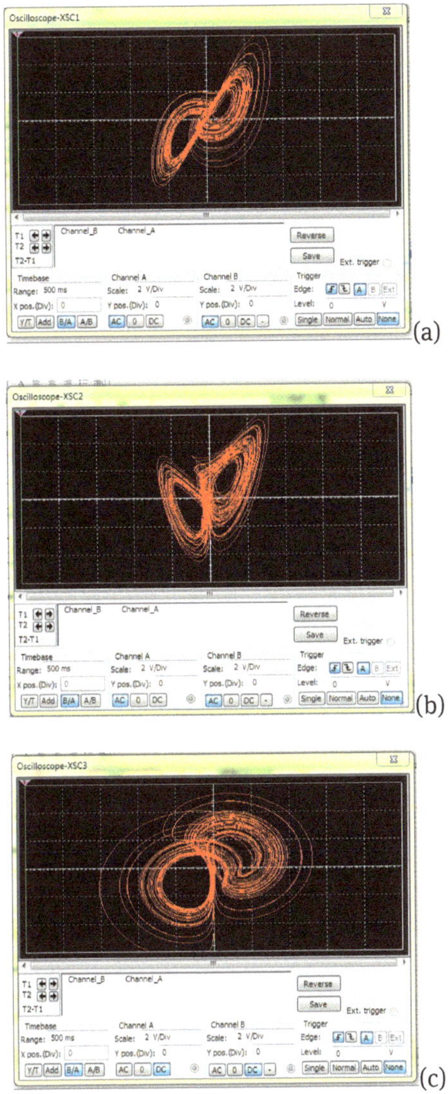

Fig. 6. Chaotic attractors of the fractional-order system (15) observed by the oscilloscope 2V/Div : (a) $x - y$ (b) $x - z$ (c) $y - z$.

Bibliography

Ahmad W.M., Sprott J.C., Chaos in fractional-order autonomous nonlinear systems. Chaos, Solitons and Fractals, (2003), 16, 339–351.

Baleanu D., Guvenc B., Tenreiro-Machado J.A., New Trends in Nanotechnology and Fractional Calculus Applications. Eds., (2010), New York: Springer.

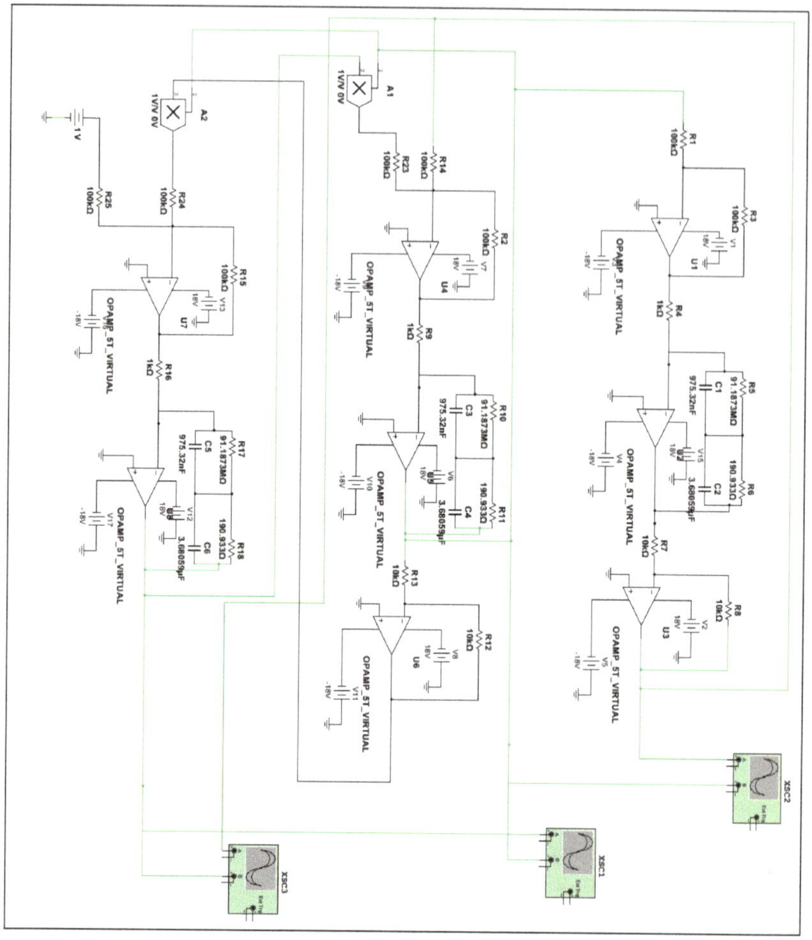

Fig. 7. Circuit diagram for the realization of the fractional-order chaotic system (15).

Cafagna D., Grassi G., Chaos in a new fractional-order system without equilibrium points. CNSNS, (2014), Vol. 19, 9, 2919–2927.

Caponetto R., Dongola G., Fortuna L., Fractional order systems: Modeling and control application. World Scientific, (2010), Singapore.

Caponetto R., Fazzino S., An application of Adomian Decomposition Method for analysis of fractional-order chaotic systems. Int. J. Bifurcation Chaos, (2013), 23, 1-7.

Caputo M., Linear models of dissipation whose Q is almost frequency independent. J. Roy. Astral. Soc., (1967), 13, 529-539.

Diethelm K., Ford N., Freed A., Luchko Y., Algorithms for the fractional calculus: a selection of numerical method. Computer Methods in Applied Mechanics and Engineering, (2005), 94, 743-773.

Diethelm K., Ford N., Analysis of fractional differential equations. Journal of Mathematical Analysis and Applications, (2002), 265, 229-248.

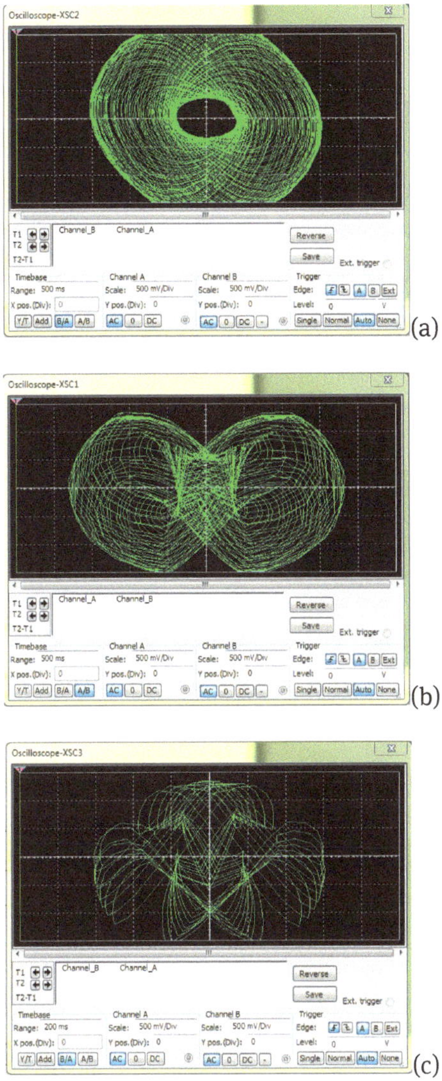

Fig. 8. Attractors of the fractional-order system (15) observed by the oscilloscope (500mV/Div): (a) $x-y$ (b) $x-z$ (c) $y-z$.

Goh S.M., Noorani M.S.M., Hashim I., Efficacy of variational iteration method for chaotic Genesio system–Classical and multistage approach. Chaos, Solitons and Fractals, (2009), 40(5), 2152-2159.

Grigorenko I., Grigorenko E., Chaotic dynamics of the fractional Lorenz system. Physical review letters, (2003), 91(3), 034-101.

Jun-Guo L., Chaotic dynamics and synchronization of fractional-order Genesio–Tesi systems. Chinese Physics, (2005), 14(8), 1517.

Hammouch Z., Mekkaoui T., Circuit design and simulation for the fractional-order chaotic behavior in a new dynamical system, Submitted.
Hammouch Z., Mekkaoui T., Control of a new chaotic fractional-order system using Mittag-Leffler stability, Preprint.
Hammouch Z., Mekkaoui T., Chaos synchronization of a fractional nonautonomous system. Nonautononmous Dynamical Systems, (2014), Vol.1, 61-71.
Hartley T.T., Lorenzo C.F., Killory H.Q., Chaos in a fractional order Chua's system. Circuits and Systems I: Fundamental Theory and Applications. IEEE Transactions, (1995), 42(8), 485-490.
Li C., Chen G., Chaos and hyperchaos in the fractional-order Rossler equations. Physica A: Statistical Mechanics and its Applications, (2004), 341, 55-61.
Liu C.X., Fractional- order Chaotic Circuit Theory and Applications. (2011), Xian: Xian Jiaotong University Press.
Lorenz E.N., Deterministic nonperiodic flow. J. Atmospheric Science. (1963), 20, 130-141.
Ma T., Guo D., Circuit simulation and implementation for synchronization, http://www.paper.edu.cn.
Matignon D., Stability results for fractional differential equations with applications to control processing. Proceedings Comp. Eng. Sys. Appl., (1996), 963-968.
Mekkaoui T., Hammouch Z., Numerical calculations of the Lyapunov exponents for variable-order fractional systems, Submitted.
Miller K.S., Rosso B., An Introduction to the Fractional Calculus and Fractional Differential Equations. Wiley, (1993), New York.
Oldham K.B., Spanier J., The Fractional Calculus. Academic Press, (1974), New York, NY, USA.
Petras I., Fractional-Order Nonlinear Systems: Modeling, Analysis and Simulation. (2011), Springer.
Pham V.T., Volos C., Jafari S., Wang X., Vaidyanathan S., Hidden hyperchaotic attractor in a novel simple memristive neural network. Opt. and Advanced Materials, (2014), Vol.8.
Podlubny I., Fractional Differential Equations: Mathematics in Science and Engineering. (1999), Academic Press, USA.
Strogatz S.H., Nonlinear Dynamics and Chaos: With Applications to Physics, Biology, Chemistry and Engineering. (1994), Perseus Books Pub.
Sun K., Wang X., Sprott J.C., Bifurcations and chaos in fractional-order simplified Lorenz system. International Journal of Bifurcation and Chaos, (2010), 20 (04), 1209-1219.
Sun, K., Wang, X., and Sprott, J. C. (2010). Bifurcations and chaos in fractional-order simplified Lorenz system. International Journal of Bifurcation and Chaos, 20(04), 1209-1219.
Sprott J.C., Elegant chaos: algebraically simple chaotic flows. (2010), World Scientific.
Sprott J.C., Simplest Chaotic Flows with Involutional Symmetries. International Journal of Bifurcation and Chaos, (2014), 24(01).
Tavazoei M., Haeri M., A necessary condition for double scroll attractor existence in fractional-order systems. Physics Letters A, (2007), 367(1), 102-113.
Wang S., Yu Y., Application of Multistage Homotopy-perturbation Method for the Solutions of the Chaotic Fractional Order Systems. International Journal of Nonlinear Science, (2012), Vol.13, No.1, 3-14.
Zaslavsky G., Hamiltonian Chaos and Fractional Dynamics. (2008), Oxford University Press.
Zhang X., Sun Q., Cheng P., Design of a four-wing heterogeneous fractional-order chaotic system and its circuit simulation. International Journal of Smart Design and Intelligent Systems, (2014), Vol.7.

E. Guariglia

Fractional Derivative of the Riemann Zeta Function

Abstract: Fractional derivative of the Riemann zeta function has been explicitly computed and the convergence of the real part and imaginary parts are studied. In particular, this fractional derivative is equal to a complex series, whose real and imaginary parts are convergent in a half-plane which depends on the order of the fractional derivative.

Keywords: Fractional derivative; Riemann zeta function; gamma function; convergence.

1 Introduction

It is well known that Riemann zeta function plays an important role both in number theory and in several applications of quantum mechanics: in particular, the Riemann hypothesis can be formulated using the quantum terminology (Pozdnyakov 2012); furthermore, its zeros are related with the Hamiltonian of a quantum mechanical system (Sierra 2010).
Fractional calculus and fractional derivatives have been increasingly studied due to the many practical and theoretical applications: especially its geometrical interpretation gives us a powerful mathematical tool to model the most advanced concepts of modern physics (Adda 1997, Podlubny 2002, Wang 2012).
Several authors have obtained the fractional derivative of complex functions, in particular in (Owa 2008) it is shown how to extend the fractional derivative to analytical functions on the unit circle $U = z \in \mathbb{C} : |z| < 1$ as a complex series

$$f(z) = z + \sum_{n=2}^{\infty} a_n z^n \tag{1}$$

This possibility seems to be unaffordable for the Riemann ζ-function, because this function can not be expressed as (1). In the following chapter, we will approach this problem by using a suitable modification of the Caputo fractional derivative as given by Ortigueira. It is known that the main advantage of the Caputo derivative is that the fractional derivative of a constant is zero, which is not true for the Riemann-Liouville derivative. In his paper (Ortigueira 2006), Ortigueira defines a generalized Caputo derivative for complex functions with respect to a given direction of the complex plane. Thus the Ortigueira fractional derivative can be considered as the direc-

E. Guariglia: Dept. of Physics "E. R. Caianiello", University of Salerno, Via Giovanni Paolo II, 84084 Fisciano, Italy, E-mail: eguariglia@unisa.it tional Caputo fractional derivative, so that the Ortigueira fractional derivative gives us the fractional derivative of a complex function in the θ-direction of the complex plane. It will be shown that Ortigueira-Caputo fractional derivative of the Riemann ζ-function can be easily computed as a complex series. The real and imaginary parts of this complex series will be studied: in particular it will be shown where these two series converge.

2 Preliminary Remarks on Riemann Zeta Function

The Riemann zeta function is defined as (Whittaker et al. 1927):

$$\zeta(s) \triangleq \sum_{n=1}^{\infty} \frac{1}{n^s}, \qquad (n \in \mathbb{N},\ s \in \mathbb{C}) \tag{2}$$

This infinite series (Cattani 2010, Voxman et al. 1981) converges for all the complex numbers such that $\Re(s) > 1$. This function can be also defined by the integral:

$$\zeta(s) = \frac{1}{\Gamma(s)} \int_0^{\infty} \frac{x^{s-1}}{e^x - 1} dx \tag{3}$$

where Γ is the gamma function (Apostol 2010), given by:

$$\Gamma(t) \triangleq \int_0^{\infty} \frac{x^{t-1}}{e^{-x}} dx \tag{4}$$

in which $t \in \mathbb{C}$ and the last integral converges if $\Re(t) > 0$ (Abramowitz et al. 1965). The Riemann zeta function can be defined only for $\Re(s) > 1$, otherwise the absolute value of n^s would be too small and the series in (2) would diverge. Riemann has proven that the function ζ defined by this series has an unique analytical continuation to the entire complex plane, excluding the point $s = 1$, where this function has a simple pole with complex residue 1 (Riemann 1859). Moreover, it can be shown that:

$$\lim_{s \to 1} (s-1)\zeta(s) = 1 \tag{5}$$

so that Riemann zeta function is meromorphic on the whole complex plane, and holomorphic for every complex s except for $s = 1$. Concerning its zeros, ζ is equal to zero at the negative even integers, i.e. $\zeta(s) = 0$ when s is $-2, -4-, 6, \ldots$: they are called *trivial zeros* (Edwards 1974, Voxman et al. 1981). Euler has also shown that (Hardy 1914, Mollin 2010):

$$\zeta(s) = \prod_{p \in \mathbb{P}} \frac{1}{1 - p^{-s}} \tag{6}$$

where $\mathbb{P}$ is the set of prime numbers (Mathai et al. 2008, Davenport 2008). Moreover, by using the last equation, we can easily see that there are no zeros in the plane regions where $\Re(s) > 1$, so that the only non-trivial zeros must belong to the strip $0 < Re(s) < 1$, also called *critical strip* (Hardy et al. 1921). Riemann conjectured that they lie on the line $\Re(s) = 1/2$, called *critical line*: this is the famous *Riemann hypothesis*. In 1914, Hardy has shown that ζ has infinitely many non-trivial zeros on the critical line (Hardy et al. 2008).

3 Derivative in the Ortigueira Sense and Caputo Derivative in Complex Plane

Recently, M. D. Ortigueira has proposed a new type of fractional derivative, defined in the complex plane, starting from the following generalization of the Cauchy's integral formula (Ortigueira 2011):

$$D^{\alpha} f(s) = \frac{\Gamma(\alpha+1)}{2\pi j} \int_{C} \frac{f(w)}{(w-s)^{\alpha+1}} dw \tag{7}$$

where C is a U shaped contour that encircles the branch cut line, while j is the imaginary unit. This operator (called the derivative in the Ortigueira sense) gives us the fractional derivative in the θ-direction of the complex plane: it is generally indicated by ${}_{O}D^{\alpha}$ and is defined as follows:

$${}_{O}D^{\alpha} f(s) = \frac{e^{j(\pi-\theta)\alpha}}{\Gamma(-\alpha)} \int_{0}^{\infty} \frac{f(xe^{j\theta}+s) - \sum_{k=0}^{n} \frac{f^{(k)}(s)}{k!} e^{jk\theta} x^{k}}{x^{\alpha+1}} dx \tag{8}$$

in which $n = \lfloor \alpha \rfloor$ is the largest integer non greater than α, while s belongs to the complex plane and $\theta \in [0, 2\pi)$. The derivative in the Ortigueira sense has many proprieties: below we discuss three of the most important (Li et al. 2009, Ortigueira et al. 2014).

1. *Consistency with order-derivative*

   If $n-1 < \alpha < n \in \mathbb{Z}^{+}$, $f(s)$ is analytic in a region that contains the Hankel contour C, and without loss of generality we can suppose $\theta \in [0, 2\pi)$: it is easy to show (Li et al. 2009) that:

$$\lim_{\alpha \to n^{-}} {}_{O}D^{\alpha} f(s) = f_{\theta}^{(n)}(s)$$

$$\lim_{\alpha \to (n-1)^{+}} {}_{O}D^{\alpha} f(s) = f_{\theta}^{(n-1)}(s)$$

where f_θ^n is the n-th directional derivative in θ-direction.

2. *Composition with integer-order derivatives*
   Let A_k be the set of complex functions $f : D \subseteq \mathbb{C} \to \mathbb{C}$ such that

$$\frac{d^k}{ds^k} \int_0^\infty f(xe^j\theta + s)dx = \int_0^\infty \frac{d^k}{ds^k} f(xe^j\theta + s)dx$$

   in which $\theta \in [0, 2\pi)$ and $k \in \mathbb{Z}^+$. If $f \in A_k$, in (Li et al. 2009) it is shown that:

$$\frac{d^k}{ds^k} \left({}_0D^\alpha f(s)\right) = {}_0D^\alpha \left(\frac{d^k}{ds^k} f(s)\right)$$

3. *Composition with itself*
   Let us give $k < \alpha < k + 1 \in \mathbb{Z}^+$, $h < \beta < h + 1 \in \mathbb{Z}^+$ and $f \in A_k$. The following equations hold true:

$$\begin{aligned}
&\text{a) } {}_0D^{-\alpha} f(s) \left({}_0D^\alpha f(s)\right) = {}_0D^\alpha \left({}_0D^{-\alpha} f(s)\right) \\
&\text{b) } {}_0D^{\alpha} f(s) \left({}_0D^{-\beta} f(s)\right) = {}_0D^{-\beta} \left({}_0D^{\alpha} f(s)\right) \\
&\text{c) } {}_0D^{\alpha} f(s) \left({}_0D^{\beta} f(s)\right) = {}_0D^{\beta} \left({}_0D^{\alpha} f(s)\right) \neq {}_0D^{\alpha+\beta} f(s)
\end{aligned}$$

   where the one holds for $f \in A_{\max[h,k]}$ (Li et al. 2009).

We recall that the classical Caputo fractional derivative of order α, of a $\mathfrak{C}^m$-differentiable function $f(t)$ is defined as (Bagley 2007):

$${}_CD^\alpha f(t) \triangleq \frac{1}{\Gamma(m-\alpha)} \int_0^t (t-x)^{m-\alpha-1} f^m(x)\, dx \tag{9}$$

with $m - 1 < \alpha < m \in \mathbb{Z}$.

Fractional derivatives in Ortigueira sense have many other proprieties (Ortigueira 2006), but for our purposes the most significant fact is how it generalizes the classical Caputo derivative (Caputo 1967, De Oliveira et al. 2014) in the complex plane.

By using the Ortigueira derivative, the α-th Caputo derivative along with the θ-direction of the complex plane is defined as:

$${}_CD^\alpha f(s) \triangleq_0 D^{\alpha-m} \left(f^{(m)}(s)\right) = \frac{e^{j(\pi-\theta)(\alpha-m)}}{\Gamma(m-\alpha)} \int_0^\infty \frac{f^{(m)}(xe^{j\theta} + s)}{x^{\alpha-m+1}} dx \tag{10}$$

where $m - 1 < \alpha < m \in \mathbb{Z}^+$ (Kilbas et al. 2006, Li et al. 2009): this operator is called Ortigueira-Caputo fractional derivative of the complex function $f(s)$.

4 Fractional Derivative of $\zeta(s)$

We now have all the mathematical tools necessary to calculate the Ortigueira-Caputo fractional derivative of the Riemann ζ-function. From (10), it follows that:

$$
\begin{aligned}
{}_C D^\alpha \zeta(s) &= \frac{e^{j(\pi-\theta)(\alpha-m)}}{\Gamma(m-\alpha)} \int_0^\infty \frac{d^m}{ds^m}\left(\zeta\left(xe^{j\theta}+s\right)\right)\frac{1}{x^{\alpha-m+1}}dx \\
&= \frac{e^{j(\pi-\theta)(\alpha-m)}}{\Gamma(m-\alpha)} \int_0^\infty \frac{d^m}{ds^m}\left(\sum_{n=1}^{\infty} n^{-s} n^{-xe^{j\theta}} x^{m-\alpha-1}\right) dx
\end{aligned}
$$

Bringing the integral sign under both derivative and summation, we obtain:

$$
{}_C D^\alpha \zeta(s) = \frac{e^{j(\pi-\theta)(\alpha-m)}}{\Gamma(m-\alpha)} \frac{d^m}{ds^m}\left(\sum_{n=1}^{\infty} n^{-s} \int_0^\infty n^{-xe^{j\theta}} x^{m-\alpha-1} dx\right) \tag{11}
$$

Let us begin by computing the integral in the right hand side (RHS) of the equation 11: by a change of variables $xe^{j\theta} = z$, we have:

$$
\begin{aligned}
\int_0^\infty n^{-xe^{j\theta}} x^{m-\alpha-1} dx &= \int_0^\infty n^{-z} e^{-j\theta(m-\alpha-\cancel{1})} z^{m-\alpha-1} \cancel{e^{-j\theta}} dz \\
&= e^{-j\theta(m-\alpha)} \int_0^\infty e^{-z\log n} z^{m-\alpha-1} dz \\
&= e^{-j\theta(m-\alpha)} \int_0^\infty e^{-x} x^{m-\alpha-1} (\log n)^{\cancel{1}+\alpha-m} \frac{dx}{\cancel{\log n}} \\
&= e^{-j\theta(m-\alpha)} (\log n)^{\alpha-m}\, \Gamma(m-\alpha)
\end{aligned}
$$

where we have performed another change of variable, i.e. $z\log n = x$; furthermore, we have used the definition of gamma function (see equation 4). Given that, from the definition of Ortigueira-Caputo fractional derivative $m-\alpha > 0$, the last RHS of the continued equality above makes sense. Substituting:

$$
\begin{aligned}
{}_C D^\alpha \zeta(s) &= \frac{e^{j(\pi-\cancel{\theta})(\alpha-m)}}{\cancel{\Gamma(m-\alpha)}} \frac{d^m}{ds^m} \sum_{n=1}^{\infty} n^{-s} \cancel{e^{-\theta(m-\alpha)}} (\log n))^{\alpha-m} \cancel{\Gamma(m-\alpha)} \\
&= e^{j\pi(\alpha-m)} \sum_{n=1}^{\infty} (\log n)^{\alpha-m} \frac{d^m}{ds^m}\left(n^{-s}\right)
\end{aligned} \tag{12}
$$

We are now ready to compute the derivative in the RHS:

$$\begin{aligned}
\frac{d}{ds}\left(n^{-s}\right) &= -n^{-s}\log n \\
\frac{d^2}{ds^2}\left(n^{-s}\right) &= (-1)^2\, n^{-s}\,(\log n)^2 \\
&\vdots \\
\frac{d^m}{ds^m}\left(n^{-s}\right) &= (-1)^m\, n^{-s}\,(\log n)^m
\end{aligned}$$

so that

$$\begin{aligned}
{}_C D^{\alpha}\zeta(s) &= (-1)^m e^{j\pi(\alpha-m)} \sum_{n=1}^{\infty} (\log n)^{\alpha-\cancel{m}}\, n^{-s} \cancel{(\log n)^m} \\
&= (-1)^m e^{j\pi(\alpha-m)} \sum_{n=1}^{\infty} \frac{(\log n)^{\alpha}}{n^s}
\end{aligned}$$

i.e.

$$ {}_C D^{\alpha}\zeta(s) = (-1)^m e^{j\pi(\alpha-m)} \sum_{n=1}^{\infty} \frac{(\log n)^{\alpha}}{n^s} \tag{13}$$

In Fig. 4 is represented the modulus of this fractional derivative, with $\alpha = 0.5$ and upper limit of sum $n = 100$. For simplicity, from now on, we will denote the fractional derivative computed above with ψ:

$$\psi \triangleq {}_C D^{\alpha}\zeta$$

in which $m - 1 < \alpha < m \in \mathbb{Z}^+$ and $\zeta = \zeta(s)$, it follows immediately that $\psi = \psi(\alpha, s)$. Let us write this function in rectangular form (Mathews et al. 2008). Since $s \in \mathbb{C} \Rightarrow s = x + jy$, it is:

$$n^{-s} = n^{-x} n^{-jy} = n^{-x} e^{-jy\log n} = n^{-x}\left(\cos(y\log n) - j\sin(y\log n)\right)$$

from which we obtain:

$$\begin{aligned}
\psi(\alpha, s = x + jy) &= (-1)^m \left(\cos(\pi(\alpha-m)) + j\sin(\pi(\alpha-m))\right)\cdot \\
&\quad \cdot \sum_{n=1}^{\infty} (\log n)^{\alpha} n^{-x}\left(\cos(y\log n) - j\sin(y\log n)\right) \\
&= (-1)^m \sum_{n=1}^{\infty} (\log n)^{\alpha} n^{-x} \Big(\cos(\pi(\alpha-m))\cos(y\log n) + \sin(\pi(\alpha-m))\cdot \\
&\quad \cdot \sin(y\log n) + j\left(\sin(\pi(\alpha-m))\cos(y\log n) - \cos(\pi(\alpha-m))\sin(y\log n)\right)\Big)
\end{aligned}$$

It follows that:

$$
\begin{aligned}
\Re(\psi(\alpha,s)) &= (-1)^m \sum_{n=1}^{\infty} (\log n)^{\alpha} n^{-x} \Big(\cos(\pi(\alpha-m))\cos(y\log n) + \sin(\pi(\alpha-m))\sin(y\log n) \Big) \\
&= (-1)^m \sum_{n=1}^{\infty} (\log n)^{\alpha} n^{-x} \frac{1}{2} \Big(\cos(\pi(\alpha-m)+y\log n) + \cos(\pi(\alpha-m)-y\log n) \\
&+ \cos(\pi(\alpha-m)-y\log n) - \cos(\pi(\alpha-m)+y\log n) \Big) \\
&= (-1)^m \sum_{n=1}^{\infty} (\log n)^{\alpha} n^{-x} \cos(\pi(\alpha-m)-y\log n)
\end{aligned}
$$

and similarly:

$$
\begin{aligned}
\Im(\psi(\alpha,s)) &= (-1)^m \sum_{n=1}^{\infty} (\log n)^{\alpha} n^{-x} \Big(\sin(\pi(\alpha-m))\cos(y\log n) - \cos(\pi(\alpha-m))\sin(y\log n) \Big) \\
&= (-1)^m \sum_{n=1}^{\infty} (\log n)^{\alpha} n^{-x} \frac{1}{2} \Big(\sin(\pi(\alpha-m)+y\log n) + \sin(\pi(\alpha-m)-y\log n) \\
&- \sin(\pi(\alpha-m)+y\log n) - \sin(y\log n-\pi(\alpha-m)) \Big) \\
&= (-1)^m \sum_{n=1}^{\infty} (\log n)^{\alpha} n^{-x} \sin(\pi(\alpha-m)-y\log n)
\end{aligned}
$$

In summary, we have shown the following property.

Theorem 37. *The Ortigueira-Caputo fractional derivative of the Riemann ζ-function is equal to:*

$$
{}_C D^{\alpha} \zeta(s) = (-1)^m e^{j\pi(\alpha-m)} \sum_{n=1}^{\infty} \frac{(\log n)^{\alpha}}{n^s} \tag{14}
$$

and its real and imaginary parts are given by:

$$
\boxed{
\begin{aligned}
\Re(\psi(\alpha,s)) &= (-1)^m \sum_{n=1}^{\infty} \frac{(\log n)^{\alpha}}{n^x} \cos(\pi(\alpha-m)-y\log n) \\
\Im(\psi(\alpha,s)) &= (-1)^m \sum_{n=1}^{\infty} \frac{(\log n)^{\alpha}}{n^x} \sin(\pi(\alpha-m)-y\log n)
\end{aligned}
} \tag{15}
$$

where $s \in \mathbb{C}$ *and* $m-1 < \alpha < m \in \mathbb{Z}^+$.

5 Convergence of $\Re(\psi(\alpha,s))$ and $\Im(\psi(\alpha,s))$

At this point we check whether $\Re(\psi(\alpha,s))$ and $\Im(\psi(\alpha,s))$ converge somewhere, using the classical comparison test and a generalization of the well known harmonic series.

Theorem 38. *(Comparison test)*
Suppose that

$$\sum_{n=1}^{\infty} a_n$$

converges and $0 \le b_n \le a_n$. *Under these hypotheses, the series*

$$\sum_{n=1}^{\infty} b_n$$

converges.

Proof. See (Stirling 2009, pp. 99-100) □

Corollary 6. *Let* $p \in \mathbb{R}$, *the following series*

$$\sum_{n=1}^{\infty} \frac{1}{n^p} \tag{16}$$

converges if $p > 1$ *and diverges if* $p \le 1$.

Proof. Just use the well known property about the convergence of the harmonic series. □

The series 16 is called p-series (Marsden et al. 1993) and if $p > 1$ it is equal to $\zeta(p)$. These two properties are essential to show the following statement.

Theorem 39. *The two real functions* $\Re(\psi(\alpha, s))$ *and* $\Im(\psi(\alpha, s))$ *converge in the half-plane:*

$$x > 1 + \alpha \tag{17}$$

Proof. Since the $sin(\cdot)$ is a bounded function in $[-1, 1]$, it is:

$$\sin(x) \le |\sin(x)| \le 1$$

therefore:

$$\frac{(\log n)^{\alpha}}{n^x} \sin(\pi(\alpha - m) - y \log n) \le \frac{(\log n)^{\alpha}}{n^x} < \frac{n^{\alpha}}{n^x} = \frac{1}{n^{x-\alpha}}$$

In the last RHS of the relations above, the last term form is that of a p-series: it converges for

$$x - \alpha > 1$$

Using the comparison test, it is clear that $\Im(\psi(\alpha, s))$ also converges in the half-plane $x > 1 + \alpha$. Similarly, it can be shown that the same holds true for $\Re(\psi(\alpha, s))$ (see fig. 1). □

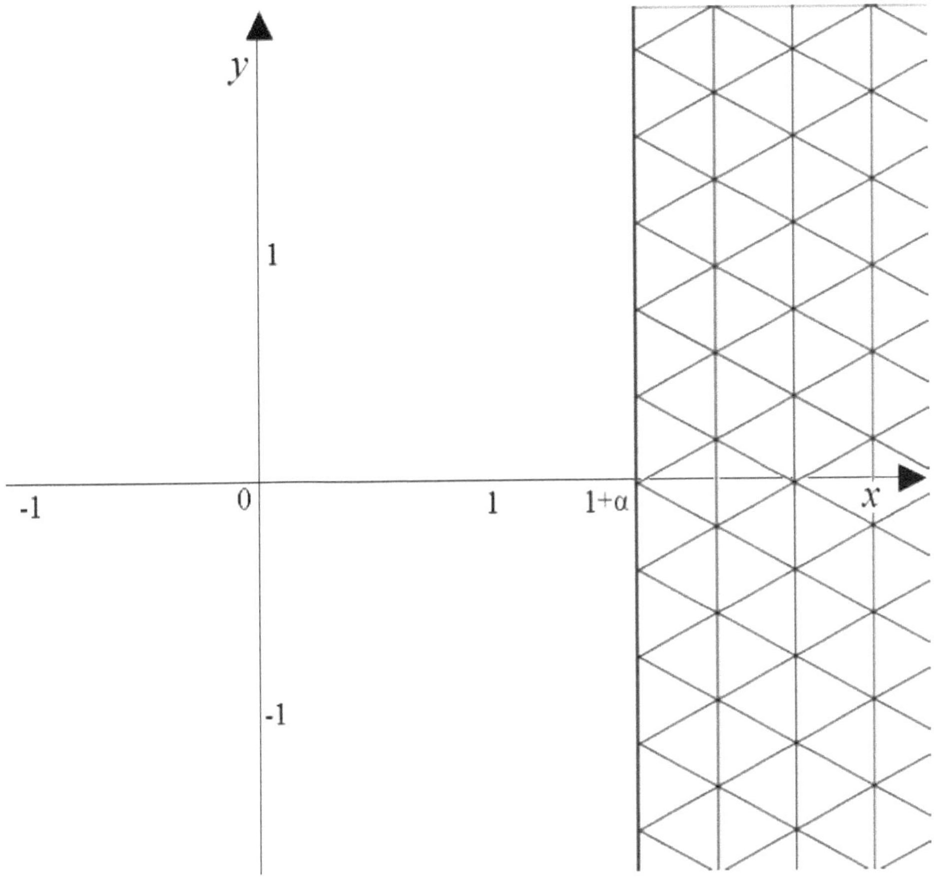

Fig. 1. *Half-plane of convergence for* $\Re(\psi(\alpha, s))$, $\Im(\psi(\alpha, s))$ *and* $\psi(\alpha, s)$*: it is interesting to note that* $\zeta(s)$ *converges for* $x > 1$ *and its* α*-order fractional derivative converges for* $x > 1 + \alpha$.

There follows that

Corollary 7. *The fractional derivative of the Riemann zeta-function converges in the half-plane* (17).

Proof. It follows from the theorem above remembering that a complex series converges if and only if both the real and imaginary parts converge (Lang 2003). □

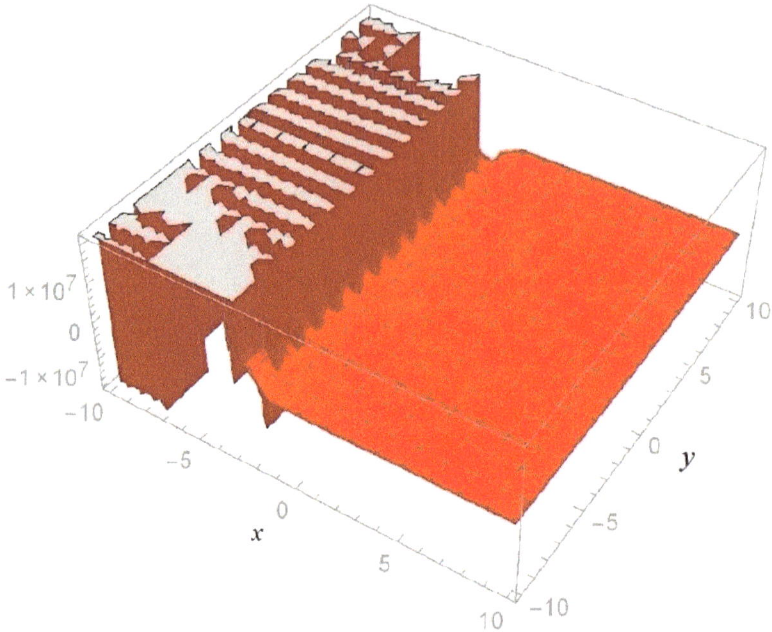

Fig. 2. *3D plot of the function* $\Re(\psi(\alpha, s))$ *with* $\alpha = 0.5$*, upper limit of the sum* $n = 100$ *and* $(x, y) \in [-10, 10]^2$*: the surface satisfies the* (17).

In Fig. 2 (respectively Fig. 3 and Fig. 4), it is plotted $\Re(\psi(\alpha, s))$ (respectively $\Im(\psi(\alpha, s))$ and $|\psi(\alpha, s)|$) with $\alpha = 0.5$ and upper bound $n = 100$: the surface behavior is in accord with the above theorem.

Conclusion

In this chapter, the fractional derivative of the Riemann zeta function was computed, using the Caputo derivative in the Ortigueira sense. It was shown where the real and imaginary parts of this new function converge.

Bibliography

Abramowitz et al. (1965) Abramowitz, M. & Stegun, I. A. (1964). *Handbook of Mathematical Functions With Formulas, Graphs, and Mathematical Tables*. Washington DC, Dover Pubblications.

Adda, F. B. (1997). *Geometric interpretation of the fractional derivative*. Journal of Fractional Calculus, 11, 21-52.

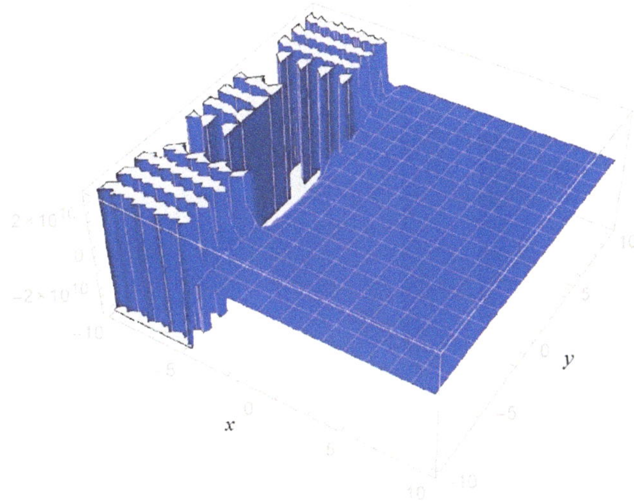

Fig. 3. *3D plot of the function* $\Im(\psi(\alpha, s))$ *with* $\alpha = 0.5$, *upper limit of the sum* $n = 100$ *and* $(x, y) \in [-10, 10]^2$*: also here, the condition* 17 *is true.*

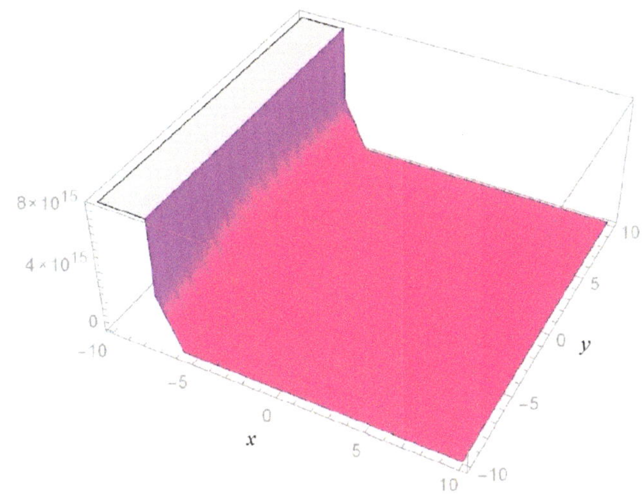

Fig. 4. *3D plot of the function* $|\psi(\alpha, s)|$ *with* $\alpha = 0.5$, *upper limit of the sum* $n = 100$ *and* $(x, y) \in [-10, 10]^2$*: it still satisfies the* (17) *and has the same shape of* $|\zeta(s)|$ (see Pegg et al. 2004).

Apostol, T. M. (2010). *Introduction to Analytic Number Theory*. New York: Springer.

Bagley, R. (2007). *On the Equivalence of the Riemann-Liouville and the Caputo Fractional Order Derivatives in Modeling of Linear Viscoelastic Materials*. Fractional Calculus and Applied Analysis, 10 (2), 123-126.

Caputo, M. (1967). *Linear Models of Dissipation whose Q is almost Frequency Independent-II*. Geophysical Journal of the Royal Astronomic Society, 13 (5), 529-539.
Cattani, C. (2010). *Fractal Patterns in Prime Numbers Distribution*. Computational Science and its Applications. Berlin/Heidelberg Springer, 6017, 164-176.
Davenport, H. (2008). *The Higher Arithmetic: An Introduction to the Theory of Numbers*. Cambridge: Cambridge University Press.
De Oliveira, E. C. & Tenreiro Machado, J. A. (2014). *A Review of Definitions for Fractional Derivatives and Integrals*. Mathematical Problems in Engineering, 6 pages.
Edwards, H. M. (1974). *Riemann's Zeta Function*. New York: Academic Press.
Hardy, G. H. (1914). *Sur les zeros de la fonction $\zeta(s)$ de Riemann*. Comptes Rendus de l'Académie des Sciences, 158, 1012-10124.
Hardy, G. H. & Littlewood, J. E. (1921). *The zeros of Riemann's zeta-function on the critical line*. Mathematische Zeitschrift, 10 (3-4), 283-317.
Hardy, G. H. & Wright, E. M. (2008). *An Introduction to the Theory of Numbers*. Oxford: Oxford University Press.
Kilbas, A. A., Srivastava, H. M. & Trujillo, J. J. (2006). *Theory and Applications of Fractional Differential Equations*. New Jersey: Springer.
Lang, S. (2003). *Complex Analysis*. New York: Springer.
Li, C., Dao, X. & Guo, P. (2009). *Fractional derivatives in complex planes*. Nonlinear Analysis: Theory, Methods & Applications, 71, 1857-1869.
Marsden, J. E. & Hoffman, M. J. (2002). *Elementary Classical Analysis*. San Francisco: W. H. Freeman and Company.
Mathai, A. M. & Haubold , H. J. (2008). *Special function for Applied Scientists*. New York: Springer.
Mathews, J. H. & Howell, R. W. (2008). *Complex Analysis for Mathematics and Engineering*. New York: Springer-Verlag.
Mollin, R. A. (2010). *Advanced Number Theory with Applications*. Boca Raton: CRC Press.
Ortigueira, M. D. (2006). *A coherent approach to non-integer order derivatives*. Signal Processing, 86 (10), 2505-2515.
Ortigueira, M. D. (2011). *Fractional Calculus for Scientists and Engineers*. London: Springer.
Ortigueira, M. D. & Tenreiro Machado, J. A. (2014). *What is a fractional derivative?*. Journal of Computational Physics, 293, 4-13.
Owa, S. (2008). *Some properties of fractional calculus operators for certain analytic functions*. Proceedings of International Symposium on New Development of Geometric Function Theory and its Applications, Bangi (Malaysia), 214-221.
Pegg, E. Jr. & Weisstein, E. W. (2004). *Seven Mathematical Tidbits*. MathWorld Headline News.
Podlubny, I. (2002). *Geometric and Physical Interpretation of Fractional Integration and Fractional Differentiation*. Fractional Calculus and Applied Analysis, 5 (4), 2002, 367-386.
Pozdnyakov, D. (2012). *Physical interpretation of the Riemann hypothesis*. arXiv:1202.2115.
Riemann, G. F. B. (1859). *Ueber die Anzahl der Primzahlen unter einergegebenen Grösse*. Monatsberichte der Königlich Preußischen Akademie der Wissenschaften zu Berlin, 10 pages.
Sierra, G. (2010). *A physics pathway to the Riemann hypothesis*. arXiv:math-ph/1012.4264.
Stirling, D. S. G. (2009). *Mathematical Analysis and Proof*. Philadelphia, Woodhead Publishing.
Voxman, W. L. & Goetschel, R. H. (1981). *Advanced Calculus: An Introduction to Modern Analysis*. New York: CRC Press.
Wang, X. (2012). *Fractional Geometric Calculus: Toward A Unified Mathematical Language for Physics and Engineering*. Proceedings of The Fifth Symposium on Fractional Differentiation and its Applications, Hohai University (China).
Whittaket, T. & Watson, G. N. (1927). *A Course of Modern Analysis*. Cambridge: Cambridge University Press.

Fethi Bin Muhammad Belgacem, Vartika Gulati, Pranay Goswami, and Abdullah Aljoujiee

A Treatment of Generalized Fractional Differential Equations: Sumudu Transform Series Expansion Solutions, and Applications

Abstract: In this chapter we solve generalized fractional differential equations by using the Sumudu transform. The equations we treat include differential equations with generalized Riemann-Liouville fractional derivatives. The operators involved are very general in nature, and cover a wide range of fractional differential equations and their solutions in terms of various functions related to Mittag – Leffler functions.

Keywords: Caputo fractional derivative, fractional differential equations; generalized Riemann-Liouville fractional derivative; Hilfer fractional derivative, Mittag-Leffler functions; Sumudu transform

1 Introduction

In the last three decades, fractional calculus became an important tool for the development and analysis of accurate models of various phenomena of nature, in diverse fields of science such as engineering, economics, material sciences and aerospace sciences. This field of mathematical analysis - which deals with investigations and applications of integrals and derivatives of arbitrary order, received considerable attention by scientists and mathematicians in numerous theoretical and applied sciences. Recently, Special families of generalized fractional derivative operator $[D_{a\pm}^{\alpha,\beta}]$ of order α and type β, were introduced and investigated (Garra et al., 2014, Hilfer and Anton 1995; Hilfer 2000; Hilfer 2002; Hilfer 2008; Hilfer). Applications of fractional calculus may be based on fractional derivatives of different kinds, (Hilfer, Luchko, and Tomovski, 2009; Mainardi and Gorenflo, 2007; Sandev and Tomovski, 2010; Srivastava and Tomovski, 2009; Tomovski 2012). The solution of generalized differential equations of fractional order is quite involved. Some analytical methods were presented some using and numerous integral based transforms, such as the popular Laplace transform (Pod-

Fethi Bin Muhammad Belgacem: Department of Mathematics, Faculty of Basic Education, PAAET, Al-Ardhia, Kuwait, E-mail: fbmbelgacem@gmail.com
Vartika Gulati: Department of Mathematics, University of Rajasthan, Rajasthan -302055, India, E-mail: vgmaths26@gmail.com
Pranay Goswami: School of Liberal Studies, Ambedkar University Delhi, Delhi -110006, India, E-mail: pranaygoswami83@gmail.com
Abdullah Aljoujiee: Mathematics Department, Al Imam Muhammed Ibn Saud Islamic University, P O Box 90189, Riyadh 11613, Saudi Arabia, E-mail: joa111@gmail.com

lubny, 1994; Podlubny, 1999), the Fourier transform method (Miller and Ross, 1993), and more recently, the Sumudu transform (Belgacem et al., 2014, Bulut, Baskonus and Belgacem, 2013a, Bulut, Baskonus and Belgacem, 2013b; Chaurasia., Dubey, and Belgacem 2012; Dubey, Goswami and Belgacem, 2012; Gupta,, Sharma, and Belgacem, 2011 ; Katatbeh and Belgacem, 2011, Tuluce, Bulut and Belgacem, 2014) that treat fractional type equations and systems by means of the Sumudu transform method.

The prevalence and summary of the state of the art capabilities and main properties of the Sumudu transform including intricate developments, can be found in the works of Belgacem, 2006a, Belgacem 2009; and Belgacem, 2010. With more than two decades after its bringing back to circulation after Watugala Re-baptism (Watugala, 1993). With less than fifty articles in circulation, of which we recommend the following articles treating integer order types of equations of systems (Asiru, 2001; Asiru, 2002; Asiru, 2003; Belgacem, Karaballi and Kalla 2003; Belgacem 2006 b; Belgacem, 2009; Belgacem 2010; Husain and Belgacem, 2007; Kılıcman, Eltayeb, and Agarwal, 2010; Rana et al. 2007; Weerakoon, 1994; Weerakoon, 1999; Zang 2007).

In the recent published literature dedicated to Smudu transform, the only tables provided are those in the works of Belgacem, Karaballi et Kalla, 2003; Belgacem and Karaballi, 2006; Katatbeh and Belgacem, 2011; Belgacem and Silambarasan, 2012.

In this chapter, we apply the Sumudu transform of the generalized fractional derivative, and use the expansion coefficients of series to derive the explicit solution to homogeneous fractional differential equations.

2 Preamble

The Riemann – Liouvville fractional derivative is defined in the (Miller and Ross, 1993), as follows:

Definition 1: The Riemann – Liouvville fractional derivative of order, α, $m-1<\alpha<m, m \in N$, is defined as the left inverse of Riemann – Liouvville fractional integral, i.e.

$$D_t^\alpha f(t) = \frac{1}{\Gamma(m-\alpha)} D^m \int_0^t \frac{f(\tau)}{(t-\tau)^{\alpha+1-m}} d\tau. \tag{1}$$

In contrast, we find that Caputo gave an alternative definition for the fractional derivative, (Caputo,1969):

Definition 2: The Caputo fractional derivative of order α, $m-1<\alpha<m, m \in N$ is given by

$$^{C}D_t^\alpha f(t) = I_t^{m-\alpha} D^m f(t) = \frac{1}{\Gamma(m-\alpha)} \int_0^t \frac{1}{(t-x)^{\alpha-m+1}} D^m f(x)\, dx. \tag{2}$$

On the other hand, (Hilfer, 2002) stated yet another version of the fractional derivative, namely:

Definition 3: The Hilfer fractional derivative or composite fractional derivative of order $0 < \alpha < 1$, and type, $0 \le \beta \le 1$, is defined by

$$D_t^{\alpha,\beta} f(t) = I_t^{\beta(1-\alpha)} D I_t^{(1-\beta)(1-\alpha)} f(t). \tag{3}$$

Furthermore, in 24 Hilfer R. (2008) extended his own definition, and termed it generalized Riemann-Liouville.

Definition 4: For $n-1 < \alpha < n, 0 \le \beta \le 1, n \in N$, The generalized Riemann-Liouville fractional derivative is defined by

$$D_t^{\alpha,\beta} f(t) = I_t^{\beta(n-\alpha)} D^n I_t^{(1-\beta)(n-\alpha)} f(t). \tag{4}$$

In the definition above, type β allows $D_t^{\alpha,\beta}$ to interpolate continuously between the classical Riemann-Liouville fractional derivative, and the Caputo fractional derivative. Clearly, the case $\beta = 0$, reverts back to yield the classical Riemann-Liouville fractional derivative:

$$D_t^{\alpha,0} f(t) = D^n I_t^{(n-\alpha)} f(t) = D_t^{\alpha} f(t), \quad 0 < \alpha < 1. \tag{5}$$

Furthermore, in the case $\beta = 1$, the Caputo fractional derivative is obtained

$$D_t^{\alpha,1} f(t) = I_t^{(n-\alpha)} D^n f(t) = {}^{C}D_t^{\alpha} f(t), \quad 0 < \alpha < 1. \tag{6}$$

Definition 5: A generalization of the Mittag – Leffler function $E_{\alpha,\beta}(z)$ is introduced by Prabhakar, 1971, as follows:

$$E_{\alpha,\beta}^{\gamma}(z) = \sum_{k=0}^{\infty} \frac{\gamma_k}{\Gamma(\alpha k + \beta)} \frac{z^k}{k!}. \tag{7}$$

where $\alpha, \beta, \gamma \in C; \Re(\alpha) > 0; \Re(\beta) > 0; z \in C$ and γ_k denotes the familiar Pochhammer symbol or the shifted factorial, since $(1)_k = \quad k! \, (k \in N_0)$, the set of whole numbers $(N_0 = N \cup \{0\} = \{0, 1, 2 \ldots\})$, and where

$$(\gamma)_k = \frac{\Gamma(\gamma + k)}{\Gamma(\gamma)} = \begin{cases} 1 & (k = 0; \gamma \in C \setminus \{0\}) \\ \gamma(\gamma+1)\ldots(\gamma+k-1) & (k \in N; \gamma \in C) \end{cases} \tag{8}$$

Definition 6: The Sumudu transform is defined over the set of functions (See for instance Watugala, 1993):

$$A = \{f(t) \,|\, \exists M, \tau_1, \tau_2 > 0, |f(t)| < Me^{|t|/\tau_j}, \; if \; t \in (-1)^j \times [0, \infty)\},$$

by

$$\tilde{G}(u) = S[f(t)] = \int_0^{\infty} f(ut) e^{-t} dt, \quad u \in (-\tau_1, \tau_2) \tag{9}$$

With the definitions above, we now recall some useful results which are directly applicable below:

$$S^{-1}\left[u^{\alpha-1}\left(1-wu^{\beta}\right)^{-\delta}\right]=t^{\alpha-1}E_{\beta,\alpha}^{\delta}\left(wt^{\beta}\right), \tag{10}$$

where, S^{-1} denotes the inverse Sumudu transform.

We can prove the result in the following way:

$$S\left[t^{\gamma-1}E_{\beta,\gamma}^{\delta}\left(\omega t^{\beta}\right)\right]=\int_{0}^{\infty}e^{-t}\left(ut\right)^{\gamma-1}E_{\beta,\gamma}^{\delta}\left(\omega\left(ut\right)^{\beta}\right)dt.$$

By using Eq. 7, we get,

$$u^{\gamma-1}\sum_{n=0}^{\infty}\frac{(\delta)_n\left(\omega u^{\beta}\right)^{n}}{n!}=u^{\gamma-1}\left(1-\omega u^{\beta}\right)^{-\delta}.$$

By applying inverse Sumudu transform, we get our required result.

$$S^{-1}\left[\frac{u^{-2}}{\left(u^{-2}+au^{-\alpha}+b\right)}\right]=\sum_{k=0}^{\infty}(-b)^{k}t^{2k+1-1}E_{-\alpha+2,2k+1}^{k+1}\left[-at^{-\alpha+2}\right] \tag{11}$$

To find inverse Sumudu transform of this function we will use result 10

$$\begin{aligned}\frac{1}{u^{2}\left(u^{-2}+au^{-\alpha}+b\right)}&=\frac{1}{\left(1+au^{-\alpha+2}\right)\left[1+\frac{bu^{2}}{\left(1+au^{-\alpha+2}\right)}\right]}=\sum_{k=0}^{\infty}\frac{\left(-bu^{2}\right)^{k}}{\left(1+au^{-\alpha+2}\right)^{k+1}}\\&=\sum_{kr=0}^{\infty}u^{2k+1-1}(-b)^{k}\left(1+au^{-\alpha+2}\right)^{-(k+1)}\end{aligned} \tag{12}$$

Then by using result 10

$$S^{-1}\left[\frac{u^{-2}}{\left(u^{-2}+au^{-\alpha}+b\right)}\right]=\sum_{k=0}^{\infty}(-b)^{k}t^{2k+1-1}E_{-\alpha+2,2k+1}^{k+1}\left(-at^{-\alpha+2}\right) \tag{13}$$

(iii)

$$S^{-1}\left[\frac{u^{-1}}{u^{-2}+au^{-\alpha}+b}\right]=\sum_{k=0}^{\infty}(-b)^{k}t^{2k+2-1}E_{-\alpha+2,2k+2}^{k+1}\left(-at^{-\alpha+2}\right) \tag{14}$$

(iv)

$$S^{-1}\left[\frac{au^{\beta(1-\alpha)-1}}{u^{-2}+au^{-\alpha}+b}\right]=\sum_{k=0}^{\infty}a.(-b)^{k}t^{\beta(1-\alpha)+2k+1}E_{-\alpha+2,\beta(1-\alpha)+2k+2}^{k+1}\left(-at^{-\alpha+2}\right) \tag{15}$$

(v)

$$S^{-1}\left[\frac{au^{\beta(2-\alpha)-1}}{u^{-2}+au^{-\alpha}+b}\right]=\sum_{k=0}^{\infty}a.(-b)^{k}t^{\beta(2-\alpha)+2k+1}E_{-\alpha+2,\beta(2-\alpha)+2k+2}^{k+1}\left(-at^{-\alpha+2}\right) \tag{16}$$

(vi)

$$S^{-1}\left[\frac{u^{-1}}{au^{-1}+u^{-\alpha}+b}\right]=\sum_{k=0}^{\infty}(-b)^k\, t^{\alpha k+\alpha-1}E^{k+1}_{\alpha-1,\alpha k+\alpha}\left(-at^{\alpha-1}\right) \tag{17}$$

(vii)

$$S^{-1}\left[\frac{u^{\beta(1-\alpha)-1}}{au^{-1}+u^{-\alpha}+b}\right]=\sum_{k=0}^{\infty}(-b)^k\, t^{\beta(1-\alpha)+\alpha k+\alpha-1}E^{k+1}_{\alpha-1,\beta(1-\alpha)+\alpha k+\alpha}\left(-at^{\alpha-1}\right) \tag{18}$$

(viii)

$$S^{-1}\left[\frac{u^{\beta(2-\alpha)-1}}{au^{-1}+u^{-\alpha}+b}\right]=\sum_{k=0}^{\infty}(-b)^k\, t^{\beta(2-\alpha)+\alpha k+\alpha-1}E^{k+1}_{\alpha-1,\beta(2-\alpha)+\alpha k+\alpha}\left(-at^{\alpha-1}\right) \tag{19}$$

Lemma 1: The Sumudu transform of generalized fractional derivative, $\left(D^{\alpha,\beta}_{a\pm}y\right)(x)$ is given as follows:

$$S\left(D^{\alpha,\beta}_{a\pm}y\right)(x)=\quad u^{-\alpha}S[y(x)](u)-\sum_{k=0}^{n-1}[u^{k-n+\beta(n-\alpha)}\lim_{x\to a+}\frac{d^k}{dx^k}(I^{(n-\alpha)(1-\beta)}_{a+}y)(x)]. \tag{20}$$

Proof. The generalized fractional differential operator is defined as

$$\left(D^{\alpha,\beta}_{a\pm}y\right)(x)=\left(\pm I^{\beta(1-\alpha)}_{a\pm}\left(D^{\alpha+\beta-\alpha\beta}_{a\pm}y\right)\right)(x). \tag{21}$$

Applying the integral operator, (I^{α}_{a+}), on both side of 21, (see for instance Podlubny I. 1994).

$$\begin{aligned}I^{\alpha}_{a+}\left(D^{\alpha,\beta}_{a+}y\right)(x)&=\left(I^{\beta(1-\alpha)+\alpha}_{a+}\left(D^{\alpha+\beta-\alpha\beta}_{a+}y\right)\right)(x)\\&=y(x)-\sum_{k=0}^{n-1}\frac{(x-a)^{k-(n-\alpha)(1-\beta)}}{\Gamma(k-(n-\alpha)(1-\beta)+1}\lim_{x\to a+}\frac{d^k}{dx^k}(I^{(n-\alpha)(1-\beta)}_{a+}y)(x)\end{aligned} \tag{22}$$

Then applying the Sumudu transform on each side of above equation we get

$$u^{\alpha}S\left(D^{\alpha,\beta}_{a+}y\right)(u)=S[y(x)](u)-\sum_{k=0}^{n-1}[\lim_{x\to a+}\frac{d^k}{dx^k}(I^{(n-\alpha)(1-\beta)}_{a+}y)(x)](u-a)^{k-(n-\alpha)(1-\beta)} \tag{23}$$

Multiplying both sides by, $u^{-\alpha}$, and taking $a=0$, we get required result.

$$S\left(D^{\alpha,\beta}_{0+}y\right)(u)=u^{-\alpha}S[y(x)](u)-\sum_{k=0}^{n-1}[u^{k-n+\beta(n-\alpha)}\lim_{x\to 0+}\frac{d^k}{dx^k}(I^{(n-\alpha)(1-\beta)}_{0+}y)(x)].$$

□

3 Main Results

In the foreword we establish the main findings of this chapter.

Theorem 1: The following generalized fractional differential equation

$$y''(t) + ay^{\alpha,\beta}(t) + by(t) = 0 \tag{24}$$

where $n-1 < \alpha < n$ and $0 < \beta \le 1$ and $a, b \in R$

with the initial conditions, $y(0) = c_0 and y'(0) = c_1 \left(I_0^{(1-\beta)(1-\alpha)} y\right)(0) = c_2$ and $\left(I_0^{(1-\beta)(2-\alpha)} y\right)(0) = c_3$, has a solution given by

$$y(t) = c_0 \sum_{k=0}^{\infty} \frac{(-b)^k t^{2k}}{k!} \sum_{r=0}^{\infty} \frac{\Gamma(r+k+1)\left(-at^{2-\alpha}\right)^r}{\Gamma[(2-\alpha)r+2k+1]r!} \tag{25}$$

$$+ c_1 \sum_{k=0}^{\infty} \frac{(-b)^k t^{2k+1}}{k!} \sum_{r=0}^{\infty} \frac{\Gamma(r+k+1)\left(-at^{2-\alpha}\right)^r}{\Gamma[(2-\alpha)r+2k+2]r!}$$

$$+ c_2 \sum_{k=0}^{\infty} \frac{(-b)^k t^{\beta(1-\alpha)+2k+1}}{k!} \sum_{r=0}^{\infty} \frac{\Gamma(r+k+1)\left(-at^{2-\alpha}\right)^r}{\Gamma[(2-\alpha)r+\beta(1-\alpha)+2k+2]r!}$$

$$+ c_3 \sum_{k=0}^{\infty} \frac{(-b)^k t^{\beta(2-\alpha)+2k+1}}{k!} \sum_{r=0}^{\infty} \frac{\Gamma(r+k+1)\left(-at^{2-\alpha}\right)^r}{\Gamma[(2-\alpha)r+\beta(2-\alpha)+2k+2]r!}$$

Proof. Applying the Sumudu Transform to Eq. 24, we obtain:

$$\begin{aligned} &u^{-2}S[y(t)] - u^{-2}y(0) - u^{-1}y'(0) + au^{-\alpha}S[y(t)]u \\ &-au^{\beta(1-\alpha)-1}(I_0^{(1-\beta)(1-\alpha)}y)(0) - au^{\beta(2-\alpha)-1}\left(I_0^{(1-\beta)(2-\alpha)}y'\right)(0) + bS[y(t)] = 0 \end{aligned} \tag{26}$$

Solving for S[y(t)] and using initial conditions we have

$$\begin{aligned} S[y(t)] &= \frac{u^{-2}}{u^{-2} + au^{-\alpha} + b}c_0 + \frac{u^{-1}}{u^{-2} + au^{-\alpha} + b}c_1 + \frac{au^{\beta(1-\alpha)-1}}{u^{-2} + au^{-\alpha} + b}c_2 + \frac{au^{\beta(2-\alpha)-1}}{u^{-2} + au^{-\alpha} + b} \\ &= 0. \end{aligned} \tag{27}$$

Now taking the inverse Sumudu transform on both sides and using results 11 to 16,

$$\begin{aligned} y(t) =& c_0 \sum_{k=0}^{\infty} (-b)^k t^{2k+1-1} E^{k+1}_{-\alpha+2,2k+1}\left(-at^{-\alpha+2}\right) \\ &+ c_1 \sum_{k=0}^{\infty} (-b)^k t^{2k+2-1_1} E^{k+1}_{-\alpha+2,2k+2}\left(at^{-\alpha+2}\right) \end{aligned}$$

$$+ c_2 \sum_{k=0}^{\infty} a.(-b)^k\, t^{\beta(1-\alpha)+2k+1} E^{k+1}_{-\alpha+2,\beta(1-\alpha)+2k+2}\left(at^{-\alpha+2}\right)$$
$$+ c_3 \sum_{k=0}^{\infty} a.(-b)^k\, t^{\beta(2-\alpha)+2k+1} E^{k+1}_{-\alpha+2,\beta(2-\alpha)+2k+2}\left(at^{-\alpha+2}\right) \quad (28)$$

$$y(t) = c_0 \sum_{k=0}^{\infty} \frac{(-b)^k t^{2k}}{k!} \sum_{r=0}^{\infty} \frac{\Gamma(r+k+1)\left(-at^{2-\alpha}\right)^r}{\Gamma[(2-\alpha)r+2k+1]\, r!}$$
$$+ c_1 \sum_{k=0}^{\infty} \frac{(-b)^k t^{2k+1}}{k!} \sum_{r=0}^{\infty} \frac{\Gamma(r+k+1)\left(-at^{2-\alpha}\right)^r}{\Gamma[(2-\alpha)r+2k+2]\, r!}$$
$$+ c_2 \sum_{k=0}^{\infty} \frac{(-b)^k t^{\beta(1-\alpha)+2k+1}}{k!} \sum_{r=0}^{\infty} \frac{\Gamma(r+k+1)\left(-at^{2-\alpha}\right)^r}{\Gamma[(2-\alpha)r+\beta(1-\alpha)+2k+2]\, r!}$$
$$+ c_3 \sum_{k=0}^{\infty} \frac{(-b)^k t^{\beta(2-\alpha)+2k+1}}{k!} \sum_{r=0}^{\infty} \frac{\Gamma(r+k+1)\left(-at^{2-\alpha}\right)^r}{\Gamma[(2-\alpha)r+\beta(2-\alpha)+2k+2]\, r!} \quad (29)$$

□

Theorem 2: The following generalized fractional differential equation

$$y^{\alpha,\beta}(t) + ay'(t) + by(t) = 0 \quad \text{where } 1 < \alpha < 2 \text{ and } 0 < \beta \le 1 \text{ and } a, b \in R \quad (30)$$

with the initial conditions $y(0) = c_0$ and $y'(0) = c_1$, $\left(I_0^{(1-\beta)(1-\alpha)}y\right)(0) = c_2$ and $\left(I_0^{(1-\beta)(2-\alpha)}y\right)(0) = c_3$, has its solution given by:

$$y(t) = ac_0 \sum_{k=0}^{\infty} \frac{(-b)^k}{k!} \sum_{r=0}^{\infty} \frac{\Gamma(r+k+1)(-a)^r\, t^{(\alpha-1)r+\alpha k+\alpha-1}}{\Gamma[(\alpha-1)r+\alpha k+\alpha]\, r!}$$
$$+ c_1 \sum_{k=0}^{\infty} \frac{(-b)^k}{k!} \sum_{r=0}^{\infty} \frac{\Gamma(r+k+1)(-a)^r\, t^{\beta(1-\alpha)+\alpha k+\alpha-1+(\alpha-1)r}}{\Gamma[(\alpha-1)r+\beta(1-\alpha)+\alpha k+\alpha]\, r!}$$
$$+ c_2 \sum_{k=0}^{\infty} \frac{(-b)^k}{k!} \sum_{r=0}^{\infty} \frac{\Gamma(r+k+1)(-a)^r\, t^{\beta(2-\alpha)+\alpha k+\alpha-1+(\alpha-1)r}}{\Gamma[(\alpha-1)r+\beta(2-\alpha)+\alpha k+\alpha]\, r!} \quad (31)$$

Proof. Applying the Sumudu transform of Eq. 30, we obtain:

$$u^{-\alpha} S[y(t)]\, u - u^{\beta(1-\alpha)-1}(I_0^{(1-\beta)(1-\alpha)}y)(0)$$
$$-u^{\beta(2-\alpha)-1}\left(I_0^{(1-\beta)(2-\alpha)'}y\right)(0) + au^{-1}S[y(t)] - au^{-1}y(0) + bS[y(t)] = 0 \quad (32)$$

Solving for $S[y(t)]$ and using the initial conditions above we get

$$S[y(t)] = a\frac{u^{-1}}{au^{-1}+u^{-\alpha}+b}c_0 + \frac{u^{\beta(1-\alpha)-1}}{au^{-1}+u^{-\alpha}+b}c_1 + \frac{u^{\beta(2-\alpha)-1}}{au^{-1}+u^{-\alpha}+b}c_2 = 0 \quad (33)$$

Now taking the inverse Sumudu transform on both sides and using results 17-19,

$$\begin{aligned} y(t) =& ac_0 \sum_{k=0}^{\infty} (-b)^k \, t^{\alpha k+\alpha-1} E_{\alpha-1,\alpha k+\alpha}^{k+1} \left(-at^{\alpha-1}\right) \\ &+ c_1 \sum_{k=0}^{\infty} (-b)^k \, t^{\beta(2-\alpha)+\alpha k+\alpha-1} E_{\alpha-1,\beta(2-\alpha)+\alpha k+\alpha}^{k+1} \left(-at^{\alpha-1}\right) \\ &+ c_2 \sum_{k=0}^{\infty} (-b)^k \, t^{\beta(1-\alpha)+\alpha k+\alpha-1} E_{\alpha-1,\beta(1-\alpha)+\alpha k+\alpha}^{k+1} \left(-at^{\alpha-1}\right) \end{aligned} \tag{34}$$

$$\begin{aligned} y(t) =& ac_0 \sum_{k=0}^{\infty} \frac{(-b)^k}{k!} \sum_{r=0}^{\infty} \frac{\Gamma(r+k+1)(-a)^r \, t^{(\alpha-1)r+\alpha k+\alpha-1}}{\Gamma[(\alpha-1)r+\alpha k+\alpha]\, r!} \\ &+ c_2 \sum_{k=0}^{\infty} \frac{(-b)^k}{k!} \sum_{r=0}^{\infty} \frac{\Gamma(r+k+1)(-a)^r \, t^{\beta(2-\alpha)+\alpha k+\alpha-1+(\alpha-1)r}}{\Gamma[(\alpha-1)r+\beta(2-\alpha)+\alpha k+\alpha]\, r!} \\ &+ c_1 \sum_{k=0}^{\infty} \frac{(-b)^k}{k!} \sum_{r=0}^{\infty} \frac{\Gamma(r+k+1)(-a)^r \, t^{\beta(1-\alpha)+\alpha k+\alpha-1+(\alpha-1)r}}{\Gamma[(\alpha-1)r+\beta(1-\alpha)+\alpha k+\alpha]\, r!} \end{aligned} \tag{35}$$

□

Theorem 3: The following equation:

$$y^{\alpha,\beta}(t) - by(t) = 0, \text{ where } 0 < \alpha < 1 \quad \text{and } b \in R \tag{36}$$

with the initial condition $y(0) = c_0$ has its solution given by

$$y(t) = c_0 \sum_{k=0}^{\infty} \frac{(-b)^k t^{\beta(1-\alpha)+\alpha+\alpha k-1}}{\Gamma\beta(1-\alpha)+\alpha+\alpha k}$$

Proof. We apply the Sumudu transform of 36, that is

$$u^{-\alpha} S[y(t)] - u^{\beta(1-\alpha)-1} I_0^{(1-\alpha)(1-\beta)} y(0) - bS[y(t)] \tag{37}$$

Solving for $S[y(t)]$

$$S[y(t)] = c_0 \frac{u^{\beta(1-\alpha)-1}}{(u^{-\alpha}-b)} = \frac{u^{\beta(1-\alpha)-1}}{u^{-\alpha}(1-bu^{\alpha})} = \sum_{k=0}^{\infty} \left(bu^{\alpha}\right)^k u^{\beta(1-\alpha)-1+\alpha} = \sum_{k=0}^{\infty} (b)^k u^{\beta(1-\alpha)-\alpha+\alpha k-1} \tag{38}$$

Now taking the inverse Sumudu of the equation above, we get:

$$y(t) = c_0 \sum_{k=0}^{\infty} \frac{(-b)^k \, t^{\beta(1-\alpha)+\alpha+\alpha k-1}}{\Gamma\beta(1-\alpha)+\alpha+\alpha k} \tag{39}$$

□

Remark 1 If $a = 0$ in Eq. 22, then the equation,

$$y^{\alpha,\beta}(t) - by(t) = 0 \qquad 1 < \alpha \le 2 \tag{40}$$

with the initial conditions $y(0) = c_0 and y'(0) = c_1$ has its solution given by

$$y(t) = c_0 \sum_{k=0}^{\infty} \frac{(-b)^k t^{\beta(1-\alpha)+\alpha+\alpha k-1}}{\Gamma\beta(1-\alpha)+\alpha+\alpha k} + c_1 \sum_{k=0}^{\infty} \frac{(-b)^k t^{\beta(2-\alpha)+\alpha+\alpha k-1}}{\Gamma\beta(2-\alpha)+\alpha+\alpha k} \tag{41}$$

The next reult treats the generalized Vibration equation which is a particular case of the previous one.

Theorem 4: A nearly simple generalized harmonic vibration equation:

$$y^{\alpha,\beta}(t) - w^2 y(t) = 0, \qquad 1 < \alpha \le 2 \tag{42}$$

with the initial conditions $y(0) = c_0 and y'(0) = c_1$ has its solution given by

$$y(t) = c_0 \sum_{k=0}^{\infty} \frac{\left(-w^2\right)^k t^{\beta(1-\alpha)+\alpha+\alpha k-1}}{\Gamma\beta(1-\alpha)+\alpha+\alpha k} + c_1 \sum_{k=0}^{\infty} \frac{\left(-w^2\right)^k t^{\beta(2-\alpha)+\alpha+\alpha k-1}}{\Gamma\beta(2-\alpha)+\alpha+\alpha k} \tag{43}$$

Proof. By putting, $b = w^2$ in Eq. 40, the proof of the statement of the theorem is completed. □

4 Conclusion

In this work we have derived solutions of some generalized and special type of fractional differential equations. These solutions are general in nature and to get them we have used the Sumudu transform technique. The Sumudu transform method turns out to be a very powerful and direct technique towards the elucidation of such applications. It is also very user friendly. Along with the references we provide below, we hope that our respected readers would find ample information about this multitasking tool, its properties, the obvious ones and the more intricate ones, as well as the various applications in fractional Calculus it was capable to successfully tackle. Any feedback or related communications would be highly welcome and taken into consideration.

Acknowledgment: F. B. M. Belgacem wishes to acknowledge the continued support of the Public Authority for Applied Education and Training Research Department, (PAAET RD), Kuwait.

The Authors wish to acknowledge the helpful support of Editor Prof. Carlo Cattani, and DGO Managing Editor Mathematics, Agnieszka Bednarczyk-Drag.

Bibliography

Asiru M. A. (2001). Sumudu Transform and the solution of integral equations of convolution type. *International Journal of Mathematical Education in Science and Technology.* 32(6). 906–910.

Asiru M. A. (2002). Further properties of the Sumudu transform and its applications. *International Journal of Mathematical Education in Science and Technology.* 33(3).441–449.

Asiru M. A. (2003). Classroom note: application of the Sumudu transform to discrete dynamic systems, *International Journal of Mathematical Education in Science and Technology.* 34(6). 944–949.

Belgacem F.B.M. (2006). Introducing and Analyzing Deeper Sumudu Properties, *Nonlinear Studies Journal* 13(1). 23–41.

Belgacem F.B.M. (2006). Applications of the Sumudu Transform to Indefinite Periodic Parabolic Problems. Proceedings of the 6th International Conference on Mathematical Problems & Aerospace Sciences. Chap. 6, (2007), 51–60, Cambridge, UK, Cambridge Scientific Publishers.

Belgacem F.B.M. (2009). Sumudu Applications to Maxwell's Equations, PIERS Online.5. 1–6.

Belgacem F.B.M, (2010), Sumudu Transform Applications to Bessel Functions and Equations, *Applied Mathematical Sciences.* 4(74), 3665 - 3686.

Belgacem F.B.M., Bulut H., Baskonus M. H., Akurk T. (2014). Mathematical Analysis of the Generalized Benjamin Kdv Equations via the Extended Trial Equations Method, *Journal of the Associations of Arab Universities for Basic and Applied Sciences.* (16). 91-100.

Belgacem, F.B.M., Karaballi, A.A. (2006). Sumudu Transform Fundamental Properties Investigations and Applications. *Journal of Applied Mathematics and Stochastic Analysis.* 2006.

Belgacem F.B.M., Karaballi A.A. and Kalla S.L. (2003), Analytical Investigations of the Sumudu Transform, and Applications to Integral Production Equations. *Mathematical Problems in Eng.* 3.103–118.

Belgacem, F.B.M., Silambarasan, R. (2012). Theory of Natural Transform. *International Journal of Mathematics in Engineering, Science and Aerospace.* 3(1). 99-124.

Bulut H., Baskonus M.H., Belgacem F.B.M. (2013) The Analytical Solution of Some Fractional Ordinary Differential Equations by the Sumudu transform Method, *Applied and Abstract Analysis.* 2013.1 - 6.

Bulut H., Baskonus M.H., and Belgacem F.B.M. (2013). Partial Fractional Differential Equation Systems solutions by using Adomian Decomposition method implementation. Proceedings of the Fourth International Conference on Mathematical and Computational Applications. (pp.138-146) Manisa, Turkey.

Caputo M. (1969)., Elasticita e Dissipazione, Zanichelli, Bologna, Italy.

Chaurasia V.B.L., Dubey R.S., Belgacem F.B.M. (2012). Fractional radial diffusion equation analytical solution via Hankel and Sumudu transforms, *International Journal of Mathematics in Engineering Science & Aerospace.* 3(2).179-188.

Dubey R.S., Goswami P., Belgacem, F.B.M. (2014). Generalized Time-Fractional Telegraph Equation Analytical Solution by Sumudu and Fourier transforms, *Journal of Fractional Calculus and Applications.* 5(2), 52-58.

Garra R., Gorenflo R., Mainardy P., Tomovski Z. (2014). Hilfer–Prabhakar Derivatives and Some Applications, arXiv:1401.6668 [math.PR].

Goswami P., Belgacem F. B. M. (2012). Fractional differential equation solutions through a Sumudu rational, *Nonlinear Studies* 19(4).591-598.

Gupta, V.G., Sharma, B., Belgacem, F. B. M. (2011). On the Solutions of Generalized Fractional Kinetic Equations, *Journal of Applied Mathematical Sciences.* 5(19) 899 – 910.

Hilfer R. and Anton L. (1995). Fractional master equations and fractal time random walks, *Phys. Rev. E* 51 pp. 848–851

Hilfer R. (ed.) (2000). Applications of Fractional Calculus in Physics. Singapore, New Jersey, London and Hong Kong. World Scientific Publishing Company.

Hilfer R. (2002). Experimental evidence for fractional time evolution in glass forming materials, *Journal of Chemical Physics*. 284. 399–408.

Hilfer R. (2008). Threefold Introduction to fractional derivatives. In R. Klages, G. Radons and I. M. Sokolov (Eds). Anomalous Transport: Foundations and Applications. (pp.17-73) Wiley-VCH Verlag, Weinheim.

Hilfer R., Luchko Y., and Tomovski Ž. (2009). Operational method for solution of the fractional differential equations with the generalized Riemann-Liouville fractional derivatives, *Fractional. Calculus Appl. Anal.* 12, pp. 299–318.

Hussain, M.G.M, Belgacem, F.B.M. (2007), Transient Solutions of Maxwell's Equations Based on Sumudu Transformation, *Journal of Progress in Electromagnetic Research (PIER)*, No. 74, pp. 273-289.

Kılıcman A., Eltayeb H., and Agarwal P. R. (2010). On Sumudu transform and system of differential equations, *Abstract and Applied Analysis*, Article ID 598702.

Katatbeh Q.K., Belgacem F.B.M. (2011). Applications of the Sumudu Transform to Fractional Diff. Equations, *Nonlinear Studies*. 18(1). 99-112

Miller K.S., Ross B. (1993), An Introduction to the Fractional Calculus and Fractional Differential Equations, Wiley, New York.

Mainardi F. and Gorenflo R. (2007). Time-fractional derivatives in relaxation processes: A tutorial survey, *Fract. Calc. Appl. Anal.* 10. 269-308.

Prabhakar, T. R. (1971). A singular integral equation with a generalized Mittag-Leffler function in the kernel, *Yokohama Math. J.*, 19. 7-15.

Podlubny I. (1994). The Laplace Transform Method for Linear Differential Equations of Fractional Order, Slovak Republic. Slovac Academy of Science.

Podlubny I. (1999). Fractional Differential Equations. New York. Academic Press.

Rana, M.A., Siddiqui, A.M., Ghori, and Q.K, Qamar, R., (2007) , Application of He's homotopy Perturbation Method to Sumudu Transform, *International Journal of Nonlinear Sciences and Numerical Simulations*, Vol.8, No.2, 185-190.

Sandev T. and Tomovski Ž. (2010). General time fractional wave equation for a vibrating string, *J. Phys. A Math. Theoret.* 43 Article ID. 055204.

Srivastava H.M. and Tomovski Ž. (2009). Fractional calculus with an integral operator containing a generalized Mittag-Leffler function in the kernel. *Appl. Math. Comput.* 211. 198-210.

Tomovski Ž. (2012)., Generalized Cauchy type problems for nonlinear fractional differential equations with composite fractional derivative operator, *Nonlinear Analysis*.

Tuluce D. S., Bulut H., Belgacem F. B. M., (2014), Sumudu Transform Method for Analytical Solutions of Fractional Type Ordinary Differential Equations, *Mathematical Problems in Engineering*

Watugala G. K. (1993), Sumudu transform: an integral transform to solve differential equations and control engineering problems, *Int. Journal of Mathematical Education in Science and Technology*. 24, 35–42.

Weerakoon, S. (1994); Applications of the Sumudu Transform to Partial Differential Equations. *Int. J. Math Edu. Sci. Technol.*, Vol.25, No. 2, 277-283.

Weerakoon, S. (1998).; Complex Inversion Formula for Sumudu Transform. *Int. J. Math Edu. Sci. Technol.*, Vol.29, No. 4, 618-621.

Zang, J. (2007); A Sumdu based algorithm for solving differential equations, *Computer Science Journal of Moldova*, vol.15, no. 3(45), 303-313.

FSC
www.fsc.org
MIX
Papier | Fördert gute Waldnutzung
FSC® C083411

Kolibri 360 GmbH
Ferdinand-Jühlke-Straße 7
99095 Erfurt, Deutschland
produktsicherheit@kolibri360.de